Geometry and Its Applications

LIMITED WARRANTY AND DISCLAIMER OF LIABILITY

ACADEMIC PRESS, INC. ("AP") AND ANYONE ELSE WHO HAS BEEN INVOLVED IN THE CREATION OR PRODUCTION OF THE ACCOMPANYING CODE ("THE PRODUCT") CANNOT AND DO NOT WARRANT THE PERFORMANCE OR RESULTS THAT MAY BE OBTAINED BY USING THE PRODUCT. THE PRODUCT IS SOLD "AS IS" WITHOUT WARRANTY OF MERCHANTABILITY OR FITNESS FOR ANY PARTICULAR PURPOSE. AP WARRANTS ONLY THAT THE MAGNETIC DISC(S) ON WHICH THE CODE IS RECORDED IS FREE FROM DEFECTS IN MATERIAL AND FAULTY WORKMANSHIP UNDER THE NORMAL USE AND SERVICE FOR A PERIOD OF NINETY (90) DAYS FROM THE DATE THE PRODUCT IS DELIVERED. THE PURCHASER'S SOLE AND EXCLUSIVE REMEDY IN THE VENT OF A DEFECT IS EXPRESSLY LIMITED TO EITHER REPLACEMENT OF THE DISC(S) OR REFUND OF THE PURCHASE PRICE, AT AP'S SOLE DISCRETION.

IN NO EVENT, WHETHER AS A RESULT OF BREACH OF CONTRACT, WARRANTY, OR TORT (INCLUDING NEGLIGENCE), WILL AP OR ANYONE WHO HAS BEEN INVOLVED IN THE CREATION OR PRODUCTION OF THE PRODUCT BE LIABLE TO PURCHASER FOR ANY DAMAGES, INCLUDING ANY LOST PROFITS, LOST SAVINGS OR OTHER INCIDENTAL OR CONSEQUENTIAL DAMAGES ARISING OUT OF THE USE OR INABILITY TO USE THE PRODUCT OR ANY MODIFICATIONS THEREOF, OR DUE TO THE CONTENTS OF THE CODE, EVEN IF AP HAS BEEN ADVISED ON THE POSSIBILITY OF SUCH DAMAGES, OR FOR ANY CLAIM BY ANY OTHER PARTY.

ANY REQUEST FOR REPLACEMENT OF A DEFECTIVE DISC MUST BE POSTAGE PREPAID AND MUST BE ACCOMPANIED BY THE ORIGINAL DEFECTIVE DISC, YOUR MAILING ADDRESS AND TELEPHONE NUMBER, AND PROOF OF DATE OF PURCHASE AND PURCHASE PRICE. SEND SUCH REQUESTS, STATING THE NATURE OF THE PROBLEM, TO ACADEMIC PRESS CUSTOMER SERVICE, 6277 SEA HARBOR DRIVE, ORLANDO, FL 32887, 1-800-321-5068. AP SHALL HAVE NO OBLIGATION TO REFUND THE PURCHASING PRICE OR TO REPLACE A DISC BASED ON CLAIMS OF DEFECTS IN THE NATURE OR OPERATION OF THE PRODUCT.

SOME STATES DO NOT ALLOW LIMITATION ON HOW LONG AN IMPLIED WARRANTY LASTS, NOR EXCLUSIONS OR LIMITATIONS OF INCIDENTAL OR CONSEQUENTIAL DAMAGE, SO THE ABOVE LIMITATIONS AND EXCLUSIONS MAY NOT APPLY TO YOU. THIS WARRANTY GIVES YOU SPECIFIC LEGAL RIGHTS, AND YOU MAY ALSO HAVE OTHER RIGHTS WHICH VARY FROM JURISDICTION TO JURISDICTION.

THE RE-EXPORT OF UNITED STATES ORIGINAL SOFTWARE IS SUBJECT TO THE UNITED STATES LAWS UNDER THE EXPORT ADMINISTRATION ACT OF 1969 AS AMENDED. ANY FURTHER SALE OF THE PRODUCT SHALL BE IN COMPLIANCE WITH THE UNITED STATES DEPARTMENT OF COMMERCE ADMINISTRATION REGULATIONS. COMPLIANCE WITH SUCH REGULATIONS IS YOUR RESPONSIBILITY AND NOT THE RESPONSIBILITY OF AP.

Geometry and Its Applications

Walter Meyer
Department of Mathematics and Computer Science
Adelphi University
Garden City, New York

San Diego London Boston
New York Sydney Tokyo Toronto

Cover image: Al Held, C-Y-I, 1978, acrylic on canvas, 114×114″, Albright-Knox Art Gallery, Buffalo, New York, Gift of Seymour H. Knox, 1979

This book is printed on acid-free paper. ∞

ACADEMIC PRESS
A division of Harcourt Brace & Company
525 B Street, Suite 1900, San Diego, CA 92101-4495, USA
http://www.apnet.com

Academic Press
24–28 Oval Road, London NW1 7DX, UK
http://www.hbuk.co.uk/ap/

Harcourt/Academic Press
200 Wheeler Road
Burlington, MA 01803
http://www.harcourt-ap.com

Library of Congress Number: 98-28224

International Standard Book Number: 0-12-493270-3

Printed in the United States of America
98 99 00 01 02 IP 9 8 7 6 5 4 3 2 1

Table of Contents

Preface

This book is for a first college-level geometry course and is suitable for mathematics majors and especially prospective high school teachers. Much of the content will be familiar to geometry instructors: a solid introduction to axiomatic Euclidean geometry, some non-Euclidean geometry, and a substantial amount of transformation geometry. However, we present some important novelties: We pay significant attention to applications, we provide optional dynamic geometry courseware for use with *The Geometer's Sketchpad,* and we include a chapter on polyhedra and planar maps. By extending the content of geometry courses to include applications and newer geometry, such a course can not only teach mathematical skills and understandings, but can help students understand the twenty-first century world that is unfolding around us. By providing software support for discovery learning, we allow experiments with new ways of teaching and learning.

The intertwined saga of geometric theory and applications is modern as well as ancient, providing a wonderful mathematical story that continues today. It is a compelling story to present to students to show that mathematics is a seamless fabric, stretching from antiquity until tomorrow and stretching from theory to practice. Consequently, one of our goals is to express the breadth of geometric applications, especially contemporary ones. Examples include symmetries of artistic patterns, physics, robotics, computer vision, computer graphics, stability of architectural structures, molecular biology, medicine, pattern recognition, and more. Perhaps surprisingly, many of these applications are based on familiar, long-standing geometric ideas — showing once again that there is no conflict between the timelessness and modernity of good mathematics.

In recent years, high school instruction in geometry has become much less extensive and much less rigorous in many school districts. Whatever advantage this may bring at the high school level, it changes the way we need to instruct mathematics students at the college level. In the first place, it makes instruction in geometry that much more imperative for all students of mathematics. In addition, we cannot always assume extensive familiarity with proof-oriented basic Euclidean geometry. Consequently, we begin at the beginning, displaying a portion of classical Euclidean geometry as a deductive system. For the most part, our proofs are in the style of Euclid — which is to say that they are not as rigorous as they could be. We do present a snapshot of some geometry done with full rigor, so that students will have exposure to that. In addition, there is a careful discussion of why full rigor is important in some circumstances and why it is not always attempted in teaching, research, or applications.

Except for Chapters 5 and 6 and parts of Chapter 7, this text requires little more than high school mathematics. Nonetheless, students need the maturity to deal with proofs and careful calculations. In Chapters 5 and 6 and parts of Chapter 7, we assume a familiarity with vectors as commonly presented in multivariable calculus. Derivatives also make a brief appearance in Chapter 5. Matrices are used in Chapter 6, but it is not necessary to have studied linear algebra in order to understand this material. All that we assume is that students know how to multiply matrices and are familiar with the associative law.

It would be foolish to pretend that this book surveys all of the major topics and applications of geometry. For example, differential geometry is represented only by one short section. I have tried to choose topics that would be most appealing and accessible to undergraduates, especially prospective high school teachers.

A good deal of flexibility is possible in selecting a sequence of topics from which to create a course. This book contains two approaches to geometry: the axiomatic and the computational. When I am teaching mostly prospective teachers, I emphasize the axiomatic (Chapters 1–4) and sprinkle in a little computational material from Chapters 5 or 6. When I have mainly mathematics majors with applied interests and others, such as computer science majors, I reverse the emphasis, concentrating on Chapters 5, 6, and 7. I find Chapter 8 works well in either type of course.

There is a lot of independence among the chapters of the book. For example, one might skip Chapters 1 through 3 since there are only a few places in other chapters (mainly Chapter 4) where there is any explicit dependence on them. An instructor can remind students of the relevant theorems as the need arises. Chapter 4 is not needed for any of the other parts. Chapter 5 can be useful in preparation for Chapter 6 only insofar as we often think of points as position vectors in Chapter 6. Chapter 7 relies on one section of Chapter 2 and one section of Chapter 5. Chapter 8 is completely independent of the other chapters. More detailed descriptions of prerequisites are given at the start of each chapter.

In writing this book, I am aware of the many people and organizations that have shaped my thoughts. I learned a good deal about applications of geometry at the Grumman Corporation (now Northrop-Grumman) while in charge of a robotics research program. Opportunities to teach this material at Adelphi University and during a year spent as a visiting professor at the U.S. Military Academy at West Point have been helpful. In particular, I thank my cadets and my students at Adelphi for finding errors and suggesting improvements in earlier drafts. Thanks are due to the National Science Foundation, the Sloan Foundation, and COMAP for involving me in programs dedicated to the improvement of geometry at both the collegiate and secondary levels. Finally, I wish to thank numerous individuals with whom I have been in contact (for many years in some cases) about geometry in general and this book in particular: Joseph Malkevitch, Donald Crowe, Robert Bumcrot, Andrew Gleason, Greg Lupton, John Oprea, Brigitte Selvatius, Marie Vanisko, and Sol Garfunkel.

Prof. Walter Meyer
Adelphi University

Supplements for the Instructor

The following supplements are available from Academic Press:

1. Answers to the even-numbered exercises
2. Instructor's guide to *The Geometer's Sketchpad* explorations that are contained in the disk that accompanies this text.

Introduction

Geometry is full of beautiful theorems, and its logical structure can be inspiring. As the poet Edna St. Vincent Millay wrote, "Euclid alone has looked on beauty bare." But beyond beauty and logic, geometry also contains important tools for applied mathematics. This should be no surprise, since the word "geometry" means "earth measurement" in Greek. As just one example, we will illustrate the appropriateness of this name by showing how geometry was used by the ancient Greeks to measure the circumference of the earth without actually going around it. But the story of geometric applications is modern as well as ancient. The upsurge in science and technology in the last few decades has brought with it an outpouring of new questions for geometers. In this introduction we provide a sampler of the big ideas and important applications that will be discussed in this book.[1]

Individuals often have preferences, either for applications in contrast to theory or vice versa. This is unavoidable and understandable. But the premise of this book is that, whatever our preferences may be, it is good to be aware of how the two faces of geometry enrich each other. Applications can't proceed without an underlying theory. And theoretical ideas, although they can stand alone, often surprise us with unexpected

[1] This introduction also appears in *Perspectives on the Teaching of Geometry for the 21st Century*, ed. V. Villani and C. Mammana, copyright Kluwer Academic Publishers b.v., 1998.

applications. Throughout the history of mathematics, theory and applications have carried out an intricate dance, sometimes dancing far apart, sometimes close. My hope is that this book gives a balanced picture of the dance at this time, as we enter a new millennium.

Axiomatic Geometry

Of all the marvelous abilities we human beings possess, nothing is more impressive than our visual systems. We have no trouble telling circles apart from squares, estimating sizes, noticing when triangles appear congruent, and so on. Despite this, the earliest big idea in geometry was to achieve truth by proof and not by eye. Was that really necessary or useful? These ideas are explored in Chapters 1 and 2.

Creating a geometry based on proof required some basic truths — which are called *axioms* in geometry. Axioms are supposed to be uncontroversial and obviously true, but Euclid seemed nervous about his parallel axiom. Other geometers caught this whiff of uncertainty and, about 2000 years later, some were bold enough to deny the parallel axiom. In doing this they denied the evidence of their own eyes and the weight of 2000 years of tradition. In addition, they created a challenge for students of this so-called "non-Euclidean geometry," which asks them to accept axioms and theorems that seem to contradict our everyday visual experience. According to our visual experience, these non-Euclidean geometers are cranks and crackpots. But eventually they were promoted to visionaries when physicists discovered that the far-away behavior of light rays (physical examples of straight lines) is different from the close-to-home behavior our eyes observe. Astronomers are working to make use of this non-Euclidean behavior of light rays to search for "dark matter" and to foretell the fate of the universe. These revolutionary ideas are explored in Chapter 3.

Rigidity and Architecture

If you are reading this indoors, the building you are in undoubtedly has a skeleton of either wooden or steel beams, and your safety depends partly on the rigidity of this skeleton (see Figure I.1b). Neither a single rectangle (Figure I.1a), nor a grid of them, would be rigid if it had hinges where the beams meet. Therefore, when we build frameworks for buildings, we certainly don't put hinges at the corners — in fact, we make these corners as strong as we can. But it is hard to make a corner perfectly rigid, so every additional safeguard is welcome. A very common safeguard, which makes a single rectangle rigid, is to add a diagonal brace. Perhaps surprisingly, if we have a grid of many rectangles, it is not necessary to brace every rectangle. The braced grid in

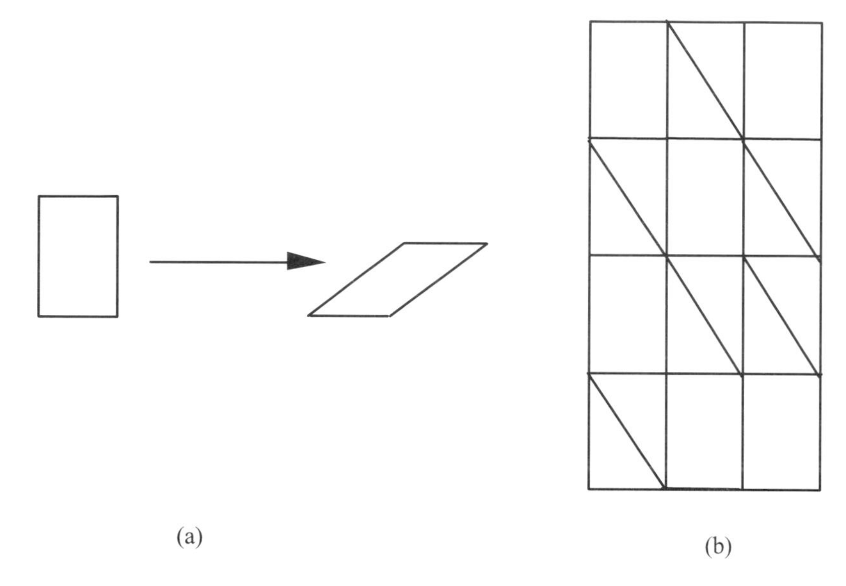

Figure I.1 (a) A hinged rectangle flexing. (b) A braced grid that cannot flex even if totally hinged.

Figure I.1b turns out to be rigid even if every corner is hinged. In Section 2.3 of Chapter 2, we work out a procedure for determining when a set of braces makes a grid of rectangles rigid even though all corners are hinged.

Computer Graphics

The impressionist painter Paul Signac (1863–1935) painted "The Dining Room" by putting lots of tiny dots on a canvas (Figure I.2a). If you stand back the tiny dots blend together to make a picture. This painting technique, called *pointillism*, was a sensation at the time, and foreshadows modern image technology. For example, if you take a close look at your TV screen, you'll see that the picture is composed of tiny dots of light. Likewise, a computer screen creates a picture by "turning on" little patches of color called *pixels*. Think of them as forming an array of very tiny light bulbs, arranged in rows and columns in the x-y plane so that each point with integer coordinates is the center of a pixel (Figure I.2b).

When a graphics program shows a picture, how does it calculate which pixels to turn on and what colors they should be? Here is a simple version of the problem: If we are given two pixels (shaded in Figure I.2b) and want to connect them with a set of pixels to give the impression of a blue straight line, which "in between" pixels should be turned

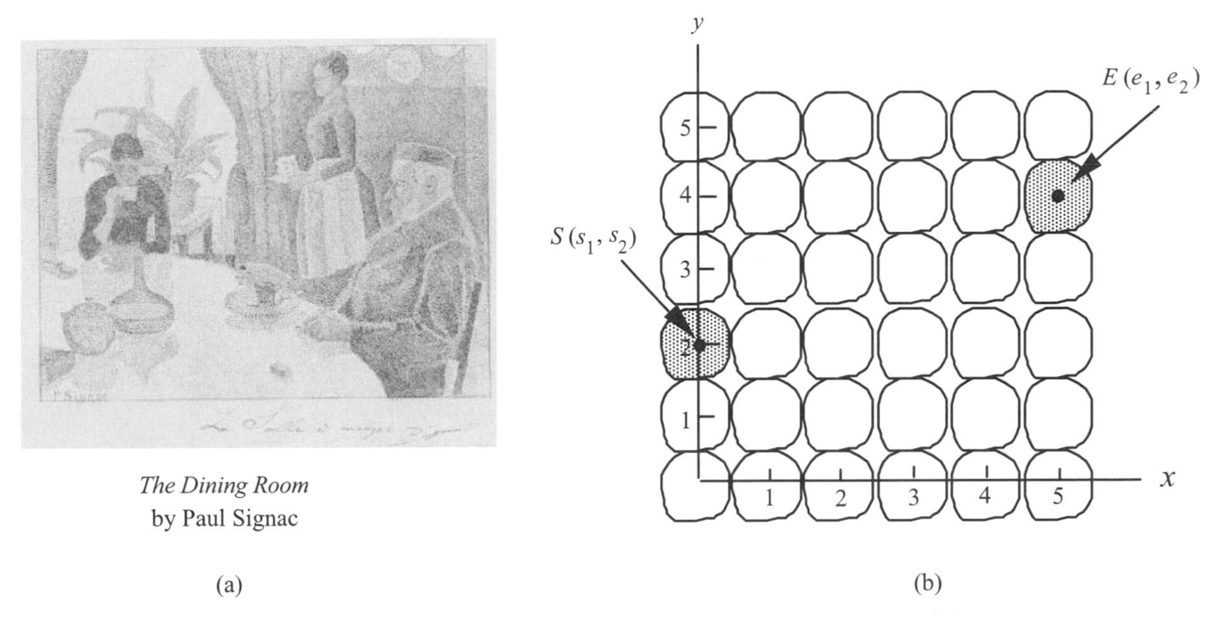

Figure I.2 Pixels in (a) art and (b) computer graphics. Courtesy of Metropolitan Museum of Art.

blue? Neither Signac nor you would have any problem with this, painting by eye, but how can the computer do it by calculations based on the coordinates of the centers of the start and end pixels? In this computer verison of the problem, the desired answer is a list of pixel centers, each center specified by x and y coordinates. Chapter 1 gives you an idea of how to create such a list.

Symmetries in Anthropology

In studying vanished cultures, anthropologists often learn a great deal from the artistic patterns these cultures produced. Figure I.3 shows two patterns you might find on a cloth or circling around a clay pot. Each of these patterns has some kind of symmetry. But what kind? What do we mean by the word *symmetry*? Some would say the bottom pattern has more symmetry than the top pattern. How can symmetry be measured? The questions we pose here for patterns are similar to ones that arise, in three-dimensional form, in the study of crystallography.

We study these questions in Chapter 4, with particular attention paid to the pottery of the San Ildefonso pueblo in the southwestern United States.

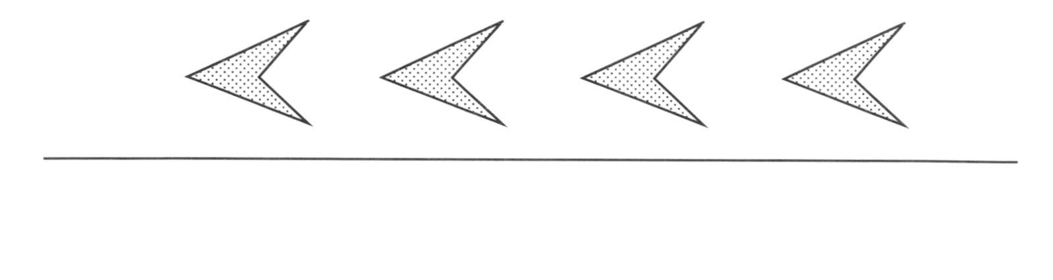

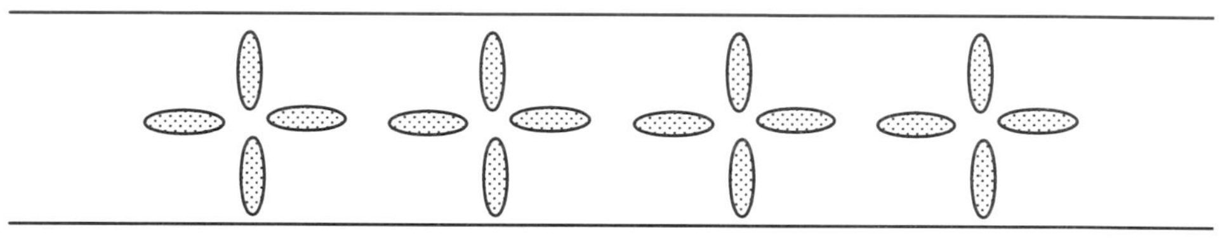

Figure I.3 Two strip patterns.

Robotics

Figure I.4 shows a robot about to drill a hole. Perhaps the hole is being drilled into someone's skull in preparation for brain surgery, or maybe it's part of the manufacture of an automobile. Whatever the purpose, the drill tip has to be in just the right place and pointing just the right way. The robot moves the drill about by changing its joint angles. If we specify the x, y, z coordinates of the drill tip, how do we calculate the values of θ_1, θ_2, and θ_3 needed to bring the drill tip to the desired point? Can we also specify the direction of the drill? These are questions in *robot kinematics*. In Section 6.5 of Chapter 6, we study the basics of robot kinematics for a two-dimensional robot. This is the same kind of mathematics used for three-dimensional robot kinematics.

Molecular Shapes

As chemistry advances, it pays more attention to the geometric shapes of molecules. In 1985 a new molecular shape was discovered, called the *buckyball*[2], that reminded chemists of the pattern on a soccer ball. The molecule consisted of 60 carbon atoms distributed in

[2] Named for the architect Buckminster Fuller, who promoted the idea of *geodesic domes*, buildings in the shape of unusual polyhedra.

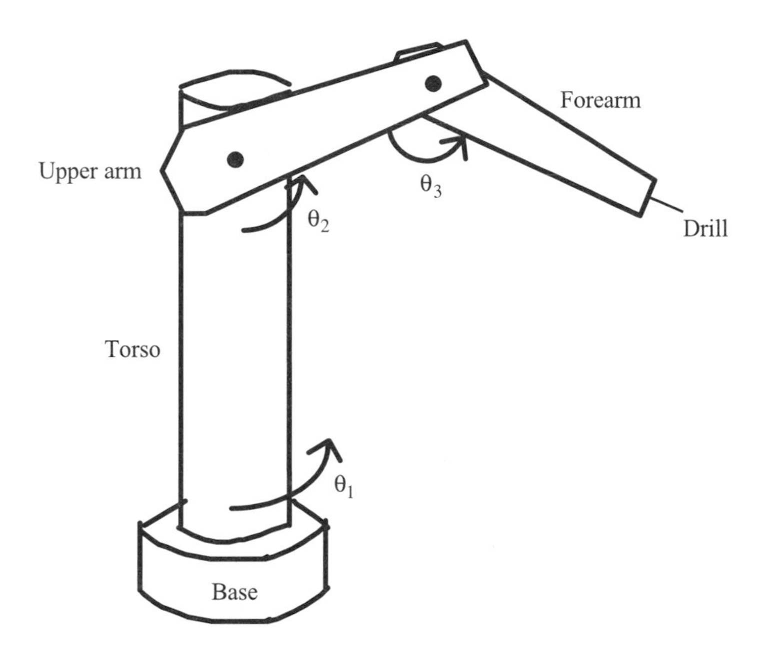

Figure I.4 A simplified version of a PUMA robot arm.

a roughly spherical shape — like the corners of the pattern on the soccer ball. Coincidentally, this pattern had been discovered not only by soccer ball manufacturers, but many centuries ago by mathematicians. Figure I.5 shows the pattern in a Renaissance drawing of a truncated icosahedron, by Leonardo da Vinci. To understand the structure of the buckyball, think of the corners of the truncated icosahedron as being occupied by carbon atoms and think of the connecting links as representing chemical bonds betweeen certain carbon atoms. Each carbon atom is connected to each of three other nearby carbon atoms with a chemical bond. This pattern contains 12 pentagons and 20 hexagons.

Once chemists discovered this molecule, they looked for other molecules involving just carbons, where each carbon is connected to exactly three others and where the pattern has only hexagons and pentagons. Each such molecule turns out to have 12 pentagons. This is not a chemical quirk! It follows from some mathematics we study in Section 8.4 of Chapter 8.

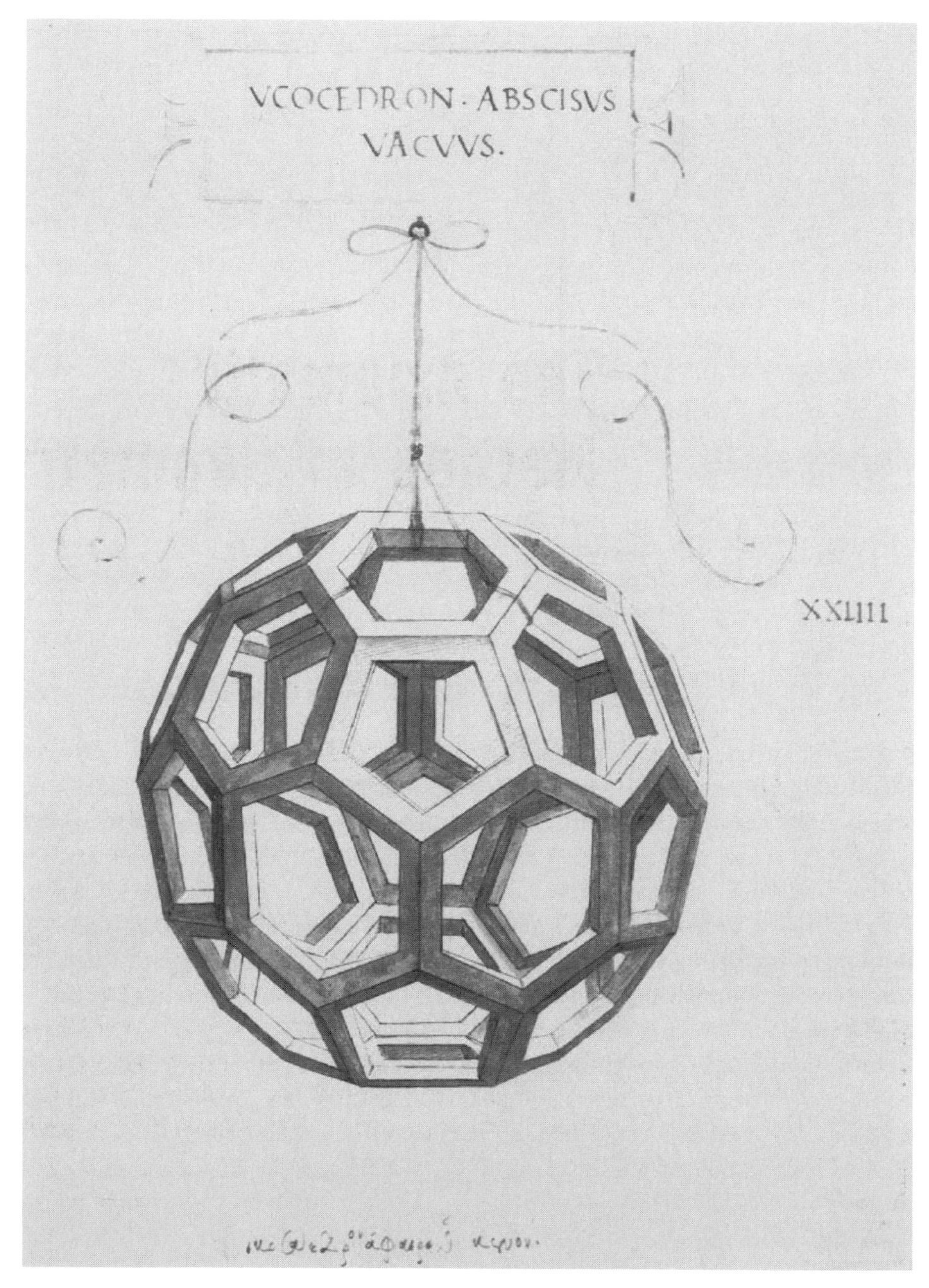

Figure I.5 The pattern of the buckyball molecule, drawn five centuries before the discovery of the molecule! Courtesy of Jerry Blow.

Chapter 1

The Axiomatic Method in Geometry

Prerequisites: High school mathematics

We human beings are at home in the physical world: We see it clearly and move about with ease, and we need no theory to do so. But our natural understanding of geometry does not serve all our needs. The axiomatic method is the most powerful method so far devised for a more complete understanding of our geometric enivronment.

The key idea of the axiomatic method is that we start with assumptions we have complete confidence in, and we reason our way to things we might not believe without a proof. Our objective in this chapter is to provide an overview of this axiomatic method, including some attention to how standards of proof have evolved over time. A short discussion of computer graphics provides an opportunity to see how the ideal concepts of geometry apply to the less ideal real world. In the last section of this chapter we display the axiom set we will use for the part of Euclidean geometry we develop in this and the next chapter.

1.1 The Aims of Axiomatic Geometry

In Euclidean geometry, as in all of mathematics, we try to discover new knowledge by applying already-known principles. This style of thinking is called *deduction*, or the *axiomatic method*, and it can also be seen in many other fields, including fields outside the sciences such as politics and religion. As an example, much of Western culture long ago selected the Ten Commandments as central moral principles, with the presumption that in the thousands of different moral dilemmas which come up in life, we can always deduce the right thing to do from the Ten Commandments. A little more recently, in the U.S. Declaration of Independence, certain truths are said to be self-evident and the document goes on to justify the independence of the American colonies as a deduction from those truths. Likewise, in our legal system, deductions are drawn from laws and evidence in order to decide individual cases. However, the use of deduction in nonmathematical fields often seems to leave lots of room for difference of opinion, something which rarely occurs in mathematics. For this reason, mathematics remains the purest and most successful example of deduction at work — an ideal to which other fields often aspire.

The idea that some scientific truths can be deduced from other more basic ones was surely known before the development of Euclidean geometry, but it certainly received its biggest boost in the sciences from its use in geometry. In this section we want to briefly ask "Why is this idea so appealing in geometry?"

Approaches to Geometry

We can discover a lot about the physical space we live in through simple visual observation. Fred Flintstone and his prehistoric buddies could tell if a ditch was too big to jump over without having gone to school. Likewise, even children who have not attended school take for granted their ability to move about without bumping into things, to judge whether they can squeeze through tight spaces or jump over ditches. (Ironically, researchers find it a tremendous challenge to get robots to make these judgments even half as well as people.)

The eye and the brain are very capable, but there is much they can't do. If Fred Flintstone wants to know exactly how wide the ditch is, he will have to measure it — say, with a tape measure. This is the second great way to learn about our physical world: Apply measuring instruments such as rulers and protractors to it. The deductive method is the third major way to learn about the physical space in which we live. This book will give you a good idea of what deductive geometry is and how we pursue it. However, many people never have need for any more than measurement, so why did mathematicians develop a deductive theory of geometry? Why isn't measurement good enough?

Here are some shortcomings of measurement as a way to learn about our physical world:

1. Measurement applies to objects and physical configurations that exist. What if we want to build something and predict its geometric nature in advance? For example, if we have plans for a house, we need to know how much surface area the outside has, so we can determine how many bundles of shingles will be needed, and so on. Information of this kind is needed to estimate the cost and decide whether it is practical to build that house.

   For another example, consider the simplified robot in Figure 1.1. We want to move the joint angles θ_1 and θ_2 to values that will center the drill tip at the point $T(2, 1)$. What angles are needed? There is no way to measure these angles until the tip is in the right place. But the most efficient way to get the tip into the right place is to know what angles to command the robot to move to.

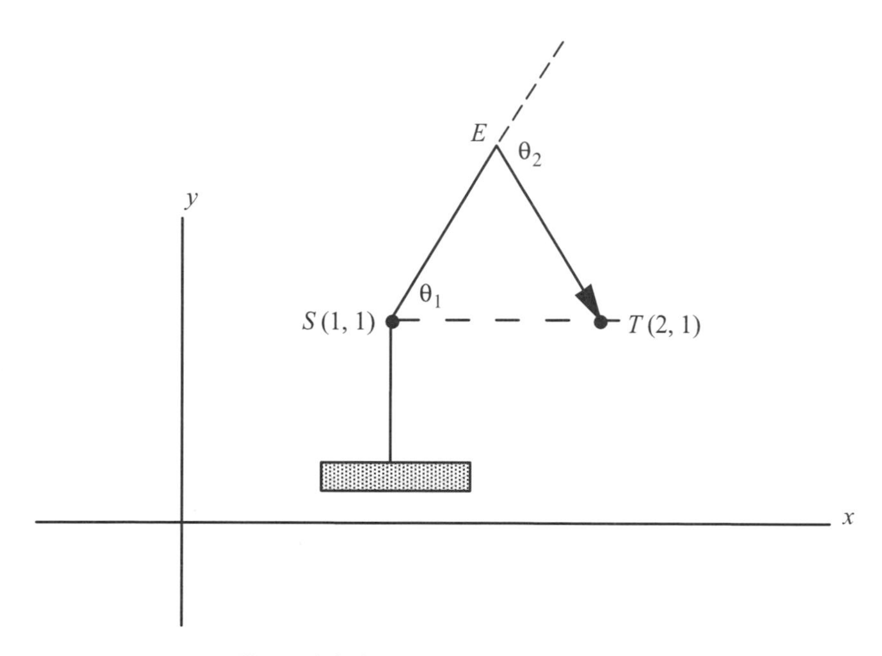

Figure 1.1 A robot with a drill tip (arrow).

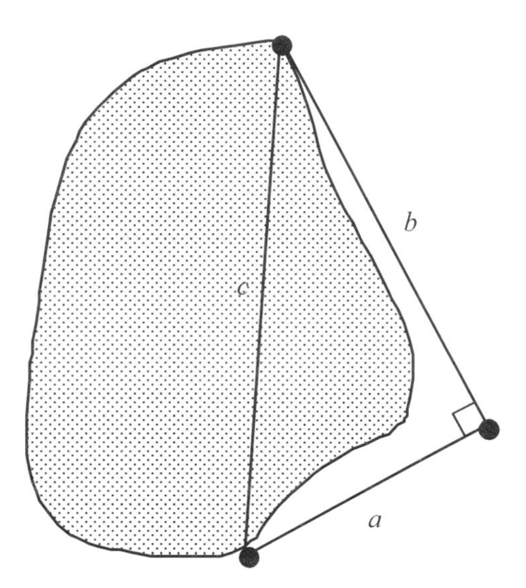

Figure 1.2 Finding the distance across a lake.

2. Another shortcoming is that we can only measure what we can get to. Suppose Fred Flintstone wants to know the distance across a lake (c in Figure 1.2) and he finds it inconvenient to swim across, dragging his tape measure along. (Can you think of a way around this problem if the triangle is a right triangle?)

As we shall see, some of the great moments in early geometry were deductions that could not be duplicated by simple measurement. A good example concerns estimating the circumference of the earth (Section 2.3 of Chapter 2). This was carried out by Eratosthenes more than 2000 years ago at a time when doing an actual measurement by walking, riding, or flying around the earth was simply inconceivable. Early geometers also estimated the distance from the earth to the moon (and distances between other astronomical bodies) before we could get to the moon.

3. Another shortcoming of measurement is that it is only approximate. Even the best instruments never give truly exact answers. For practical purposes, we often don't need much exactness, but sometimes we do. Suppose the robot of Figure 1.1 is carving an artificial hip joint out of plastic that will be used to replace a diseased hip joint in a patient. The new joint needs to be just the right size so that the patient will be able to walk without

pain for many years. To move the drill tip with the necessary precision, the corresponding sequence of angular values needs to be computed very precisely.

4. Measurements tell us what is true, but not why. If we measure the width of a ditch, it seems fairly accidental what the number comes out to. Likewise, any relationships about sides of a triangle that we discover by measurement, even if we happen to get the correct relationships, will merely seem like unrelated facts if we discover each relationship by measurement. For example, we might measure a few triangles and discover a few that had right angles. In these triangles we might discover that

$$c^2 = a^2 + b^2.$$

Here c is the length of the hypotenuse and a and b are the lengths of the two legs (see Figure 1.2). Is it just an accident that right triangles display this numerical relationship? As we know, the numerical relationship can be deduced from the fact that a triangle has a right angle. The proof shows us that the facts are connected. Being able to see connections confirms our hope that our world is understandable and not just a bunch of senseless accidents.

5. Measurement only tells us about the particular objects we measured. What we really would like is information that pertains to *all* objects of a certain kind. For example, suppose we believe that there is a relation between the three sides of a right triangle, such as the right triangle in Figure 1.2. We know, from deductive geometry, what that relationship is, but let's pretend we don't know and try to find a relationship by measurement. Well, we can't measure the actual triangle of interest because one side stretches across the lake. What if we looked at other triangles that we could measure? If we looked at a right triangle with sides 3, 4, and 5 we might suppose the relationship is that the hypotenuse is the longer side plus one-third the shorter $[5 = 4 + (1/3)(3)]$. This is also seen to be true in the right triangle with 6, 8, and 10 as its sides. If we stopped with these two triangles we wouldn't know that this relationship is false for most right triangles. The deductive approach in geometry tries to produce truths that can be asserted for *all* figures of a certain type (all triangles, all right triangles, all pentagons, etc.).

Summarizing all of this, we can provide the following rough definition of the aims of deductive geometry: Deductive geometry tries to determine truths of geometry which are *exact* and applicable to *all figures*, even those we can't physically measure or which exist only in our imaginations. It shows how some truths can be deduced from others, so it is as much about the *relationships of geometric truths* as it is about the truths themselves.

Achieving Certainty in Deductive Geometry

To understand how we try to achieve reliable geometric knowledge through the deductive method, let us compare the following two statements:

ROSCOE RECHARGES

Roscoe the robot is located at a point S whose coordinates on a factory floor are (1, 2). Roscoe needs to travel to $G(4, 6)$. He has enough power in his battery to go 4.9 units. But SG, the distance from S to G, is 5. Roscoe needs to recharge his batteries first.

EUCLID'S EXTENDIBILITY AXIOM

If A and B are any two points, then the segment $\overline{AB}$ can be extended arbitrarily far on either side of segment $\overline{AB}$, making a new line segment of any desired length.

The first statement, Roscoe's assertion that $SG = 5$, is far from obvious. If you hadn't studied geometry before this, and if there were no expert around to ask, you might have no confident opinion about whether it was true or false. However, it takes no special education to feel certain about Euclid's extendibility axiom, even though it is claimed to be true for all pairs of points, including points on Venus or Jupiter where we have never been.[1] One of the main motivations for deductive geometry is to show how things we are not sure of, like Roscoe's distance calculation, can be deduced from a set of statements anyone would be confident about, like the one about the extendibility of line segments. These easy-to-believe statements are called *axioms*.[2] We make no attempt to prove our axioms. A proof of an axiom would have to be based on some principles. But then we could ask for a proof of those principles, and so on. We need to start somewhere. The statements we start with are called *axioms* and we use them to prove other statements called *theorems*.

We shall return to Roscoe's distance calculation in the exercises of Section 2.3 in the next chapter. By that point, we will have established all the theorems we need to justify Roscoe's calculation.

---

[1] When astronauts visited the moon for the first time in 1969, they didn't bother to check whether the extendibility principle was true there. They spent their limited time collecting moon rocks and hopping about demonstrating the low lunar gravity. Did they make a bad choice about how to spend their time?

[2] These are sometimes also called *postulates*.

Applying Geometry

Once we set out our axioms (Section 1.3), in no time at all we will see that we are led to think about points as being "infinitely small." More precisely, no matter how small a physical thing you might imagine, be it a pencil point or an atom, or something even smaller, a point must be smaller than that. Obviously then, a point is not a physical thing, but an ideal concept existing in our minds. Some concepts imagined by our minds, like ghosts and goblins under the bed, have no value for dealing with the practicalities of everyday life. However, the ideal concepts of geometry are extremely useful.

For example, an airline working out routes between various cities may think of the cities as points even though cities are quite large. What makes this reasonable is the fact that, in relation to the whole country, which is the area of interest for finding air routes, a city is small. Another way to justify it is to suppose that the point on the globe labeled "Chicago" is actually representing an infinitely small point at the center of that city.

For another example, imagine a traffic control application. Now we need to think of the city not as a point, but as the entire area of interest. In a street map of this city, we might consider intersections where roads cross to be points — not because an intersection is small in any absolute sense, but because, for practical purposes, we can ignore its size when thinking about the entire city. Alternatively, think of the point as representing the infinitely small center of the intersection.

When we turn our attention to lines and line segments, various physical examples come to mind. One of the first is a ruler. But how was the ruler made straight? And how could we check that it really is straight (without using another ruler)? It would be nice to have examples of line segments that don't provoke such awkward questions.

Perhaps the handiest way to make a physical example of a line segment is to take a skinny piece of string and stretch it tight — to get the shortest path from one end to the other. However, it is impossible to stretch a string totally tight. As long as the string has mass and there is some gravitational attraction on this mass, the string will sag a little, no matter how tight we pull it. For very short strings, the sag won't be noticeable, but it is there all the same. Surveyors, who often stretch chains from one point to another, sometimes have to take this into account.

Since antiquity, physicists have thought of light rays in a vacuum as traveling along straight lines, and much of physics is based on this assumption. Light rays in the near vacuum of outer space are often considered the best example of line segments. But when light rays pass into the atmosphere, or into water, or pass through the lenses of your glasses, they don't behave like the ideal lines of geometry.

As you can see, there seems to be nothing in nature that is absolutely straight enough to be a "true" line segment. As with points, we must think about a line segment as an ideal concept, which is approximated by physical examples such as light rays, rulers, and stretched strings. In this way of thinking, a theory of geometry is a model for the physical world, not an exact description of it. In this book, we present two models of geometry: For most of the book we deal with Euclidean geometry, but non-Euclidean geometry makes a brief appearance in Chapter 3.

APPLICATION: Scan Converting a Line Segment

Applying Euclidean geometry on a computer graphics screen is an interesting challenge because here the closest thing we have to points are little patches (Figure I.2 of the Introduction) of phosphor-based chemicals that can be turned on in various colors. These patches, called *pixels*, are very small and a typical screen might have over 600 rows and 600 columns of pixels. These pixels may have somewhat irregular shapes, and are jammed together with little or no space between them. Despite this, it is convenient to reduce the sizes of the pixels to small perfect circles, as in Figure 1.3, in order to make a neater picture. The pixels are evenly spaced so that, if we place a coordinate system on top of them, we find the pixel centers occupying all the points with integer x and y coordinates. What happens if we pick two pixels and ask the computer to show a segment connecting the two? What pixels should be lit up to show the segment?

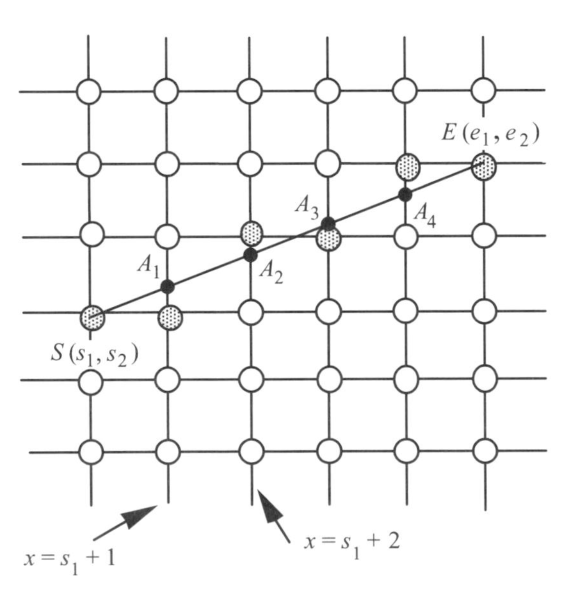

Figure 1.3 Vertical grid line crossing algorithm (pixels reduced).

If we are solving this problem pictorially, we can lay a ruler on the picture, connecting the center of the start pixel $S(s_1, s_2)$ and the center of the end pixel $E(e_1, e_2)$, and we can pick out some pixels near the ruler line. But the computer has to solve this problem in a purely numerical way, without a picture or ruler. Input to the problem is the coordinates of the pixel centers, S and E. The desired output is a list of "in between" pixels, each one identified by the coordinates of its center point. The problem of finding this list is called *scan converting* a line segment.

Various methods for scan conversion have been proposed. Here is one, based on elementary ideas you have learned before, which gives reasonable output when $|\text{slope of } \overline{SE}| \leq 1$. (After you study it, see if you can determine what could go wrong if $|\text{slope of } \overline{SE}| \geq 1$.)

Vertical Grid Line Crossing Algorithm (applicable if $|\text{ slope of } \overline{SE}| \leq 1$):

1. Find the equations of the vertical grid lines between S and E. In Figure 1.3, these would be $x = s_1 + 1$, $x = s_1 + 2$, $x = s_1 + 3$, $x = s_1 + 4$.
2. We next find the intersections of segment $\overline{SE}$ with these vertical lines, the points A_1, A_2, etc.
3. For each point A_i, we find the closest pixel center on the same vertical line by rounding up or down, whichever is more appropriate.

EXERCISES

**Marks challenging exercises.*

Use your knowledge of geometry, as you learned it in high school, to solve these problems. In some cases, what is wanted is intelligent discussion rather than any particular answer.

Approaches to Geometry

1. Create four right triangles by picking one leg to be length 1 inch and the other leg to have a randomly chosen larger length, b. You don't actually have to do any drawing. Then calculate the hypotenuse by the Pythagorean theorem. Now apply the "one-third rule" (hypotenuse is the longer leg plus one-third the shorter) and see how close this comes to being correct. Measure the percentage error by:

$$\%\ \text{error} = 100\% \times (\text{true hypotenuse} - \text{one-third rule hypotenuse})/\text{true hypotenuse}.$$

2. For the previous exercise, express the percentage error in terms of b.

3. In Exercise 1, how big do you think the percentage error could possibly be in absolute value? You can attack this as a calculus problem, using the formula found in Exercise 2.

4. In Exercise 1, if we confine ourselves to creating small triangles, say, $1 \leq b \leq 3$, can we detect that the "one-third rule" is false if we can only measure with a ruler to the nearest quarter of an inch (every measurement is rounded to the nearest quarter inch)? Furthermore, whenever a hypotenuse h is calculated (by either rule), the answer is rounded off to the nearest quarter inch. You can answer these questions by actually drawing and making measurements. Alternatively, make a table in which b varies from 1 to 3 in steps of a quarter inch and the hypotenuse is calculated (with rounding as described) two ways.

5. Suppose you had to seed a soccer field that is presently bare earth. You would like to know how much seed to buy so you don't have to go back to the store too often. Could you do this without measurement and without theory? If so, how? Could you do it easier with measurement and theory? How? How could you find out how much ground a box of seed will cover?

6. Suppose $SE = ET = 1$ in Figure 1.1 and $S = (1, 1)$ as shown. By studying triangle SET determine if there are any choices for θ_1 and θ_2 that will result in the tip T being at (2, 1) as shown. How many choices are there?

7. An experimenter is raising a thermometer into the atmosphere on a kite to radio back measurements at different heights. Normally there is an altimeter to measure the altitude, but it is broken today. You instead mark off lengths on the kite string, so you can measure how much string has been unwound. An assistant walks to a spot just under the kite (so that the triangle from kite to assistant to you has a right angle at the assistant) and then measures her distance to you. Illustrate this with a labeled figure and then explain with a formula how you could calculate the altitude of the kite. (Assume the kite string stretches from you to the kite in a straight line.)

Achieving Certainty

8. What makes a statement about geometry easy to believe? Is it conceivable that a person could find Roscoe's assertion easier to believe than the extendibility axiom? Explain.

9. From your recollection of geometry, do you remember any theorems that were surprising to you when you first learned them? Give one or more examples. Do you remember any theorems that seemed obvious? Give one or more examples. Why would your time be taken up by theorems that don't tell you anything you did not already know?

10. When we see an elephant standing on one foot in the circus it is pretty amazing, but it is not a very practical way for an elephant to stand. But when mathematicians selected axiom systems for geometry, they often tried to have as few axioms as possible. Is it really a good idea for all of geometry to rest on the smallest possible base of assumed truths? Or is this analogous to the elephant showing off his tricks?

Scan Conversion of a Line Segment

11. In Figure 1.3, if $S = (5, 12)$, $E = (10, 14)$, find the coordinates of A_1, A_2, A_3, A_4.

12. In Figure 1.3, if $S = (6, 20)$, $E = (10, 18)$, find the coordinates of all the A_i. Then find the centers of the pixels to light up.

13. In Figure 1.3, find a formula for the y coordinate of A_i in terms of i, s_1, s_2, e_1, e_2.

14. As you can see in Figure 1.3, the centers of the chosen (shaded) pixels don't necessarily lie exactly on a straight line. There are precisely three slopes for which these centers will lie on a straight line when we carry out the vertical grid line crossing algorithm. Determine what these slopes are and draw the pictures to illustrate.

15. Draw a figure to illustrate why the vertical grid line crossing algorithm presented here is not a good one for lines where $|\text{slope}| > 1$.

16. Devise a variant of the vertical grid line crossing algorithm that performs well if $|\text{slope}| > 1$.

17. (a) Given the center of a pixel, say, (s_1, s_2), give formulas in terms of s_1 and s_2 for the centers of the four pixels that are neighbors of the one containing (s_1, s_2).

    (b) Here is a proposal for a scan conversion algorithm. First find the centers of the neighbors of the start pixel. If any of them is E, no intermediate pixels should be lit. Otherwise, find the neighbor pixel where the distance from its center point to E is shortest and light up that pixel. In case of a tie, pick at random. Now continue this process. Carry this out when the start and end pixels are centered at (3, 2) and (6, 3). As you do this, keep track of the number of additions, subtractions, and multiplications you need to do. (Scan conversion may have to be done many thousands of times to create a picture—methods that use a lot of arithmetic are not preferred.) Compare it to the amount of arithmetic done in the vertical grid line algorithm.

(c) Will the pixels produced by the distance algorithm described in part (b) be the same as the ones produced by the vertical grid line crossing algorithm?

*18. Let y_i be the y coordinate of A_i. Let r_i be y_i rounded to the nearest integer (if y_i has 0.5 as its fractional part, round up). Define $d_i = y_i - r_i$.

(a) Show how you can find r_{i+1} from r_i and d_{i+1} and $m = (e_2 - s_2)/(e_1 - s_1)$, the slope of $\overline{SE}$.

(b) Give an algorithm for finding d_{i+1} from d_i and m.

(c) Combine your answers to parts (a) and (b) to form an algorithm for finding all the centers $(s_1 + i, r_i)$ of the pixels to light up.

(d) Modify your algorithm so that it contains only integer quantities.

1.2 Proofs in Axiomatic Geometry

As we have seen, we need axioms to start on the process of proving theorems—so it appears that listing our axioms should be our next order of business. But just as a tool is designed based on how it will be used, we have chosen our axiom set to fit a modern style of proof. Before discussing our axioms in the next section, we pause to discuss how the concept of proof has evolved in geometry.

In the ancient world a good deal of geometric knowledge was accumulated by a variety of cultures, including the Babylonians, Chinese, Egyptians, and the Hindus, beginning as early as 2000 B.C. In all of these cultures, up until about 600 B.C., arguments were sometimes offered for geometric assertions. For example, in the *Sulvasutras*, which date from 600 B.C. or earlier in India, there is a hint of a proof of a special case of the Pythagorean theorem. In around 1000 B.C. a Chinese work called the *Jinzhang Suanshu* offered a sketch of a proof of this theorem in somewhat greater generality. However, neither of these works presented comprehensive proofs for every proposition. It appears that many propositions were accepted because they seemed obvious to the eye or because they had been checked by measurement in special cases.

As one might expect, without rigorous standards of proof, a lot of errors and crude approximations were accumulated along with correct principles. For example, it was common to compute the circumference of a circle by multiplying the diameter by 3. This approximate formula even made its way into the Old Testament (I Kings, 7:23). For another example, the Egyptians were accustomed to computing the area of any quadrilateral, even if it was not a rectangle, by the formula used for rectangles.

Around 300 B.C., a Greek scholar named Euclid published a geometry book called the *Elements*, which forever put an end to the old style of doing geometry. This great work was actually the capstone of work done by many geometers over the preceding few hundred years, beginning with Thales (c. 640–546 B.C.). Not only were proofs required in this approach to geometry, but the standard of proof was demanding enough — "rigorous" enough, to use the modern jargon — that this standard was accepted until late in the nineteenth century. Proofs were required to be carried out by the axiomatic method — something we'll soon see in detail. This approach to geometry resulted in a significant expansion of securely established geometric knowledge.

During the nineteenth century, it became clear that the Euclidean standards of proof were not as high as they could be: Steps in proofs were often accepted because they seemed obvious to anyone looking at the picture. A new standard arose, one we'll call *symbolic rigor.* This style of geometry rejects the evidence of our eyes in doing proofs and demands that every step be justified by an axiom or previous theorem. A consequence of this is that the symbols of a proof (the words and mathematical notations) must be convincing by themselves. If a picture accompanies the proof, as it usually does, it is only a crutch.

Interestingly, the symbolic approach to geometry has some roots in the Euclidean era, even though it did not really blossom until the work of geometers such as Moritz Pasch (1843–1930) and David Hilbert (1862–1943). For example, Aristotle (384–322 B.C.) proclaims: "The geometer bases no conclusion on the particular line which he has drawn being that which he has described, but [he refers to] what is *illustrated* by the figures." Euclid's extendibility axiom (previous section) provides a more concrete example of the early desire to replace visual intuition with verbal reasons in proofs. This axiom was included in the *Elements* because, in many proofs, it is necessary to extend a line segment and Euclid wanted to have an axiom to justify this step. But suppose the axiom had been left out and, whenever a line segment was extended in a proof, no reason was given to justify it. Who would object? A reader who believed that a proof only had to be visually persuasive would surely not object. By including this axiom, it appears that Euclid was aspiring to the symbolic standard and was not satisfied with visual acceptability.

But Euclid was inconsistent. Here is a "proof" that depends on something being visually obvious, which is typical of proofs found in Euclid's *Elements*. It would be regarded as flawed according to the standards of symbolic rigor.

ASSERTION 1

If two sides of a triangle are congruent, say, $AB = BC$ in Figure 1.4, then the angles opposite these sides are congruent, $m\angle A = m\angle C$.

PROOF

Let M be the point where the angle bisector of angle B meets segment $\overline{AC}$. Now examine triangles BAM and BCM. These are congruent by "side-angle-side." Thus $\angle A$ is congruent to $\angle C$ since corresponding parts of congruent triangles are congruent. ■

For many years, this proof appeared in textbooks and was considered a totally convincing proof. However, mathematicians with the highest standards of rigor would say the proof has at least one step we have not justified: how do we know that the angle bisector crosses line $\overleftrightarrow{AC}$? And if it does, how do we know the crossing M is between A and C? Visually speaking, these assertions seem obvious. But if we are aiming

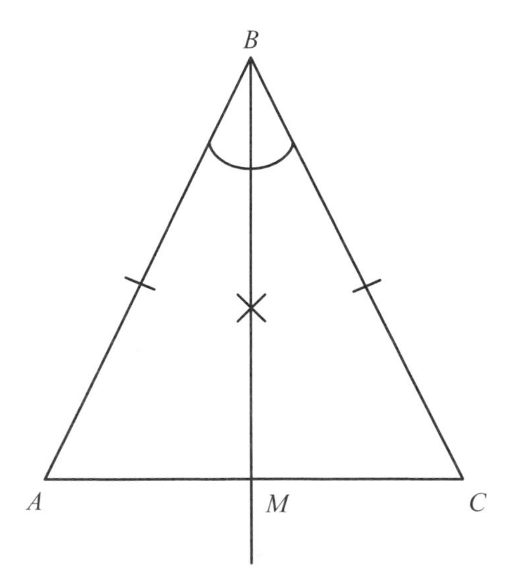

Figure 1.4 Base angles of an isosceles triangle are equal.

for symbolic rigor, the standard of evidence is higher than "It looks like it to me, at least in this case." Of course this doesn't mean that the theorem should be rejected. A devotee of symbolic rigor would try to fill the gap in the proof or look for another proof. For this theorem, other proofs are available; we give one in the next chapter.

But there were other theorems for which no airtight proofs could be found using the axioms set forth by Euclid. It became necessary to add some axioms to Euclidean geometry that Euclid had never contemplated. We return to this matter in the next section when we set forth our axiom system.

If you thought that the argument given for Assertion 1 was convincing, and if you thought that time spent on showing M exists and lies between A and C is time wasted, then you may doubt whether symbolic rigor has any merit. Here are some things in its favor.

Our eyes sometimes deceive us. Just as the magician's hand is quicker than the eye, once in a while an invalid proof based on an inaccurate picture can fool the eye. Here is an example of a ridiculous assertion with a "proof" in which it is hard to find the fault.

ASSERTION 2

Every point inside a circle is actually on the circumference of the circle (Figure 1.5).

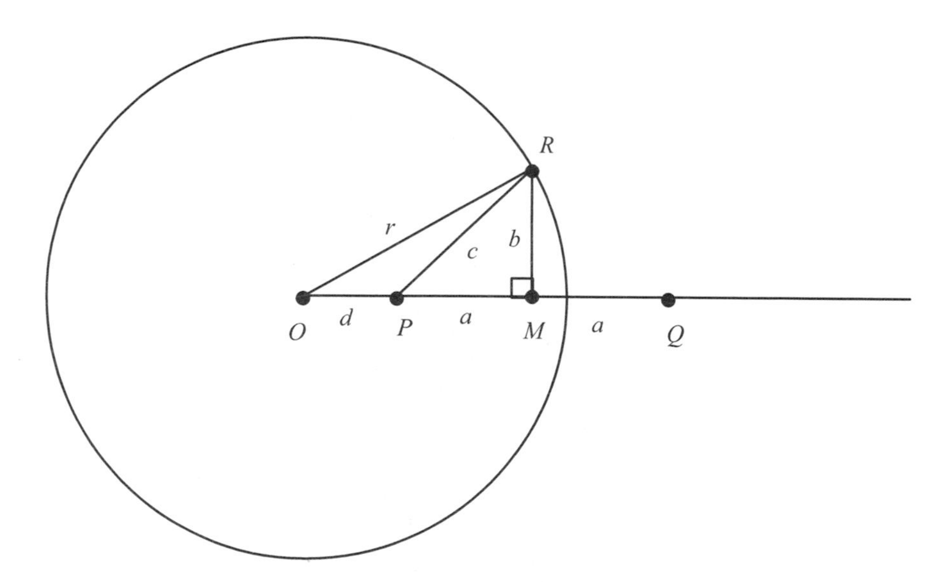

Figure 1.5 The inside of a circle is really on the boundary.

PROOF

Let P be any point inside the circle of radius r centered at O. On ray $\overrightarrow{OP}$, find point Q so that $OQ = r^2/OP$. Choose M as the midpoint of $\overline{PQ}$. Erect a perpendicular to $\overline{OQ}$ at M, meeting the circle at R.

1.	$r^2 = d(d + 2a)$	Construction of Q.
2.	$= (a + d)^2 - a^2$	Algebra
3.	$r^2 + a^2 = (a + d)^2$	Transposing in 2.
4.	$r^2 + a^2 + b^2 = (a + d)^2 + b^2$	Adding b^2 to both sides.
5.	$a^2 + b^2 = c^2$	Pythagorean theorem
6.	$(a + d)^2 + b^2 = r^2$	Pythagorean theorem
7.	$r^2 + c^2 = r^2$	Substituting 5 and 6 in 4.
8.	$c = 0$	Algebra applied to 7.
9.	$P = R$	Step 8 says their distance apart is 0. ■

What we have "proved" is obviously incorrect, but where is the flaw? Our eyes have been fooled by an inaccurate picture. A more accurate picture would show that M lies outside the circle, so R does not even exist. Thus steps 5 and 6 of our "proof" are invalid since they deal with triangles that don't exist. An adherent of symbolic rigor would not accept step 5 unless a proof could be given for M being inside the circle. Obviously, no such proof can be given.

People do not always agree on what is obvious or correct. For example, during the time when the Greek geometers were listing the extendibility axiom as self-evident, Greek cosmologists were asserting that the entire material world, including the earth, sun, moon, planets, and stars, was contained in one large sphere. There was nothing beyond that sphere and no way to move outside that sphere. The universe had an outer wall. In this view of things, a line segment with an endpoint on that sphere could not be extended beyond that point.

The contradiction between the extendibility axiom and the "one large sphere" view of the universe did not seem to bother the Greeks too much. It does play a role in one of the sadder episodes of European intellectual history though. In the year 1600 Giordano Bruno was driven in an oxcart through the streets of Rome to be burned at the stake in a public square. The path was lined with onlookers who shouted to him to recant his heretical thoughts. What were these thoughts that so enraged the authorities of the day? Bruno was a philosopher, well ahead of his time, and one of his ideas was to take seriously Euclid's notion that line segments could be extended. This appeared to make the universe infinite. Bruno went further to believe that, since the universe went on forever, there would be other worlds like the earth that contained intelligent life. Thus, he was perhaps the first believer in extraterrestrial intelligence. This was flatly contradictory to prevailing religious beliefs about the creation of the human race and the earth.

Modern scientists no longer argue whether the entire universe is inside some big sphere — no one today believes in such a sphere. However, other disagreements have arisen about the geometric nature of the universe. Do this mental experiment. Imagine an ant crawling around the equator of a large sphere, like the earth. At any point the ant can always crawl more, so it seems that an extendibility property of sorts is true in the ant's world. But, whether the ant can detect this or not, eventually it will circle back around to where it started. So this is not the sort of extendibility Euclid meant. A similar conclusion would be true if the ant were crawling on a large donut. There are no walls to stop a crawl, but the "world" of the donut, like that of the sphere, is not infinite in size. Some physicists maintain that our universe is a sort of three-dimensional version of a sphere or donut in the sense that there are no walls that a light ray would bump into as it hurtles through space, but the light ray wouldn't "go to infinity" because the universe is not infinitely large. In some way or other, the light ray would "go around in circles." These scientists claim that we are unable to notice this strange shape of our universe because the universe is large and our experience of it is limited. By analogy, think of the ant on the sphere. If the sphere is very large, how could the ant tell that it was not an infinite plane?

Since people do not agree on what is "obviously" true, we need to justify each step of a proof — even those that seem obvious to us — by recourse to assertions that have been agreed on in advance, namely, the axioms.

There are problems where neither looking at pictures (nor measuring) can help. One such example was the famous problem of trisecting an arbitrary angle using ruler and compass. The Greeks understood how to cut any given angle in half — this problem has an easy solution often studied by beginning students in geometry (see the exercises). It seemed logical that one also ought to be able to use the ruler and compass to cut an angle into three equal parts. However, geometers were unable to find a way to carry out the trisection for 2000 years. Staring at figures and measuring their parts was not the least bit helpful. In the nineteenth century it was demonstrated that trisecting any given angle with ruler and compass was impossible. Of course, this does not mean that there are no trisecting rays for the angle in question. What it means is that the allowable tools, ruler and compass, are inadequate to construct those rays. This is a bit analogous to discovering that you cannot drive to Europe from North America. It doesn't mean that Europe does not exist — it merely means that the network of roads, bridges, and tunnels won't get you there. Returning to the matter of trisection with ruler and compass, not only was studying a picture not helpful in finding a trisection method, it was also useless in understanding why no such construction could exist. The proof of impossibility used abstract algebra, a theory which is notable for proceeding in a purely symbolic way, with no help from pictures.

Without symbolic rigor, people sometimes disagree about whether a proof is right. There is at least one important episode in the history of mathematics where the shortcomings of the Euclidean standard of proof led to uncertainty about an important issue in the minds of some of the best mathematicians. During a period of 29 years, the eminent mathematician Adrien Marie Legendre (1752–1833) published numerous "proofs" that one of Euclid's axioms, one concerning parallel lines (which we will encounter in the next section), could actually be proved from the other axioms. Finding such a proof had long been a sort of holy grail of geometry, easily as famous as that of trisecting an angle. Legendre's attempts at proof were not widely accepted. This standoff between Legendre and his doubters would have been avoided if symbolic rigor had made its appearance during Legendre's time. We return to this issue in the next two chapters.

Geometric theorem proving software requires symbolic rigor. An interesting side effect of symbolic rigor, probably not anticipated by the mathematicians of yesteryear who originally promoted it, is the fact that computer software systems for proving geometric theorems are based on manipulating symbols in the computer's memory. For this reason, an understanding of symbolic rigor is essential for those wishing to devise such software. This is ironic, because symbolic rigor has often been criticized as complicated to the point of being impractical, whereas computers are considered very practical.

Surprisingly, despite these numerous arguments in favor of symbolic rigor, it is little used in research, and in applications of geometry. It makes its appearance mainly in teaching some advanced geometry courses. Some reasons for this include the following:

1. *Seeing is believing.* Mathematicians aspire to transcend this proverb, but we are only human. It is no accident that the word *theorem* comes from the Greek word *theorein*, which means *to look at*.

2. *Few or no theorems have had to be recalled.* If the grading standard for a course is raised, more students will probably fail. Analogously, because symbolic rigor is a tougher standard of proof than the Euclidean standard, one might expect that some geometric assertions that had been widely regarded as correctly proven by the Euclidean standard would turn out to be not provable when the symbolic standard of rigor is used. This never happened to any significant degree. What has happened is that new, longer proofs have been proposed for the old, familiar theorems of Euclid.

3. *Symbolic rigor often leads to long, complicated proofs of painfully obvious statements.* This makes geometry time consuming, never a good thing for people who want to make progress quickly. It also makes geometry hard to learn for all but the most advanced students.

As our last word on this awkward issue of how much rigor to insist on, the following quotation from Heath shows that the issue is age-old and was already being disputed in Euclid's time. This quotation pertains to Euclid's proof of the following theorem, which many would regard as not requiring proof at all.

✯ THEOREM: *The Triangle Inequality*

In any triangle, each side is shorter than the sum of the other two side lengths.

In Sir Thomas Heath's *A History of Greek Mathematics*, he reports (Figure 1.6):

> It was the habit of the Epicureans, says Proclus, to ridicule this theorem [the triangle inequality] as being evident to even an ass and requiring no proof, and their allegation that the theorem was "known" even to an ass was based on the fact that, if fodder is placed at one angular point and the ass at another, he does not, in order to get his food, traverse the two sides of the triangle but only the one separating them ... Proclus replies truly that a mere perception of the truth of a theorem is a different thing from a scientific proof of it and a knowledge of the reason why it is true.

Figure 1.6 Jack forgets the triangle inequality.

We'll have a short look at symbolic rigor in the next section as we begin proving theorems in Euclidean geometry. After that section, we return to the more relaxed standard established by Euclid and still used in most research, applications, and teaching.

EXERCISES

**Marks challenging exercises.*

1. (a) Using the notation of Figure 1.5, verify line 2 of the proof and prove that $(d+a)^2 = r^2 + a^2$. (*Hint*: Don't use the picture, which is, after all, wrong.)

   (b) Use part (a) to prove $d + a \geq r$, proving that M is not inside the circle.

2. (a) Suppose ABC is a triangle with $m\angle B = 90^\circ$. Does the following construction create a trisecting ray? Assume that each construction step can be carried out and that each intersection point actually does exist.

   (i) Extend $\overline{BC}$ to a segment $\overline{BD}$, which is twice as long.
   (ii) With your compass opened to length BD, place the point at C and draw a circle, which meets line $\overleftrightarrow{AB}$ at E where A is between B and E. What are the angles of triangle BEC?
   (iii) Drop a perpendicular from B to line $\overleftrightarrow{EC}$, meeting it at F. What is $m\angle FBC$?

   Is $\overrightarrow{BF}$ a trisecting ray?

   (b) If you believe this construction creates a trisecting ray, why does this not contradict our discussion of trisection in this section?

3. Here is a construction of the angle bisector of $\angle ABC$. Do you think it meets the standard of symbolic rigor? Explain.

   (i) Draw a circle of any radius centered at B, meeting $\overleftrightarrow{BA}$ at E and $\overleftrightarrow{BC}$ at F.
   (ii) Draw circles of the same radius, centered at E and F, and let G be an intersection of these circles.

   Triangles BEG and BFG are congruent by side-side-side, and the measures of $\angle EBG$ and $\angle FBG$ are equal because these angles are corresponding parts.

4. Criticize the construction of a triangle with two right angles shown in Figure 1.7. Let two circles meet in points A and B. Let $\overline{AC}$ and $\overline{AD}$ be diameters of these two circles from A, C being on one circle and D on the other. Let E be the point

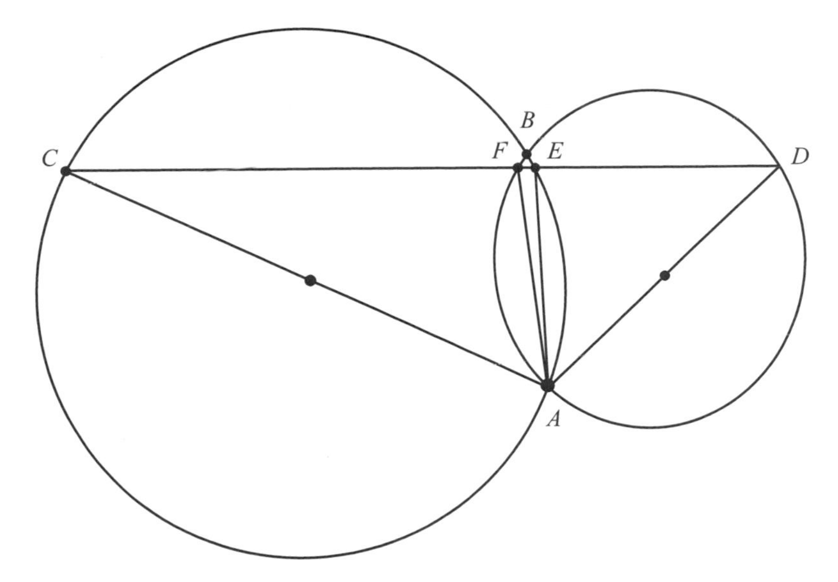

Figure 1.7 Does triangle *FEA* have two right angles?

where $\overleftrightarrow{CD}$ meets the circle containing C. Let F be the point where $\overleftrightarrow{CD}$ meets the circle containing D. Because $\angle AEC$ is inscribed in a semicircle, it is a right angle. Likewise $\angle AFD$ is a right angle. Thus triangle AFE has two right angles.

1.3 Axioms for Euclidean Geometry

In this section, we describe the axiom set we will use for our study of Euclidean geometry. (Some of these axioms are also used in the two non-Euclidean geometries described in Chapter 3.) We then carry out rigorous proofs of the first few deductions from this axiom set. If time is limited, establishing the spirit of the enterprise is perhaps more important than studying each axiom and theorem intensively.

Our axiom set is a descendant of the five axioms provided by Euclid, but there are some differences. There are two main reasons why we do not use Euclid's axioms as he originally gave them:

1. Euclid phrases his axioms in a way that is hard for the modern reader to appreciate.
2. It has been necessary to add axioms to Euclid's set in order to be able to give rigorous proofs of many Euclidean theorems.

A number of individuals, and at least one committee, have taken their turn at improving Euclid's axiom set; notably, David Hilbert in 1899, G. D. Birkhoff in 1932, and the School Mathematics Study Group (SMSG) working during the 1960s. Even though these axiom sets differ from one another, and from Euclid's, they all lead to the well-known theorems in Euclid's *Elements*. Consequently, we say that they are all axiom sets for Euclidean geometry. The axioms we list in this section for our use are a minor rewording of the SMSG axiom set.

As we embark on our study of axiomatic Euclidean geometry, you will be asked to consider proofs of some statements that seem obvious and many that you have learned before. To enter into the spirit of our study, you must put aside what you have learned before or find obvious. In earlier sections we have relied on some geometry you have studied before. But in this section, our proofs are constructed only from the axioms we are about to list and any theorems we have previously proved from those axioms. Keep in mind that our objective in our axiomatic discussion of Euclidean geometry is not to learn the most facts of geometry, but to learn about the logical structure of geometry and to practice the art of proof.

The Objects of Geometry: Points, Lines, and All That

However, whether we emphasize facts or logical structure, geometry is about geometric figures. One would suppose that a deductive study of geometry would begin by carefully defining the objects geometry deals with. Indeed, Euclid attempts to do just that by telling us, for example, that a point is "that which has no part." But to understand this, we need to know the meaning of *part*. We could look "part" up in a dictionary and find more words, which we could also look up, and so on. Logically speaking, there is no end to this. We'll get around this by leaving *point* and *line* and *plane* as *undefined terms* in our axiomatic system.

This appears to ignore the fact that, as a practical matter, we all have a pretty good idea of what these words mean. But there is no contradiction here. Geometry exists as a logical system (in which *point* is undefined), but also as a practical social enterprise — something we all do together and talk about and show each other pictures of. In geometry as a social enterprise, we have no trouble reaching a common understanding of *point*. For example, a teacher says the word *point* and draws a very tiny dot on the blackboard. Many experiences like this convey the customary interpretation of what a point is. So, it turns out there is no harm done in leaving it undefined in our axiomatic system.

Not every term in geometry needs to be left undefined. For example, we will shortly define line segment in terms of points and lines. In fact, every other term in geometry, from altitude to zonotope, can be defined in terms of our three

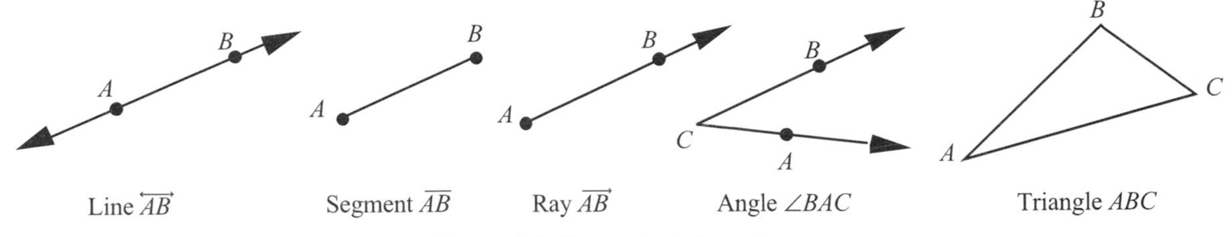

Figure 1.8 The cast of characters.

undefined terms. The most important geometric figures are shown in Figure 1.8. But here is something you might not expect: We will need axioms to make some of these definitions.

Axioms for Points on Lines

AXIOM 1: *The Point–Line Incidence Axiom*

A line is a set of points. Given any two different points, there is exactly one line which contains them.

We denote the line connecting A and B by $\overleftrightarrow{AB}$. Our first theorem about lines uses a style of proof called *proof by contradiction* or *indirect proof*. It is based on the idea that the truths of Euclidean geometry do not contradict one another. In particular, if you reason correctly on statements that are true, then you can never deduce a statement that contradicts another statement known to be true. If you do find a contradiction, then one of the statements you have been reasoning from must be false. In our proof we will make a supposition and show it leads to a contradiction. This proves the supposition false.

☆ THEOREM 1.1

Two lines intersect in at most one point.

PROOF

Suppose lines L and M contain the two points A and B. Then A and B would have two lines containing them, violating Axiom 1. This contradiction shows that our supposition that L and M contain two points in common must be false. ■

Our next axiom is just a mathematical way of saying what everyone who has ever used a ruler will find familiar: A line "comes with" a set of numerical markings that we can use for calculations and proofs.

AXIOM 2: *The Ruler Axiom*

For any line, there exists a 1:1 correspondence f between the points of the line and the real numbers. This means:

1. Every point A on the line has a number $f(A)$ associated with it.
2. Different points have different numbers associated with them.
3. Every number, positive, negative, and zero, has some point with which it is associated.

The function f is called the *ruler function* for that line. The number $f(A)$ is called the *coordinate* of A.

This axiom allows us to use the properties of the real numbers to find out things about lines. For example, there are infinitely many real numbers, so we must have infinitely many corresponding points on a line. The ruler axiom also allows us to define the key geometric ideas of distance and betweenness.

DEFINITIONS

Let A, B, and C be three points on a line and f be the ruler function for that line.

1. We say B is *between* A and C if either $f(A) < f(B) < f(C)$ or $f(C) < f(B) < f(A)$. We write $A\text{-}B\text{-}C$ to indicate that B is between A and C. $C\text{-}B\text{-}A$ has the same meaning as $A\text{-}B\text{-}C$.
2. The *segment* from A to B, denoted $\overline{AB}$, is defined to be the set consisting of A, B, and all points X where $A\text{-}X\text{-}B$.
3. The *distance* from A to B is defined to be $|f(B) - f(A)|$ and we denote this distance by AB. Note that if A and B designate the same point $AB = 0$.
4. If $AB = CD$ then the segments $\overline{AB}$ and $\overline{CD}$ are called *congruent*.

☆ THEOREM 1.2

1. $AB = BA$.
2. If $A\text{-}B\text{-}C$ then $AB + BC = AC$.
3. If A, B, C are three different points on a line, exactly one of them is between the other two.

Proof

1. $AB = |f(B) - f(A)| = |-[f(A) - f(B)]| = |f(A) - f(B)| = BA$.
2. There are two cases. First, suppose $f(A) < f(B) < f(C)$. Then

$$\begin{aligned} AB + BC &= |f(B) - f(A)| + |f(C) - f(B)| \\ &= f(B) - f(A) + f(C) - f(B) \\ &= f(C) - f(A) \\ &= |f(C) - f(A)| \\ &= AC \end{aligned}$$

   We leave the second case where $f(C) < f(B) < f(A)$, as an exercise.

3. Let a, b, c be the coordinates of A, B, C according to the ruler function for the line on which they lie. It is a well-known fact about numbers that, out of three different numbers, exactly one can lie between the other two. Consequently, by the definition of what it means for one point to be between two others, the result follows. ■

The following theorem was assumed as an axiom by Euclid. If Euclid had included our ruler axiom among his axioms, then, of course, he would not have needed to assume what we are about to prove.

☆ THEOREM 1.3: *Extendibility*

If A and B are any two points, then the segment $\overline{AB}$ can be extended by any positive distance on either side of segment $\overline{AB}$ (Figure 1.9).

Proof

Let $e > 0$ be the amount of extension wanted and let's say we want to extend past B to a point C so that B is between A and C and $BC = e$. Let a and b be the real numbers $f(A)$ and $f(B)$ under the ruler function for the line $\overleftrightarrow{AB}$.

Case 1: $a < b$ (Figure 1.9). Then define $c = b + e$. By part 3 of the ruler axiom, there is a point C which corresponds to the number c. C is the point we want since:

(a) B is between A and C (since $a < b < b + e$)

(b) $BC = |f(C) - f(B)| = |(b + e) - b| = |e| = e$ since $e > 0$. To extend on the other side of the segment, find the point corresponding to $a - e$.

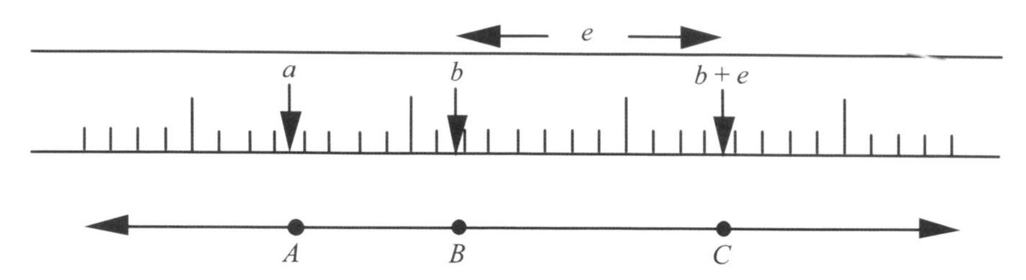

Figure 1.9 The ruler function helps extend segment $\overline{AB}$.

CASE 2: $a > b$. We leave this as an exercise. ■

✯ THEOREM 1.4: *The Midpoint Theorem*

Every segment has a midpoint. That is, for any points A and B, there is a point M on segment $\overline{AB}$ so that $AM = AB/2$.

PROOF

We leave this as an exercise. ■

The midpoint theorem suggests that we think of points as infinitely small. If they had any positive size, a line segment would be a bit like a necklace (Figure 1.10). If the number of points were even, there would not be one exactly in the middle as required by Theorem 1.4.

DEFINITION

If A and B are distinct points, the *ray* from A through B, denoted $\overrightarrow{AB}$, is the set of all points C on line $\overleftrightarrow{AB}$ such that A is not between B and C. We call A the *endpoint* of the ray.

The negative phrasing of this definition is sometimes awkward, so the following theorem is sometimes handy. Its proof is based on one of our previous results. Can you find it?

✯ THEOREM 1.5

$\overrightarrow{AB}$ consists of segment $\overline{AB}$ together with all points X where A-B-X. ■

Figure 1.10 A segment $\overline{AB}$ of six "fat" points would not have a midpoint.

DEFINITION

Let A, B, and C be points which are not on the same line. $\overrightarrow{AB} \cup \overrightarrow{AC}$ is called the *angle BAC* and denoted $\angle BAC$ (Figure 1.8). We may also denote this angle $\angle CAB$.

When we think of an angle we often think of it as the space between the rays that border it. Our definition of angle does not try to capture that idea. We need a separate definition of the *interior* of an angle, and we base it on the next axiom.

Separation

AXIOM 3: *Pasch's Separation Axiom for a Line*

Given a line L in the plane, the points in the plane which are not on L form two sets, H_1 and H_2, called *half-planes*, so that:

(a) If A and B are points in the same half-plane, then $\overline{AB}$ lies wholly in that half-plane.

(b) If A and B are points not in the same half-plane, then $\overline{AB}$ intersects L.

H_1 and H_2 are also called *sides* of L. L is called the *boundary line* of H_1 and H_2.

Notice that the half-planes mentioned in Pasch's axiom do not contain their boundary line. They are sometimes referred to as *open* half-planes for this reason.

DEFINITION

Let A, B, and C be three noncollinear points, as in Figure 1.11. (This means there is no line which contains all three of them.) Let H_B be the half-plane determined by $\overleftrightarrow{AC}$ which contains B. Let H_C be the half-plane determined by $\overleftrightarrow{AB}$ which contains C. The *inside* or *interior* of $\angle BAC$ is defined to be $H_B \cap H_C$.

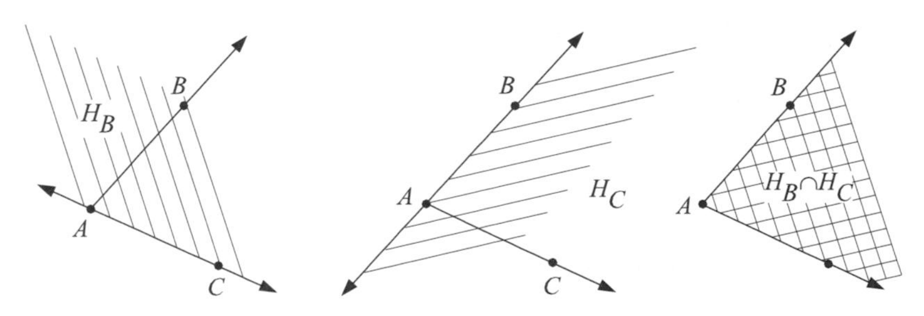

Figure 1.11 An angle and its interior.

Pasch's axiom was only added to the axiom set for Euclidean geometry in the late nineteenth century when geometers became aware that, for many geometric figures, there was no way to rigorously define the inside or outside of the figure, much less prove theorems about the insides and outsides. For example, if you had asked Euclid to prove that if a line contains a point on the inside of an angle then it must cross at least one of the rays making the angle, he would have been unable to do so. He would undoubtedly have been unconcerned about this, thinking this theorem to be too obvious to bother with.

✯ THEOREM 1.6

If a ray $\overrightarrow{AB}$ has endpoint A on line L, but B does not lie on L, then all points of the ray, except for A, lie on the same side of L as B.

PROOF

The proof is indirect. Assume there is a point C on the ray so that C and B are on opposite sides of L. By Pasch's axiom, $\overline{BC}$ crosses L at some point. This must be A since $\overline{BC} \subset \overleftrightarrow{AB}$ and, by Theorem 1.1, $\overleftrightarrow{AB}$ crosses L in just one point, namely, A. Since A is not B or C, the fact that A is in $\overline{BC}$ means B-A-C. But this means C is not in $\overrightarrow{AB}$ by the definition of a ray. ■

Triangles play a starring role in geometry, and now it is time to define them. Let A, B, C be three points that are not collinear. In that case, we define the *triangle* ABC to be $\overline{AB} \cup \overline{BC} \cup \overline{AC}$. Can you see how to define the interior of a triangle?

Suppose we have a triangle ABC and we extend side AC to D, thereby creating an exterior angle $\angle BCD$ as in Figure 1.12. Pick any point M on $\overline{BC}$ that is not B or C, then extend $\overline{AM}$ past M to a point E. Will E be in the inside of the exterior angle $\angle BCD$? Our visual intuition says yes, but if we want the highest degree of rigor, we need a proof. Here it is, but with reasons for some steps left out for you to supply.

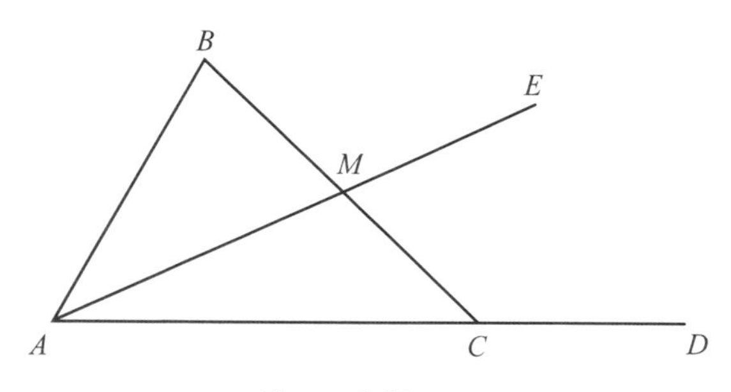

Figure 1.12

★ THEOREM 1.7

If A, B, and C are not collinear and

1. A-C-D
2. B-M-C
3. A-M-E

then E is in the interior of $\angle BCD$.

PROOF

According to our definition of the interior of an angle, we need to show two things:

(a) that E is on the same side as D of line $\overleftrightarrow{BC}$, and

(b) that E is on the same side as B of line $\overleftrightarrow{CD}$.

(a) A and D are on opposite sides of $\overleftrightarrow{BC}$. (Why?) A and E are on opposite sides of $\overleftrightarrow{BC}$. (Why?) Thus, we have shown both E and D to be on the opposite side from A of line $\overleftrightarrow{BC}$. Therefore, E and D must be on the same side of $\overleftrightarrow{BC}$ as we wished to prove.

(b) By hypothesis, B is not on $\overleftrightarrow{AC}$, so B is not on $\overleftrightarrow{CD}$ since $\overleftrightarrow{AC}$ and $\overleftrightarrow{CD}$ are the same line. Thus, by the previous theorem, $\overrightarrow{CB}$ lies wholly on the B side of $\overleftrightarrow{CD}$ (except for C). But B-M-C means M is on $\overrightarrow{CB}$ and so M and B lie on the same side of $\overleftrightarrow{CD}$. Likewise, E and M lie on the same side of $\overleftrightarrow{AC}$. Thus, E and B lie on the same side of $\overleftrightarrow{CD}$ as we wished to prove. ■

Axioms for Measuring Angles

We have spoken of angles, but not about measuring them. To fill this gap, we come now to a group of axioms that does for angles what the ruler axiom does for lines. We might refer to them, as a group, as the *protractor axioms.*

AXIOM 4: *The Angle Measurement Axiom*

To every angle, there corresponds a real number between 0° and 180° called its *measure* or *size*. We denote the measure of $\angle BAC$ by $m\angle BAC$.

DEFINITION

If $m\angle BAC = m\angle PQR$ then we say $\angle BAC$ and $\angle PQR$ are *congruent* angles.

AXIOM 5: *The Angle Construction Axiom*

Let $\overrightarrow{AB}$ lie entirely on the boundary line L of some half-plane H. For every number r where $0° < r < 180°$ there is exactly one ray $\overrightarrow{AC}$ where C is in H and $m\angle CAB = r$.

AXIOM 6: *The Angle Addition Axiom*

If D is a point in the interior of $\angle BAC$, then $m\angle BAC = m\angle BAD + m\angle DAC$.

DEFINITION

If A-B-C, and D is any point not on line $\overleftrightarrow{AC}$, then the angles $\angle ABD$ and $\angle DBC$ are called *supplementary* (Figure 1.13a).

AXIOM 7: *The Supplementary Angles Axiom*

If two angles are supplementary, then their measures add to 180°.

Now consider two lines crossing at X, making four angles as in Figure 1.13b. Each angle has two neighboring supplementary angles and one that is "across" from it. For example, $\angle AXB$ is across from $\angle A'XB'$. An angle and the one across from it are said to be *vertical angles*[3] or to form a *vertical pair.* A technical definition goes like this: If A and A' are points on one line where A-X-A' and B and B' are on the other line with B-X-B, then $\angle AXB$ and $\angle A'XB'$ are vertical angles. In addition $\angle AXB'$ and $\angle A'XB'$ are vertical angles. Notice that an angle cannot be vertical by itself—it is only vertical in relation to another.

✯ THEOREM 1.8: *Vertical Angles Theorem*

Vertical angles are congruent.

Proof

We use the notation of Figure 1.13. $\angle AXB$ and $\angle BXA'$ are supplementary. Likewise, $\angle BXA'$ and $\angle A'XB'$ are supplementary. These facts, together with Axiom 7, give the equations

$$m\angle AXB + m\angle BXA' = 180°,$$
$$m\angle A'XB' + m\angle BXA' = 180°.$$

Subtracting one equation from the other, we deduce $m\angle AXB = m\angle A'XB'$. ■

---

[3] This terminology really makes no sense. It is traditional so we use it—but with apologies.

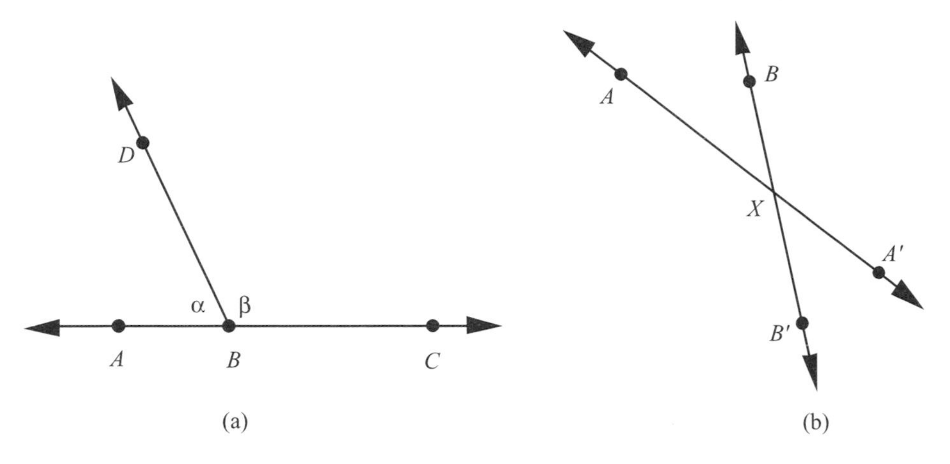

Figure 1.13 (a) Supplementary angles: $\alpha + \beta = 180°$. (b) $\angle AXB$ and $\angle A'XB'$ are vertical angles; $\angle BXA'$ and $\angle B'XA$ are vertical angles.

Axioms for Congruence and Parallelism

The next two axioms are real workhorses of Euclidean geometry so we just list them here, saving a more extensive discussion for the next chapter.

AXIOM 8: *The Side-Angle-Side (SAS) Congruence Axiom*

If

1. one angle of a triangle is congruent to a certain angle of a second triangle, and if
2. one side forming the angle in the first triangle is congruent to a side forming the congruent angle in the second triangle, and if
3. the remaining side forming the angle in the first triangle is congruent to the remaining side forming the congruent angle in the second triangle,

then the triangles are congruent.

AXIOM 9: *Euclid's Parallel Axiom*

Given a point P off a line L, there is at most one line in the plane through P not meeting L.

Axioms for Three-Dimensional Geometry

The axioms we have given so far describe matters in one single plane. This is adequate for most of our work since plane geometry is our main objective. But for completeness and for some brief applications of three dimensions that appear in this book, we now ask what needs to be assumed to deal with the third dimension. First, all the previously mentioned axioms are still true with the understanding that they hold for all the planes in the three-dimensional space. For example, Pasch's separation axiom needs to be understood as holding for every plane which contains L. Likewise, the parallel axiom needs to be understood as holding not just for "the plane," but for any plane. But, in addition to these reinterpretations, we need some extra axioms.

AXIOM 10: *Point–Plane Incidence Axiom*

A plane is a set of points. Given any three points, there is at least one plane containing them. If the points are not collinear, there is exactly one plane passing through them.

AXIOM 11: *Plane Intersection Axiom*

Two distinct planes either don't intersect or intersect in a line.

AXIOM 12: *Nontriviality*

(a) Every plane contains at least three points which are not all on the same line.

(b) There exist at least four points which are not all in the same plane.

AXIOM 13: *Line–Plane Incidence Axiom*

If two points of a line are in a plane, then the line lies entirely in that plane.

AXIOM 14: *Pasch's Separation Axiom for a Plane*

Given a plane N, the points which are not on N form two sets, S_1 and S_2, called *half-spaces*, with these properties:

(a) If A and B are points in the same half-space, then $\overline{AB}$ lies wholly in that half-space.

(b) If A and B are points not in the same half-space, then $\overline{AB}$ intersects N.

S_1 and S_2 are also called *sides* of N. N is called the *boundary plane* of S_1 and S_2.

Axioms for Area and Volume

Even without a logical study of geometry, we all know that area concerns itself with the inside of a figure — for example, a polygonal region. But the proper definition of a polygonal region requires a bit of work. Exercises 16 and 17 at the end of this section will give an idea of what is involved. In three dimensions, there are similar difficulties in defining the insides of solids. We'll skip the details and get right to the axioms, which, fortunately, are very simple.

AXIOM 15

To every polygonal region there corresponds a definite positive real number called its *area*.

AXIOM 16

If two triangles are congruent, then they have the same area.

AXIOM 17

Suppose that the polygonal region R is the union of two polygonal regions R_1 and R_2 which intersect at most in a finite number of segments and individual points. The area of R is the sum of the areas of R_1 and R_2: $\text{area}(R_1 \cup R_2) = \text{area}(R_1) + \text{area}(R_2)$.

AXIOM 18

The area of a rectangle (a four-sided figure whose angles all measure 90°) is the product of the length of the base and the length of the height.

Although we do not use them in this book, for completeness we finish with the following axioms concerning volume:

AXIOM 19

The volume of a rectangular parallelepiped is equal to the product of the length of its altitude and the area of its base.

AXIOM 20: *Cavalieri's Principle*

Suppose two solids and a plane are given. Suppose also that every plane which is parallel to the given plane either does not intersect either solid or intersects both in planar cross-sections with the same area. In that case, the solids have the same volume.

EXERCISES

**Marks challenging exercises.*

Axioms for Points on Lines

1. Which of the three properties of f in the ruler axiom ensures that a line is infinite in extent?

2. Which of the three properties of f in the ruler axiom ensures that for any two different points A and B, $AB \neq 0$.

3. Suppose we defined distance as $AB = [f(B) - f(A)]^2$. Which facts about distance proved Theorems 1.2, 1.3, and 1.4 are no longer true, and which are still true?

4. Suppose the absolute value were dropped from the definition of distance. Would parts 1 and 2 of Theorem 1.2 still be true? Explain.

5. Suppose we change the definition of what it means for A-B-C by changing the strict inequalities ($<$) to nonstrict inequalities ($\leq$). If we keep our definition of segment unchanged, will the nature of segments be any different from before? What about the nature of rays?

6. Suppose we took Theorem 1.1 as an axiom and deleted Axiom 1 from our set of axioms. Could you prove that two points determine exactly one line?

7. Prove case 2 of part 2 of Theorem 1.2.

8. Prove case 2 of Theorem 1.3.

9. Prove the midpoint theorem (Theorem 1.4). [*Hint*: How is $f(M)$ related to $f(A)$ and $f(B)$?]

10. Which previous theorem is needed to establish the characterization of rays in Theorem 1.5? Can you give a proof of Theorem 1.5?

    In each of the next four exercises, you are asked to prove two sets are equal to one another. A common way to proceed is with two steps: Show that if X is any point in the first set, then it also lies in the second; then show that if X lies in the second, it also lies in the first.

11. Prove that $\overrightarrow{AB} \cap \overrightarrow{BA} = \overline{AB}$.

12. Prove that $\overrightarrow{AB} \cup \overrightarrow{BA} = \overleftrightarrow{AB}$.

13. Prove that if A-B-C then $\overline{AB} \cup \overline{BC} = \overline{AC}$.

14. Prove that if A-X-B and A-B-C then A-X-C.

*15. There is nothing in the axioms that directly states that the ruler functions for the different lines in the plane need to be related in any way. One might try to choose each function independently of the others. Here is a line of thought that undermines that idea. Suppose we had a set of ruler functions that made a consistent theory of Euclidean geometry; this means that it is not possible to deduce a theorem that contradicts an axiom or another theorem. Now take a single line L and replace its ruler function f_L by a new one g_L defined by $g_L(P) = 2f_L(P)$ for each point P on L. Show that this will allow you to deduce a contradiction to one of the axioms.

Separation

16. Give a definition of the interior of a triangle.

17. Suppose we define a quadrilateral like this: If A, B, C, D are any four points where no subset of three of them is collinear, then the *quadrilateral ABCD* is $\overline{AB} \cup \overline{BC} \cup \overline{CD} \cup \overline{DA}$. Figure 1.14 shows three types of quadrilaterals.
    (a) Can you propose a definition for each of these categories; that is, what property or combination of properties puts a quadrilateral in one of these categories instead of another?
    (b) Can you prove that your categories don't overlap (no quadilateral is in more than one)? Can you prove that every quadrilateral is in one of your categories?
    (c) Propose a definition for the interior of a quadrilateral (or, if you like, give a separate definition for each category).

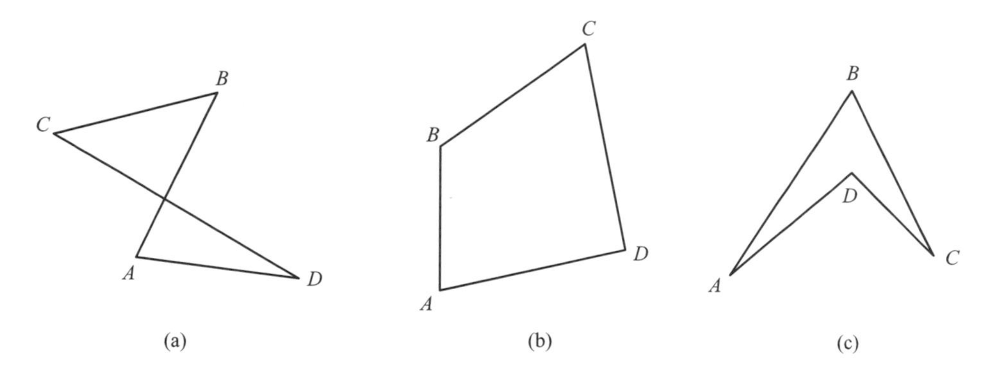

Figure 1.14 (a) Self-intersecting. (b) Not self-intersecting and convex. (c) Not self-intersecting and not convex.

18. Fill in the missing steps in Theorem 1.7 (respond to all the "Why's" in the proof.)

19. Prove that if A, B, and C are not collinear and B-D-C then D is in the interior of $\angle BAC$.

20. Suppose A, B, and C are noncollinear points and L is a line that does not contain any of them. Prove that if L intersects one of the segments of the triangle ABC then it intersects a second one.

21. Prove that, in triangle ABC, $\overleftrightarrow{AB}$ has no points in common with the triangle except for those of $\overline{AB}$.

*22. Prove that if a line L contains a point D on the inside of an angle $\angle BAC$ then L must intersect at least one of the rays making the angle.

Chapter 2

The Euclidean Heritage

Prerequisites: Sections 1.1 and 1.2 of Chapter 1. Depending on one's attitude toward rigor, and the needs of the students, Section 1.3 of Chapter 1 may also be desirable.

Our main objective in this chapter is to study how some key theorems in Euclidean geometry arise from the axioms of Chapter 1. To make quicker progress, we relax slightly the degree of rigor used in the last section of Chapter 1 to a standard closer to that found in Euclid's *Elements*.

A starring role will be played by the SAS axiom and, in the third section, Euclid's parallel axiom. Students with a firm grasp of how proofs are constructed in geometry may go quickly through this chapter. Although the theorems of this chapter are very old, our applications span the range from ancient to modern.

2.1 Congruence

To say that figures are congruent means, in nontechnical language, that they have the same size and shape even though they may be in different positions (Figure 2.1). There is a lot more to say to make this precise and useful, so in this section we will flesh out the concept of congruence with a detailed theory for triangles in the plane.

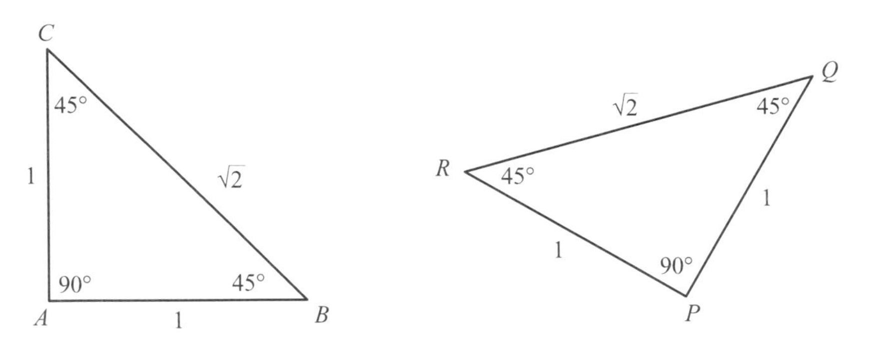

Figure 2.1 Two congruent triangles.

But all this attention to triangles shouldn't obscure the fact that the concept also applies to figures other than triangles. It is entirely reasonable to ask whether two quadrilaterals are congruent and to ask what evidence would convince us of it. The triangle theory can help with quadrilaterals, pentagons, and so on. You might like to think about whether it helps us with curved figures.

It is quite possible that the human mind does some kind of three-dimensional congruence checking in order to recognize familiar people or objects. If you are sitting in front of your computer, how do you know it is your computer and not the microwave oven? The computer has a certain size and shape, which is somehow recorded in your mind. Even if you are seeing it from a new angle right now — maybe an angle you have never seen if it has tipped over — you can somehow compare the current view to the remembered image and tell that they differ just by position.

This recognition of objects is an important problem (only partly solved at the time of this writing) in developing computer vision systems for robots and other uses. Solving this problem will probably require a variety of approaches — perhaps the color of the object needs to be taken into account in figuring out what it is — but the geometry of congruence may turn out to be part of the story.

Congruence and Correspondences

We need a better definition of congruence than "having the same size and shape." To understand the definition of congruent triangles, let us first examine triangles ABC and PQR in Figure 2.1. If we fasten our attention on any angle of ABC, say, the angle at A for the sake of argument, there is an angle in the other triangle that has the same size — angle P. If we also match up B with R and C with Q we have a complete $1:1$ correspondence of vertices where corresponding angles are congruent. So part of what it means for triangle ABC to be congruent to triangle PQR is that we can match up vertices so that vertices which correspond in the matching have congruent angles. If, in addition, we look at any pair of vertices in triangle ABC, say, A and B for example, the length of the side they determine is the same as the length of the side determined by the corresponding vertices — PR in this case. Summarizing:

DEFINITION

Two triangles are *congruent* if it is possible to find a 1 : 1 correspondence from the vertices of one triangle to the vertices of the other so that:

1. Corresponding angles are congruent.
2. Sides whose endpoints correspond are congruent.

Conditions 1 and 2 are abbreviated "corresponding parts of congruent triangles are congruent." It is important to note that the two triangles in the definition might be the same (we didn't say "two different triangles"). This is relevant in the proof of Theorem 2.1. A key point here is that you have to find the right 1 : 1 correspondence (match-up) if you want to demonstrate that the triangles are congruent.

EXAMPLE 2.1

In Figure 2.1 there are two 1 : 1 correspondences that show congruence:

$$A \to P,\ B \to Q,\ C \to R$$

$$A \to P,\ B \to R,\ C \to Q$$

But $A \to Q$, $B \to P$, $C \to R$ is not a congruence. ■

We often show a congruence without the arrow notation; instead we show the correspondence by the order in which we list the vertices. For example, if we say that triangle ABC is congruent to PQR we mean that A (the first vertex) matches up with P (also first on its list), B with Q, and C with R. In Figure 2.1 we might also say that triangle ABC is congruent to triangle PRQ, but we would not say that triangle ABC is congruent to triangle QPR.

Our definition of congruence suggests a strategy for finding the 1 : 1 correspondence which shows that two given triangles are congruent: Match up vertices according to the sizes of the angles. Once we have a 1 : 1 correspondence of vertices this creates a 1 : 1 correspondence among the sides. If the triangles are truly congruent, corresponding sides are congruent. Notice that checking whether a given 1 : 1 correspondence is a congruence seems to require checking six equalities, since there are six parts (sides and angles) to a triangle. The SAS axiom (Axiom 8 in Chapter 1) says the work can be cut down. Here is the SAS axiom again.

The SAS Axiom

AXIOM 8. *The Side-Angle-Side (SAS) Congruence Axiom*

1. If one angle of a triangle is congruent to a certain angle of a second triangle, and
2. if one side forming the angle in the first triangle is congruent to a side forming the congruent angle in the second triangle, and
3. if the remaining side forming the angle in the first triangle is congruent to the remaining side forming the congruent angle in the second triangle,

then the triangles are congruent.

EXAMPLE 2.2: *Why We Should Believe the SAS Axiom*

In Figure 2.2, suppose $m\angle A = m\angle A'$ and $AB = A'B'$ and $AC = A'C'$. (Note that we have used solid lines in the figure to indicate these known parts.) It seems obvious to the eye that the given information implies that $BC = B'C'$ and that $m\angle B = m\angle B'$ and $m\angle C = m\angle C'$. As an informal argument for this conclusion, imagine moving triangle ABC so that vertex A falls on vertex A' and so that $\overline{AB}$ falls on top of $\overrightarrow{A'B'}$ and $\overline{AC}$ falls on top of $\overrightarrow{A'C'}$. Clearly B, in its new position, is the same point as B'. Likewise, C, in its new position, is the same point as C'. In other words the two triangles coincide exactly. Now the motion of the first triangle doesn't change any of its lengths or the measures of any angles. Consequently, the parts of the original ABC are congruent to the corresponding parts of $A'B'C'$—that is, the triangles are congruent.

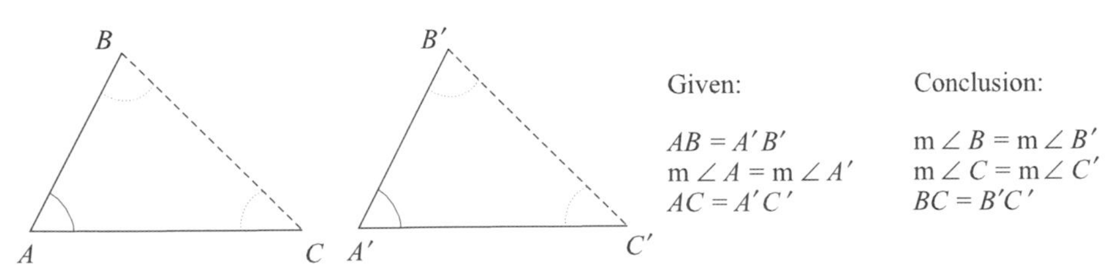

Figure 2.2 Side-angle-side axiom (given parts in solid lines).

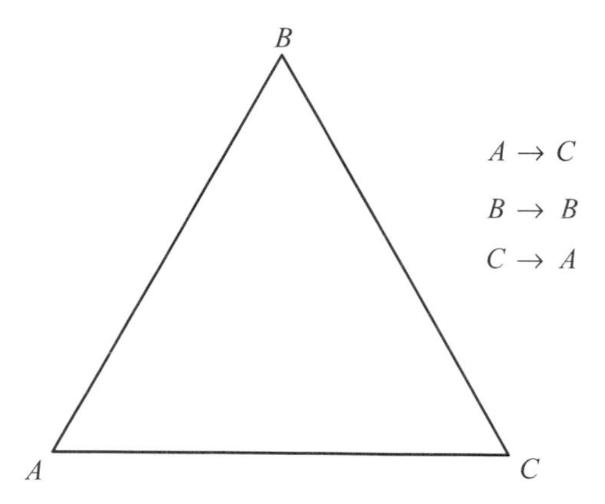

Figure 2.3 Base angles of an isosceles triangle are congruent.

The argument just given seems a lot like a proof of the SAS axiom, but we have not labeled the SAS principle a theorem, which we would do if we had a proof of it. The reason our argument is not a proof is that it relies on the assumption that figures can be moved without changing the sizes of their sides and angles. There is no reason that such an axiom could not be added to our set of axioms—it is after all quite believable. However, it is common to exclude ideas of motion from the axioms of Euclidean geometry (but not from our intuitions about geometry) and we follow this approach. ■

Here is our first deduction from the SAS axiom: the well-known theorem that base angles of isosceles triangles are congruent. Recall that a triangle is an *isosceles* triangle if it has two congruent sides (Figure 2.3).

☆ THEOREM 2.1

If two sides of a triangle are congruent, say, $AB = BC$ in Figure 2.3, then the angles opposite these sides are congruent, $m\angle A = m\angle C$.

Proof

The idea is to show that $A \to C$, $B \to B$, $C \to A$ is a congruence of the triangle to itself. Under this 1:1 correspondence, $\angle B$ corresponds to itself, and is clearly congruent to itself, so we have the "angle" part of an SAS proof. For $\overline{BA}$ the corresponding side is $\overline{BC}$ and these are congruent by hypothesis. For $\overline{BC}$ the corresponding side is $\overline{BA}$ and these are congruent by the same hypothesis. By SAS, the correspondence is a congruence. Consequently $m\angle A = m\angle C$ since these are corresponding angles. ■

Our proof of Theorem 2.1 sometimes strikes readers as odd, perhaps even invalid, since it deals with a congruence of a triangle with itself. Perhaps for this reason, this proof has rarely appeared in geometry texts. It is true that most often when we deal with congruence we are dealing with different triangles, but there is nothing about our definition of congruence or the SAS axiom that requires this. This proof seems first to have been found by Pappus around 300 A.D. There is a story that a computer program which was designed to find proofs of theorems in geometry came up with this proof rather than any of the proofs commonly included in geometry texts.

EXAMPLE 2.3: *Combining SAS and the Vertical Angles Principle*

We will prove that if the diagonals of a quadrilateral bisect each other, the opposite sides are congruent (Figure 2.4).

Proof

In Figure 2.4 we use SAS to show that triangle ABC is congruent to triangle EDC, with the $1:1$ correspondence $A \to E$, $B \to D$, $C \to C$. We have $AC = CE$ and $BC = CD$ by hypothesis. We also have $\alpha = \beta$ since they are measures of vertical angles. Thus, SAS implies that the triangles are congruent. Then corresponding sides are congruent, so $AB = DE$. A similar proof will show $BE = AD$. ■

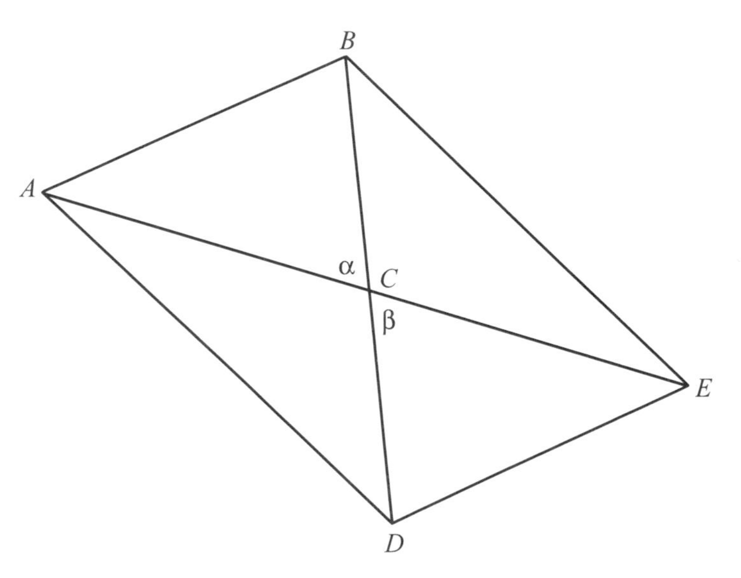

Figure 2.4 Diagonals of a quadrilateral bisecting each other.

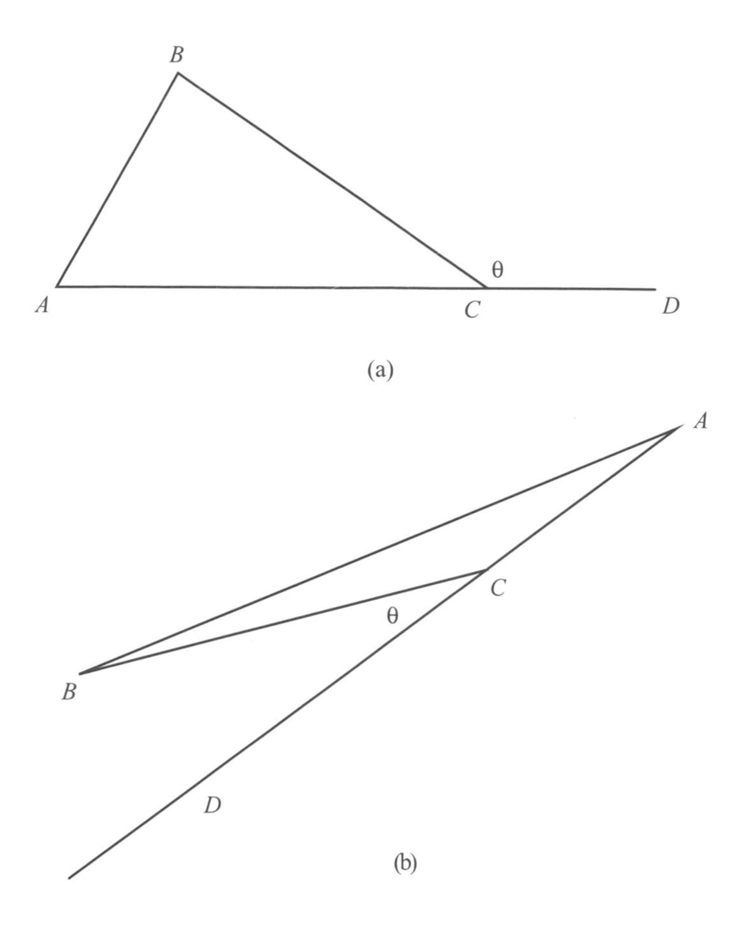

Figure 2.5 Illustrating the exterior angle inequality.

If we have a triangle, such as ABC in Figure 2.5a, and we extend side $\overline{AC}$ by some unspecified amount past C to D, then $\angle BCD$, whose measure is θ, is called an *exterior angle* of the triangle. We could make another exterior angle at C by extending $\overline{BC}$. There are two exterior angles per vertex, which happen to be vertical angles; six exterior angles in all for the whole triangle. The angles at A, B, and C are called *interior* angles. The ones not touching C, namely, $\angle A$ and $\angle B$, are called *remote* from the exterior angle labeled θ. In Figure 2.5a, it seems clear that the exterior angle labeled θ is greater than either of the remote interior angles $\angle A$ and $\angle B$. But in other pictures, such as Figure 2.5b, the inequality seems less obvious. We would like to be convinced of this inequality without drawing hundreds of triangles and measuring their angles. So here is a proof.

★ THEOREM 2.2: *The Exterior Angle Inequality*

An exterior angle of a triangle is greater than any of its remote interior angles (Figure 2.6).

PROOF

We show only $m\angle B < \theta$. The proof that $m\angle A < \theta$ is almost the same. Let M be the midpoint of segment $\overline{BC}$. We do not have to actually construct this midpoint using instruments such as a ruler and compass—it is enough to know that it exists. Now imagine segment $\overline{AM}$ extended an equal amount to a point E. Finally, connect E to C, thereby splitting the exterior angle θ (perhaps not evenly).

Our strategy now is to show that triangles ABM and ECM are congruent. (You'll see in a minute how this helps us with the exterior angle.) Because α and β are vertical angles, they are congruent Next we deal with segments. $BM = MC$ by construction and likewise $MA = ME$ by construction. By the SAS axiom, we have the congruence we wanted. Consequently, $m\angle B = m\angle ECM$. But $m\angle ECM < \theta$. Therefore, $m\angle B < \theta$. ■

There is one lapse from complete rigor in the proof just given. We need to prove (not just observe visually) that E is situated in such a way that $m\angle ECM < \theta$ (see Exercise 8 at the end of this section).

In thinking about this theorem, you might be tempted to give a different proof, like this:

$$m\angle A + m\angle B + m\angle C = 180^\circ \text{ so}$$
$$m\angle B = 180^\circ - m\angle A - m\angle C < 180^\circ - m\angle C = \theta.$$

This would be based on your recollection that the sum of the measures of the angles of a triangle is 180°. We have not yet proved this principle about the "angle sum" of a triangle,

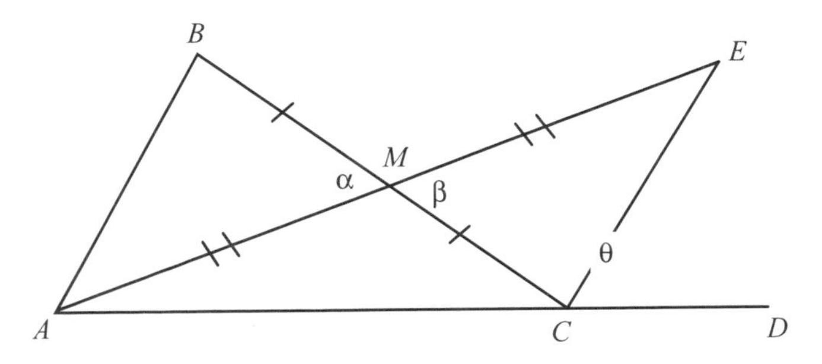

Figure 2.6 Proving the exterior angle inequality.

so such a proof for the exterior angle inequality would be incorrect at this point. In our chapter on non-Euclidean geometry, we will find it important that we have proved the exterior angle inequality without relying on anything about the angle sum of a triangle.

Other Congruence Principles

The exterior angle inequality is useful in establishing another congruence principle, the *angle-angle-side* principle. If we go around a triangle, such as ABC in Figure 2.7 (but ignore segment $\overline{DC}$), visiting an angle ($\angle A$), the next angle ($\angle B$), then the side we come to after the last visited angle ($\overline{BC}$), and we do the same in another triangle in such a way that the parts visited in the two triangles are the same in size ($m\angle A = m\angle A'$, $m\angle B = m\angle B'$, $BC = B'C'$), then the triangles are said to have the angle-angle-side relationship. We abbreviate this as AAS. This implies that the triangles are congruent.

✯ THEOREM 2.3: *The AAS Congruence Principle*

Triangles with the AAS relationship are congruent (Figure 2.7).

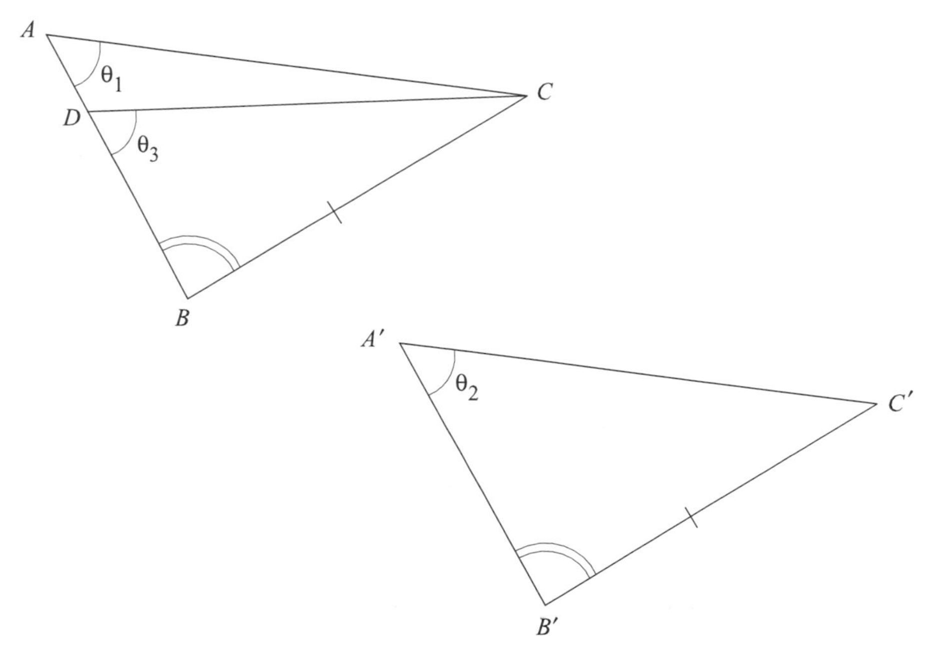

Figure 2.7 The AAS congruence principle.

Proof

If we knew that $AB = A'B'$, we could get the congruence immediately by SAS. So the tack we will take is to show that $AB \neq A'B'$ leads to a contradiction and is therefore impossible. There are two cases: $AB < A'B'$ and $AB > A'B'$. We show that $AB > A'B'$ leads to a contradiction and then leave the other case as an exercise.

Because we are assuming that $AB > A'B'$, if we lay off a length BD along BA congruent to $B'A'$, D falls between A and B. Connect D to C. Triangles DBC and $A'B'C'$ are congruent by SAS. Therefore $\theta_2 = \theta_3$. (Why?) But since $\theta_2 = \theta_1$, we have $\theta_3 = \theta_1$. This contradicts a previous theorem. (Which one?) ■

Here are two more congruence principles whose proofs we leave for the exercises.

☆ THEOREM 2.4: *The SSS Congruence Principle*

If there is a 1 : 1 correspondence between two triangles in which corresponding sides are congruent, then the correspondence is a congruence. ■

☆ THEOREM 2.5: *The ASA Congruence Principle*

If two angles of one triangle and the side between them are congruent, respectively, to two angles and the side between them in a second triangle, the triangles are congruent. ■

APPLICATION: The Carpenter's Level

You can make a simple but useful carpenter's level by making an isosceles triangle *ABC*, as in Figure 2.8 and marking the midpoint *M* on the base. Now hang a weight from a long string attached to *B*. When the weight stops swinging, it will be pointing vertically. Now rotate the triangle carefully, without letting the weight swing, till the string passes across the point *M*. You can use SSS to prove that $m\angle BMA = m\angle BMC$, which implies that $\overline{AC}$ is horizontal. We leave the details to you.

APPLICATION: Thales's Estimates of Distance

Thales, one of the earliest of the Greek geometers, suggested the following procedure for estimating the distance of a ship at sea (Figure 2.9). Mark your initial position *A* on the shore and walk at a 90° angle to the line of sight to the ship. After a short walk, drive a spike into the ground at *B*. Continue walking an equal amount in the same direction to *C*. Turn so your new direction is at a 90° angle to your previous direction. Now walk till you reach a point where the ship (*E*) and the stake (*B*) line up as you look at them.

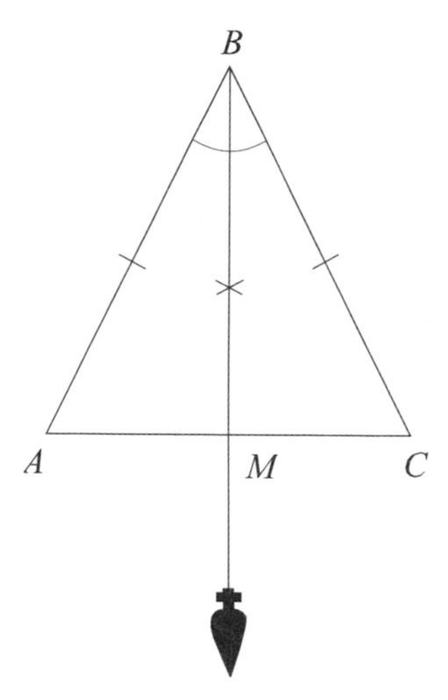

Figure 2.8 Isosceles triangles and a simple carpenter's level.

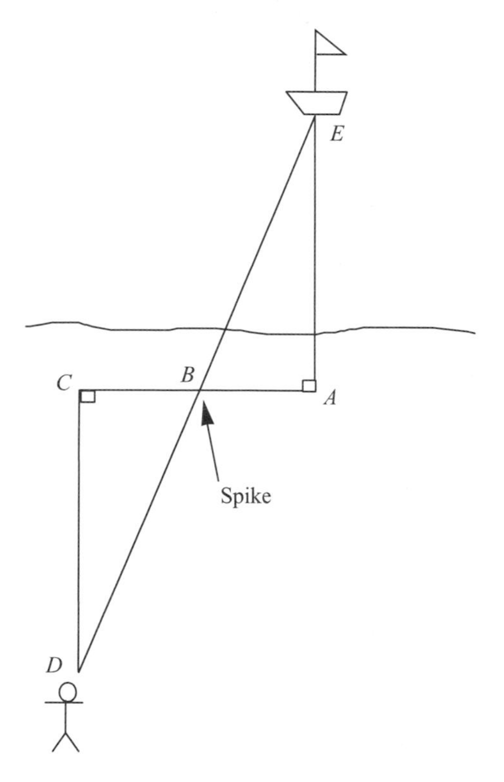

Figure 2.9 Using ASA to estimate distance *AE*.

If we could conclude that triangles EAB and DCB are congruent, then we could conclude that $DC = EA$, and then all we have to do is measure DC. These triangles will indeed be congruent because of the ASA congruence principle: $m\angle A = m\angle C = 90°$, $AB = BC$, $m\angle EBA = m\angle ABC$ (vertical angles).

Both of our applications illustrate the fact that showing triangles to be congruent is usually a tool to some other end. In the carpenter's level, we needed to be sure two angles were congruent. In the distance estimation, it was two segments we cared about. The congruence of two triangles is hardly ever the primary objective.

EXERCISES

**Marks challenging exercises*

Congruence and Correspondences

1. List all the possible 1 : 1 correspondences between the two triangles of Figure 2.1. Which of these is a congruence?

2. Show an example of two triangles where there are at least three different 1 : 1 correspondences showing congruence.

3. Let ABC and PQR be two triangles and suppose both of the following correspondences are congruences:

   (i) $A \to P$, $B \to Q$, $C \to R$

   (ii) $A \to P$, $B \to R$, $C \to Q$

   Explain why ABC must be isosceles.

4. If there are two different 1 : 1 correspondences that demonstrate congruence between triangles ABC and PQR, must the triangles be isosceles? If not, give an example. If so, give a proof. (*Hint:* Study the previous exercise.)

5. If there are three different 1 : 1 correspondences that demonstrate congruence between triangle ABC and a second triangle, must the triangles be equilateral? If not, give an example. If so, give a proof.

The SAS Axiom

6. Suppose $\overline{AB}$ and $\overline{CD}$ are diameters of a circle (chords that pass through the center). Show that $AC = BD$.

7. Let $ABCD$ be a quadrilateral where $\angle B$ and $\angle C$ are right angles and where $AB = CD$. Show that the diagonals are congruent.

8. There is a lack of rigor in the proof of the exterior angle inequality. When we connect C to E, how do we know this segment is in the interior of the exterior angle? And then how do we know that $m\angle BCE < \theta$? Can you resolve this problem with the help of the theorems and one of the axioms about angles in Section 1.3 of the previous chapter?

9. In Example 2.3, suppose we tried to show congruence with the 1 : 1 correspondence $A \rightarrow D$, $B \rightarrow E$, $C \rightarrow C$. Would that work? Explain your answer.

10. Complete the proof in Example 2.3 by showing that $BE = AD$.

11. In the exterior angle inequality, show that $\theta > m\angle A$. (*Hint*: There is another exterior angle at C.)

12. Prove that if two sides of a triangle are not equal in length, then the angles opposite are unequal in measure in the same order. For example, if $AB > AC$, then prove $m\angle C > m\angle B$. (*Hint*: Let D be a point between A and B so that $AD = AC$. Now combine two of the theorems of this section.)

13. Prove that if two angles of a triangle are not equal in measure, then the sides opposite are unequal in measure in the same order. For example, if $m\angle C > m\angle B$, then prove $AB > AC$. (*Hint*: Use the previous exercise.)

14. Prove the triangle inequality: If P, Q, and R are not on a line, then $PQ + QR > PR$. (*Hint*: Extend $\overline{PQ}$ beyond Q to S so that $QS = QR$. Try to use the result of the previous exercise.)

Other Congruence Principles

15. In the proof of Theorem 2.3, we left out the case where $AB < A'B'$. Show that this inequality would lead to a contradiction.

16. Answer the questions in parentheses in the proof of Theorem 2.3.

17. Prove the ASA principle of congruent triangles. (*Hint*: Try to get another pair of sides congruent so you can apply SAS. Consider a proof by contradiction. Read the proof of Theorem 2.3 for an analogous strategy.)

18. Prove that if a triangle has two angles congruent, then the sides opposite those angles are congruent.

19. Complete the following proof of the SSS principle of congruent triangles. Let the triangles be ABC and $A'B'C'$ with $AB = A'B'$, $BC = B'C'$, $CA = C'A'$. Find a point D on the other side of $\overleftrightarrow{AC}$ from B where $m\angle CAD = m\angle A'$ and $m\angle ACD = m\angle C'$. Triangles ACD and $A'C'B'$ are congruent by ASA. Now we try to show that triangles ABC and ADC are congruent. Connect B to D and try to show $m\angle B = m\angle D$. (How? Why does that help?)

*20. Show by example that SSA is not a correct congruence principle.

Other Exercises

21. How would you determine whether the two triangles in the following table are congruent? Describe your plan of attack with enough detail that a computer could be programmed to carry it out.

Triangle 1	Triangle 2
$A(1, 4)$	$P(-2, 0)$
$B(-3, 5)$	$Q(1, 2)$
$C(-1, 2)$	$R(-1, 4)$

22. Devise a ruler and compass construction for finding the midpoint of a given segment $\overline{AB}$. Prove that it works.

23. We all know that a circle crosses a line at most twice. How can we prove this? Show that if a circle crossed some line three times we could deduce a contradiction.

24. Show that if we have two quadrilaterals so that a side, a neighboring angle, the next side, the next angle, and the next side in one quadrilateral are congruent, respectively, to a side, a neighboring angle, the next side, the next angle, and the next side in another quadrilateral, then the quadrilaterals are congruent. (More briefly: SASAS implies congruence.)

25. Show that if we have two quadrilaterals so that a side, a neighboring angle, and the next three sides in one quadrilateral are congruent, respectively, to a side, a neighboring angle, and the next three sides in one quadrilateral, then the quadrilaterals are congruent. (More briefly: SASSS implies congruence.)

*26. Is it true that if you have any five parts of one quadrilateral congruent, respectively, to five corresponding parts of another, then the quadrilaterals are congruent? (The previous exercises give two instances where this is true.)

27. Prove that if all sides of a quadrilateral are congruent, then the diagonals are perpendicular. Show by a drawing that the converse is not true.

Applications

28. Jack makes a carpenter's level as described in this section. But he gets some of the instructions wrong. He makes the isosceles triangle and marks the midpoint of the base, but instead of tying the string to the vertex B, he ties the string to point M. Can he use his level? If there is a method he can use, prove that it works.

*29. Suppose there are two ships at sea and you need to find a method for determining the distance from one ship to another. Can you add something to Thales's method to get this additional information?

30. In Thales's distance estimation method, explain whether or not it is really necessary to have $m\angle EAB = 90°$. If it is not necessary, explain how you could do it differently. If it is necessary, explain why.

2.2 Perpendicularity

When two lines cross so that all the four angles are equal (i.e., each is 90°, called a *right angle*), then we call the two lines *perpendicular*. This is one of the most important concepts in geometry and has inspired similar concepts outside of geometry. We will see its power in one classical application in physics, which has proved its value over the centuries, and in a second very modern application currently attracting much attention in computer science.

✯ THEOREM 2.6

1. If B is equidistant from A and C, then B is on the perpendicular bisector of segment $\overline{AC}$ (Figure 1.4 of Section 1.2 in the previous chapter).

2. Conversely, if B is on the perpendicular bisector of segment $\overline{AC}$, then B is equidistant from A and C.

Proof

1. Let M be the midpoint of $\overline{AC}$. Triangles ABM and CBM are congruent by SSS. Consequently, the angles at M are right angles and $\overline{AM}$ is part of the perpendicular bisector of $\overline{AC}$.

2. We leave the proof of the converse as an exercise. ■

You should be aware that, although parts 1 and 2 of the previous theorem use mostly the same words, they are actually saying things that are logically different. We say one is the converse of the other. Statements of the form "If P then Q" and "If Q then P" are called *converses* of one another. If we know one of them is true, we do not automatically know the other is true as well. Each needs a separate proof. As an example consider the statements:

- If all sides of a quadrilateral are congruent, the diagonals are perpendicular.
- If the diagonals of a quadrilateral are perpendicular, then all sides are congruent.

One of these is true while the other is not. (Can you see which is true and give a proof? For the one which is not true, can you draw an example that demonstrates that?)

Given a point P lying off a line L, many lines can pass through P to meet L, and they create various angles with L (Figure 2.10). Will there be a line that makes right angles with L? Could there be two such lines? Our intuition tells us there will be exactly one, but in axiomatic geometry we try to deduce this even though we have little or no doubt about it.

We can readily rule out two perpendiculars, because if there were two as in Figure 2.10, we would have a contradiction to the exterior angle inequality (Theorem 2.2 of the previous section). (The exterior angle $\angle AEP$ has the same measure as its remote interior angle $\angle EFP$, since both are right angles.)

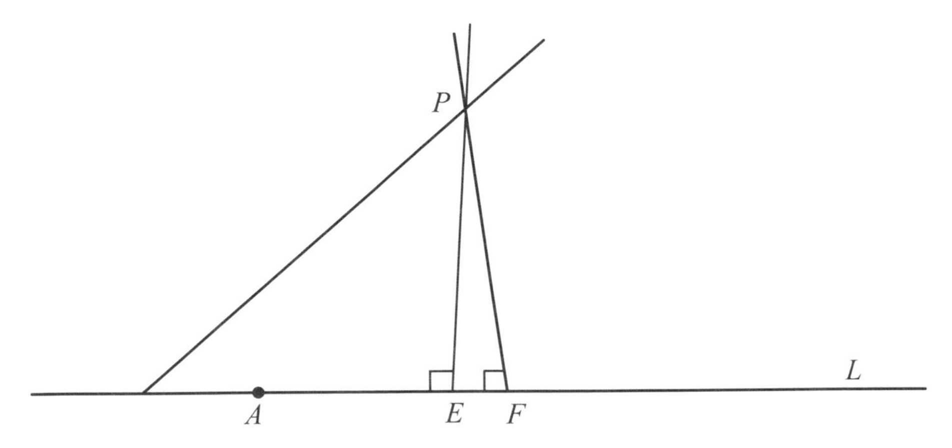

Figure 2.10 Are two perpendiculars possible from P?

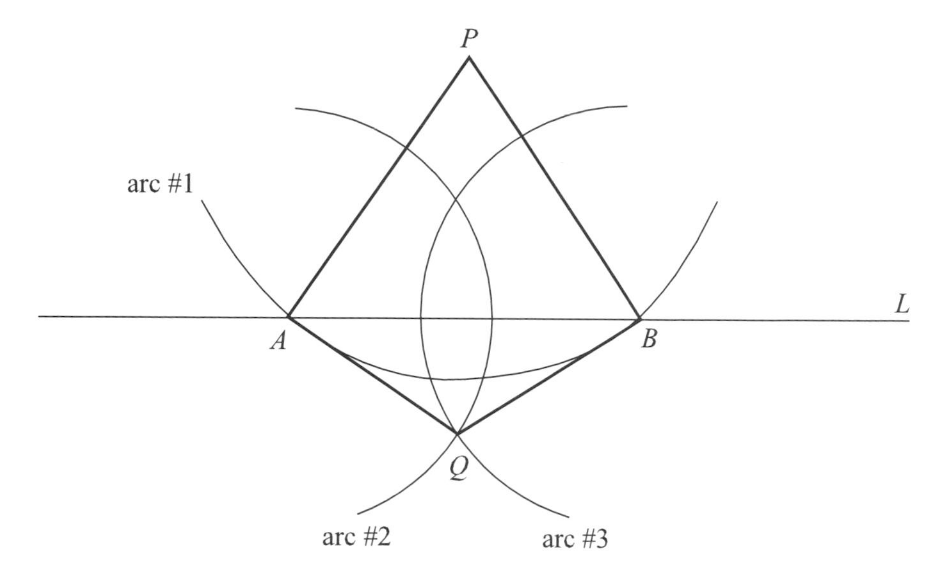

Figure 2.11 Construction for dropping a perpendicular.

To show that there is at least one perpendicular to L passing through P, we give a method of constructing one with a ruler and compass (Figure 2.11).

Constructing the Perpendicular Bisector $\overleftrightarrow{PQ}$ to line L.

1. Place the point of the compass at P, open the compass enough so that it reaches the line L, and swing an arc (arc #1 in Figure 2.11), meeting L at two points. We label these points A and B. Note that

$$PA = PB. \tag{2.1}$$

2. Now place the point of the compass at A and swing an arc (see the arc labeled arc #2 in Figure 2.11).

3. Move the point to B, and keeping the compass open to the same amount, swing another arc (arc #3), crossing arc #2 at Q. Note that

$$QA = QB. \tag{2.2}$$

To show that $\overleftrightarrow{PQ}$ is perpendicular to line L, we make use of Theorem 2.6. By Eq. (2.1), P has the same distance to A and B. Likewise, by Eq. (2.2) Q is equidistant to A and B. Thus $\overleftrightarrow{PQ}$ is perpendicular to line L.

APPLICATION: Fermat's Least Time Principle

When light rays bounce off a flat mirror, the angle between an incoming ray and the mirror is congruent to the angle between the outgoing ray and the mirror. This is also true of radio and TV signals and every form of electromagnetic radiation. Fermat's least time principle attempts to explain this behavior. In preparation for explaining this principle, let's consider the somewhat whimsical burning tent adventure (Figure 2.12). You have just returned from a hike and see that your tent is burning. There happens to be a stream nearby, with a conveniently straight shape. For some odd but fortunate reason, you have been carrying an empty bucket, so you need only rush to the stream, rush to the tent, and put out the fire. But which point on the stream should you aim for? Obviously, you want to find the point that makes your total trip to the stream

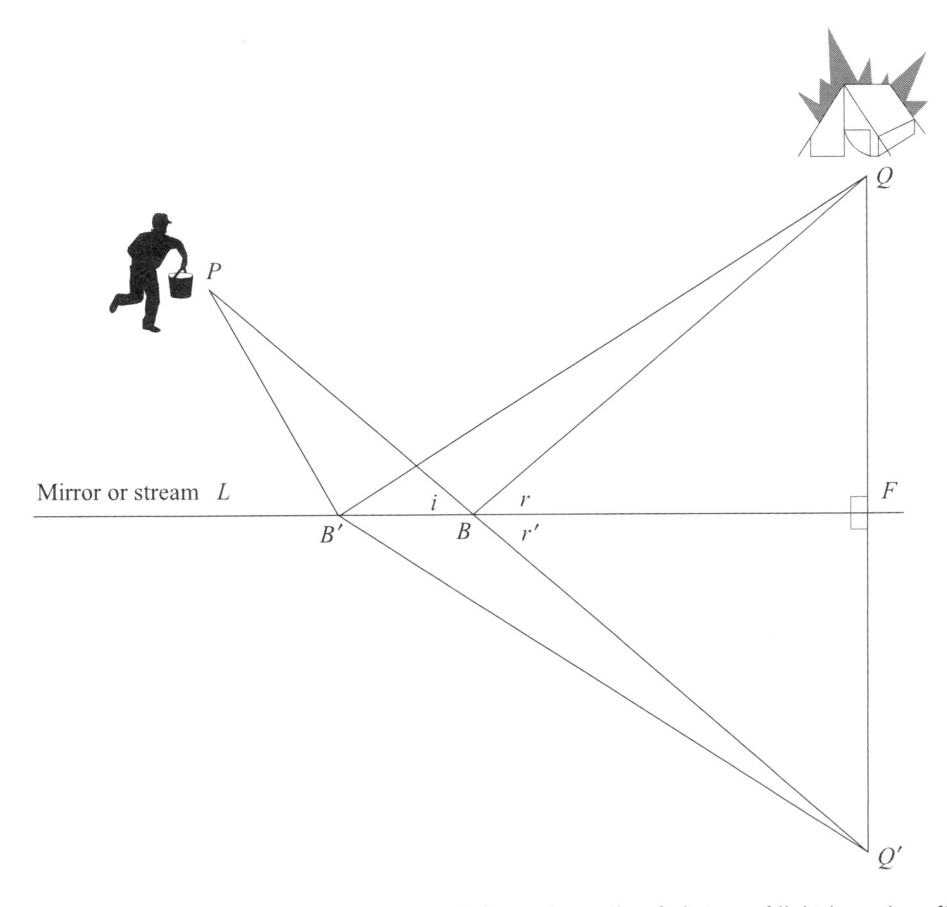

Figure 2.12 The "burning tent" or actual and alternative paths of photons of light bouncing off a mirror.

and then to the tent the shortest. We prove that the best point for you to aim at is the point where the angle between your incoming path and the stream is congruent to the angle between your outgoing path and the stream. Light rays bounce off a mirror the same way, only here we call the angles the *angle of incidence* (that is the incoming angle, $\angle i$ in Figure 2.12) and the *angle of reflection* ($\angle r$).

✯ THEOREM 2.7

If P and Q are on the same side of line L, the shortest two-segment path from P to Q via line L is one where the angle of incidence has the same measure as the angle of reflection (Figure 2.12).

PROOF

Our first step is to drop a perpendicular from Q to line L at F and extend this an equal distance on the other side of L to Q'. Now connect P to Q' with a line segment and let B be the point where it crosses L. We prove that B is the best bounce point. Let B' be any other bounce point.

Triangles $QB'F$ and $Q'B'F$ are congruent by SAS ($QF = Q'F$, the right angles at F are congruent, and $B'F = B'F$). Consequently,

$$B'Q = B'Q'. \tag{2.3}$$

Similarly, triangles QBF and $Q'BF$ are congruent. Consequently,

$$BQ = BQ'. \tag{2.4}$$

Equations (2.3) and (2.4) imply that if we wish to compare the path that bounces at B' with the one that bounces at B, we can compare the broken line $PB'Q'$ with the straight segment PBQ'. But the triangle inequality tells us that PBQ' is shorter. Thus, B is a better bounce point than B'.

It remains now to show $m\angle r = m\angle i$. From the congruence of the triangles QBF and $Q'BF$ we have $m\angle r = m\angle r'$, and from the vertical angles theorem we have that $m\angle r' = m\angle i$. From these two equations we get $m\angle r = m\angle i$. ■

Back in the seventeenth century, Pierre Fermat realized that Theorem 2.7 means that the bouncing behavior of light rays can be deduced from the assumption that they behave according to the *principle of least time*. In simple terms this means that if you look at any point Q to which the light ray has gotten after the bounce, the path from P to Q via the mirror is the quickest (i.e., shortest) possible. It is as if the light ray knows where it wants to get to and then solves the burning tent problem for that destination. Notice that we say *as if*. It is obviously ridiculous to suppose that a light ray has a destination in mind and then solves a math problem to get there the fastest. So it is a bit mysterious that the behavior of a light ray can be predicted from the least time principle.

Figure 2.13 A radio telescope with parabolic reflector to collect signals from space. Courtesy of National Radio Astronomy Observatory.

An interesting sidelight on this equal angles principle is its use for creating "dishes" for TV reception from satellites and in creating radio telescopes (Figure 2.13). TV signals and electromagnetic emissions from stars display the equal angles principle just as light does. Furthermore, if such rays hit a mirror that is curved, the equal angles principle still applies with just a small variation: The angle between an incoming ray and the tangent to the mirror at the bounce point is congruent to the angle between the outgoing ray and the tangent. This can be used to prove the following remarkable fact. If we take a curve called a *parabola* and spin it on its axis of symmetry, we get a surface, called a *paraboloid*, with a useful property: All incoming rays that are parallel to the rotation axis get bounced so that they pass through one fixed point, called the *focus* of the paraboloid. This makes a paraboloid ideal for concentrating incoming signals. The idea is to point the dish in the direction of the incoming rays and place a receiver at the focus where all the signals come together, creating a combined signal of maximum strength at the focus.

APPLICATION: Voronoi Diagrams

A fire is reported at a certain address. Which of the dozens of fire stations in a large city should send firefighters to the scene? Obviously, the closest station should respond. How do we know which is closest? Suppose we had a map on which the sites of the fire stations are indicated with dots and where the fire can be located by its address. We could apply a ruler to the map, measuring the distance from the fire to each station. An alternative is to divide the map up into regions, each region containing one fire station and consisting of all the points closer, or as close, to that station than to any other. Mark the region boundaries on the map. When we locate the fire on the map, we immediately see in which region it is located with no need for measurements. Each such region is called a *Voronoi region.* The diagram consisting of all such regions is called the *Voronoi diagram* corresponding to the given sites (fire stations).

A chain of pizza parlors receives an order from a certain address. Which of the many pizza parlors in the chain should bake and deliver the order? This is obviously the same type of problem as the fire station problem just mentioned; in fact, there are many applications of this sort for Voronoi diagrams.

Finding the Voronoi diagram is closely related to the idea of a perpendicular bisector of a segment. Say we had two sites, S_1 and S_2, and let L be the perpendicular bisector of $\overline{S_1 S_2}$. By part 1 of Theorem 2.6 any point off L is closer to one site than the other. In Figure 2.14, the left side of the perpendicular bisector, plus L itself, is the Voronoi region for S_1, which we denote $V(S_1)$. The right side of L, together with L itself, is the Voronoi region for S_2, $V(S_2)$.

Now let's look at three sites. Drawing the Voronoi diagram accurately for three sites will involve drawing the perpendicular bisectors of each of the three sides of triangle $S_1 S_2 S_3$. To do the drawing accurately requires that we be aware of the following surprising theorem.

✮ THEOREM 2.8

The perpendicular bisectors of the three sides of a triangle meet at a single point (Figure 2.15).

PROOF

Let I be the point of intersection of L_2 and L_3. We wish to show that I is on L_1 as well. Because I is on L_2, its distances to S_1 and S_3 are equal by part 2 of Theorem 2.6.

$$IS_1 = IS_3.$$

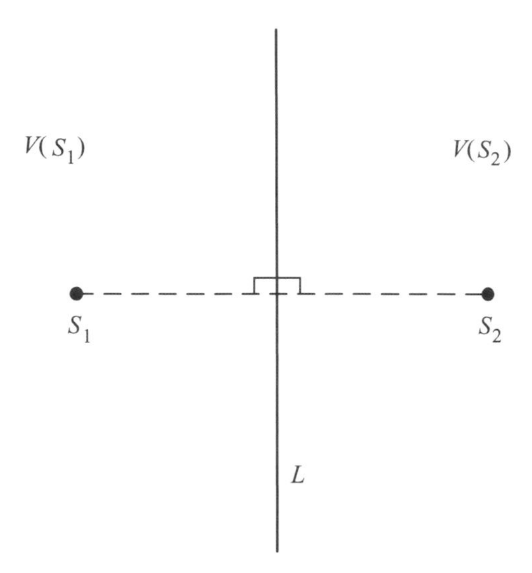

Figure 2.14 The perpendicular bisector separates the Voronoi regions for two sites.

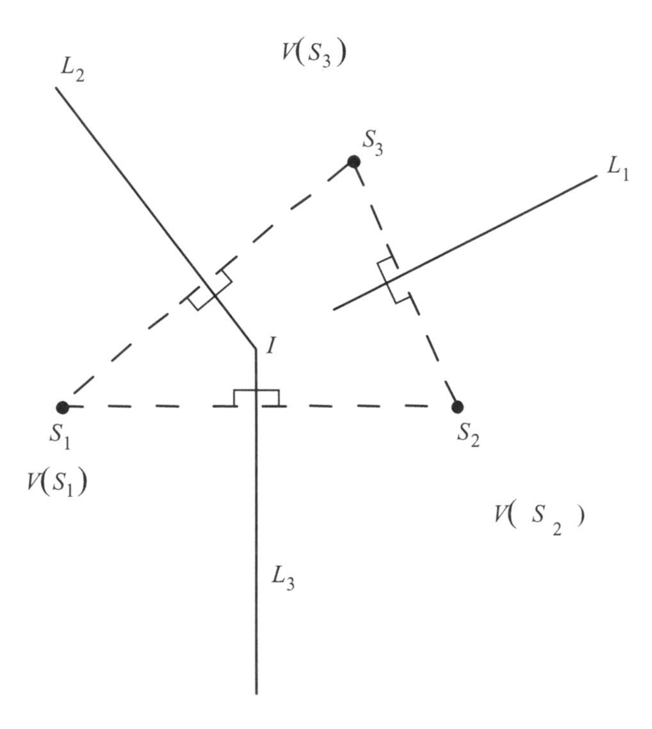

Figure 2.15 Will L_1 pass through I?

Because I is on L_3, the same reasoning gives

$$IS_1 = IS_2.$$

Comparing these equations, we deduce that $IS_2 = IS_3$. But if a point is equidistant from S_2 and S_3, it is on the perpendicular bisector of the segment connecting them (part 1 of Theorem 2.6). Thus, I is on L_1, the perpendicular bisector of $\overline{S_2S_3}$. ■

So how would you find the Voronoi diagram for four sites? One approach is to remove one site and use the preceding method for three sites. Then add the removed site back and figure out what to adjust. To begin, we leave out S_4 and build the Voronoi diagram for the first three sites, as shown in Figure 2.16a. Now when we add S_4 back in, we ask ourselves if any points of $V(S_3)$ should actually be in $V(S_4)$. To find out, we connect S_3 to S_4 and draw the perpendicular bisector of this segment. This creates a wedge (shaded portion of Figure 2.16b) of points presently in $V(S_3)$ that needs to be placed in $V(S_4)$ instead. In a similar way, we find a wedge-shaped part (not shown) of $V(S_2)$ that we need to add to $V(S_4)$. However, when we do the construction for S_1, we find that none of $V(S_1)$'s points need to be transferred to $V(S_4)$. We wind up with the Voronoi diagram shown in Figure 2.16c.

This is a general approach that will work to handle any number of sites. For example, if you had five sites, remove two and create the Voronoi diagram. Restore one site and adjust the diagram. Restore the last site and adjust the diagram again.

APPLICATION:

Example 2.4 Pattern Recognition Using Feature Space

A common problem in computer vision is to take a picture of an object and determine in which category it belongs. This is commonly called *pattern recognition.* As an example imagine a security robot that patrols various rooms in an art museum. From time to time it needs to know which room it is in. If we human beings had to solve this problem, we might look at the nature of the paintings in the room. If we see George Washington and other men in powdered wigs we would conclude that we are in the Early American room. Robot vision systems aren't that good yet at recognizing what's in a painting, but they can readily estimate the lengths and widths of paintings. Because each room has a distinctive combination of painting sizes, the robot's problem can be solved by giving it the list of painting sizes for each room—all carefully measured with great accuracy once and for all. For simplicity, let's say there are just four paintings in the whole museum, as shown in Table 2.1.

Now let's say the vision system is trying to determine the painting in the camera's field of view. We'll call this the *query painting.* The vision system determines the length and width of the query painting. This measurement of length and width via the camera and associated software is an interesting process, but all we wish to say about it here is that it is very inexact for a variety of reasons. The light may be bad, leading

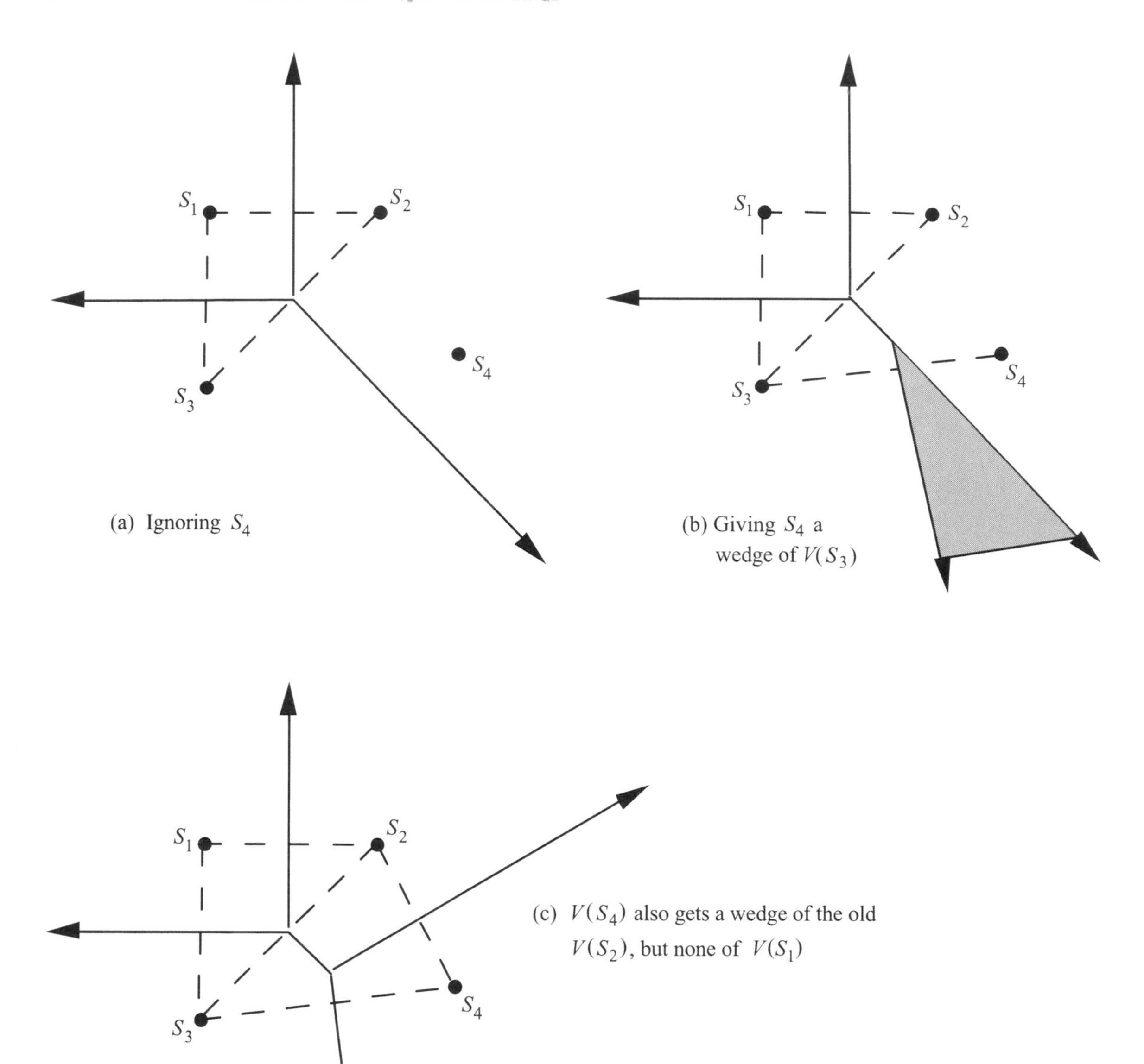

(a) Ignoring S_4

(b) Giving S_4 a wedge of $V(S_3)$

(c) $V(S_4)$ also gets a wedge of the old $V(S_2)$, but none of $V(S_1)$

Figure 2.16 Building a Voronoi diagram.

TABLE 2.1 PAINTING FEATURES

Painting	Length (inches)	Width (inches)	Point in Feature Space
A	40	40	(40, 40)
B	60	50	(60, 50)
C	80	30	(80, 30)
D	70	20	(70, 20)

to an image that is hard to analyze. Or light-colored painting frames may blend into the background color of the wall, making it hard for the robot to find the boundaries of the painting. For these and other reasons, the vision system's best guess about length and width of the query painting may be wrong and won't correspond exactly to any of the paintings, which have been measured exactly. What the computer needs to do is to find the closest of the four exactly measured paintings to the query painting.

Let's turn this into a Voronoi diagram problem. Imagine that we human beings are going to solve the problem with paper, pencil, and drawing instruments. Begin by plotting a point on graph paper (Figure 2.17) for each of the four paintings in Table 2.1, the x coordinate being the length and the y coordinate being the width. Our graph paper, with points interpreted as real or potential paintings, is called the *feature space* for this recognition problem. Next draw the Voronoi diagram of these four sites

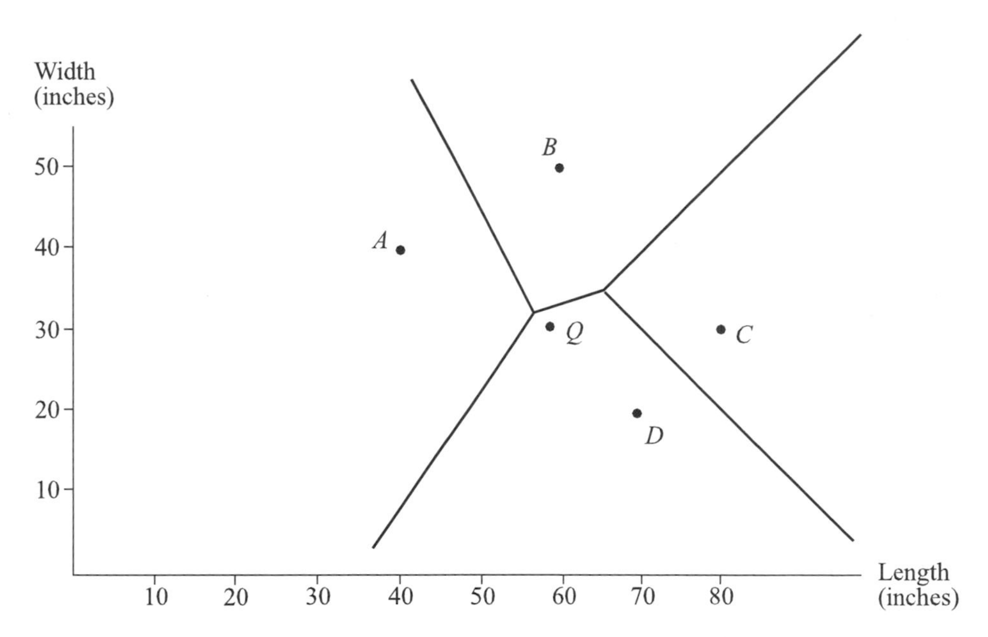

Figure 2.17 Feature space for paintings, divided into Voronoi regions.

(paintings) in feature space. Now suppose the best guess for the query painting is the point $Q(58, 31)$. Plot this query point and see in which Voronoi region it lies. We would call this approach the *Voronoi diagram algorithm.*

It is instructive to compare this to an alternative approach we will call the *distance formula algorithm.* In this approach, we don't use a Voronoi diagram. We simply calculate the distance from the query point to each of the four sites using the distance formula. For example, let's say the query painting is estimated to have length l_q and width w_q, so it would be represented by the point (l_q, w_q). To find the distance of this query painting to painting B (see Table 2.1), which would be represented by point (60, 50), we would use the formula:

$$\text{distance} = \sqrt{(l_q - 60)^2 + (w_q - 50)^2}.$$

Use similar formulas to get the distances to paintings A, C, and D. Pick the painting (site) that is closest and declare that to be the query painting.

The distance formula algorithm seems quicker, if we are working by hand, than the Voronoi diagram algorithm because drawing a Voronoi region is time consuming, whereas four times through the distance formula is not so bad with a hand calculator. On the other hand, suppose you had to do 100 query points. Now the Voronoi diagram approach starts to look better since you only have to draw the Voronoi diagram once. Then for each query point, you just have to plot the point and then visually observe in which Voronoi region the query point is located. Each plot plus visual observation is quick and easy. If you were using the distance formula algorithm, you would have to calculate 400 distances.

Of course, a computer vision system is not going to draw the Voronoi diagram on paper and plot points. Instead it works out a numerical description of the Voronoi diagram. Despite this difference in the way humans and computers would deal with the problem, the Voronoi diagram algorithm done by a vision system will have this in common with a human being doing a drawing: It is going to be quicker than the distance formula algorithm if there are many points. This is far from obvious; in fact, it is a key insight in the subject of *computational geometry.* See, for example, the book by O'Rourke (1994).

If the paintings had a third important numerical feature that differed from painting to painting, such as the overall lightness or darkness of the colors in the painting, our feature space could use this as well and be three dimensional. A feature space can be a useful tool in computer vision recognition problems. These spaces can often have very high dimension. Imagine how many measurements you would need so that different human faces would be different points in a feature space for faces.

EXERCISES

**Marks challenging exercises.*

1. Prove part 2 of Theorem 2.6.

2. Suppose you had started on the proof of Theorem 2.6 by proving part 2 first. Use this and Figure 2.18 to give an alternate proof of part 1. (*Hint:* Also use indirect proof and the triangle inequality.)

3. In the construction for dropping a perpendicular, do *all* the arcs have to have the same radius? (Check whether this is required in the proof.) Do *any* have to?

4. Here is a construction for erecting a perpendicular to a line at a given point M.

   (i) Put the point of your compass at M and swing an arc, cutting the line at A and B.

   (ii) From A and B, using the same radius each time, swing two arcs, and let C be a point of intersection of these arcs.

   (iii) Connect C to M.

   Prove that this works.

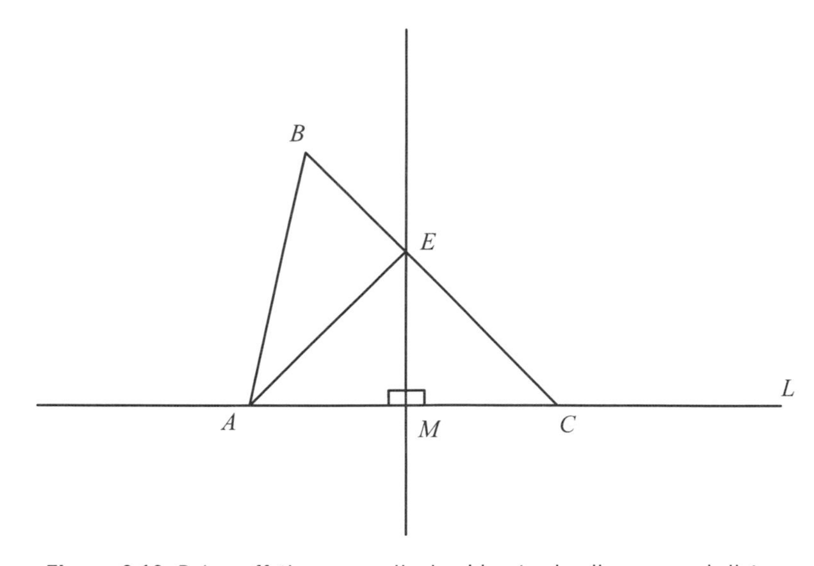

Figure 2.18 Being off the perpendicular bisector implies unequal distances.

Fermat's Least Time Principle

5. In Theorem 2.7, is there anything special about B' that is required to make the proof work?

6. In Theorem 2.7, fill in the steps which prove that triangles QBF and $Q'BF$ are congruent.

7. Suppose light went faster after bouncing off a mirror (see Figure 2.12). If it still were to behave according to the least time principle, would the angle of incidence measure less than, equal to, or greater than the measure of the angle of reflection?

Voronoi Diagrams

8. A researcher is studying the radio stations in an area. Each station has one transmitting tower and all have equal power. The further a home or car is from a station, the weaker the signal. Generally speaking, listeners have a preference for the station whose signal is strongest. The researcher has made the Voronoi diagram with the towers as sites. What is the significance of the Voronoi regions? Phrase your answer in a way that makes the significance for radio listening clear.

9. Find the Voronoi diagram when there are four sites and they are at the corners of a rectangle.

10. The two configurations of sites shown in Figure 2.19 were obtained by taking the corners of a square and moving one point along the diagonal toward or away from the opposite corner.

    (a) Make an accurate drawing (use a ruler, but estimate distances and right angles by eye) of the Voronoi diagram for Figure 2.19a.

    (b) Make an accurate drawing of the Voronoi diagram for Figure 2.19b.

11. Draw the Voronoi diagram of a set of five sites arranged in the pattern of a regular pentagon, that is, a set of five points arranged at intervals of 72° around the circumference of a circle.

12. If we have three sites, there is a point that all Voronoi regions have in common (e.g., in Figure 2.15 this is the point I). On the other hand, if there are more than three sites, as we can see from Figure 2.16, there may not be a point that all Voronoi regions have in common. Draw a set of six sites so that there will be a point that all six Voronoi regions have in common.

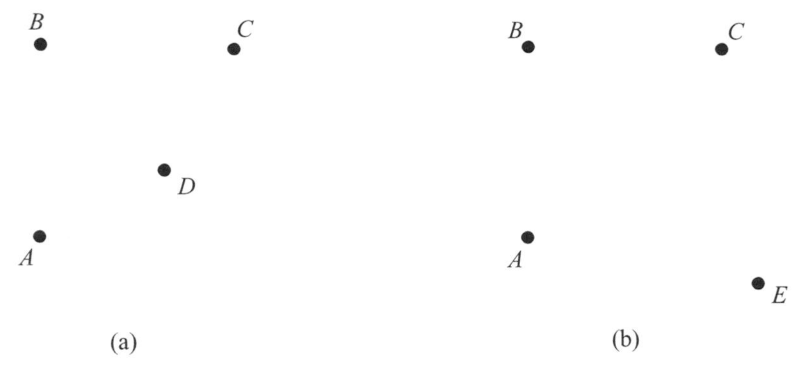

Figure 2.19

13. The vision system in Example 2.4 has estimated a query painting's length as 65 inches and width as 40 inches.

    (a) Use the Voronoi diagram algorithm and Figure 2.17 to decide which painting this is most likely to be.

    (b) Answer the same question using the distance formula algorithm.

14. A washer used in machinery is characterized by its outer diameter and the diameter of its hole. Table 2.2 shows washer types present in a bin. A computer vision system observes a washer and determines (inexactly) that the outer diameter is 13.5 millimeters and the diameter of its hole is 9.25 millimeters. Use the distance formula algorithm to determine which type it probably is. Now use the feature space algorithm (use graph paper). Do you get the same results?

15. A query painting (see Example 2.4) has a length estimated as 85 inches. This is longer than the longest of the paintings (painting C) in Table 2.1. Can we conclude therefore that the closest of the sites is painting C? Or do we need to find the width and use either the Voronoi diagram algorithm or the distance formula algorithm as in the previous exercise? Explain your answer with one or more examples.

16. If we have four sites, S_1, S_2, S_3, S_4, we define $H_{1j} = \{X \mid XS_1 \leq XS_j\}$ for $j = 2, 3, 4$.

    (a) What significance can you attach to H_{12}? Explain in words and illustrate with a picture.

    (b) What significance can you attach to $H_{12} \cap H_{13} \cap H_{14}$? Explain in words and illustrate.

TABLE 2.2 WASHER FEATURES

Type	Diameter of Hole (mm)	Outer Diameter (mm)
A	5	10
B	4	16
C	10	15
D	7	13
E	9	11

17. If we have n sites, $S_1, S_2, S_3, \ldots, S_n$, and we define H_{1j} as in the previous exercise, what significance can you attach to $H_{12} \cap H_{13} \cap \ldots \cap H_{1n}$?

18. Let H be a half-plane whose boundary line is L. If you choose two arbitrary points in H, can you be sure that the segment connecting them lies wholly in H? Justify your answer.

DEFINITION

If U is a set with the property that, no matter how you choose points A and B in U, the segment $\overline{AB}$ lies wholly in U, then U is said to be *convex*. (If your answer to the previous exercise was yes, you would say H is convex.)

19. Let H and H' be two half-planes, corresponding to lines L and L'. Prove that $H \cap H'$ is convex.

*20. Let U and V be any two convex sets. Prove that $U \cap V$ is convex.

21. Explain why a Voronoi region $V(S_i)$ is convex. (*Hint:* Examine previous exercises.)

22. If A and B are two points in $V(S_1) \cap V(S_2)$ and if C lies on the segment $\overline{AB}$, show that C belongs to $V(S_1) \cap V(S_2)$ also.

*23. Suppose we have n sites $S_1, S_2, \ldots, S_n$. Some pairs of sites will have Voronoi regions that have points in common, while other pairs may not. Prove that if S_1 is closer to S_2 than to any other S_i then $V(S_1) \cap V(S_2) \neq 0$. (*Hint:* Keep the triangle inequality in mind.)

*24. Suppose $\overline{AB}$ is the segment that two Voronoi regions $V(S_1)$ and $V(S_2)$ have in common. Is it always the case that the segments $\overline{AB}$ and $\overline{S_1S_2}$ intersect?

25. Fire engines cannot go in a straight line to a fire — they need to follow streets that usually have a rectangular pattern as illustrated in Figure 2.20. Shade the streets, or parts of streets, which are closer to S_1 than S_2. (Two such streets

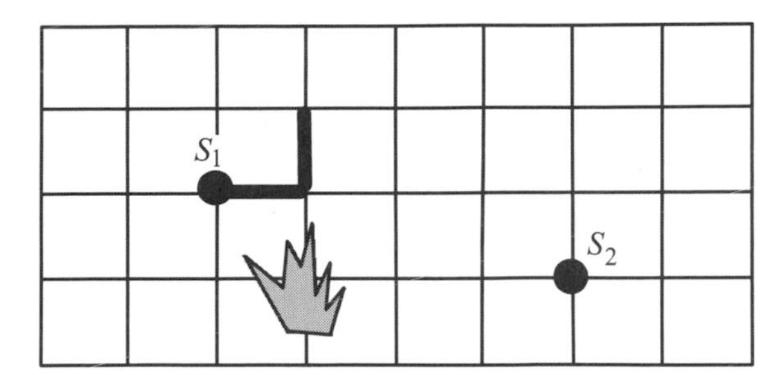

Figure 2.20 Two sites and a fire.

are already shaded as an example.) In this problem, distance means driving distances rather than straight-line distances. For example, the distance from the corner with the fire to S_1 is two blocks, not $\sqrt{1^2 + 1^2}$.

26. Draw the Voronoi diagram based on driving distances rather than straight-line distances for the three fire stations in Figure 2.21. This problem should be done in the spirit of the previous exercise. Use bold shading for $V(S_1)$, wiggly shading for $V(S_2)$ and leave the streets, or parts of streets, in $V(S_3)$ unshaded. Ignore points which are not on streets.

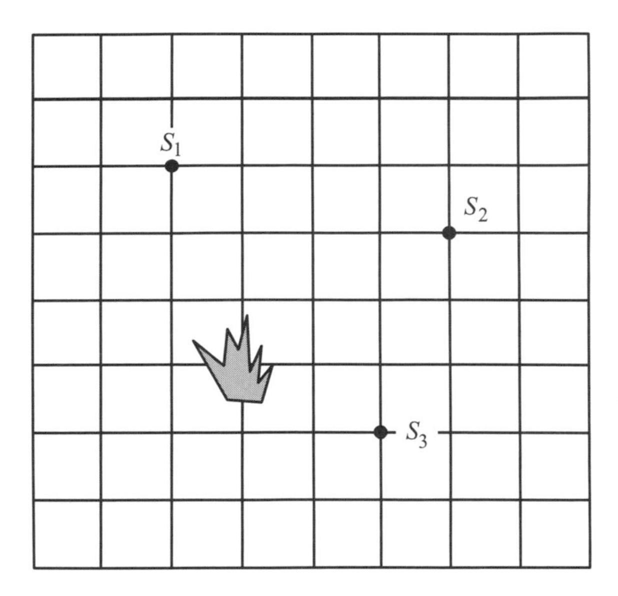

Figure 2.21 Three fire stations. What streets should they serve?

27. (a) Suppose we are given a Voronoi diagram for three sites, as in Figure 2.15 except, for simplicity, take the sites to be the vertices of an equilateral triangle. Now suppose the sites and their connecting lines are erased. Is it possible to find the sites from I and the lines L_1, L_2, L_3? Or might there be a second set of three sites that gives the same Voronoi diagram?

    (b) What if the triangle is not equilateral, but any triangle?

*28. In a Voronoi diagram, there will always be Voronoi regions that are unbounded ("go to infinity"), and usually some that are bounded. Is there a simple way to tell (simpler than making the Voronoi diagram that is) which sites have bounded regions and which have unbounded regions?

2.3 Parallelism

The Theory of Parallel Lines in a Plane

Lines in a plane that do not meet are called *parallel*. Are there any such things? It is one thing to make a definition, for example that a hoofed animal with one twisting horn is a unicorn, but it is another thing to find an example of what you have defined.

Let's put the question a little more precisely. Suppose we have a line L and a point P off of L. Many lines pass through P and lie in the plane of P and L. How many never cross L? Our first approach to this problem is to show how we can construct one such parallel to L through P. This means the answer to the question cannot be zero as long as we subscribe to our axiom set. Our construction is based on dropping and erecting perpendiculars as described in the previous section.

> **Construction of a Parallel to Line L through Point P**
>
> 1. Drop a perpendicular from P to L. Call this perpendicular line M.
> 2. Erect a perpendicular to M at P. Call this line N.

☆ THEOREM 2.9

Given a line L and point P off it, if we construct line N as just described, N is parallel to L (Figure 2.22).

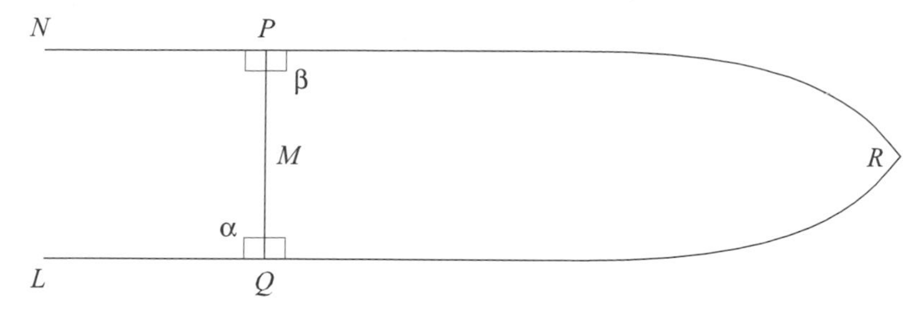

Figure 2.22 Lines with a common perpendicular are parallel.

Proof

We will give an indirect proof, in which we assume that the lines are *not* parallel and show that this leads to a conclusion which contradicts a previous theorem. This contradiction forces us to conclude that the lines are parallel.

Suppose the lines cross at a point R. Then triangle PQR has an exterior angle α measuring 90° and one of its remote interior angles, β, also measures 90°. This contradicts the exterior angle inequality (Theorem 2.2 of Section 2.1 in this chapter). ■

(Figure 2.22 shows a difficulty in illustrating indirect proofs. We made the assumption that the lines cross, later shown to be false, but we want to illustrate the assumed crossing in order to help visualize the proof. The picture we draw is bound to be unbelievable.)

So now we know that there is at least one parallel to L through P. Could there be more than one? Our intuition tells us that the answer is no. Consequently, it is quite natural to try to prove this. Euclid and his predecessors, working more than 2000 years ago, were unable to find such a proof. As a result, Euclid assumed as an axiom what he couldn't prove. His axiom about parallels was equivalent to our more modern version:

AXIOM 9: *Euclid's Parallel Axiom*

Given a point P off a line L, there is at most one line through P parallel to L.

Geometers who followed in the footsteps of Euclid were never happy assuming this assertion, because they hoped it would be possible to prove it as a theorem from the other axioms. One reason we might be uneasy with this axiom is that we prefer not to assume something with which we have little experience. For example, in Figure 2.22 imagine placing a line N' through P, which makes an angle of 89.999999999999999° to $\overline{PQ}$. This would be barely distinguishable from the perpendicular line N. Even though you might believe that N' will eventually cross line L, to see if it really does you might have to follow the constructed line for miles. This is not the kind of experiment we do very often. Even if you did carry out such an experiment, you might get it wrong; you could draw the line (or stretch the string or follow the light ray) inaccurately and your answer would be useless.

Because geometers were uneasy with assuming Euclid's parallel axiom, and were by and large unable to substantiate it with accurate experiments, they tried very hard to prove the assertion from the other axioms so that its status could be changed from axiom to theorem. This endeavor went on for more than 2000 years—surely one of the longest running stories in mathematics and science. Finally, in the nineteenth century geometers were able to prove that no such proof could ever be devised. (A more detailed account of this fascinating story is found in the next chapter and in books on the foundations of geometry.[1]) Following Euclid, in this section we assume Euclid's parallel axiom.

We turn now to the question of how we can determine whether two lines are parallel. Given two lines L and M (Figure 2.23), we call a third line a *transversal* if it cuts both lines. At each crossing point we have four angles. Two of them are said to be *interior* since they lie between the lines L and M. In Figure 2.23, α_3 and α_4 are interior at the crossing with M and β_1 and β_2 are interior at the crossing with L. The other angles are not interior. Interior angles, one at L and the other at M, are called *alternate interior* angles if they lie on opposite (alternate) sides of the transversal. In Figure 2.23, α_3 and β_2 have the alternate interior relationship. Likewise α_4 and β_1 are alternate interior angles.

Two angles made by a transversal across lines L and M are said to be *corresponding angles* if

(a) one is made by L and the transversal and the other by M and the transversal,

(b) the angles are on the same side of the transversal, and

(c) one is between L and M and the other is not.

For example, α_1 and β_1 are corresponding. We have labeled the angles in Figure 2.23 so that angles with the same subscript are corresponding.

☆ THEOREM 2.10

(a) If two lines L and M (Figure 2.24) are parallel and cut by a transversal, then alternate interior angles (for example, α and β) are congruent.

(b) If two lines are cut by a transversal and alternate interior angles are congruent, then the lines are parallel.

---

[1] See, for example, Cederberg (1989), Prenowitz and Jordan (1965), and Trudeau (1987).

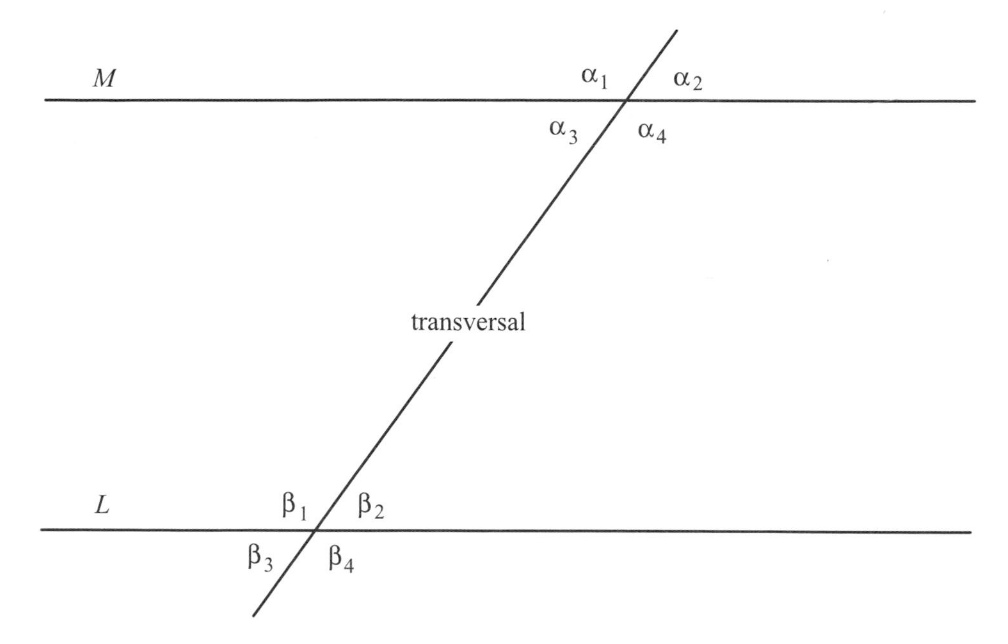

Figure 2.23 Alternate interior angles made by a transversal to two lines.

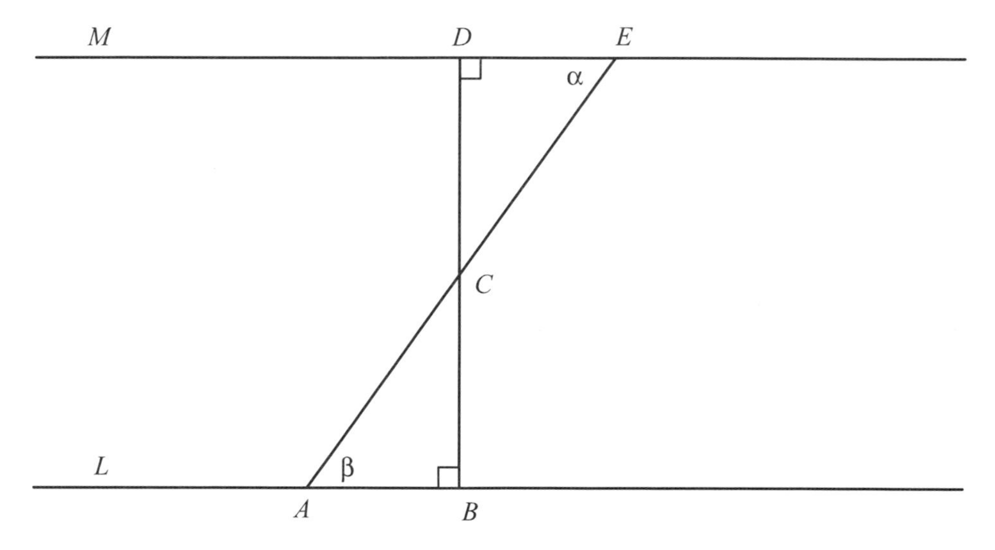

Figure 2.24 The alternate interior angles theorem.

PROOF

First we prove part (a) where the lines are given to be parallel. Let C be the midpoint of transversal $\overline{AE}$. From C drop a perpendicular to line L, meeting it at B. Now extend this perpendicular back till it reaches M at D. (Can you explain why there will be a crossing of the perpendicular with M?) What angle does $\overline{BD}$ make with M? Although we have labeled this a right angle in Figure 2.24, this needs proof. We know by Theorem 2.9 that if we were to draw a perpendicular to $\overline{BD}$ at D we would obtain a parallel to L. But this has to be line M since M is parallel to L and Euclid's parallel axiom says there cannot be two parallels to L through D. Because M is perpendicular to $\overline{BD}$, $\angle BDE$ is a right angle. We can now assert that triangles ABC and EDC are congruent by AAS. (The right angles at D and B are congruent, the vertical angles at C are congruent, and $CE = CA$ by construction.) Thus, $\alpha = \beta$ since these are corresponding parts of the congruent triangles. This concludes the proof of part (a).

We leave the proof of part (b) to the reader as an exercise. ■

Notice (Figure 2.23) that when two angles are corresponding, then an angle which is vertical to one of them is alternate interior to the other. This suggests that the previous theorem has consequences for corresponding angles. Here are some questions you should think about. If two lines are parallel and cut by a transversal, what can you say about corresponding angles? If two lines are cut by a transversal, what information about a pair of corresponding angles would allow you to conclude that the lines are parallel?

In many examples and applications of geometry, we encounter four-sided figures where the opposite sides are parallel. Such figures are called *parallelograms*. The most important theorem about parallelograms is that their "opposite parts" are congruent. Here is a more precise statement and the proof.

☆ THEOREM 2.11

If $ABCD$ is a parallelogram (Figure 2.25), then:

(a) Opposite sides are congruent.

(b) Opposite angles are congruent.

PROOF

For simplicity we will just show $m\angle A = m\angle C$. (The same type of proof can be adapted to show $m\angle B = m\angle D$.) Likewise, we only show $BC = AD$.

The standard technique for proving parts congruent in geometric figures is to show that they are corresponding parts of congruent triangles. The parallelogram has no

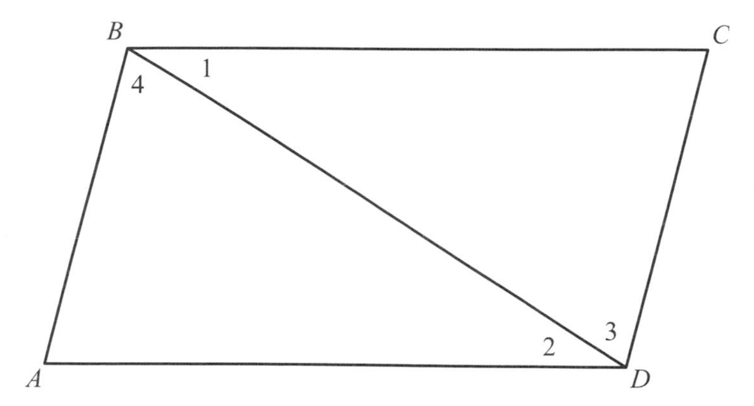

Figure 2.25 Opposite parts of a parallelogram are congruent.

triangles to start with so we introduce some by drawing a diagonal $\overline{BD}$. We show that triangles ABD and CDB are congruent by ASA: $m\angle 1 = m\angle 2$ since these are alternate interior angles of the parallel lines $\overleftrightarrow{AD}$ and $\overleftrightarrow{BC}$; $m\angle 3 = m\angle 4$ since these are alternate interior angles of the parallel lines $\overleftrightarrow{AB}$ and $\overleftrightarrow{CD}$; and, finally, $\overline{BD}$ is common to both triangles. Now ASA gives us the congruence we want. $\angle A$ and $\angle C$ are corresponding parts and therefore have the same measure. Likewise, $\overline{AD}$ and $\overline{BC}$ are corresponding parts so $AD = BC$. ■

✯ THEOREM 2.12

A quadrilateral $ABCD$ (Figure 2.25) where

1. no side intersects another side, and
2. opposite sides are congruent

will have its opposite sides parallel (i.e., it will be a parallelogram).

Proof

Triangle ABD is congruent to triangle CDB by SSS (with the correspondence $A \to C$, $B \to D$, $D \to B$) so

$$m\angle 1 = m\angle 2.$$

These are alternate interior angles of lines $\overleftrightarrow{BC}$ and $\overleftrightarrow{AD}$, cut by transversal $\overleftrightarrow{BD}$. By part (b) of Theorem 2.10, the congruence of these angles implies that $\overleftrightarrow{BC}$ is parallel to $\overleftrightarrow{AD}$. In a similar way we show $m\angle 3 = m\angle 4$, so that $\overleftrightarrow{AB}$ is parallel to $\overleftrightarrow{CD}$. ■

There is something about this proof we have left as a challenge for you to think about. Why do we need condition 1 of the hypothesis? And what does it mean for a quadrilateral to have a side that intersects another side? (Draw an example, or look at Exercise 17 of Section 1.3 in Chapter 1.) We do not use condition 1 in the proof, so why not delete it? Or could it be that the theorem does not hold for *self-intersecting* quadrilaterals? If so, we should have used condition 1 and our proof is not rigorous.

We come, finally, to one of the most central theorems of Euclidean geometry, one which says nothing at all about parallelism in its statement but which cannot be proved without somehow using Euclid's parallel axiom. In this theorem, the phrase *angle sum* means the sum of the measures of the three angles of the triangle.

✯ THEOREM 2.13

The angle sum of any triangle ABC is 180° (Figure 2.26).

Proof

Place a line L through B, parallel to $\overleftrightarrow{AC}$. The idea of the proof is to show that adding the measures of the angles of the triangle is the same as adding the measures of the angles at B: $m\angle 1 + m\angle 2 + m\angle 3$. By Theorem 2.10,

$$m\angle A = m\angle 1,$$
$$m\angle C = m\angle 3.$$

Thus $m\angle A + m\angle B + m\angle C = m\angle 1 + m\angle 2 + m\angle 3 = 180^\circ$. ■

We leave the proof of the following useful theorem as an exercise. It follows partly from the previous theorem and by consideration of isosceles triangles.

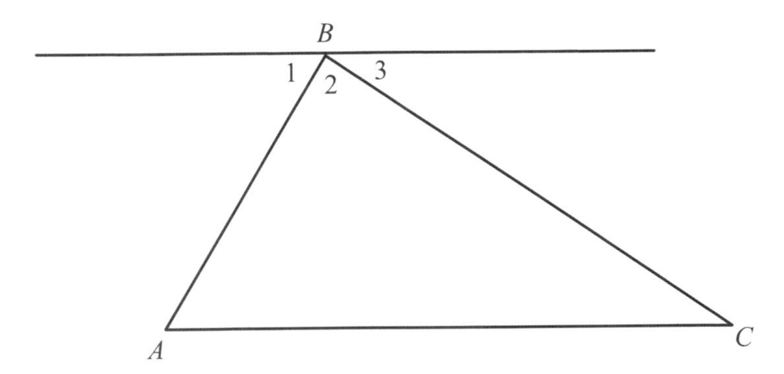

Figure 2.26 The angle sum of a triangle.

★ THEOREM 2.14

(a) If P, Q, and R lie on the circumference of a circle and the center C lies on $\overline{PR}$ (in this case we say $\angle PQR$ is *inscribed in a semicircle*), then $m\angle PQR = 90^\circ$.

(b) If P and R are fixed points, then the locus of all points Q where $m\angle PQR = 90^\circ$ is a circle whose center is the midpoint of $\overline{PR}$.

APPLICATION: The Circumference of the Earth

Eratosthenes (280 – 195 B.C.), one of the leading mathematicians of antiquity, was able to use Theorem 2.10 to estimate the circumference of the earth. Figure 2.27 shows his method.

Eratosthenes began with the fact that the distance from the city of Alexandria (A) to the city of Syene (S) is 5000 stadia, which is about 500 miles. He allegedly made this measurement of circular arc $\widehat{AS}$ by following a camel caravan from one city to the other and counting the steps taken by one particular camel. Now he needed to know what fraction of the whole way around the earth that 500 miles is. As we now describe, Eratosthenes found that $\widehat{AS}$ is 1/50 of the whole circumference, so he got his answer by multiplying 500 miles by 50 to obtain 25,000 miles, remarkably close to the modern estimate.

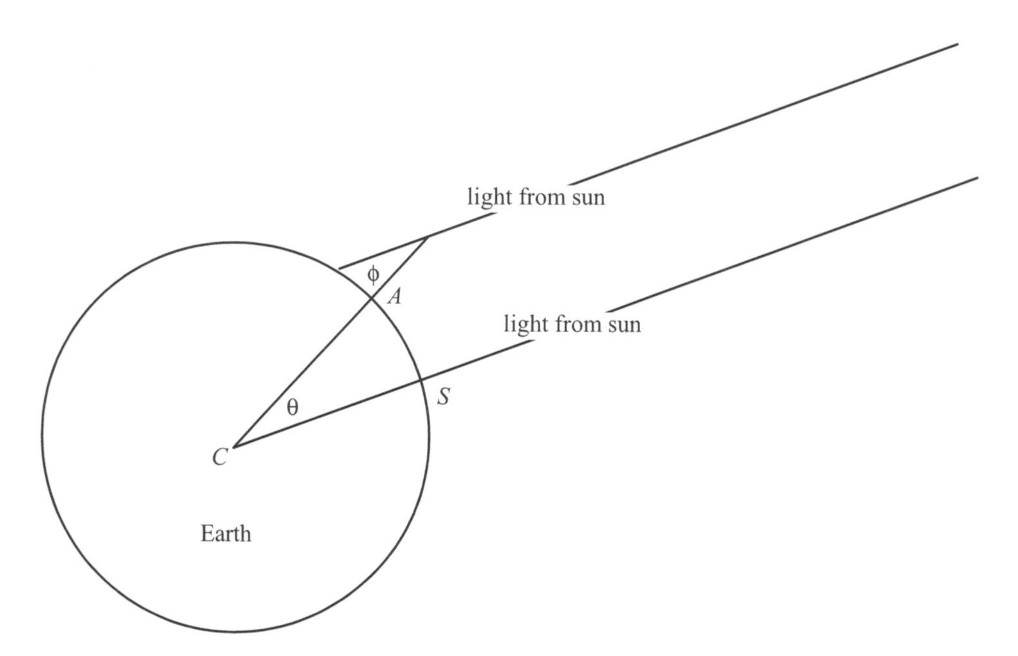

Figure 2.27 Measurement of the circumference of the earth.

That fraction Eratosthenes needed is related to the measure θ of the central angle as follows:

$$(\text{length of arc } \widehat{AS})/\text{circumference} = \theta/360. \tag{2.5}$$

Now it remained to estimate θ. Of course, this couldn't be done directly since we cannot get to the center of the earth to make the measurement.

To work around this, Eratosthenes started with the fact that on a certain day of the year, at a certain time of day, the sun shone directly down a vertical well in Syene. This was determined by noticing that no part of the water at the bottom was in shadow. Since the well was vertical, it also meant that the beam of light illuminating the bottom would, if extended, pass through C, the center of the earth. At Alexandria, there was a vertical pole. Because it was vertical, if we could somehow extend its line, this line would also pass through the center of the earth. Eratosthenes reduced the problem to a planar one by slicing the earth with the plane determined by the lines $\overleftrightarrow{CS}$ and $\overleftrightarrow{AC}$.

Now Eratosthenes made a few approximations. He thought of the sun as a point and the light beams from the sun to the well and the pole as lines. Because these lines cross at the sun, they are not parallel, but Eratosthenes wanted to think of them as parallel anyhow. The justification for this is that the sun is so far away that the directions of these lines are almost the same. On the assumption of parallelism, we have that $\theta = \phi$ [part (a) of Theorem 2.10].

After making this assumption, all Eratosthenes had to do was to measure ϕ at the pole in Alexandria — an easy task that yielded the value 7.2° — and plug that value in as θ in the fraction $\theta/360$ he is looking for. Thus, he obtained

$$(\text{length of arc } \widehat{AS})/\text{circumference} = \theta/360 = 7.2/360 = 1/50, \tag{2.6}$$

$$\text{circumference} = (50)(\text{length of arc } \widehat{AS}) = (50)(500) = 25{,}000 \text{ miles}. \tag{2.7}$$

This is amazingly close to the modern estimate of 24,860 miles.

Rigidity of Frameworks

Suppose we take four sticks and hinge them together at the ends. Will we be able to change the shape of this quadrilateral by bending the hinges? Figure 2.28 suggests that, for a rectangle, the answer is yes, it can *flex*. You can use your hands to build a flexible quadrilateral that is not a rectangle. Put your two thumbs together at their tips and then put your two forefingers together at their tips. You can open and close this quadrilateral by opening and closing the joints where your thumbs meet their neighboring forefingers.

It is convenient to simplify things by replacing sticks or fingers with line segments and keeping in mind that all of the angles are hinged. To keep things simple, we assume that the hinges are such that the only flexings they allow result in a four-sided figure that

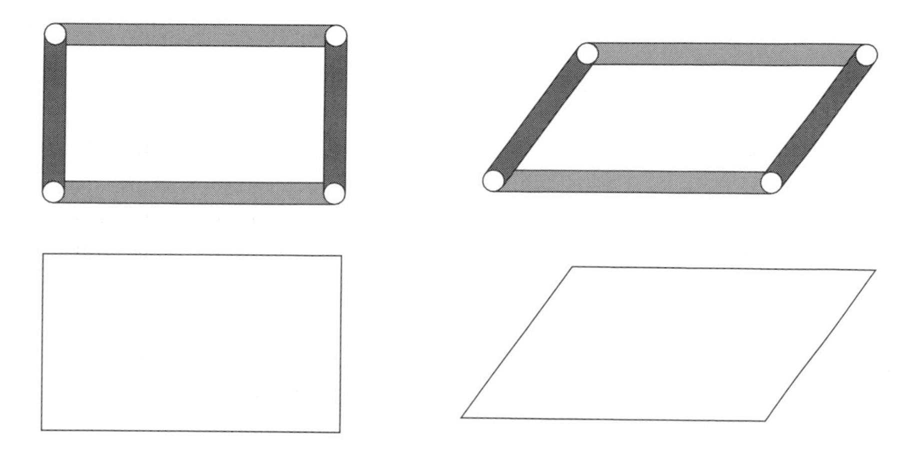

Figure 2.28 A quadrilateral that flexes at its hinges.

lies in a plane. (If you try, you can easily make your thumbs and forefingers create a nonplanar configuration — the kind we want to rule out.)

Just because a set of line segments has hinges where the segments meet does not mean that the framework can flex. For example, a triangle with three hinges cannot flex. The proof is indirect: Suppose we could flex the triangle to create a second triangle. Since we are not changing any lengths, each side of the second triangle can be matched up with a congruent side of the original. Therefore, SSS tells us that the triangles are congruent. This means that all the angles are congruent. In short, you can't change any of the hinged angles of a triangle. We say it is *rigid.* A framework of lines with hinges where the lines meet is called rigid if none of the hinge angles can be changed.

The framing of a house typically consists of pieces of wood called studs, nailed together to form a grid of rectangles (see Figure 2.29). This becomes a kind of skeleton on which the inner and outer walls are attached. As we have seen in Figure 2.28, a rectangle is not, by itself, rigid, so something else has to be done to keep it from flexing. Of course, there are no hinges at the corners, but if just one nail is used where two pieces of wood meet at a corner, that is hardly much better than a hinge. Various aids to rigidity come to the rescue, such as extra nails and the plywood sheets nailed onto the rectangles. Most interesting for us, a brace can be nailed from one corner to the opposite corner, making two triangles out of the rectangle.

With this as motivation, let's consider the mathematical problem of making a grid of squares rigid by adding diagonals. Our objective is to add enough braces to ensure that none of the squares can change its shape. Perhaps surprisingly, it is not necessary to brace every square. Figure 2.29a shows a set of braces that confers rigidity on

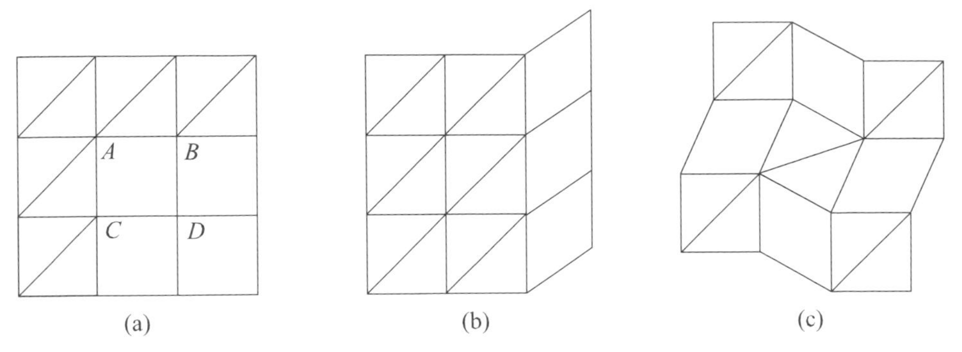

Figure 2.29 (a) A rigid braced grid; (b) and (c) Braced grids that have flexed.

the grid (you'll see how to prove this soon.) But the number of braces used is not the key factor. Figure 2.29b shows a bracing that uses more braces than in 2.29a, but the grid has flexed away from its "all square" shape. The lack of rigidity in Figure 2.29b is an instance of the following fairly obvious fact:

If there is a row (or column) with no braces, the grid is not rigid.

Figure 2.29c shows that a grid can flex even if every row contains a braced square and every column contains a braced square.

Now we'll describe the *segment marking algorithm* for determining whether a given bracing makes a grid of squares rigid. In describing the method, we use the phrase *originally vertical* to describe segments that are vertical in the "all square" shape of the grid — the shape before any flexing has taken place. All the other segments are *originally horizontal.* If flexing has occurred, some of the originally vertical (or originally horizontal) segments may no longer be vertical (or horizontal). But it won't interfere with our search for a flexing if we agree not to move the originally vertical segment in the upper left corner. This segment is shown in boldface in the grids of Figure 2.30 and in the first grid of Figure 2.31.

In some flexings of a braced grid of squares, some originally vertical segments may remain vertical, and some originally horizontal segments may remain horizontal. More interestingly, with some bracings, there may be originally vertical segments that *must* remain vertical for any conceivable flexing of the grid. Likewise, some originally horizontal segments must remain horizontal in any possible flexing. These segments, which must remain in their original directions, are called *fixed.* The segment marking algorithm consists of looking for these and marking them in bold. If we find that every segment has been boldfaced, then no square can change its shape and the braced grid is rigid.

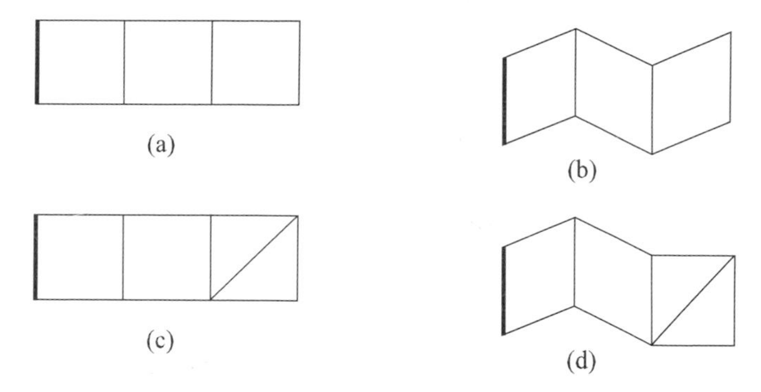

Figure 2.30 Illustrating our principles.

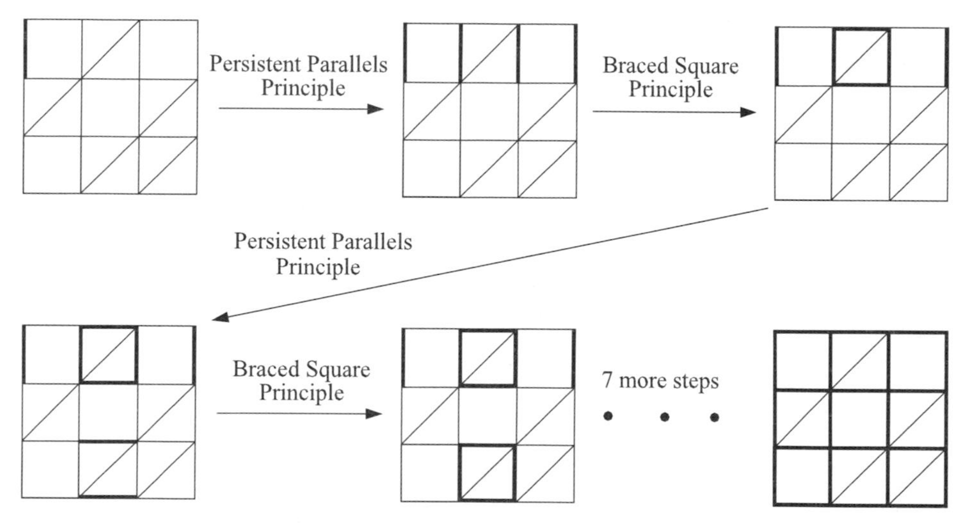

Figure 2.31 The segment marking algorithm checking for rigidity.

The search for fixed segments uses two principles. Theorem 2.12 implies that cells remain parallelograms no matter how they have flexed, as long as the flexing is not so severe as to produce a self-intersection of the cell boundary. But a quadrilateral that starts as a square cannot intersect itself, so we have:

PERSISTENT PARALLELS PRINCIPLE

In any row of the grid, regardless of whether it has flexed or not, all the originally vertical sides of cells in that row are parallel.

In any column of the grid, regardless of whether it has flexed or not, all the originally horizontal sides of cells in that column are parallel.

THE BRACED SQUARE PRINCIPLE

If a square is braced, it is rigid. If one pair of opposite sides is vertical, then the other sides are horizontal and vice versa.

To illustrate the persistent parallels principle, consider the single row of squares in Figure 2.30a). Recall that the left side of the first square is fixed in a vertical direction. As we see from Figure 2.30b, a good deal of flexing is possible. However, as the persistent parallels principle tells us, each cell has two vertical sides. The originally horizontal sides are no longer horizontal, but each remains parallel to its opposite side.

Now suppose we have a brace as in Figure 2.30c. Flexing is still possible, as we see in Figure 2.30d, but the braced square principle and the persistent parallels principle together ensure that the last cell's sides are vertical and horizontal even after the first two cells have flexed.

We illustrate how the segment marking algorithm uses these principles in Figure 2.31. Begin by applying the persistent parallels theorem to the first row, taking advantage of the fact that we have agreed to keep the boldface segment at the extreme left vertical. We mark the other segments to show that they remain vertical even if flexing were to occur. Then we look for a braced square in which to apply the braced square principle—in this case the second square in the first row. We mark the horizontal segments of that cell. We go on alternating the two principles until we have either marked every segment or we cannot apply the principles anymore. If every segment has been marked as a fixed segment, then the framework is rigid. The converse is also true (though less obvious): If one or more segments remain unmarked when we are done, then the famework is not rigid.

EXERCISES

**Marks challenging exercises.*

Theory of Parallel Lines in a Plane

1. Prove part (b) of Theorem 2.10.

2. Prove that the diagonals of a parallelogram bisect each other.

3. Prove that if the diagonals of a quadrilateral bisect each other, then the quadrilateral is a parallelogram.

4. Prove that the sum of the angles of a quadrilateral is 360°. What can you say about the angle sum of a polygon with n sides?

5. (a) Prove that if two lines are parallel and cut by a transversal, then corresponding angles are congruent.

   (b) Prove that if two lines are cut by a transversal and two corresponding angles are congruent, then the lines are parallel.

6. Suppose triangles ABC and $A'B'C'$ are such that $\overleftrightarrow{AB}$ is parallel to $\overleftrightarrow{A'B'}$, $\overleftrightarrow{BC}$ is parallel to $\overleftrightarrow{B'C'}$ and $\overleftrightarrow{CA}$ is parallel to $\overleftrightarrow{C'A'}$. Prove that $m\angle ABC = m\angle A'B'C'$.

7. Prove that if a quadrilateral has diagonals which are congruent and bisect one another, then each angle of the quadrilateral is a right angle. (*Hint:* Show all angles are congruent and use a previous exercise.)

8. Prove that an exterior angle of a triangle has measure equal to the sum of the measures of the two interior angles remote from it.

9. Prove that if opposite angles of a quadrilatral that does not intersect itself are congruent, then the quadrilateral is a parallelogram.

10. In Theorem 2.11, if we just wanted to show $m\angle A = m\angle C$, we wouldn't need to draw the diagonal. We could use Theorem 2.10. Can you write this proof?

11. In Theorem 2.11, how would you show $m\angle B = m\angle D$?

12. Let $A_1A_2A_3A_4A_5A_6$ be a hexagon where opposite sides are parallel (e.g., A_1A_2 is parallel to A_4A_5, etc.). Prove that opposite angles are congruent (e.g., $m\angle A_1 = m\angle A_4$, $m\angle A_2 = m\angle A_5$, etc.).

13. With the hypotheses of the previous exercise, can we prove that opposite sides of the hexagon are congruent? If so, prove it. If not, explain by example why the opposite sides might not be congruent.

*14. (a) Show by example that Theorem 2.12 is not correct if you remove hypothesis 1 and allow the quadrilateral to be self-intersecting (e.g., $\overline{AD}$ intersects $\overline{BC}$).

   (b) Explain why the proof given does not work if the quadrilateral is self-intersecting.

   (c) Find a rigorous proof for Theorem 2.12. (This is a challenge.)

15. Prove Theorem 2.14. Keep in mind that for part (b) one thing you need to show is that if a point R is not on the circle described, then $m\angle PQR \neq 90°$.

16. Let point A have x coordinate x_A and let B have x coordinate x_B (Figure 2.32). Let M be the midpoint of $\overline{AB}$ and suppose the x coordinate of M is x_M. Prove that $x_M - x_A = x_B - x_M$.

17. Figure 2.33 illustrates a method of proving the Pythagorean theorem, the fact that in a right triangle with legs of lengths a and b and hypotenuse c, $c^2 = a^2 + b^2$. The idea is to take the given right triangle and assemble four rotated copies of it into a big square whose area we express two ways: first in terms of the side length of the square, second as the sum of five pieces. We equate the expressions for the two areas and simplify to get $c^2 = a^2 + b^2$. To assemble, first we turn a copy of the triangle 1 so its hypotenuse is horizontal. Next turn a copy so its shorter side fits along the longer side of 1 with no gaps or overlaps (this would be 2 in Figure 2.33). Here are some details for you to work out:

    (a) Prove that when you fit triangles 1 and 2 together, you get a right angle at their common corner. (The same type of proof should establish that the other corners are right angles.)

    (b) Prove that the corners of the inner figure (5) are right angles (making it a rectangle).

    (c) Determine the lengths of the sides of the inner rectangle in terms of a, b, and c.

    (d) Get the two expressions for the area of the big square and equate them and simplify.

*18. We return to Roscoe's decision (Section 1.1 of the previous chapter). In Figure 2.34, we have marked the information about the coordinates of S and G as lengths of segments on the x and y axes and we have indicated which angles are known to be right angles. Nothing is known about unmarked angles, but maybe you can prove some useful facts. Can you show that $SG'' = 3$ and that $G''G = 4$? Can you show that $\angle SG''G$ is a right angle so that the Pythagorean theorem can be applied to find SG?

Circumference of the Earth

19. In Eratosthenes's calculation of the circumference of the earth, how would you measure ϕ? You may assume you have an angle-measuring device such as a protractor. Be specific about how you would use it or any other equipment.

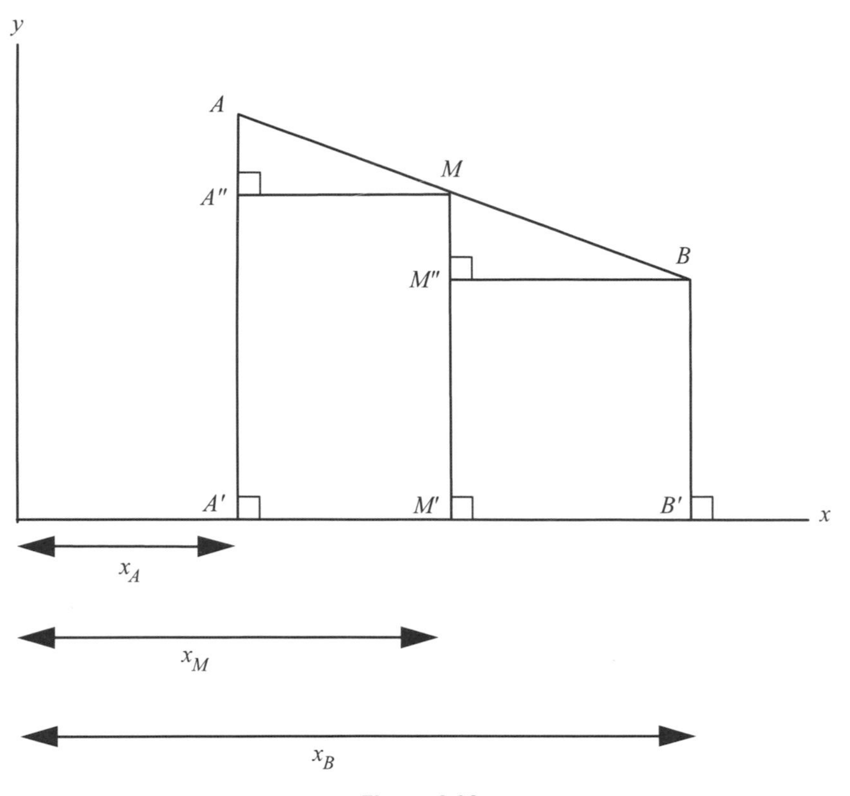

Figure 2.32

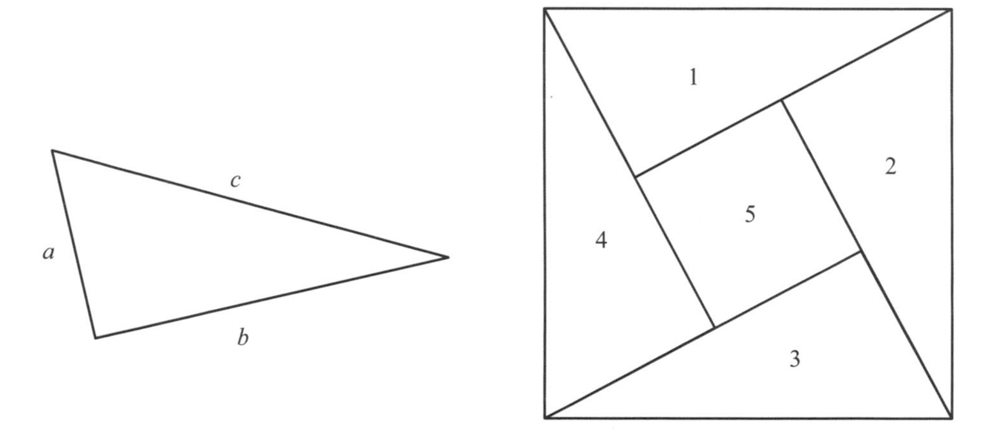

Figure 2.33 Proving the Pythagorean theorem.

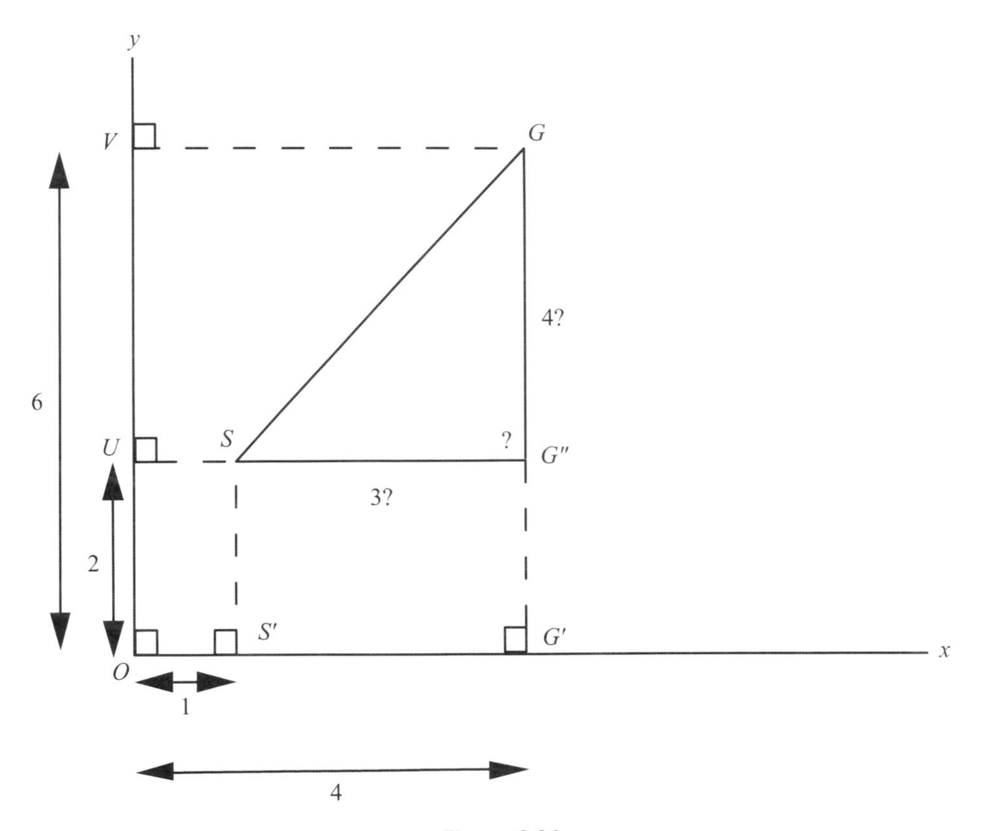

Figure 2.34

20. Could you do Eratosthenes's measurement and calculation as a term project? If not, what obstacles stand in the way?

21. Suppose Eratosthenes was willing to consider that the light rays from the sun were not really parallel after all but met at an angle of ψ at the sun, which is assumed to be a point. Say, for the sake of a trial calculation, $\psi = 1^\circ$. Show how this could be taken into account in the calculation of θ. Why didn't Eratosthenes do something like this?

Parallel and Skew Lines in Three Dimensions

In three-dimensional space, lines that do not meet are not necessarily called parallel. To be parallel, they must also lie in the same plane. Lines that are not in the same plane and do not meet are called skew.

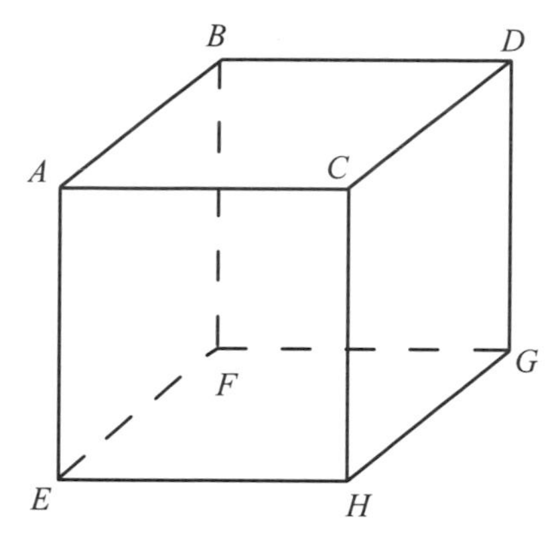

Figure 2.35

22. Among the lines determined by the edges of the cube of Figure 2.35, are there any lines that are skew to $\overleftrightarrow{AB}$? If so, list them. How many parallels to $\overleftrightarrow{AB}$ are there?

23. Among the lines determined by the edges of the tetrahedron of Figure 2.36, are there any lines that are skew to $\overleftrightarrow{AB}$? If so, list them. How many parallels to $\overleftrightarrow{AB}$ are there?

24. Consider the lines determined by the diagonals of the four-sided faces of the cube in Figure 2.35. (For example, the segment $\overline{AD}$ is such a diagonal and the line $\overleftrightarrow{AD}$ it determines is one of the lines under consideration now.) Among these lines, are there any lines that are skew to $\overleftrightarrow{AD}$? If so, list them. How many parallels to $\overleftrightarrow{AD}$ are there?

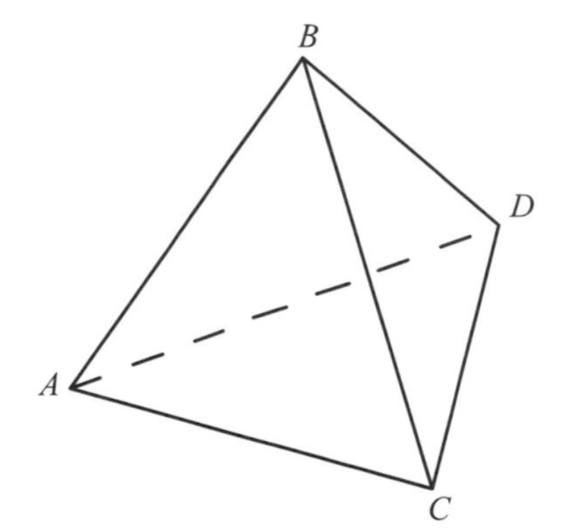

Figure 2.36

Rigidity

25. Apply the segment marking algorithm to Figure 2.29a.

26. Apply the segment marking algorithm to Figure 2.29b.

27. Apply the segment marking algorithm to Figure 2.29c.

28. Determine whether the braced grid of squares shown in Figure 2.37 is rigid using the segment marking algorithm. Make a series of drawings to show how our two principles lead to segment markings.

29. Show, by a careful drawing, a flexed version of the braced grid shown in Figure 2.38.

*30. In Figure 2.37, suppose we remove the brace from the square in the fourth row and third column. Apply the segment marking algorithm. Is the grid rigid? If not, draw it in a flexed shape.

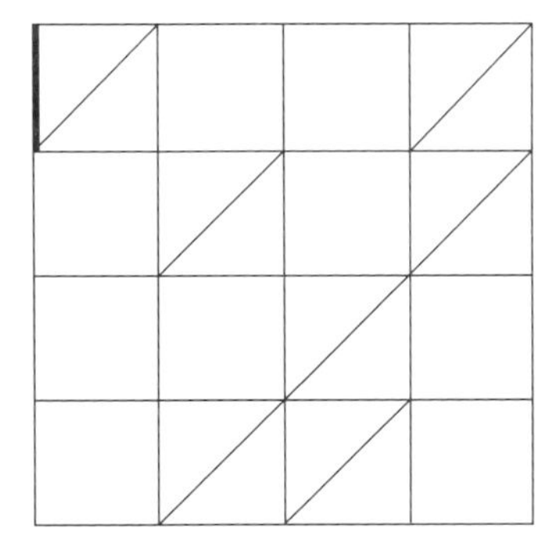

Figure 2.37

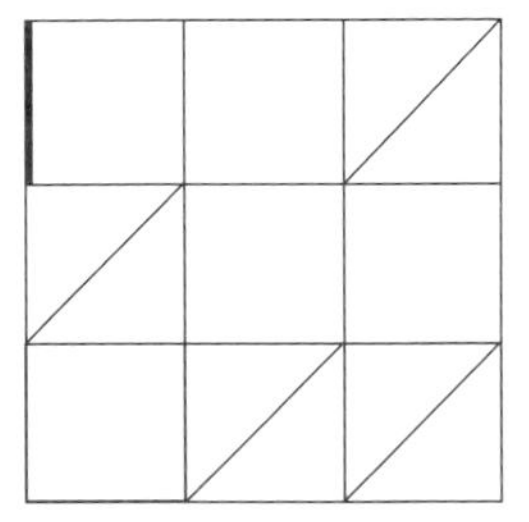

Figure 2.38

31. Here is another way to argue that the braced grid of Figure 2.29a is rigid. Three of the angles at A must remain right angles because they are parts of braced squares. Therefore, the fourth angle is also a right angle. Therefore, the center square is as good as braced, so we can draw a dotted diagonal in it to show this. Can you continue and use this right angle principle at other angles? Can you demonstrate rigidity this way?

*32. Construct an example of a braced grid that can be shown to be rigid by the segment marking algorithm, but where the method of right angles, described in the previous exercise, does not work.

33. Prove that, in a grid of squares with m rows and n columns, it is possible to brace the grid with $m+n-1$ braces. (Show a pattern of bracings.)

*34. Prove that, in a grid of squares with m rows and n columns, if the number of braces is $\leq m+n-2$, then the segment marking algorithm will not mark all the segments. (*Hint:* there are $m+n$ rows and columns that need to be marked.)

35. Consider a 2×2 grid of four rectangles which are not squares. Assume there are no braces. Show that you can flex this grid in such a way that at least one of the rectangles becomes self-intersecting.

2.4 Area and Similarity

Area formulas are among the most practical fruits of geometry for everyday applications. But the theory of area, perhaps surprisingly, also serves as the basis for the theory of similar triangles. Our objective in this section is to derive some well-known area formulas from our area axioms (Axioms 15 to 20 in Section 1.3 of Chapter 1) and then to apply the formula for the area of a triangle to obtain some central theorems about similar triangles.

Area

✮ THEOREM 2.15

The area of a parallelogram is the product bh, where b is the length of the base, and the height h is the length of the altitude dropped to the base from the opposite side (Figure 2.39).

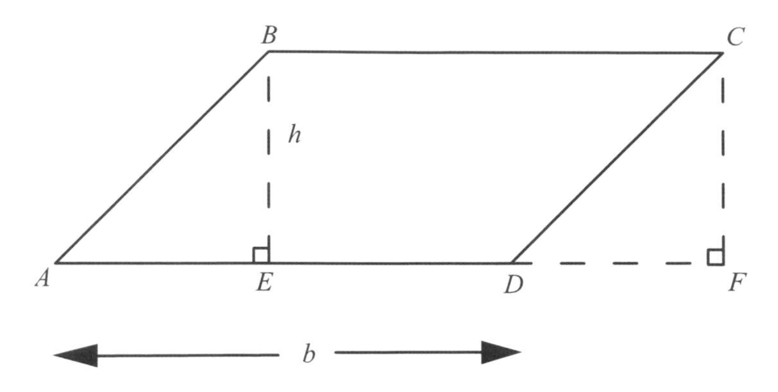

Figure 2.39 The area of a parallelogram.

Proof

In Figure 2.39 we need to show area$(ABCD) = (AD)(BE)$. From the vertices B and C, which are not on base $\overline{AD}$, drop perpendiculars to the line of the base, meeting it at E and F. Triangles ABE and DCF are congruent by AAS. (Both triangles are right triangles; $\mathrm{m}\angle BAE = \mathrm{m}\angle CDF$ since the lines $\overleftrightarrow{AB}$ and $\overleftrightarrow{CD}$ are parallel; and $AB = DC$ since opposite sides of a parallelogram are congruent.) We now express the area of trapezoid $ABCF$ by two different decompositions, using Axiom 17, after which we apply Axiom 18.

$$\begin{aligned}\text{area}(ABCF) = \text{area}(ABE) + \text{area}(BCFE) &= \text{area}(ABCD) + \text{area}(DCF)\\ \text{area}(ABE) + (EF)h &= \text{area}(ABCD) + \text{area}(DCF).\end{aligned}$$

Because the two triangles in the previous equation are congruent, their areas are equal by Axiom 16 and can be cancelled from the equation. In addition, because $AE = DF$ (corresponding parts of congruent triangles), $EF = ED + DF = ED + AE = AD = b$, so we arrive at

$$bh = \text{area}(ABCD).$$

■

Our proof has the shortcoming that it is tailored to a particular picture. Can you draw a parallelogram for which a different approach is needed? (See the exercises at the end of this section.)

There is an interesting mathematical issue lurking under the terminology of the area formula in the previous theorem: Which side is the base? In fact, either side may be taken as the base (with the altitude taken perpendicular to it). It is far from obvious why the area should come out the same no matter which side is taken as the base. However, if the axioms of geometry are consistent, this must be the case.

✯ THEOREM 2.16

The area of a triangle is $\frac{1}{2}bh$, where b is the length of any side and the height h is the length of the altitude to that side from the opposite vertex (Figure 2.40).

PROOF

Suppose triangle ABD is the given triangle and we choose $\overline{AD}$ as the base. Draw a line through B, parallel to the base, and then another line to complete a parallelogram $ABCD$. Triangles ABD and CDB are congruent (can you supply the reasons?) and have the same area by Axiom 16. Thus the area we seek is half the area of the parallelogram, which is bh by the previous theorem. Thus the area of the triangle is $\frac{1}{2}bh$. ∎

Figure 2.40 reminds us that the altitude (the perpendicular dropped from B) of a parallelogram or a triangle can fall outside the polygon.

DEFINITION

A *trapezoid* is a quadrilateral in which one pair of opposite sides is parallel. These sides are called the *bases* of the trapezoid and their lengths are denoted b_1 and b_2. The perpendicular distance from one base to the other is called the *height*, and its length is denoted h (Figure 2.41).

✯ THEOREM 2.17

The area of a trapezoid is $\frac{1}{2}(b_1 + b_2)h$.

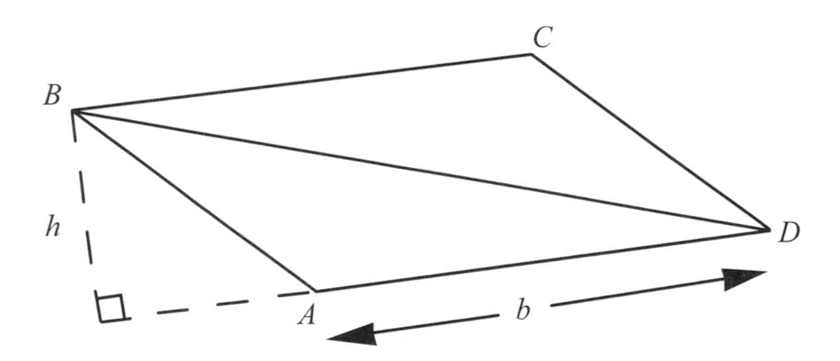

Figure 2.40 The area of a triangle.

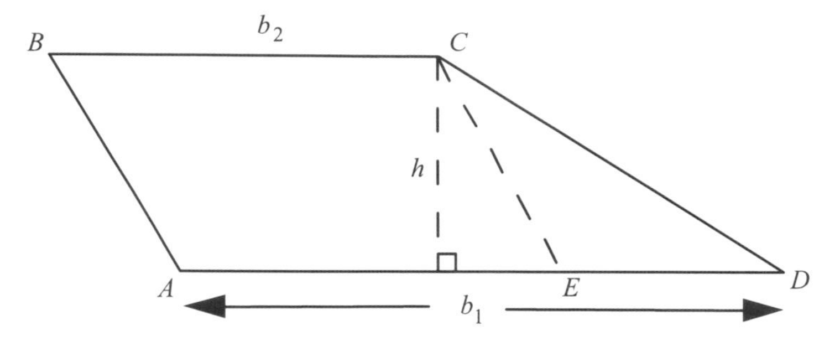

Figure 2.41 The area of a trapezoid.

Proof

Divide the trapezoid into a parallelogram and a triangle as shown in Figure 2.41. Because opposite sides of a parallelogram are congruent, $AE = b_2$ and so $ED = b_1 - b_2$. By Axiom 17,

$$\begin{aligned}\text{area}(ABCD) &= \text{area}(ABCE) + \text{area}(CED)\\ &= (AE)h + \frac{1}{2}(ED)h\\ &= b_2h + \frac{1}{2}(b_1 - b_2)h\\ &= \frac{1}{2}(b_1 + b_2)h.\end{aligned}$$

■

The following example is a useful preparation for the calculations in Theorem 2.18.

EXAMPLE 2.5

We find the area of the trapezoid created as follows. From the point $A(3, 4)$ drop a perpendicular to the x axis, meeting it at A'. From the point $B(5, 7)$ drop a perpendicular to the x axis, meeting it at A'. The area of trapezoid $ABB'A'$ is $[(4 + 7)/2](5 - 3) = 11$.

Now suppose we don't know the numerical coordinates but we denote A as (x_A, y_A) and B as (x_B, y_B). If we know $y_A, y_B > 0$ and $x_B > x_A$, we obtain the formula $[(y_A + y_B)/2](x_B - x_A)$. ■

Suppose we are given the coordinates of the vertices of a triangle and we want the area. The formula $\frac{1}{2}bh$ is not too helpful unless we can find some easy way of finding h. It would be nice to have an area formula that only depends on the coordinates of A, B, and C. Here it is:

✯ THEOREM 2.18

The area of triangle *ABC* (Figure 2.42) in terms of the coordinates of its vertices is

$$\left|\frac{1}{2}(x_A y_B - x_B y_A + x_B y_C - x_C y_B + x_C y_A - x_A y_C)\right|. \tag{2.8}$$

Proof

Figure 2.42 shows

1. our triangle;
2. two trapezoids underneath the triangle, *ABED* and *EBCF*;
3. and a large trapezoid, *ACFD*.

If we subtract the two lower trapezoids from the large one, we get the area of the triangle.

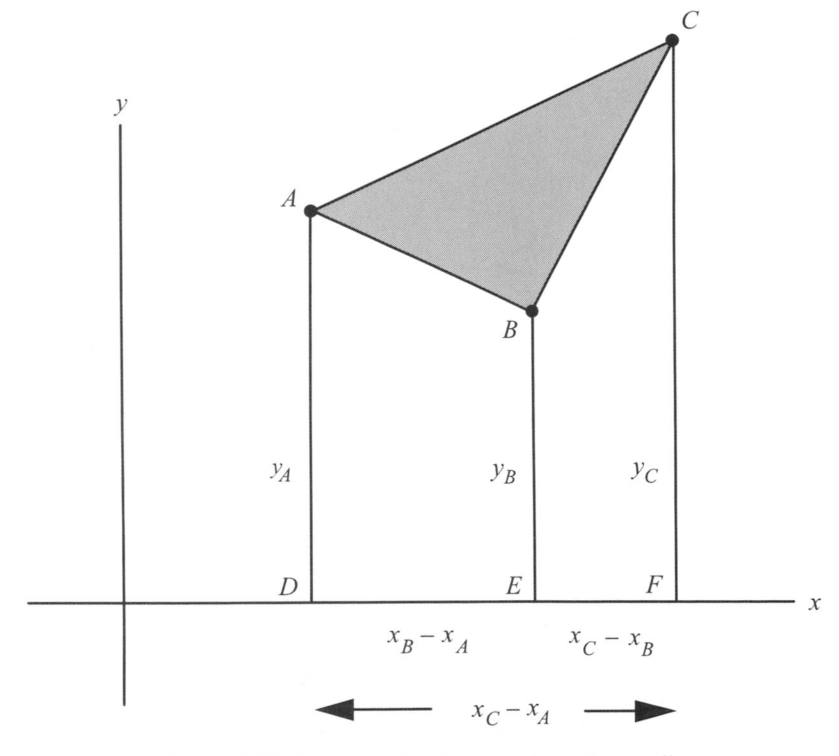

Figure 2.42 The area of a triangle from its coordinates.

As in Example 2.5, the trapezoid areas can be found conveniently from the coordinates of A, B, and C as follows:

$$\text{area}(ACFD) = \frac{1}{2}(y_A + y_C)(x_C - x_A) = \frac{1}{2}(y_A x_C - y_A x_A + y_C x_C - y_C x_A),$$
$$\text{area}(ABED) = \frac{1}{2}(y_A + y_B)(x_B - x_A) = \frac{1}{2}(y_A x_B - y_A x_A + y_B x_B - y_B x_A),$$
$$\text{area}(BCFE) = \frac{1}{2}(y_B + y_C)(x_C - x_B) = \frac{1}{2}(y_B x_C - y_B x_B + y_C x_C - y_C x_B).$$

Subtracting the second and third areas from the first, simplifying, and rearranging gives:

$$\frac{1}{2}(x_A y_B - x_B y_A + x_B y_C - x_C y_B + x_C y_A - x_A y_C). \tag{2.9}$$

But Formula (2.8) includes an absolute value sign, suggesting that Formula (2.9), which we just derived, might give the negative of the area for some configurations of points. Nothing about our proof suggests that this could happen. The reason is that we based the proof on particular positional relationships among the points A, B, and C. We leave to the reader the task of showing that Formula (2.9) might be the negative of the area for certain configurations and the absolute value really needs to be there. ■

Readers who have studied determinants may check that the Formula (2.9) just derived is identical with

$$\begin{vmatrix} x_A & y_A & 1 \\ x_B & y_B & 1 \\ x_C & y_C & 1 \end{vmatrix}.$$

When we are dealing with areas in everyday life, we usually don't know (or need) the coordinates of the vertices and so Formula (2.8) is not useful. However, it is useful in any piece of computer software where areas are wanted. Nearly all software that deals with geometry represents a polygonal figure in terms of the coordinates of its vertices. If the area of a polygon is wanted, it is first divided into triangles (how this is done by the software is itelf an interesting problem) and then Formula (2.8) is applied to each triangle. An example would be CAD/CAM software, which is used by engineers to design machine parts, buildings, and physical objects of all kinds. Say you have just used such a system to design the floor plan of a house. You will need to know the area of the floor in order to estimate the cost of covering it with a rug. When you pull down the menu choice for "area" the computer will probably use Formula (2.8) thanks to the programmer who knew this formula.

Here is an application of the use of Formula (2.8) which may be surprising: You can use it to check if three points are collinear. Suppose you are writing a computer game

"Save the Titanic" in which a player needs to guide a ship through a field of icebergs. In a typical play of the game, the ship has a current location $S(x_S, y_S)$ and wants to get to New York at $N(x_N, y_N)$. There is an iceberg at $I(x_I, y_I)$. If the player chooses to go straight, will there be a collision? In order for the game software to be able to answer the player, you need to write a software module that checks if one point is on the segment determined by two others. How should you do it?

Part of the task is to check if the three points are collinear. One approach to this is to compare the slopes of $\overleftrightarrow{SN}$ and $\overleftrightarrow{SI}$ and see if they are equal. There are at least two problems with this: If one or both segments are vertical there is no slope (there would be a zero denominator in the slope calculation and if the software attempted the division, it would crash). Even if the slopes are not exactly vertical but nearly vertical, there might still be a problem. It is conceivable the slopes may be so large that the computer can only get approximate values for them. If these approximate values are not equal we still can't be sure that the points are not collinear — maybe the true values are equal. There are various ways around this problem. The simplest, according to some computational geometers, is to forget about slopes and test whether the area of the triangle determined by the three points is 0 or not.

Similar Figures

Saying that two polygons are similar is the mathematical way of saying that they have the same shape even though they may be different sizes. One is a scaled up or scaled down version of the other. Here's how we make this precise.

DEFINITION

Two polygons are *similar* if there is a 1 : 1 correspondence f between their vertices so that:

1. Corresponding angles are congruent [for any vertex A, $m\angle A = m\angle f(A)$].
2. Corresponding side lengths have the same ratio. This means there is a *ratio of similarity* r with the property that if A and B are any two neighboring vertices on the first polygon then $AB/f(A)f(B) = r$.

EXAMPLE 2.6

In Figure 2.43, rectangle $ABCD$ is similar to rectangle $XYZW$ with correspondence $A \to X$, $B \to Y$, $C \to Z$, $D \to W$. [In the language of our definition, $f(A) = X$, $f(B) = Y$, $f(C) = Z$, $f(D) = W$.] This is because all the angles are right angles and

$$\frac{AB}{XY} = \frac{BC}{YZ} = \frac{CD}{ZW} = \frac{DA}{WX} = \frac{1}{2}.$$

Figure 2.43 Similar and nonsimilar figures.

It is important to get the correspondence right: $A \to Y$, $B \to Z$, $C \to W$, $D \to X$ does not demonstrate the similarity of these rectangles. Under this correspondence, we do have corresponding angles equal, but we do not have each ratio of a side to its corresponding side the same. For example $AB/YZ = 1$ while $BC/ZW = 0.25$.

Finally, notice that square $PQRS$ is not similar to rectangle $ABCD$ no matter what correspondence you try. ■

✯ THEOREM 2.19

In triangle ABC (see Figure 2.44) if D and E lie on sides $\overline{AB}$ and $\overline{CB}$ in such a way that $\overleftrightarrow{DE}$ is parallel to $\overleftrightarrow{AC}$, then $BA/BD = BC/BE$.

Proof

In Figure 2.44a consider the shaded triangle DEB and the triangle AEB consisting of the shaded plus striped triangles. Both triangles have the same altitude EF. Thus,

$$\text{area}(AEB) = \frac{1}{2}(BA)(EF),$$
$$\text{area}(DEB) = \frac{1}{2}(BD)(EF).$$

Dividing gives

$$\frac{BA}{BD} = \frac{\text{area}(AEB)}{\text{area}(DEB)}. \tag{2.10}$$

We can make the same type of argument, using Figure 2.44b to show

$$\frac{BC}{BE} = \frac{\text{area}(BDC)}{\text{area}(DEB)}. \tag{2.11}$$

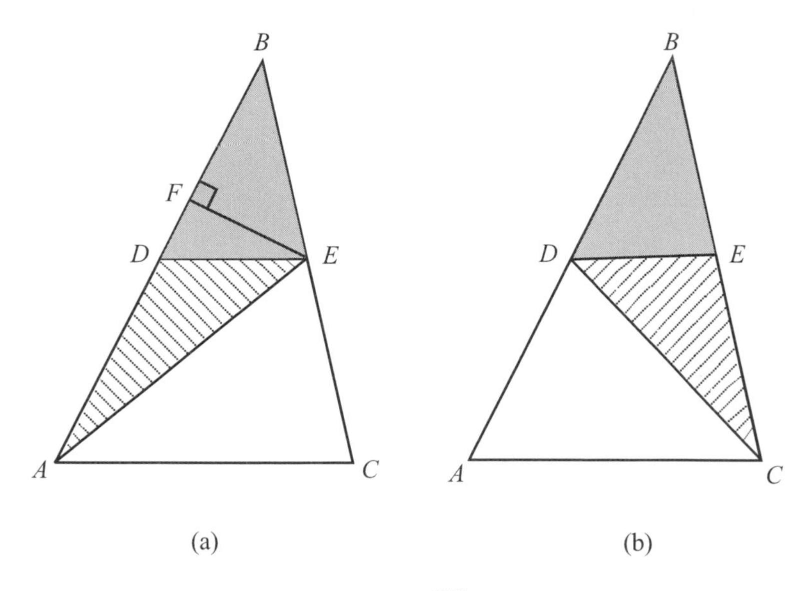

Figure 2.44 A parallel cutoff $\overline{DE}$ makes a similar triangle.

We now show that the right sides of Eqs. (2.10) and (2.11) are the same by showing area(AEB) = area(BDC). Since these triangles (look at Figure 2.44b for triangle BDC) both include the shaded triangle, it is enough if we show that the two striped triangles have the same area. For both striped triangles we can use $\overline{DE}$ as a base. But the perpendicular segments from A and C to line $\overleftrightarrow{DE}$ have the same length since $\overleftrightarrow{DE}$ is parallel to $\overleftrightarrow{AC}$. Because the bases and heights are the same, the areas are the same.

Because the right sides of Eqs. (2.10) and (2.11) are equal, so are the left sides, which is what we wished to prove. ■

In the previous theorem, because $\overleftrightarrow{DE}$ is parallel to $\overleftrightarrow{AC}$, $m\angle BDE = m\angle BAC$ and $m\angle BED = m\angle BCA$ (corresponding angles are equal). Thus, we are almost able to conclude that the shaded triangle DBE and triangle ABC are similar. However, we would still need to show that DE/AC is the same ratio as the other pairs of corresponding sides. This will emerge shortly as Theorem 2.22.

✫ THEOREM 2.20

In triangle ABC if D and E lie on sides $\overline{AB}$ and $\overline{CB}$ in such a way that $BA/BD = BC/BE$, then $\overleftrightarrow{DE}$ is parallel to $\overleftrightarrow{AC}$.

PROOF

We assume that $\overleftrightarrow{DE}$ is not parallel to $\overleftrightarrow{AC}$ and obtain a contradiction. In that case let F be a point on $\overline{CB}$ so that $\overleftrightarrow{AC}$ and $\overleftrightarrow{DF}$ are parallel. By the previous theorem, and by the hypothesis:

$$\frac{BA}{BD} = \frac{BC}{BF},$$

$$\frac{BA}{BD} = \frac{BC}{BE}.$$

The lengths in these equations are identical except for BF and BE, so we can deduce that $BF = BE$. But this means that $F = E$. Thus $\overleftrightarrow{DE}$ and $\overleftrightarrow{DF}$ are the same line and $\overleftrightarrow{DE}$ is parallel to $\overleftrightarrow{AC}$. ■

✯ THEOREM 2.21: *AAA Criterion for Similarity*

Let ABC and $A'B'C'$ be triangles where $m\angle A = m\angle A'$, $m\angle B = m\angle B'$, $m\angle C = m\angle C'$. Then the triangles are similar with $A \to A'$, $B \to B'$, $C \to C'$ (see Figure 2.45).

PROOF

If $A'B' = AB$ then the triangles are congruent (by ASA) and then they are similar as well. Thus we assume $A'B' < AB$ (if $A'B' > AB$ then interchange the triangles in the proof that follows). Find point D on $\overline{AB}$ so that $BD = A'B'$. Now find a point E on $\overline{BC}$ so that

$$\frac{BC}{BE} = \frac{BA}{BD}. \tag{2.12}$$

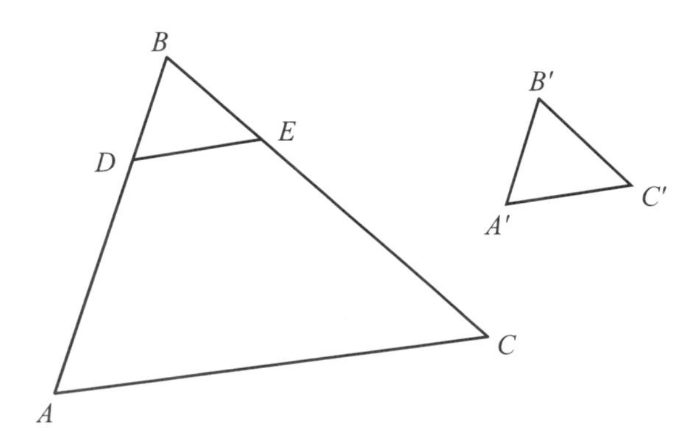

Figure 2.45 AAA criterion for similarity.

By Theorem 2.20, $\overleftrightarrow{DE}$ is parallel to $\overleftrightarrow{AC}$. Thus, by the corresponding angles of parallel lines principle and the hypothesis, $m\angle BDE = m\angle A = m\angle A'$. Thus triangles DBE and $A'B'C'$ are congruent (ASA). Using the corresponding parts principle, $BE = B'C'$. Substituting this and $BD = B'A'$ in Eq. (2.12) gives

$$\frac{BC}{B'C'} = \frac{BA}{B'A'}. \tag{2.13}$$

We can repeat the same proof, constructing a copy of triangle $A'B'C'$ in corner A and obtain

$$\frac{AC}{A'C'} = \frac{BA}{B'A'}. \tag{2.14}$$

Equations (2.13) and (2.14) together show that the proportionality requirements in the definition of similarity are satisfied. ■

✯ THEOREM 2.22: *SAS Criterion for Similarity*

Let ABC be a triangle (see Figure 2.46) and let $m\angle B = m\angle B'$ and $BA/B'A' = BC/B'C'$. Then triangles ABC and $A'B'C'$ are similar.

PROOF

Let D lie on $\overline{BA}$ and E on $\overline{BC}$ so that $BD = B'A'$ and $BE = B'C'$. Substituting these into the ratios mentioned in the hypothesis gives $BA/BD = BC/BE$. From Theorem 2.20 we see that $\overleftrightarrow{AC}$ is parallel to $\overleftrightarrow{DE}$. Therefore, the corresponding angles $\angle BDE$ and $\angle BAC$ are congruent, as are $\angle BED$ and $\angle BCA$. Consequently, we can apply Theorem 2.21 to conclude that triangles BDE and BAC are similar. However, triangle BDE is congruent to triangle $B'A'C'$ (SAS criterion for congruence). Thus, triangles $B'A'C'$ and BAC are similar. ■

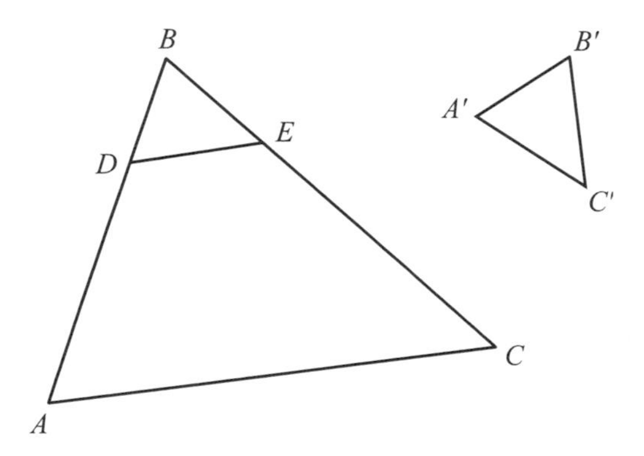

Figure 2.46 SAS criterion for similarity.

We close this section with what is undoubtedly the most famous theorem in geometry, a theorem which provoked a crisis in Greek philosophy, a theorem which yielded numbers so inscrutable as to be termed "irrational," a theorem encrusted with myth and legend, a theorem about which an entire book[1] was written containing hundreds of proofs, a theorem for which an original proof was given by the American president James Garfield — in short, a marvelous theorem.

✯ THEOREM 2.23: PYTHAGORAS

If a right triangle (Figure 2.47) has legs of length a and b and hypotenuse of length c, then $a^2 + b^2 = c^2$.

PROOF

Drop a perpendicular from the vertex of the right angle to the hypotenuse, dividing it into segments of length x and $c - x$. Triangles BCD and BAC are both right triangles and share $\angle B$. Thus $m\angle A = m\angle BCD$ and so the triangles are similar by Theorem 2.21. The lower part of Figure 2.47 shows the corresponding sides. Consequently:

$$\frac{a}{c - x} = \frac{c}{a}. \tag{2.15}$$

Likewise, triangles CDA and BCA are similar and so:

$$\frac{b}{c} = \frac{x}{b}. \tag{2.16}$$

Cross-multiply, simplify, and eliminate cx between these equations and we get the result. ■

EXERCISES

**Marks challenging exercises.*

AREA

1. For the parallelogram shown in Figure 2.48, would the proof of Theorem 2.15 (based on the construction in Figure 2.39) still work? If not could you modify it?

[1]Elisha Scott Loomis's *The Pythagorean Proposition* (1968).

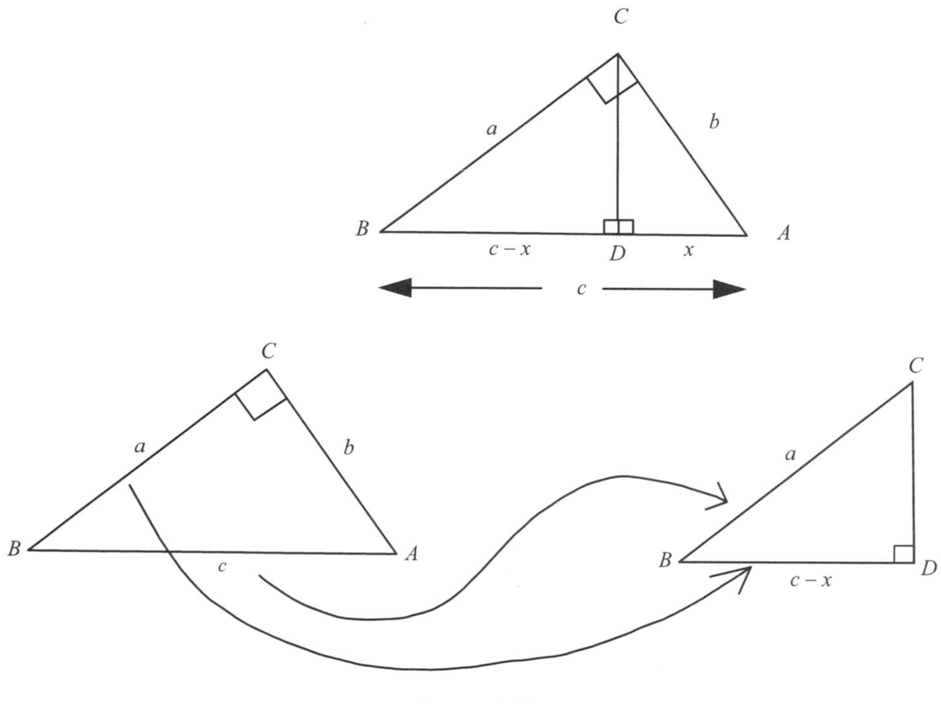

Figure 2.48

2. Would the proof method of Theorem 2.17, based on Figure 2.41, work for the trapezoid of Figure 2.49, or would you have to modify the construction and/or the proof in some way?
3. Find a proof of Theorem 2.17 that involves putting two copies of the trapezoid together to make a figure whose area formula has been previously derived.
4. In the proof of Theorem 2.16, why are triangles *ABD* and *CDB* congruent?

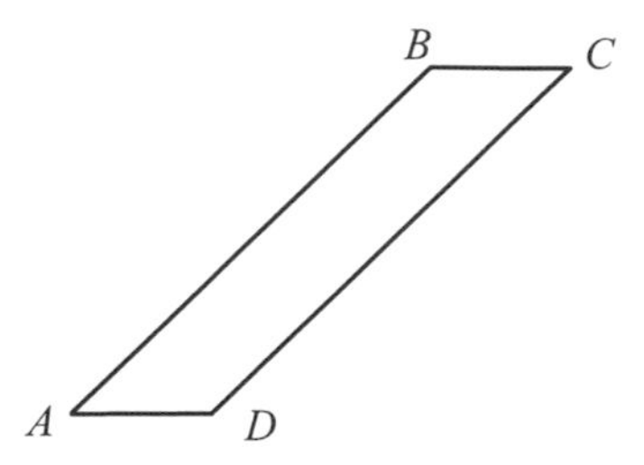

Figure 2.47 Similar triangles and the theorem of Pythagoras.

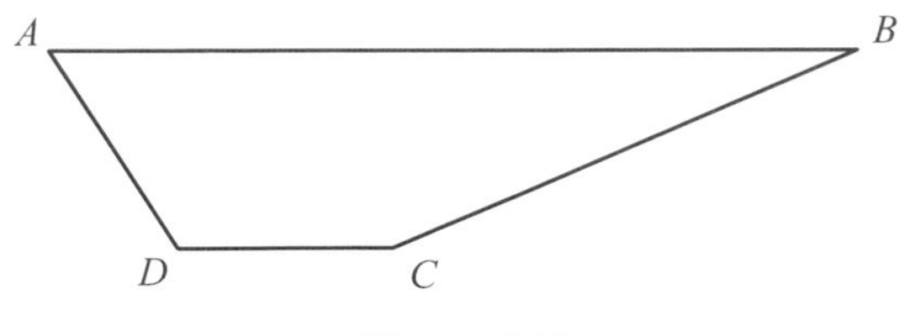

Figure 2.49

5. Is it true that the area of a quadrilateral is $\frac{1}{2}(b_1h_1 + b_2h_2)$, where b_1 and b_2 are the lengths of any two adjacent sides and h_1 and h_2 are the lengths of altitudes to those sides from the vertex which does not lie on either of those sides? If it is not true, give an example to show this. If it is true, prove it.

6. Let ABC be a right triangle whose legs have lengths a and b. Show that the area is $\frac{1}{2}ab$ directly from the axioms—without using any of the theorems of this section.

7. Find an alternate proof of Theorem 2.16 that relies on the result of the previous exercise.

8. Show that if a parallelogram has all sides equal (i.e., is a rhombus) then its area is one-half the product of the lengths of the diagonals.

* 9. Prove that if r is the radius of the circle inscribed in a triangle (the circle inside the triangle that is tangent to all the sides) and p is the perimeter of the triangle, then the area is $\frac{1}{2}pr$. (You will need to recall a fact we haven't discussed about tangents to circles.)

10. Suppose $ABCD$ is a square and E is the point where the diagonals meet. Let L be any line through E. Show that L divides the square into two equal areas.

11. Find the area of the triangle determined by the points $A(2, 4)$, $B(4, 7)$, and $C(6, 5)$ using Formula (2.8). If you had ignored the absolute value sign would you have a positive or negative answer?

12. Revisit the calculation of the previous exercise and explain why a negative answer would arise if you ignored the absolute value sign. One way you can explore this is to keep A and C fixed and try various locations for B, say, (4, 8), (4, 7), (4, 6), (4, 5), etc.

13. Suppose a quadrilateral $ABCD$ has these coordinates for its vertices: $A(0, 0)$, $B(b, 0)$, $C(b, h)$, $D(0, h)$. What do you expect the area to be? See whether Formula (2.8) gives the same answer: Divide the figure into two triangles by a diagonal and apply Formula (2.8) twice to find the area.

14. Suppose A and B are on the x axis at $(a, 0)$ and $(a + b, 0)$ with $b > 0$. Suppose C is at $(0, h)$ where $h > 0$. What do you expect the area of the triangle ABC to be? What does Formula (2.8) give?

15. Let $A = (1, 3)$ and $B = (2, 5)$. Use Formula (2.8) to find an equation for the locus of points $C(x, y)$ where the area of triangle ABC is 0.

16. Suppose a trapezoid $ABCD$ has opposite sides $\overline{AB}$ and $\overline{CD}$ horizontal. Find a formula for the area in terms of the coordinates of A, B, C, and D.

Similarity

17. In the proof of Theorem 2.19 we made use of the fact that if two lines L and M are parallel, and if A and B are any two points on L, the perpendiculars dropped from A and B to M have the same length. Explain where this was used; which two triangles did we prove to have equal areas using this principle? Prove this principle.

18. Suppose one triangle is similar to another and the ratio of corresponding sides is r. What can you say about the ratio of the areas? Prove your contention.

19. Complete the proof of Theorem 2.21 by deriving Eq. (2.14).

20. Suppose $A_1A_2A_3A_4$ and $A_1'A_2'A_3'A_4'$ are two quadrilaterals where $m\angle A_i = m\angle A_i'$ for each i and suppose

$$\frac{A_1A_2}{A_1'A_2'} = \frac{A_2A_3}{A_2'A_3'} = r.$$

Show that all other corresponding sides have ratio r as well—the quadrilaterals are similar.

21. Prove the SSS similarity criterion for triangles: If ABC and $A'B'C'$ are such that $AB/A'B' = BC/B'C' = CA/C'A'$, then the triangles are similar.

22. Prove that if triangles ABC and $A'B'C'$ are similar and the ratio of corresponding sides is r, then the ratio of the lengths of the angle bisectors of angles A and A' is also r.

23. In our proof of the theorem of Pythagoras, state the correspondence between vertices, which is the similarity that yields Eq. (2.16).

24. If an equilateral triangle has a side of length s, find the length of the altitude of the triangle.

25. If a right triangle has an angle of 30° and the third angle is 60°, find the ratio of the shorter to the longer leg.

Chapter 3

Non-Euclidean Geometry

Prerequisites: Chapter 1 and the first three sections of Chapter 2.

What if we reject one or more of Euclid's axioms? In this chapter, we outline the surprising consequences of replacing Euclid's parallel axiom with axioms that contradict it. The path we are exploring was considered so disturbing in the early eighteenth century that the most eminent mathematician of his day, C. F. Gauss, feared to write about it. What we have in this chapter is nothing less than some alternative theories of geometry. When these theories were first proposed, there were no known applications. Today there are applications and some are literally out of this world.

3.1 Hyperbolic and Other Non-Euclidean Geometries

In this section we examine the consequences of denying Euclid's parallel axiom of Section 1.3 of Chapter 1, while keeping the other axioms we have used so far. If we deny that there is at most one parallel to a line through a given point, this means there is more than one. (Recall that Theorem 2.9 of Section 2.3 of Chapter 2 rules out the possibility of no parallels.) Specifically, we make the following *replacement* for Euclid's parallel axiom:

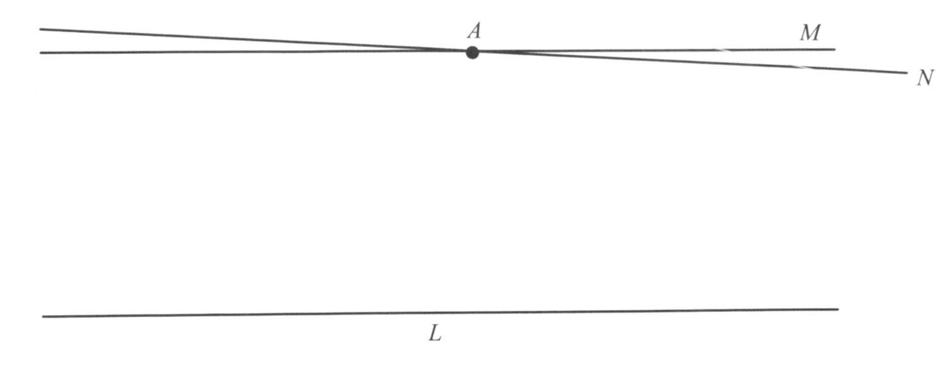

Figure 3.1 Can there be two lines through a parallel to *L*?

HYPERBOLIC PARALLEL AXIOM

Given a line *L* and a point *A* off the line, there is more than one line through *A* which does not meet *L* (Figure 3.1).

On first glance, the hyperbolic parallel axiom seems to be an incorrect description of the way points and lines behave in the real world around us. This leads to a rather important question: Why have mathematicians concerned themselves with this axiom? We'll take up this question at the end of this section, after we have seen some consequences of the axiom.

DEFINITION

The axiom set for planar hyperbolic geometry consists of axioms 1–8, area axioms 15–17, and the hyperbolic parallel axiom (taking the place of the Euclidean parallel axiom). The phrase *hyperbolic geometry* refers to this set of axioms and all the theorems that follow from it. Hyperbolic geometry is an example of a *non-Euclidean geometry*.

Because Euclidean geometry and hyperbolic geometry have different axiom sets, we expect that they have different theorems. As an example, in Euclidean geometry, as we showed in Section 2.3 of the previous chapter, every triangle has an angle sum of exactly 180°. We will see that in hyperbolic geometry we can prove that there is a triangle with an angle sum of less than 180°.

However, Euclidean geometry and hyperbolic geometry have many axioms in common—for example, that two points determine one line and the axiom about SAS guaranteeing congruence. Consequently, all the theorems in Chapter 2 that precede Section 2.3 of Chapter 2, for example, the exterior angle inequality and the theorems on congruence, are present in both geometries. However, in this section we cannot use any of the theorems of Section 2.3 of Chapter 2, with the notable exception of Theorem 2.9 (which did not depend on Euclid's parallel axiom).

✯ THEOREM 3.1

If α and γ are the measures of any two angles of a triangle (Figure 3.2), $\alpha + \gamma \leq 180°$.

PROOF

Extend one of the rays forming γ (either of the two will do) and let θ be the supplementary angle to γ so that $\theta = 180 - \gamma$. θ is an exterior angle of the triangle and so $\alpha \leq \theta$ by the exterior angle inequality (Theorem 2.2 of Section 2.1 in Chapter 2). Therefore, $\alpha \leq 180 - \gamma$, from which the result follows. ■

✯ THEOREM 3.2: SACCHERI-LEGENDRE

Given any triangle *ABC*, there is another triangle with the same angle sum, but with one of its angles no greater than $\frac{1}{2}\text{m}\angle A$.

PROOF

We use the same construction and the same diagram as in the exterior angle inequality (see Theorem 2.2 and Figure 2.6 of Section 2.1 in the previous chapter). We'll show that triangle *AEC* is the other triangle we want.

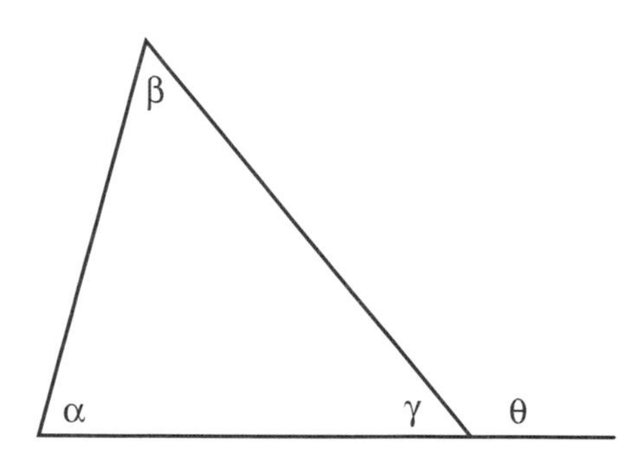

Figure 3.2 $\alpha + \gamma \leq 180°$.

Recall that the construction for the exterior angle inequality resulted in congruent triangles BAM and CEM with the correspondence $B \rightarrow C$, $A \rightarrow E$, $M \rightarrow M$. At least one of the angles $\angle BAM$ and $\angle MAC$ has measure $\leq (1/2)\text{m}\angle A$. $\text{m}\angle E = \text{m}\angle BAM$ because these are corresponding parts of congruent triangles. Therefore, at least one of the angles $\angle E$ and $\angle MAC$ has measure $\leq (1/2)\text{m}\angle A$.

To check the equality of the angle sums in triangles ABC and AEC, first note the following equalities involving corresponding parts of the congruent triangles BAM and CEM:

$$\text{m}\angle B = \text{m}\angle ECM,$$

$$\text{m}\angle BAM = \text{m}\angle E.$$

As a result

$$\begin{aligned}\text{angle sum of triangle } ABC &= \text{m}\angle MAC + \text{m}\angle BAM + \text{m}\angle B + \text{m}\angle BCA \\ &= \text{m}\angle MAC + \text{m}\angle E + \text{m}\angle ECM + \text{m}\angle BCA \\ &= \text{m}\angle MAC + \text{m}\angle E + \text{m}\angle ECA \\ &= \text{angle sum of triangle } AEC. \quad \blacksquare\end{aligned}$$

By applying the previous theorem over and over again, we could get a triangle with the original angle sum and in which one of the angles is less than some previously chosen target number. For example, suppose we have a target number of 3° and we start with a triangle ABC with angle sum s and $\text{m}\angle A = 16^\circ$. The first application of the theorem gives a new triangle $A'B'C'$ with $\text{m}\angle A' \leq 8^\circ$. One more application gives a triangle $A''B''C''$ with $\text{m}\angle A'' \leq 4^\circ$. Since our target number is 3°, we just need one more application giving a triangle $A'''\ B'''\ C'''$ with the same angle sum and $\text{m}\angle A''' \leq 2^\circ$. This idea of making an angle as small as we want, but with no change to the angle sum, becomes an essential part of the proof of the next theorem. Note that the inequality ($\leq$) in our next theorem is not strict, which is not inconsistent with Euclidean geometry where all triangles have angle sum 180°.

✯ THEOREM 3.3

The angle sum of any triangle is less than or equal to 180°.

PROOF

The proof is indirect, so we start with the assumption that there is a triangle ABC which has angle sum greater than 180°. We'll show this leads to a contradiction.

Because the angle sum is greater than 180°, there is a positive value p so that

$$m\angle A + m\angle B + m\angle C = 180° + p.$$

Now by repeated application of Theorem 3.2, we can find a triangle $A^{*}B^{*}C^{*}$ where

$$m\angle A^{*} + m\angle B^{*} + m\angle C^{*} = 180° + p,$$
$$m\angle A^{*} < p.$$

From these two relationships, we deduce that $m\angle B^{*} + m\angle C^{*} > 180°$, which contradicts Theorem 3.1. ■

The previous theorem does not say that all triangles have the same angle sum — but it doesn't rule that out either.

☆ THEOREM 3.4: STRONG EXTERIOR ANGLE INEQUALITY

If θ is the measure of an exterior angle of a triangle, and α and β are measures of the remote interior angles, then $\alpha + \beta \leq \theta$ (see Figure 3.2).

PROOF

Let γ be the interior angle of the triangle supplementary to θ. The previous theorem tells us

$$\alpha + \beta + \gamma \leq 180°.$$

But since θ and γ are supplementary, $\gamma = 180° - \theta$. Using this to substitute for γ into the inequality gives

$$\alpha + \beta + 180° - \theta \leq 180°,$$

from which the result follows. ■

The next theorem is the point we have been driving at with the previous results: the existence of a triangle with angle sum strictly less than 180°. Note carefully that the inequality in this theorem is strict ($<$ and not $\leq$), so it really is saying something strikingly different from Euclidean geometry. Notice also that this is the first theorem in this section in which we use the hyperbolic parallel axiom.

The proof we will give is motivated by a rather simple idea which, nonetheless, requires some technicalities to carry out. Here is an informal sketch. Figure 3.3 shows a construction that starts with any line L, draws a perpendicular to it at B and then a perpendicular, M_{perp}, to that line at A. By Theorem 2.9 of Section 2.3 in Chapter 2 (which does not rest on Euclid's parallel axiom for its proof), we know M_{perp}

is a parallel to L. But the hyperbolic parallel axiom asserts that there is another line, say, M, parallel to L. It is not necessary to construct this line — it is enough to know that it exists. Let δ be the measure of the angle between M and M_{perp}.

Pick a point C on L and consider the triangle ABC as C is moved to the right, through positions $C_1, C_2, \ldots$. It appears as if the measures of the angles at the various positions of C (γ, γ_1, γ_2, etc.) get as close to 0 as we want, just by moving C far enough to the right. Eventually this angle makes a negligible contribution to the angle sum of triangle ABC, so we can ignore it. (You can't really, and we'll give a more careful argument in our formal proof.) $m\angle B$ remains 90° regardless of where C is. The angle at A never measures more than $90^\circ - \delta$. Ignoring the angle at C, the angle sum is $\leq 90^\circ + 90^\circ - \delta$.

If we were doing Euclidean geometry, we could prove there is a fatal flaw in this argument: As we move C to the right, C will eventually lie on M and then "cross over" to lie within the angle of size δ between M and M_{perp}. When this happens, we couldn't say that the angle at A is not more than $90^\circ - \delta$, so our proof breaks down. But in hyperbolic geometry, this cross-over doesn't occur because M doesn't cross L. If there were a cross-over position for C, where it lies on M, then C would be a point on both M and L, which is impossible because these lines don't cross. Thus, the flaw doesn't exist in hyperbolic geometry. Here is a more formal version of this argument.

✯ THEOREM 3.5

In hyperbolic geometry, there is a triangle whose angle sum is strictly less than 180° (Figure 3.3).

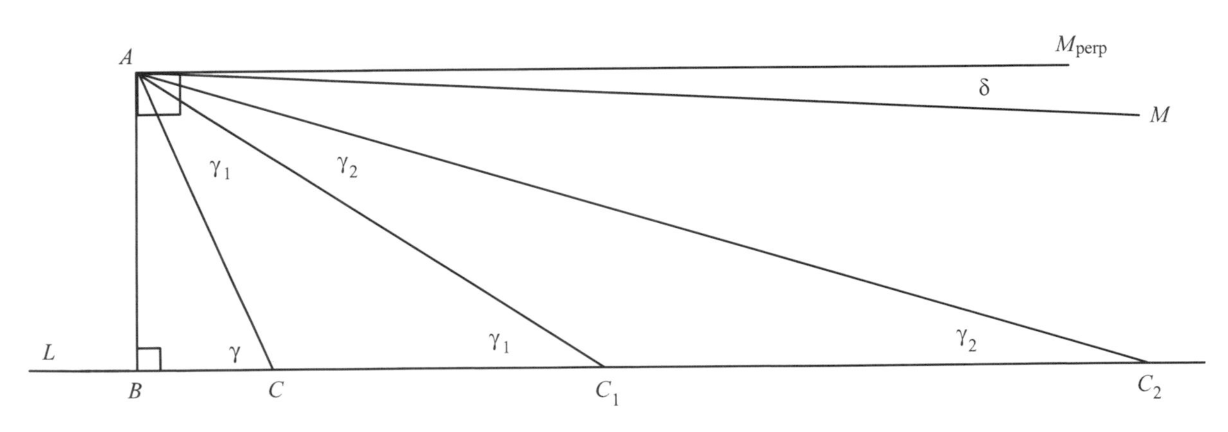

Figure 3.3 A triangle with angle sum less than 180°.

PROOF

Let L be a line, B a point on it, and $\overleftrightarrow{AB}$ perpendicular to L. We know that the line M_{perp}, which is perpendicular to $\overleftrightarrow{AB}$ at A, is parallel to L. But since there is more than one parallel, there is a second parallel line M and this line makes an angle of $90° - \delta$ with $\overleftrightarrow{AB}$ on one side of $\overleftrightarrow{AB}$. Here $\delta > 0$. On that same side of $\overleftrightarrow{AB}$ pick any point C on L to form a triangle ABC.

Let's abbreviate $\angle BCA$ as γ. Let C_1 be a point on L so that $AC = CC_1$. Because ACC_1 is an isosceles triangle, it has congruent base angles. Let's call their measure γ_1. By Theorem 3.4

$$\gamma \geq \gamma_1 + \gamma_1,$$

which implies

$$\gamma_1 \leq (1/2)\gamma. \tag{3.1}$$

Now find C_2 so that $AC_1 = C_1C_2$. Just as before, we can show

$$\gamma_2 \leq (1/2)\gamma_1.$$

Combining this with Eq. (3.1) gives

$$\gamma_2 \leq (1/4)\gamma = (1/2^2)\gamma.$$

If we do this construction n times, determining points $C_1, C_2, \ldots, C_n$, we obtain

$$\gamma_n \leq (1/2^n)\gamma. \tag{3.2}$$

The right side of this inequality can be made as small as we like just by making n large enough. In particular, we can choose n so that

$$(1/2^n)\gamma < \delta$$

so that Eq. (3.2) implies

$$\gamma_n < \delta. \tag{3.3}$$

Now we add up the angle sum of triangle ABC_n

$$\text{angle sum} = 90° + \angle BAC_n + \gamma_n.$$

But $\angle BAC_n < 90° - \delta$, and in view of Eq. (3.3) we get

$$\text{angle sum} < 90° + (90° - \delta) + \delta = 180°. \quad \blacksquare$$

The previous theorem does not tell us that every triangle has angle sum strictly less than 180°. This is true in hyperbolic geometry but we shall not prove it.

✯ THEOREM 3.6

In hyperbolic geometry, every triangle has angle sum strictly less than 180°. ■

The History of Non-Euclidean Geometry

Hyperbolic geometry arose from the desire of geometers to prove the Euclidean parallel axiom based on the other axioms of Euclidean geometry. This was an attempt to reduce to a minimum the axioms needed to support geometry. (In mathematics as in a court of law or everyday life, arguments based on many unproven assumptions — in our case, the axioms — are more suspect than arguments based on just a few.)

One way of trying to prove the Euclidean parallel axiom is to assume its opposite and attempt to derive a contradiction. One of the first attempts at doing this was made by the Italian Gerolamo Saccheri (1667–1733). Saccheri deduced one consequence after another from the hyperbolic geometry axiom set, and finally he reached a consequence that seemed so bizarre to him that he claimed to have discredited the hyperbolic parallel axiom. But in this he made a fundamental mistake. *A statement that seems to be disturbing, or to defy common sense, or to contradict something proven with a different axiom set (like Theorem 3.5) is not necessarily a contradiction.*

The first mathematicians who avoided this fundamental mistake of confusing the strange with the contradictory were the German Karl Friederich Gauss (1777–1855), the Russian Nikolai Lobachevsky (1793–1856), and the Hungarian Johann Bolyai (1802–1860). Gauss, the leading mathematician of his day, developed many consequences of the hyperbolic geometry axiom set. It appears that he had an open mind about whether our physical space is best described by hyperbolic geometry, for he is said to have carried out a careful mesurement of the angle sum of a large triangle whose vertices were three different mountain peaks. The measured angle sum was so close to 180° that it seemed quite possible that the discrepancy was due to errors in the measurement process. Gauss did not publish any of the consequences he derived from the hyperbolic parallel axiom for fear of ridicule and damage to his esteemed reputation.

Nikolai Lobachevsky, working in relative obscurity at the University of Kazan in Russia, began working on hyperbolic geometry in about 1823. Perhaps it is fortunate for him and for us that he had no great reputation (in those days) to protect, and was willing to give a public lecture on hyperbolic geometry in 1826, and then to write about it intermittently in the years thereafter. Today his reputation as a courageous and insightful founder of hyperbolic geometry is secure.

At about the same time as Lobachevsky started thinking about hyperbolic geometry, Bolyai was doing the same during his student days in Vienna. Instead of regarding his theorems (such as Theorem 3.5) with distaste, Bolyai wrote "I have discovered such magnificent things that I myself am astounded at them." He sent a copy of his results to his father, Wolfgang, who encouraged him to publish them soon. As Wolfgang Bolyai put it, "Many things have an epoch in which they are found at the same time in several places, just as the violets appear on every side in the Spring." The elder Bolyai's words could not have been more true. Not only was Lobachevsky hard at work on hyperbolic geometry at about the same time, but, when Wolfgang sent Johann's work to Gauss, the great man wrote back:

> If I commenced by saying that I am unable to praise this work (by Johann), you would certainly be surprised for a moment. But I cannot say otherwise. To praise it would be to praise myself. Indeed, the whole contents of the work, the path taken by your son, the results to which he is led, coincide almost completely with my meditations, which have occupied my mind partly for the last thirty or thirty-five years.

Gauss goes on to explain that he did not intend to publish his own results because he feared they would be misunderstood.

The development of hyperbolic geometry, although slow to catch on, was a trail-blazing event that stimulated other kinds of tinkerings with Euclid's axioms, leading to topics such as *single elliptic geometry, double elliptic geometry,* and "curved spaces" studied in the *differential geometry* of Bernhard Riemann (1826–1866). These geometries are collectively known as *non-Euclidean geometries.* In the next section we study *spherical geometry*, a variety of double elliptic geometry that is particularly easy to visualize and appreciate.

As geometers developed these exotic non-Euclidean geometries, many hoped and expected that in due course contradictions would be found in them. This would show that Euclid's geometry was the only reasonable theory of the physical space in which we live. As time went by and no contradictions were found, mathematicians began to suspect that these non-Euclidean geometries might be internally *consistent*, that is, have no contradictions among their theorems and axioms.

Finally, in the late nineteenth century it was shown, by Felix Klein (1849–1925) and Henri Poincaré (1854–1912), that if a contradiction were ever to be found in hyperbolic geometry, it would also be possible to find a contradiction in Euclidean geometry (and vice versa). Similar equivalences were found between Euclidean geometry and the other non-Euclidean geometries. Although we are not certain that Euclidean geometry is consistent, all scientists hope and assume that it is. Consequently, we hope and assume that hyperbolic geometry and its various non-Euclidean brethren are consistent as well.

Although the non-Euclidean geometries are as consistent as Euclidean geometry, they are a lot stranger. We demonstrated an unexpected theorem in hyperbolic geometry earlier in this section, but the properties of lines in some of the other non-Euclidean geometries seem even more unusual than in hyperbolic geometry. For example, double elliptic geometry denies the incidence axiom that two points determine exactly one line. In addition, the ruler axiom is denied. Lines have the property that if you travel on one long enough you'll come back around to your starting point without ever changing direction to reverse your path.

APPLICATION: Astronomy

Modern physicists welcome this because they believe that the incidence axiom and the ruler axiom are not actually true of light rays. One type of evidence they cite is the recently discovered "gravitational lenses" (see Figures 3.4 and 3.5). A gravitational lens is a massive object (see the macho in Figure 3.5), generally composed of invisible "dark matter," which bends light. If this mass lies near the line of sight between us and a star, light from the star may reach us in various ways, bending around the macho. This means we can see the same star by looking in various directions. Photos of the star will show a ring

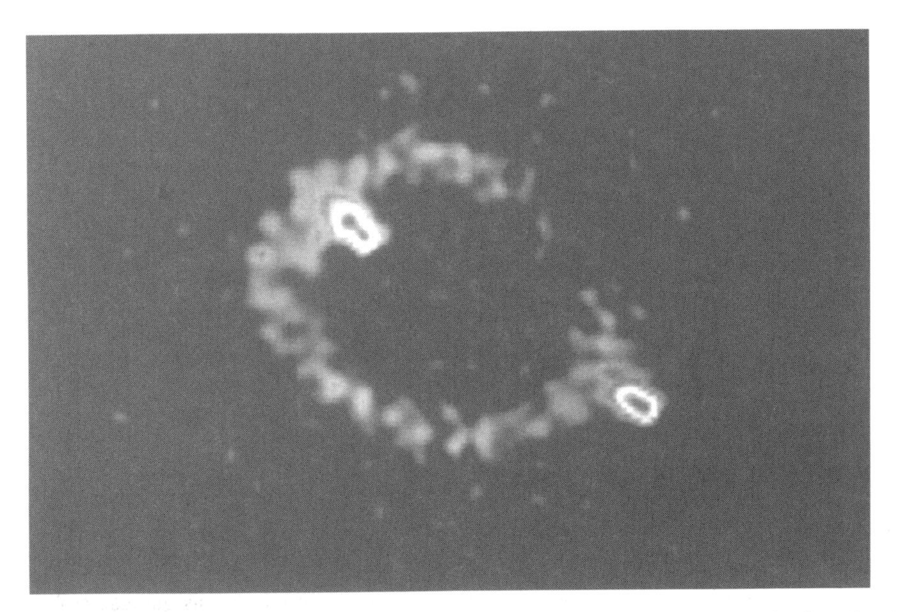

Figure 3.4 A star (MG 1131 + 0456) appearing as a ring, known as *Einstein's ring*, due to a gravitational lens. Courtesy of National Radio Astronomy Observatory.

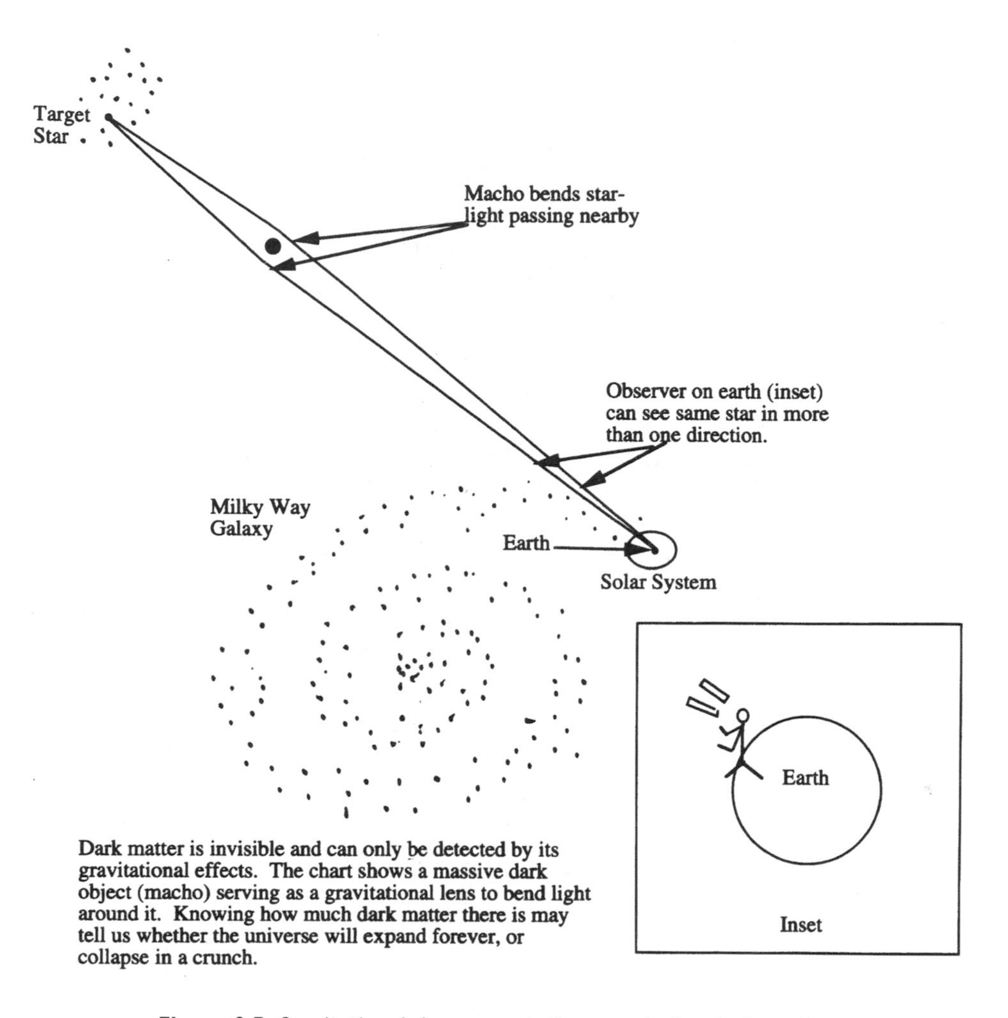

Figure 3.5 Gravitational lenses and the search for dark matter.

of light rather than a point of light (Figure 3.4.) If we regard light rays as traveling along lines, then a gravitational lens creates many lines with the same two points in common. In 1996 an international team of astronomers, working from an observatory in Australia, examined millions of target stars in the Large Magellanic Cloud (the nearest galactic neighbor to our own galaxy, the Milky Way) looking for evidence of gravitational lensing. Seven examples were found. This research was part of a campaign by astronomers to find out how much dark matter there is in the universe. The answer to this question could tell us whether the universe will eventually stop its expansion and collapse to a tiny hot lump, or whether it might expand forever.

If having lines that intersect twice seems unacceptable, we might try to get out of it by asserting that light rays are not really behaving like lines. This argument would claim that "real" lines really do behave as Euclid's axioms say they do. But then what are these "real" lines physically? Denial of "linehood" to light rays would leave us with a theory of geometry missing one of its most important applications. Meanwhile, physicists would be clamoring for a theory of geometry that does apply to light rays.

But, if Euclidean geometry is a poor description of light rays and stretched strings in our universe, why do we teach it and why does it seem so much more appealing to our visual common sense than non-Euclidean geometries? We teach Euclidean geometry because, when we deal with geometric figures like triangles and rectangles that are no larger than our earth, the discrepancy between Euclidean geometry and more accurate theories of geometry is so small that it cannot be measured by ordinary means. As we have seen, when Gauss measured the angle sum of a large triangle, the results were consistent with Euclidean geometry. For "small-scale" problems ranging from chemistry to carpentry and including even sending rockets into outer space, Euclidean geometry is entirely reliable. We only need non-Euclidean theories for cosmological problems.

Even with the foregoing explanations, when all is said and done, it is still strange to think that the large-scale behavior of lines may be different from our everyday experience of them on earth. This would not be the first time we have had to change our ideas about the physical world. For example, believers in the flat-earth theory had to give way when it became clear that the earth was round. Nonetheless, it is hard to keep an open mind about things we feel sure of.

EXERCISES

**Marks challenging exercises.*

1. In hyperbolic geometry show how to construct a quadrilateral whose angle sum is strictly less than $360°$. Do not rely on Theorem 3.6.

2. What does Theorem 3.6 imply about the angle sum of an arbitrary quadrilateral? A pentagon? A polygon with n sides?

3. In hyperbolic geometry show that if two triangles are similar (have corresponding angles congruent) then they are congruent. (*Hint:* Show how the larger triangle can be subdivided into a triangle congruent to the smaller one and a quadrilateral. What is the angle sum of the quadrilateral?)

4. In hyperbolic geometry, if ABC is a triangle with angles measuring α, β, γ, define the *defect* of ABC as $d(ABC) = 180° - (\alpha + \beta + \gamma)$. Let ABC be a triangle and D be a point lying between A and B. Show that $d(ABC) = d(ACD) + d(BCD)$.

5. In hyperbolic geometry show that if ABC is a right triangle, we can create a new triangle, not necessarily a right triangle, with twice the defect (see previous Exercise) as d(ABC). (*Hint:* If this exercise were about area instead of defect how would you do it?)

6. We would like to find a definition of the defect of a quadrilateral, d($ABCD$), in hyperbolic geometry that would allow us to assert that if we divide a quadrilateral into two quadrilaterals by a line drawn from one side to the opposite side, we could say that the defects of the pieces sum to the defect of the original quadrilateral. Find an appropriate definition of d($ABCD$). (See the previous exercises about the defect of a triangle.)

7. Using the definition you found in the previous exercise, determine whether the defects "add up" if you divide a quadrilateral into two triangles by drawing a diagonal. What if you divide a quadrilateral into a triangle and a quadrilateral by connecting a vertex to a point within a side?

8. Why did Gauss pick such a big triangle to measure the angle sum of? (*Hint:* Think about some of the previous exercises.)

Exercises 9–15 deal with Saccheri quadrilaterals. *A Saccheri quadrilateral (two are shown in Figure 3.6) is one in which a* base *makes right angles with its two neighboring sides and these neighboring sides have the same length called the* height. *The side opposite the base is called the* summit, *and the angles touching the summit are called* summit angles. *In many cases we ask you to give a proof without using either the Euclidean or hyperbolic parallel axiom or any of their consequences. These theorems are true in both Euclidean geometry and hyperbolic geometry. In other cases, you will need one or the other parallel axiom, as indicated.*

9. Prove that the diagonals of a Saccheri quadrilateral are congruent ($AC = BD$ in the left quadrilateral in Figure 3.6). Do not use either the Euclidean or hyperbolic parallel axiom or any of their consequences.

10. (a) Prove that the two summit angles of a Saccheri quadrilateral are congruent. Do not use either the Euclidean or hyperbolic parallel axiom or any of their consequences.

    (b) In Euclidean geometry, show that the summit angles of a Saccheri triangle are not only congruent but are right angles.

11. Suppose we have two Saccheri quadrilaterals whose heights are congruent and whose bases are congruent, as in Figure 3.6. Show that $\alpha = \alpha'$ and then $\beta = \beta'$. Do not use either the Euclidean or hyperbolic parallel axiom or any of their consequences.

12. Suppose we have two Saccheri quadrilaterals whose heights are congruent and whose bases are congruent, as in Figure 3.6. Show that $AC = A'C'$. Do not use either the Euclidean or hyperbolic parallel axiom or any of their consequences.

13. Suppose we have two Saccheri quadrilaterals whose heights are congruent and whose bases are congruent, as in Figure 3.6. Show that the summits are congruent and $\mathrm{m}\angle D = \mathrm{m}\angle D'$. Do not use either the Euclidean or hyperbolic parallel axiom or any of their consequences, but look for previous exercises that may help.

14. Suppose the summit of a certain Saccheri quadrilateral equals the base in length. Prove that this contradicts the hyperbolic parallel axiom. (*Hint:* Look for previous exercises that may help.)

*15. In hyperbolic geometry, prove that the summit of a Saccheri quadrilatral is longer than the base. (*Hint:* You can use some of the previous theorems. Jam some Saccheri quadrilaterals whose heights are equal together with bases on a line and look to make use of the following extension of the triangle inequality: The straight line path from A to B is shorter than any other.)

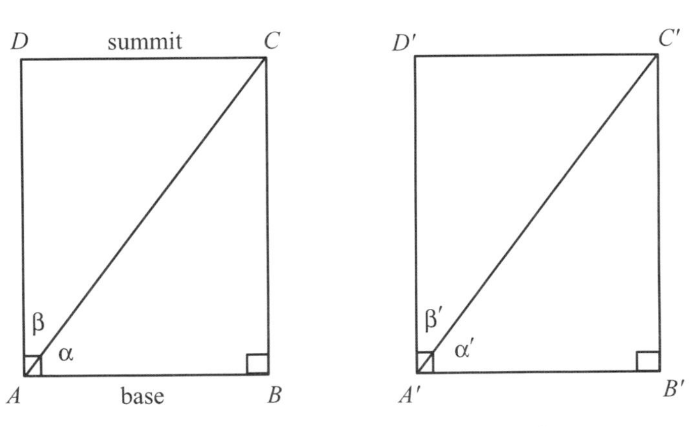

Figure 3.6 Two Saccheri quadrilaterals.

3.2 Spherical Geometry: A Three-Dimensional View

As you might suppose from the name, spherical geometry is the study of the surface of a sphere. As with Euclidean geometry, points play a key role. But there are no straight lines on the surface of a sphere. Instead, we study shortest paths from one point to another and we study certain circles lying on the sphere which are composed of shortest paths. Of course, we are not talking about shortest paths that cut through the inside of the sphere. The point of view to take is that of a ship sailing on the ocean or an aircraft following the curve of the earth's surface.

In the plane, a shortest path from A to B is the segment $\overline{AB}$, something we know a good deal about. But what is a shortest path on the sphere like? How does it bend and curve? To answer these questions, we will take full advantage of our knowledge of spheres and planes in three dimensions. (In the following section, we sketch a set of axioms that describes the shortest paths on the sphere and allows us to deduce facts about spherical geometry without any knowledge of three-dimensional geometry.)

The sphere we study is an object in Euclidean three-dimensional space, so all the axioms of Section 1.3 in Chapter 1 apply.

Shortest Paths and Geodesics

To study shortest paths on the sphere we need, first, to remind ourselves of the definition of a sphere, and then we need a foot in the door — something to make shortest paths on the sphere seem related to things we already understand. So here is the definition and the theorem that serve as our foot in the door.

DEFINITION

A *sphere of radius* R is the locus of all points in three-dimensional space that have distance R from one particular point called the *center*.

☆ THEOREM 3.7

If M_{AB} is a shortest path from A to B on the surface of a sphere S, then there is a plane P that passes through the center of the sphere with the property that $M_{AB} \subset S \cap P$.

PROOF

We omit the proof of this theorem. In the exercises we show you how to use some calculus techniques to prove it. ■

It is not hard to see that a plane which slices through the center of a sphere cuts the sphere in a circle. After all, each point of the intersection, since it is on the surface of the sphere, has the same distance to the center of the sphere. A planar figure in which each point has the same distance to some central point is, by definition, a circle.

DEFINITION

A circle on the surface of the sphere that arises by slicing the sphere with a plane through the center C of the sphere is called a *great circle*. Another name for a great circle is a *geodesic*.

In view of our definitions, Theorem 3.7 can be restated as follows.

☆ THEOREM 3.8: *The Great Circle Theorem*

Each shortest path on a sphere is an arc of a great circle. ■

There are circles on the sphere which are not great circles. For example, if you take a plane that does not pass through the center of the sphere, it can be proved that it will cut the sphere in a circle. Such a circle is called a *small circle*.

Great circles and small circles are commonly used in geography and navigation. On a globe, you will find circles marked that pass through both North and South poles (Figure 3.7). These are great circles called *longitude* circles or longitude lines. *Latitude* circles (Figure 3.8) are created by cutting the earth with planes that are perpendicular to the segment connecting the North and South poles. Among the circles of latitude, only the equator is a great circle. All the other latitude circles are created by planes which do not contain the center of the sphere. They are small circles.

Each point of the sphere has an *opposite* point. If you have ever wondered where you would come out if you dug a hole straight down into the earth, straight through the center and out the other side, then you already understand the idea. Mathematically speaking, the opposite of a point A on the sphere is the point you get when you connect A to the sphere's center and extend the segment till it meets the sphere surface again. We denote it as $-A$. For example, if N is the North pole, then $-N$ is the South pole. A useful fact about opposite points is that when two geodesics cross, each intersection point is the opposite of the other. (Can you prove this?)

Now suppose we have two points A and B on a geodesic G. There are two circular arcs along G which connect A and B, and the shorter of these paths must be the shortest path from A to B. We'll denote it by $\widehat{AB}$. But what if $B = -A$ (for example, say we are dealing with the North and South poles)? In this case each of the two

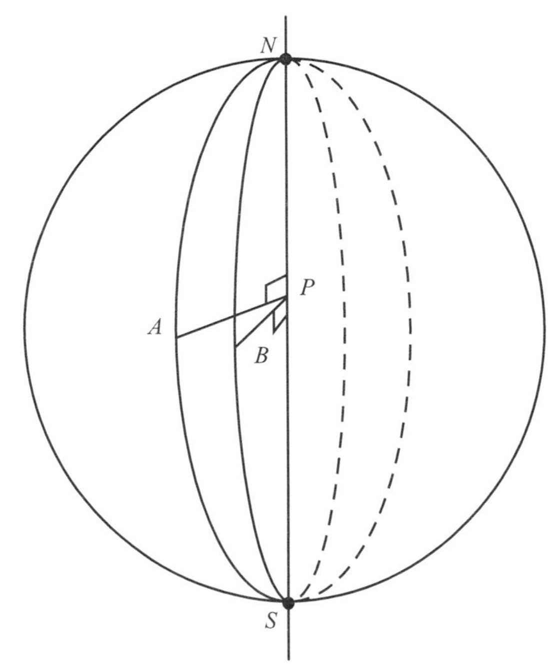

Figure 3.7 Two longitude circles and the angle $\angle APB$ between them.

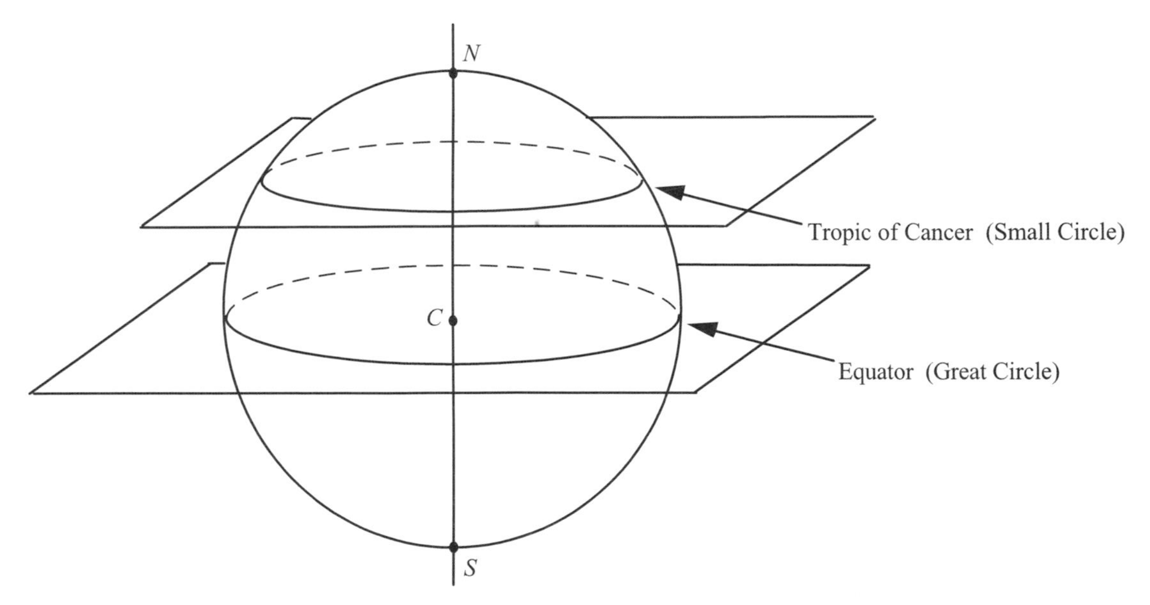

Figure 3.8 Planes perpendicular to NS creating latitude circles.

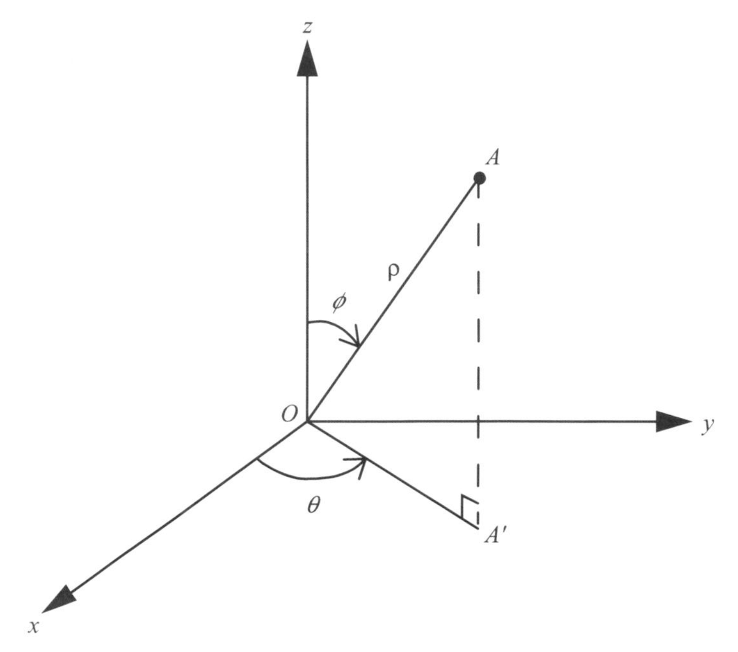

Figure 3.9 Spherical coordinates.

arcs has the same length — there is no single path shorter than all others. In fact, there are more than two paths which are tied for the honor of being shortest. There are many planes we can pass through A and $-A$ and they create an infinity of shortest paths (think of the longitude lines on the earth). In such cases where A and B are opposite, the notation $\widehat{AB}$ is ambiguous and more information is needed to specify which path is under discussion.

A point A on a sphere whose center is at the origin is often more conveniently described by *spherical coordinates* (ρ, θ, ϕ) rather than its (x, y, z) Cartesian coordinates (Figure 3.9). ρ is the distance OA, which will be the radius of the sphere, R. To get θ, drop a perpendicular from point A down onto A' in the x-y plane and take the radian measure of the angle from the positive x axis to $\overrightarrow{OA'}$. (As usual, the direction from the positive x axis to the positive y axis is the positive direction for angle measurement.) We define ϕ to be the radian measure of the angle from the positive z axis to ray $\overrightarrow{OA}$.

EXAMPLE 3.1 Spherical Coordinates

(a) What is the set of all points where ρ and θ are constant but ϕ varies?
(b) What is the set of all points where ρ and ϕ are constant but θ varies?

SOLUTION

To help us describe these sets, let's suppose the North pole is on the positive z axis.

(a) If $\rho = 0$ this set is just the origin. If $\rho > 0$ then we have a longitude circle on a sphere of radius ρ, such as those shown in Figure 3.7.

(b) If $\rho = 0$ this set is just the origin. If $\rho > 0$ then we have a latitude circle on a sphere of radius ρ, such as those shown in Figure 3.8, unless $\phi = 0°$ or $180°$, in which case we have either the North or South pole. ■

Because geodesics are created by slicing planes through the center of the sphere, the properties of planes give rise to some key facts about geodesics. Those properties we need for the following theorems are given in Axioms 10 through 14 of Section 1.3 in Chapter 1.

☆ THEOREM 3.9

Any two geodesics G_1 and G_2 have exactly two points A and B in common.

PROOF

Let P_1 be the slicing plane that creates G_1 and let P_2 be the slicing plane that creates G_2. These planes have the center of the sphere in common so they do intersect. By Axiom 11, they intersect in a line L. $G_1 \cap G_2 = (P_1 \cap S) \cap (P_2 \cap S) = (P_1 \cap P_2) \cap S = L \cap S$. But a line through the center of a sphere crosses the sphere exactly twice (see Exercise 2 at the end of this section), so $L \cap S$ consists of two points. ■

Angles and Spherical Triangles

In Euclidean plane geometry lines make angles where they cross, so we should ask whether geodesics in spherical geometry do too. In calculus we learn that when two curves cross at a point, the angle between these curves is defined as the angle between the tangent vectors to the curves (**u** and **v** in Figure 3.10). Of course, a curve has two tangent vectors at a point—each one pointing opposite to the other. Consequently, there are four possible angles (equal in pairs because of the vertical angle theorem) where curves meet. To keep our figure simple, we just show the smaller angle, θ. This smaller angle is what we mean by the angle between the geodesics.

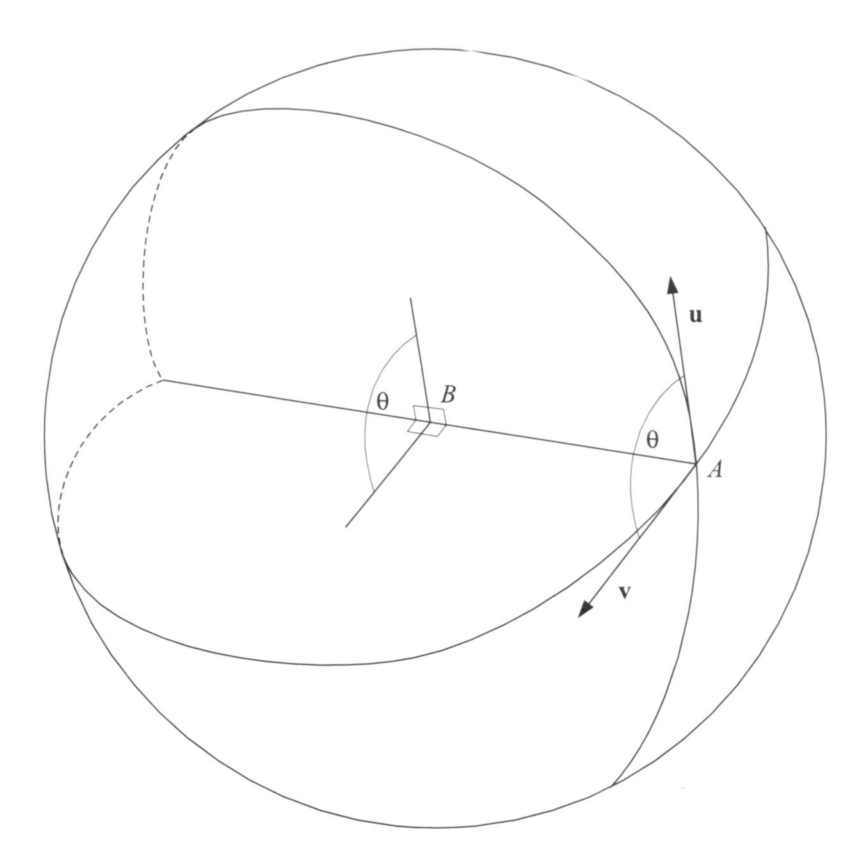

Figure 3.10 Two ways of measuring an angle between two geodesics.

There is a second equivalent way of thinking about the angle between two geodesics. Consider the planes that create these geodesics by slicing the sphere. Choose any point on the line where the planes meet, for example, B in Figure 3.10. Within each plane, erect a perpendicular ray to the line of intersection at B. Notice that, in each plane, there are two opposite directions we can pick for the ray. Again, for simplicity our figure shows just one pair of rays making the smaller angle. This angle θ between these rays in the slicing planes is the angle between the geodesics.

DEFINITION

Suppose A, B, and C are points on the sphere such that no geodesic contains all of them. The spherical triangle ABC is defined to be $\widehat{AB} \cup \widehat{BC} \cup \widehat{CA}$.

EXAMPLE 3.2: *A Spherical Triangle with Two Right Angles*

Let P_e be the plane passing through the equator of the earth (Figure 3.11). Let P_1 be a second plane, perpendicular to P_e. This second plane will create a geodesic that passes through the North and South poles — a longitude circle. Let A_1 be one of the two points where this geodesic cuts the equator. Because the planes are perpendicular, all the angles between the two geodesics are 90°. Now take plane P_1, which passes through the line from North to South poles, and spin it on that line a bit, making a new plane P_2. This new plane — also making a longitude circle — will also be perpendicular to the equatorial plane and so we have more right angles. Let A_2 be one of the two points

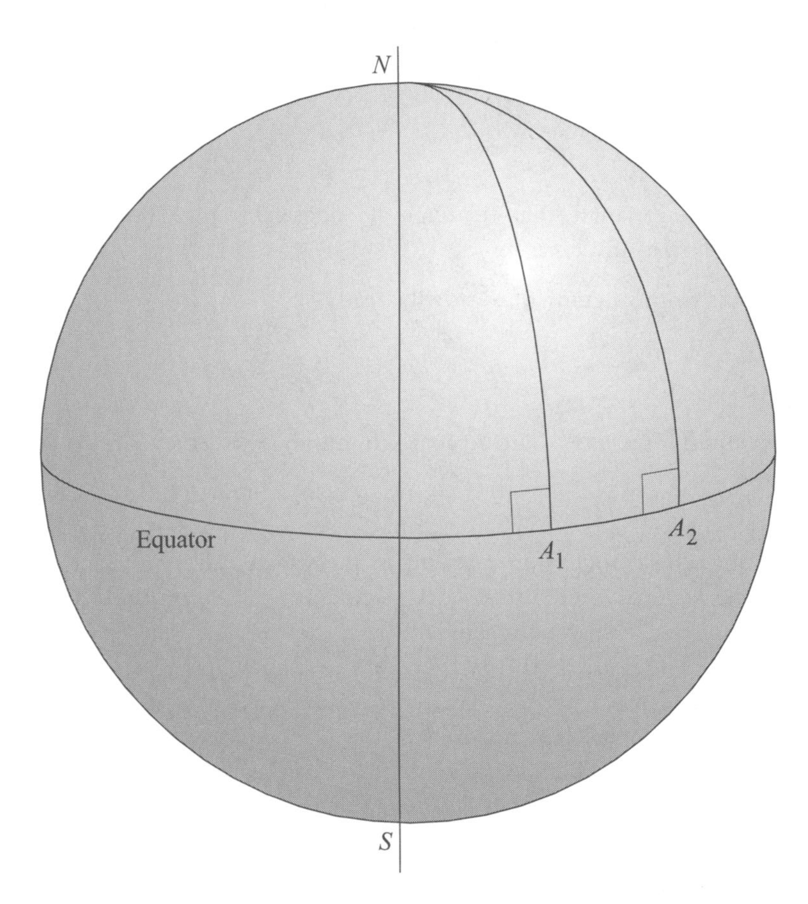

Figure 3.11 A spherical triangle with two right angles.

where this geodesic cuts the equator. Notice that the spherical triangle A_1NA_2 is a triangle with two right angles and, therefore, an angle sum of greater than 180°, something which does not happen with triangles in the plane. How much more than 180° can it be? ■

Suppose we attempt to create a spherical triangle as follows (see Figure 3.11): shortest path $\widehat{A_1N}$, shortest path $\widehat{NA_2}$, but in place of shortest path $\widehat{A_1A_2}$ we take the arc that goes around the equator the "long way" from A_1 to A_2. One side of the figure we have created is not a shortest path so this fails to meet our definition of a triangle. But why do we insist on shortest paths for the sides of a triangle? One reason is that if we did not, we could create a triangle whose boundary intersected itself. Can you draw an example of such a three-sided figure?

Our next major goal is to prove that *every* spherical triangle has an angle sum of greater than 180° (π if we measure angles in radians). This is called the *excess theorem*. We need a few preliminary results first. The next one should be obvious so we will skip the proof.

✯ THEOREM 3.10

If ABC is a spherical triangle, then the opposite spherical triangle $(-A)(-B)(-C)$ has the same shape, size, and area.

Here is a result from calculus that we will need:

✯ THEOREM 3.11

The area of a sphere is $4\pi R^2$. The area of a hemisphere is $2\pi R^2$.

The stage on which our proof of the excess theorem takes place is Figure 3.12—studying it well is half the battle of understanding the proof. Spherical triangle ABC is the one whose angle sum we want to investigate. We have turned the sphere so this triangle faces us and so that A, B, $-A$, and $-B$ are on the great circle that divides the front (visible) part of the sphere from the back (invisible) part.[1] The dotted parts of the great circles are on the back side of the sphere. Spherical triangle $AB(-C)$, on the back side of the sphere, is hatched because it plays a role in the proof and we need to remember that it is not part of the front hemisphere.

Two great circles on the sphere divide the surface of the sphere up into four *lunes*—so called because of their crescent moon shape. For example, in Figure 3.12, there are four lunes meeting at C. One of them is bounded by $CA(-C)BC$. (The hatched

---

[1] Technically, the figure shows a parallel projection of the sphere onto the plane of the paper.

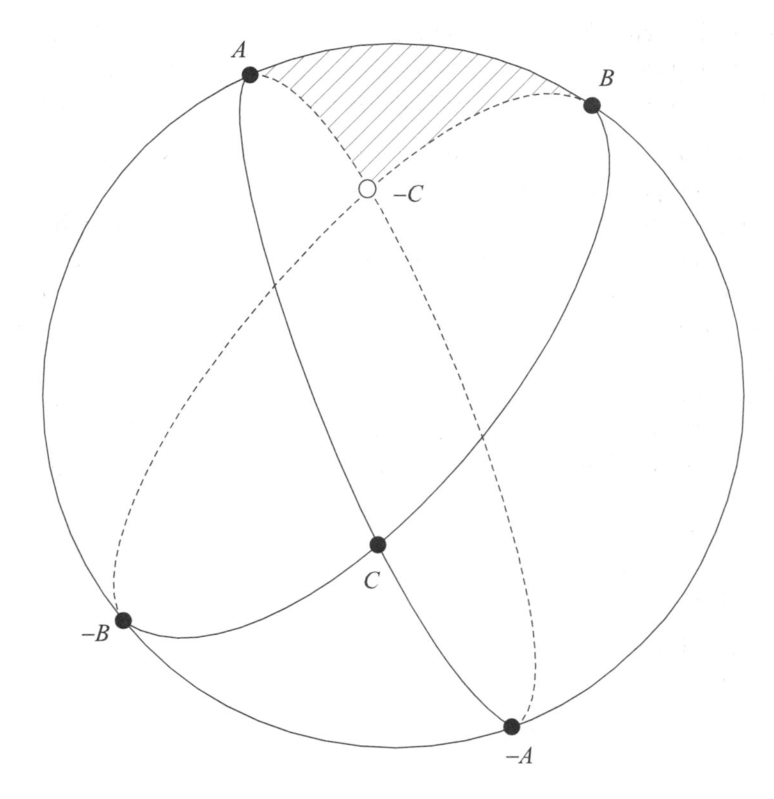

Figure 3.12 Visualizing the proof of the excess theorem.

spherical triangle is part of this lune.) The boundary of a lune consists of two shortest paths which are halves of the great circles creating the lune. The corners of a lune are opposite points. The size of a lune is conveniently conveyed by the measure of the corner angle between the great circles.

✯ THEOREM 3.12

A lune whose corner angle is θ radians has area $2\theta R^2$.

PROOF

We can imagine a lune at C being generated in the following way. Take the semicircle $CB(-C)$ and rotate it around the diameter connecting C to $(-C)$. As it rotates, it sweeps out a lune. Initially the lune is just a semicircle and has 0 area, but by the time we have rotated through π radians (180°) we have created a hemisphere with area $2\pi R^2$. If we stop at some intermediate number of radians θ, we have a fraction of this, namely, $(\theta/\pi)(2\pi R^2) = 2\theta R^2$. ■

☆ THEOREM 3.13: *The Excess Theorem*

In spherical triangle ABC (Figure 3.12),

$$m\angle A + m\angle B + m\angle C = \pi + \frac{\text{area}[ABC]}{R^2},$$

where all angles are in radians and R is the radius of the sphere.

PROOF

At each vertex of the spherical triangle ABC there is one lune which contains the spherical triangle. We call these the lune at A, the lune at B, and the lune at C. Our strategy is to add the area of those lunes and compare it with the front-facing hemisphere and some spherical triangles.

$$\text{area}[\text{lune at } A] + \text{area}[\text{lune at } B] + \text{area}[\text{lune at } C]$$
$$= \text{area of hemisphere} + 2\text{area}[ABC] - \text{area}[(-A)(-B)C] + \text{area}[AB(-C)].$$

The last two areas are equal by Theorem 3.10. Using Theorems 3.11 and 3.12 to make substitutions we get

$$2R^2\, m\angle A + 2R^2\, m\angle B + 2R^2\, m\angle C = 2\pi R^2 + 2\text{area}[ABC].$$

Dividing by $2R^2$ gives the formula we want. ■

Notice that one consequence of the excess theorem is that you can find the area of a triangle from its angles provided you know the radius of the sphere — there is no need to measure any of the side lengths.

APPLICATION: Map Making

Since the earth is nearly a sphere, it is common practice to make reasonably accurate scale models of it in the form of globes. When you look at a globe and notice that Greenland is narrower than the United States, you can conclude that the real Greenland really is narrower than the United States. Globes don't lie.

But globes are not as convenient as maps, which can be folded or rolled up or published in books. This "minor" issue of convenience leads to a major mathematical question of how to create a map that accurately shows the shapes of countries and bodies of water on the earth. In Figure 3.13 we show a

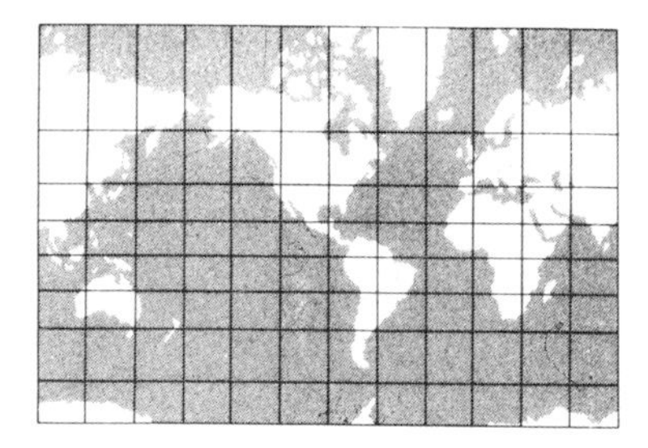

Figure 3.13 Mercator projection map of the earth (Wilford 1982). Courtesy of Vintage Books.

Mercator projection map which shows Greenland about as wide as the United States, a definite distortion of reality. Mercator projection maps, in which areas near the poles get excessively stretched out, are not the only types of maps. There are stereographic projections (studied in Section 8.3 of Chapter 8), central projections, conical projections, and dozens of others. Each has some type of distortion. Ideally we would like a map that scales each distance on the earth down by the same amount. More precisely:

PERFECT MAP REQUIREMENT

A planar map of the earth is *perfect* if there is a single scale number s such that whenever you measure the distance between any two points A and B on the earth along the shortest path $\widehat{AB}$, and you measure the straight line distance $A'B'$ between the images of A and B in the map, we have

$$\text{length}(\widehat{AB}) = s(A'B').$$

✯ THEOREM 3.14

There is no perfect map of the earth's surface (Figure 3.14).

Proof

In this theorem, we assume that the earth is a perfect sphere. (Of course, this is not exactly true, but to deal with the exact situation would take us too far afield.) We provide an indirect proof, assuming that there is a perfect map, and then derive a contradiction. Let A, B, and C lie on the equator, separated by 1/4 of the equator (Figure 3.14). Let c be the circumference of the earth, by which we mean the common circumference all great circles share. The shortest paths connecting A, B, and C to the North pole have length $c/4$, just as do the shortest paths $\widehat{AB}$ and $\widehat{BC}$. Now consider the corresponding points in a perfect map lying in the plane, A', B', C', N'. Since

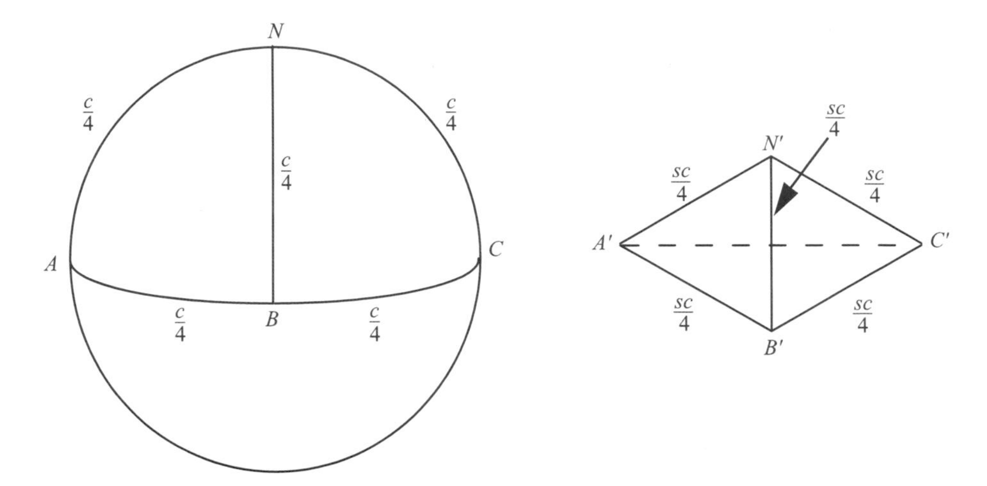

Figure 3.14 There is no perfect map of a sphere.

the map is perfect, each of the sides of the triangles in Figure 3.14 has length $sc/4$, so the triangles are equilateral and all angles measure 60°. We can readily calculate by the usual methods of Euclidean geometry (see the exercises at the end of this section) that $A'C' = (sc/4)\sqrt{3}$. This means that the length of $\widehat{AC} = (c/4)\sqrt{3}$. But, by inspecting Figure 3.14, we can easily see that the length of is $\widehat{AC}$ is $c/2$. This contradiction proves there is no perfect map. ■

EXERCISES

**Marks challenging exercises.*

Shortest Paths and Geodesics

1. Let P be any plane that cuts the sphere S (not necessarily passing through the center of the sphere). Explain why $P \cap S$ is a circle. (*Hint:* Drop a perpendicular from the center of the sphere to the plane.)

2. Use the ruler axiom (Section 1.3) and your knowledge of numbers to prove that a line passing through the center of a sphere cuts the sphere exactly twice.

3. What is the set of all points in three-dimensional space where the spherical coordinate ϕ is constant but ρ and θ vary?

4. What is the set of all points in three-dimensional space where the spherical coordinate θ is constant but ρ and ϕ vary?

5. What is the set of all points in three-dimensional space where the spherical coordinate ρ is constant but ϕ and θ vary?

6. On a sphere, draw a three-sided figure where each side is part of a geodesic and whose boundary intersects itself.

*7. Shortest Paths Lie in Great Circles: Let A_0 and A_1 be points on a sphere of radius R. We wish to show that the shortest path from A_0 to A_1 lies along a great circle. There is no harm in assuming that A_0 and A_1 have the same θ value. (We could always rotate the sphere so that our two given points have that relationship.)

   (a) Express the length of the arc of the great circle connecting A_0 and A_1 in terms of R, $\phi(0)$, and $\phi(1)$. You may assume $\phi(0) < \phi(1)$.

   (b) Prove that the Cartesian coordinates (x, y, z) of a point are related to the spherical coordinates by

$$\begin{aligned} x &= \rho \sin\phi \cos\theta, \\ y &= \rho \sin\phi \sin\theta, \\ z &= \rho \cos\phi. \end{aligned} \tag{3.4}$$

   (c) A curve from A_0 to A_1 can be parameterized by the functions $\theta(t)$ and $\phi(t)$ where $[R, \theta(0), \phi(0)]$ and $[R, \theta(1), \phi(1)]$ are the points A_0 and A_1. The length of a curve C where the Cartesian coordinates are given parametrically as $x(t)$, $y(t)$, $z(t)$ is

$$\text{length}(C) = \int_0^1 \sqrt{x'(t)^2 + y'(t)^2 + z'(t)^2}\, dt.$$

   Show that if we switch to spherical coordinates, this becomes:

$$\text{length}(C) = R\int_0^1 \sqrt{(\sin^2\theta)\theta'(t)^2 + \phi'(t)^2}\, dt.$$

   (d) Use the preceding formula to show that length$(C) \geq$ the length of the arc of the great circle that connects A_0 and A_1.

Angles and Spherical Triangles

8. Let G_1 and G_2 be two geodesics. Explain why the points in which they cross are opposites.

9. In Example 3.2, how much larger than 180° can you make the angle sum?

10. (a) On a sphere of radius 5, what is the area of a spherical triangle that has angles with these measures: $m\angle A = 1.4$ radians, $m\angle B = 1.6$ radians, $m\angle B = 2.5$ radians?

    (b) On a sphere of radius 3, what is the area of a spherical triangle that has angles with these measures: $m\angle A = 90^\circ$, $m\angle B = 80^\circ$, $m\angle B = 50^\circ$?

11. Explain why for any spherical triangle there is a hemisphere that completely contains it.

12. What does the previous exercise suggest about the largest angle sum a spherical triangle can have?

13. Suppose you were looking for a formula like the one in Theorem 3.13, but applying to spherical quadrilaterals instead of spherical triangles.

    (a) What do you think the formula would be?

    (b) Can you prove your conjecture?

14. The three shortest paths making up a spherical triangle divide the sphere into two regions, an "inside" of the triangle and an "outside." But which region is the inside and which is the ouside? More specifically, give a definition of the "inside."

15. Show an example of a spherical triangle where the exterior angle inequality does not hold; that is, there is an exterior angle that is smaller than or equal to one of its remote interior angles. (See Theorem 2.2 of Section 2.1 in the previous chapter.)

Map Making

16. Prove that $A'C' = (sc/4)\sqrt{3}$ in Figure 3.14.

17. Suppose there were a perfect map of a sphere. Let A, B, C be points on the sphere with B lying on the shortest path $\widehat{AC}$. Let A', B', and C' be the images under the perfect map. Show that B' is collinear with A' and C' and lies between them. (*Hint:* What relationship exists among the distances between the points A, B, and C that would not exist if B did not lie on $\widehat{AC}$?)

18. Assuming the result of the previous exercise, can you provide an alternative ending to the proof of Theorem 3.14, an ending that does not concern itself with the distance $A'C'$ and the length of $\widehat{AC}$?

3.3 Spherical Geometry: An Axiomatic View

Imagine a colony of intelligent ants living on the surface of a sphere. They have no experience of the third dimension — as far as they are concerned the sphere surface is all there is to the universe. They have no ability to see things inside or outside the sphere surface. If you tell them that there is an "up" or "down" direction in addition to the forward, backward, left, and right they are familiar with, they will regard you as a crackpot. And, of course, they don't know they live on a sphere, because the very idea of a sphere, with a center and having a curvature and so on, is impossible for them to imagine, just as it is impossible for us to visualize a fourth dimension.

Being intelligent creatures, and always wanting to get places in a hurry, they become interested in shortest paths in their world. To find a shortest path from A to B, they stretch a string tight along the sphere surface from A to B. In many cases, they paint along the shortest path to form a convenient set of roadways. They discover that if they extend a shortest path far enough it comes back on itself and they give the name *geodesic* to this larger geometric figure.

The ants discover that any two geodesics always intersect in two points. The first time this was discovered, it was regarded as a fluke. Other geodesics were checked and it still held up. Still there were naysayers among the ants who pointed out that there were still other geodesics that hadn't been checked. It took a lot of checking to convince everyone that any two geodesics intersected twice. This little controversy among the ants seems amusing to us because in the previous section we have demonstrated, by using three-dimensional geometry, that two geodesics always intersect twice. Alas, this route is not open to the ants because they don't understand the third dimension.

Another principle they eventually accept is that SAS guarantees congruence for spherical triangles. Their congruence principle reads just like the SAS congruence axiom for Euclidean plane geometry (Section 2.1 of the previous chapter) with one exception: Where the plane geometry axiom says "triangle," the ants' axiom says "spherical triangle."[2]

[2] Of course, the ants wouldn't call them *spherical* triangles as we do, since they have no idea what a sphere is.

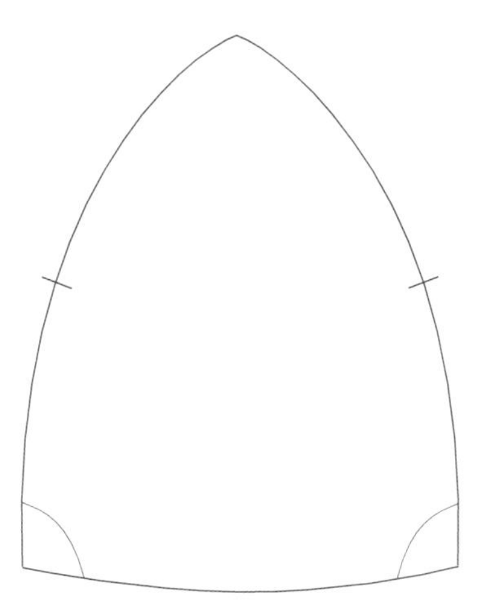

Figure 3.15 Base angles of an isosceles spherical triangle have equal measure.

Let's look in on Antland's leading thinkers as they are hard at work trying to justify another principle: that base angles of an isosceles spherical triangle are equal (Figure 3.15). What they are going to do is paint a few thousand isosceles triangles of various sizes and shapes and measure the base angles to great accuracy. They just barely get started when an ant named Emma gets the clever idea of not doing any measurements at all, but deducing the new principle from the SAS congruence axiom. Her proof is the very same one we gave for the plane geometry version of the theorem (Theorem 2.1 of Section 2.1 in the previous chapter), except for the following substitution:

$$\text{triangle} \rightarrow \text{spherical triangle}.$$

We may think of this as a "translation" of one proof into another by means of the substitution. It may seem surprising that this would work, but it is really very simple to understand (simple for us, but not for the ants — they have never heard of plane geometry). The only thing used in the plane geometry proof is the plane geometry SAS axiom. If we change "triangle" to "spherical triangle" in the proof, we still have a valid argument because the spherical version of SAS has been accepted as an axiom by the ants. But what emerges from this translated argument is the plane geometry theorem with "triangle" replaced by "spherical triangle," namely, that base angles of isosceles spherical triangles are equal in measure.

Back in Antland Emma's clever idea leads to a revolutionary plan: Choose a basic set of principles to establish by measurement. If they are well chosen then maybe everything else the ants want to know can be deduced from these with no more measurement needed. Thus the ants invent the axiomatic method for their geometry. This little fable is, of course, inspired by how human beings invented axiomatic Euclidean plane geometry.

Although we investigated spherical geometry using the tools of three-dimensional geometry in the previous section, we will now do it axiomatically the ants' way. To start with, we need undefined terms.

UNDEFINED TERMS OF SPHERICAL GEOMETRY *Point* and *geodesic.*

Here is a description of part of the spherical geometry axiom set,[3] arranged for easy comparison with the axiom set of Euclidean geometry.

Axioms with Versions in Both Spherical Geometry and Euclidean Plane Geometry

The SAS axiom is not the only Euclidean plane geometry axiom that has a "translated" version present in the axiom set for spherical geometry. Here are some others along with the "dictionary" needed to translate one axiom into its look-alike in the other geometry.

point → point
line → geodesic
triangle → spherical triangle
segment → shortest path
angle → angle

SPHERICAL GEOMETRY AXIOM 1: *The SAS Congruence Axiom.*

SPHERICAL GEOMETRY AXIOMS 2–5

All the *axioms about angles* (axioms 4 through 7 of Section 1.3 in Chapter 1).

SPHERICAL GEOMETRY AXIOM 6: *Pasch's Separation Axiom for a Geodesic*

Given a geodesic G, the points not on G form two sets, H_1 and H_2, called *sides* of G, so that

(a) if A and B are points in the same side, then the shortest path connecting A and B lies wholly in that side, and

[3] For more details see the discussion of *double elliptic geometry* in Gans (1973).

(b) if A and B are points not in the same side, then the shortest path connecting A and B intersects G.

When speaking the language of three dimensions, we would call the sides of a geodesic *hemispheres*. But we avoid this term to emphasize that we are avoiding the three-dimensional view.

SPHERICAL GEOMETRY AXIOMS 7–9

These are the first three of the Euclidean area axioms (axioms 15–17 of Section 1.3 in Chapter 1).

Any theorem in Euclidean plane geometry which has a proof that depends only on the Euclidean versions of axioms 1–9 can be translated into a theorem in spherical geometry simply by translating the proof. Thus we get some theorems of spherical geometry "for free." One example of this is the following theorem. We leave the task of creating a translated proof as an exercise for you (see Exercise 17 of Section 2.1 in the previous chapter).

☆ THEOREM 3.15

If two spherical triangles have the ASA relationship, then they are congruent.

Axioms in Euclidean Geometry with No Translated Versions in Spherical Geometry

AXIOM 1: *The Point-Line Incidence Axiom*

In Euclidean plane geometry, this axiom asserts that given any two different points, there is exactly one line which contains them. But in spherical geometry, sometimes there are two geodesics passing through two given points (for example, the North and South poles). Therefore, this axiom is not present in spherical geometry.

AXIOM 2: *The Ruler Axiom*

In Euclidean plane geometry, the ruler axiom implies that a line extends indefinitely in two directions. But if you follow a geodesic in spherical geometry far enough in one direction, you will return to your starting point. Therefore, this axiom is unsuitable for spherical geometry.

AXIOM 9: *Euclid's Parallel Axiom*

Not present in spherical geometry.

AXIOMS 10–14

The axioms dealing with the *third dimension*. Not present in spherical geometry.

AXIOMS 18–20

These axioms deal with area and volume are not present in spherical geometry.

Axioms in Spherical Geometry But Not in Euclidean Geometry

There is a body of axioms usually phrased in the language of topology that we just describe loosely (to avoid a long excursion into topology). Among them are the following:

SPHERICAL GEOMETRY AXIOM 10: *Existence of Shortest Paths (loosely stated)*

Any two points A and B can be joined by paths (see Figure 3.16b for an example), each of which has a positive number as its length. Among these paths there will be one or more that are shortest of all such paths joining A and B.

From a purely logical point of view, Spherical Geometry Axiom 10 might well have been at the head of our list. It justifies any discussion of path lengths, which is needed to speak of the idea of a shortest path, which is needed to define a spherical triangle (the sides are shortest paths), without which the SAS axiom is meaningless.

SPHERICAL GEOMETRY AXIOM 11: *Topology of Geodesics (loosely stated)*

Geodesics are *simple closed curves*. Being closed means that if you start anywhere and travel along a geodesic you will come back to your starting point, as in Figures 3.16a and 3.16c. Being simple means that you will not cross your path until you come back to the starting point — not as in Figure 3.16a.

Nontopological axioms in spherical geometry include these:

SPHERICAL GEOMETRY AXIOM 12: *Geodesics Intersect Twice*

If G and G' are two geodesics, there are precisely two points on both of them.

According to Spherical Geometry Axiom 12, geodesics are quite different from Euclidean lines, some of which never intersect at all. For this reason, spherical geometry is classified as a non-Euclidean geometry.

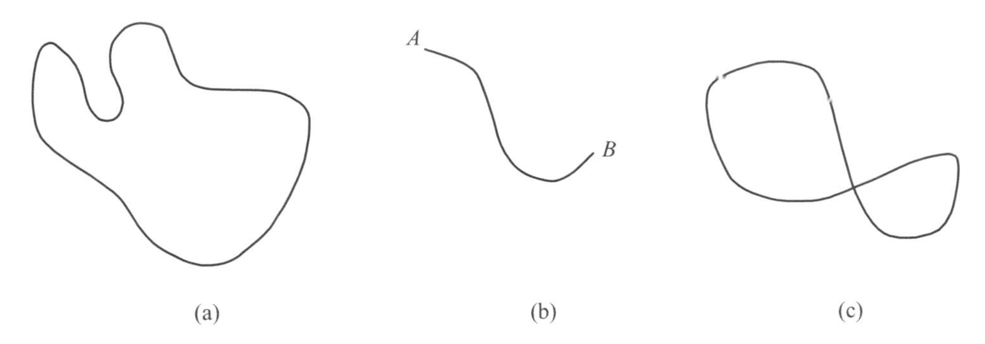

Figure 3.16 (a) Simple closed curve. (b) Path from *A* to *B*. (c) Nonsimple closed curve.

SPHERICAL GEOMETRY AXIOM 13: *Geodesics Contain Shortest Paths*

If A and B are any two points and G a geodesic containing them, then at least one of the two paths into which A and B divide G is a shortest path from A to B.

DEFINITION

If A and B are on geodesic G and the paths into which A and B divide G have the same length (and are, by Spherical Geometry Axiom 13, both shortest paths), then we say the points are *opposite* on G. In simple language, opposite points divide a geodesic in half. It should be clear that point A can only have one opposite point on a geodesic. Another point to notice is a subtlety in the definition: Conceivably, B could be opposite A on geodesic G but not opposite A on another geodesic G' that passes through A and B. Actually, this does not happen, but this fact needs proof, and the proof needs to wait until we have another axiom.

SPHERICAL GEOMETRY AXIOM 14: *Points Determine Geodesics*

1. Given any two points A and B, there is always at least one geodesic G containing them.
2. If A and B are points on geodesic G and are not opposite on G, then G is the only geodesic containing A and B.

✯ THEOREM 3.16

Suppose A and B are opposite on G and also lie on geodesic G'. Then they are opposite on G' as well.

Proof

If B is not opposite A on G' then by part 2 of Spherical Geometry Axiom 14 there is only one geodesic containing A and B. But we have two, G and G'. This contradiction shows that B is opposite A on G'. ■

The theorem just proved allows us to speak of a point being opposite another without any reference to a geodesic containing them.

✯ THEOREM 3.17

Let G and G' be any two geodesics and let A and B be their two crossing points. Then A and B are opposite.

Proof

If they were not opposite, then we would have a contradiction to part 2 of Spherical Geometry Axiom 14. ■

✯ THEOREM 3.18

Geodesics all have the same length.

Proof

Let G and G' be two geodesics and let A and B be the two points where they cross. By Theorem 3.17, A and B are opposite one another. A and B divide each geodesic into two paths—four paths altogether. By the definition of opposite, each of these is a shortest path. Because they all have the shortest possible length, they all have the same length. Each geodesic's length is twice this length. ■

DEFINITION

If A and B are opposite points, and S_1 and S_2 are shortest paths connecting them, but S_1 and S_2 are not part of the same geodesic, then we call $S_1 \cup S_2$ a *bilateral* (see the solid curves connecting A and B in Figure 3.18). S_1 and S_2 are its *sides* and A and B are its *corners*.

We may think of a bilateral informally as the boundary of what we called a *lune* in the previous section. But notice that we have not defined *lune* in our formal axiomatic development of spherical geometry, and to keep our exposition short, we don't intend to. You might like to think about how to phrase the definition of a lune. Keep in mind that you can only use terms already defined in our axiomatic treatment of spherical geometry, or taken as undefined. At this stage, these terms are *point, geodesic, path, shortest path, sides of a geodesic, angle, opposite, bilateral, side of bilateral, corner of bilateral.* Of course, you can also use the common language of all mathematical disciplines such as *union, intersection, set membership,* the idea of a point being "on" a geodesic (which is really just set membership), etc.

Our last objective in this section is to prove that the corner angles of a bilateral are equal in measure (Theorem 3.20), but first we need the following result whose proof we leave as an exercise.

☆ THEOREM 3.19

If A and B are opposite points on geodesic G, and X and Y are a different pair of opposite points on G, then each of the two paths from A to B on G must contain exactly one of X and Y. (The situation is as shown in Figure 3.17a, not as in Figure 3.17b.)

☆ THEOREM 3.20

The corner angles of a bilateral have the same measure (Figure 3.18).

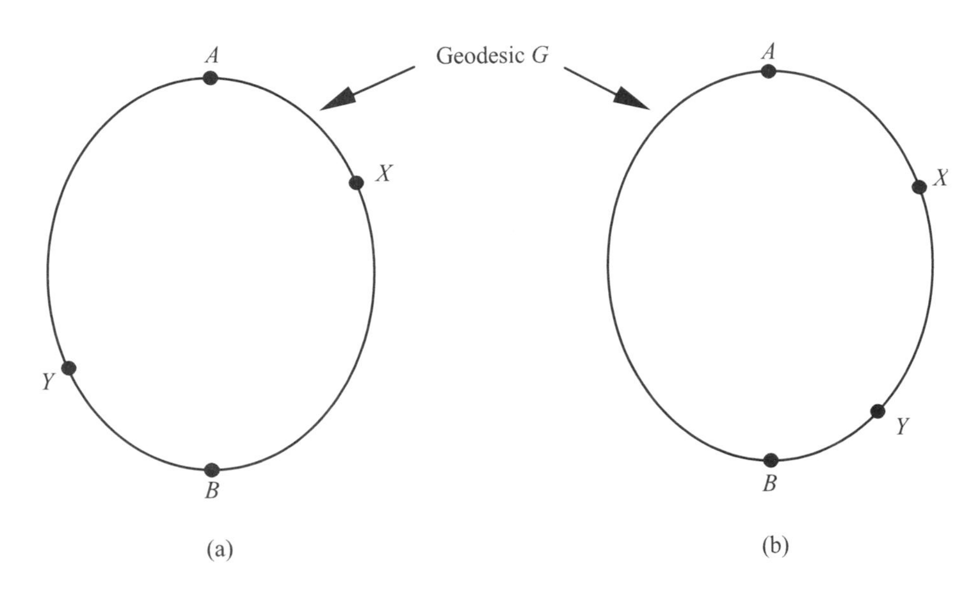

Figure 3.17 Pairs of opposite points on a geodesic separate one another.

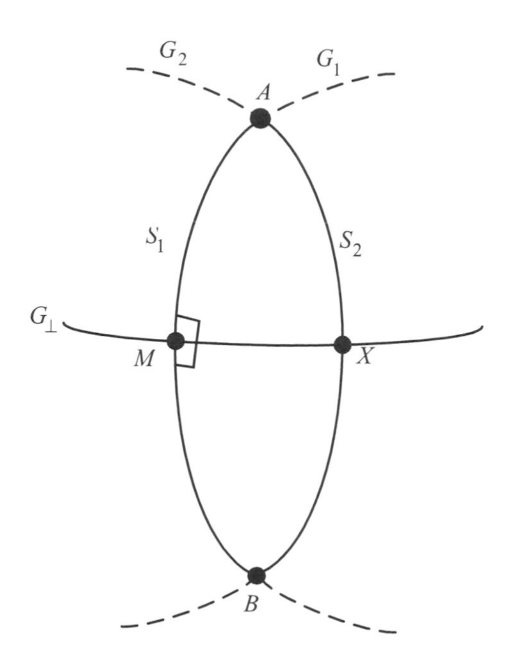

Figure 3.18 Cutting a bilateral into two congruent spherical triangles.

Proof

Our strategy is to cut the bilateral into two congruent spherical triangles. Call the sides of the bilateral S_1 and S_2, and let G_1 and G_2 be the geodesics containing those sides. Let M be the midpoint of S_1. Let $G_\perp$ be the geodesic passing through M which is perpendicular to S_1. $G_\perp$ does not contain A or B (Why?) so it crosses G_2 at two other points X and Y, which are opposite on G_2 according to Theorem 3.17. By the previous theorem, one of these points — say, X — lies on S_2. Spherical triangles AMX and BMX are congruent by SAS. The corner angles of the bilateral are corresponding parts under this congruence. Therefore, they have the same measure. ■

Spherical geometry contains many other fascinating theorems. There is, for example, a branch of the subject called spherical trigonometry, which is of daily use to navigators of ships and aircraft. However, we stop here (for more, consult Gans's (1973) chapter on double elliptic geometry in his book *An Introduction to Non-Euclidean Geometry*) and turn briefly to a "big picture" question: Why would anyone want to prove theorems about spherical geometry from its axiom set? Isn't it easier to stand outside the world of the ants, visualizing the sphere in three dimensions and using the well-known machinery of Euclidean geometry as we did in the previous section?

The value of the ant-like axiomatic approach is more than just mental gymnastics. It can prepare us for the task of understanding the structure of the three-dimensional universe we inhabit. Physicists have assembled much evidence that light rays do not behave exactly like the lines of three-dimensional Euclidean geometry (see, for example, our discussion of gravitational lenses in the previous section). They contend that Euclidean geometry gives an extremely good approximation for "small" figures (anything that can be drawn on earth for example), but when considering the most distant galaxies we need some new non-Euclidean theory.

It appears that we are in the same position with respect to our three-dimensional world as the ants are in relation to their sphere surface: The way the world seems is not the way it is. Just as the ants cannot step outside their world to see its curving spherical nature, we cannot step outside our universe to see it "from the outside." The exercise of doing spherical geometry from its axioms—from the "inside," as the ants would do it—is a useful warmup for devising a new non-Euclidean theory of our three-dimensional universe.[4]

EXERCISES

**Marks challenging exercises.*

The following exercises should be done from the axioms of spherical geometry, not by realizing what geodesics and shortest paths really are as objects on the sphere in three dimensions. (Write axiomatically like an ant.) But, of course, you can get the idea for your axiomatic work by visualizing the sphere in three dimensions. (Plan like a person.)

1. Prove Theorem 3.19.

2. Prove that if A is opposite B and C is any third point, then there is a geodesic that contains all three points.

3. Explain why $G_\perp$ does not contain A or B in the proof of Theorem 3.20.

4. Prove that $G_\perp$ crosses S_2 at right angles at X (see the proof of Theorem 3.20 and Figure 3.18).

5. We did not define rays in our axiomatic development of spherical geometry. Can you supply a definition? Or shouldn't we make the attempt?

---

[4] A task that has not yet been accomplished.

6. We did not define angles (nor are they undefined) but we have spoken of them. We need to give a definition or Spherical Geometry Axioms 2–5 will be meaningless. Can you suggest a definition?

7. Define lune.

8. Suppose two spherical triangles, ABC and $A'B'C'$, have the ASA relationship as follows: $m\angle A = m\angle A'$, $AB = A'B'$, $m\angle B = m\angle B'$. Show that they are congruent. (*Hint*: See if the proof of the analogous theorem in Euclidean plane geometry can be adapted to the spherical situation. See Exercise 17 of Section 2.1 in the previous chapter.)

9. (a) Prove that if a spherical triangle ABC has $m\angle A = m\angle B$, then the sides opposite these angles are also congruent to one another, i.e., $AC = BC$. (Consider the previous exercise.)

   (b) Let G be any geodesic and A and B points on it. Let G_A be a geodesic perpendicular to G at point A. Let G_B be a geodesic perpendicular to G at point B. Let C be a point on both G_A and G_B. Show that the length of shortest path $\widehat{CA}$ is the same as the length of shortest path $\widehat{BC}$.

10. Let $A_1, A_2, \ldots, A_n$ be points on a geodesic G such that the shortest paths $\widehat{A_1A_2}, \ldots, \widehat{A_{n-1}A_n}$ are all equal in length. Let G_i be the geodesic that crosses G perpendicularly at A_i. Let C be a point where G_1 and G_2 cross. Prove that C lies on each of the G_i and that all the shortest paths $\widehat{CA_i}$ are equal in length.

Chapter 4

Transformation Geometry I: Isometries and Symmetries

Prerequisites: High school geometry such as in Chapter 2

Symmetric patterns call attention to themselves. We seem to have a built-in appreciation for symmetry. But could you give a short definition of symmetry in words? It turns out you need some mathematics to establish a clear idea of symmetry. In this chapter, we establish a theory of symmetry in the Euclidean plane by connecting it to the concept of an isometry — a transformation that preserves distances. The familiar rotations and translations are examples of isometries, but there are others as well. One of our goals is to have a catalog of all possible isometries in the Euclidean plane. This will allow us to derive a catalog of the kinds of symmetries that strip patterns can have. We illustrate the theory with the pottery found in a Native American settlement in the southwestern United States.

4.1 Isometries and Their Invariants

One of our key objectives in this chapter is to provide a mathematically precise way of talking about symmetry — a key concept in art and science. The strip pattern (Crowe, 1986) at the top of Figure 4.1 shows a pattern we would agree has some kind of symmetry.

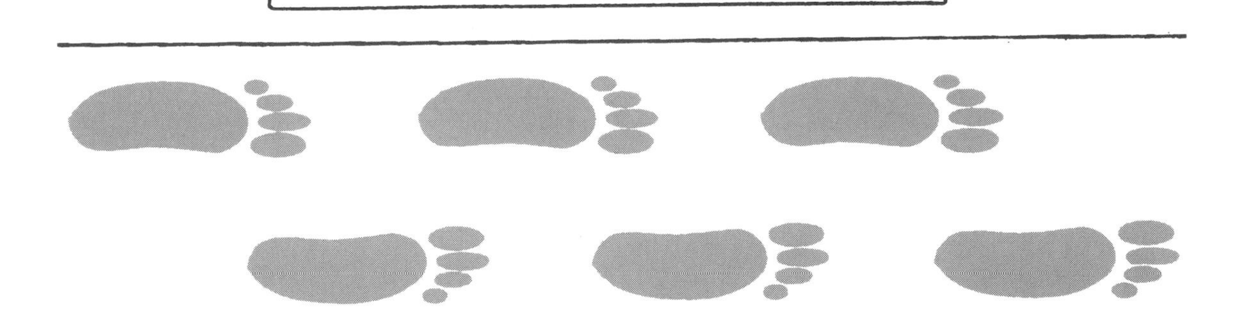

Figure 4.1 What symmetries do these three strip patterns have?
Bird pattern from Tana and Tana (1988); Middle pattern from Christie (1969). Courtesy of Dover Publications.

But what kind? What do we mean by the word *symmetry*? Compare the middle pattern of Figure 4.1. Some would say this figure has more symmetry than the top pattern. Do you agree? How can symmetry be measured? Finally, compare the top and bottom patterns. Each has a kind of symmetry. Do they have the same symmetry? We cannot answer any of these questions unless we have a mathematically precise idea of what symmetry means.

Our study of symmetry will involve the concept of an *isometry*, the central idea of this section. Roughly speaking, when an object has been moved without change of size or shape we say it has undergone an isometry. In the top example in Figure 4.1 you

Figure 4.2 A snowflake (Bentley and Humphries, 1962). Courtesy of Dover Publications.

can see a number of instances of a bird, each instance the same size and shape but with different placement. The existence of these multiple copies of the same design in the bird pattern of Figure 4.1 contributes to the impression of symmetry. The symmetry of the snowflake in Figure 4.2 seems related to the fact that one "arm" is the same size and shape as any other. In this section, we set aside these matters about symmetry so we can first establish an understanding of isometries.

The geometry we use to study isometries and symmetries is that of the Euclidean plane (Chapters 1 and 2), not that of the previous chapter. For example in this section we make use of the fact that, through a point not on a given line, there is a unique parallel to that line. We also use the fact that, from a point not on a line, we can drop exactly one perpendicular to that line.

What Is an Isometry?

Whenever we move an object without changing size and shape, when we compare "before and after" snapshots, the distance between any two points of the object before the motion is the same as the distances between the points in their new positions. For example, if you slide a piece of paper on a table, the distance between the opposite corners is the same before and after the slide. This idea allows us to give a definition of isometry in terms of functions, usually called transformations in geometry, currently the single most unifying concept in mathematics.

DEFINITION

Suppose we have some formula or rule that associates each point of a two-dimensional space with another point of a two-dimensional space. This formula or rule is called a *transformation* of the plane. The transformation is called an *isometry*[1] if, for every pair of points P and Q, the distance between their images $P' = S(P)$ and $Q' = S(Q)$ is the same as the distance between the points:

$$P'Q' = PQ. \tag{4.1}$$

(Throughout this section, if $P, Q, R, \ldots,$ are points then $P', Q', R', \ldots,$ are the images under the isometry being considered.)

EXAMPLE 4.1: A GALLERY OF ISOMETRIES

1. *Translation.*

If we move each point of the plane by the same distance and in the same direction, this is a particular kind of isometry called a *translation*. In Figure 4.3, P is translated to P'. Any other point Q moves along a segment of equal length and direction to Q'. If we were to apply a translation just to the points of a particular object, we could visualize this as pushing the object along a table top without turning it.

2. *Rotation around a point.*

Another important example of an isometry is a rotation through an angle of θ around a center of rotation C. To find the image of a point P, imagine P connected to C with a fixed-length rod hinged at C. Turn the rod counterclockwise around C through θ if $\theta > 0$, and clockwise through $|\theta|$ if $\theta < 0$. Figure 4.4 shows two points P and Q being rotated through -90°. Rotation through θ around C is denoted $R_C(\theta)$, or perhaps just $R(\theta)$ when there is no need to call attention to the center of rotation.

3. *Reflection in a line* L.

For any point P, drop a perpendicular to L (Figure 4.5). Let F be the foot of this perpendicular. Extend $\overline{PF}$ past L to a point P' where $FP' = FP$. Associate P' to P. This function is called *reflection* in line L. The word *reflection* is used because of the analogy with a mirror. When we look in a mirror, it seems there is another copy of us behind the mirror; in the same way, P has a copy of itself on the other side of the line L. We denote reflection in L by M_L, where the M stands for mirror. This way of thinking affects

[1] Some authors call this a *rigid motion* or just a *motion*.

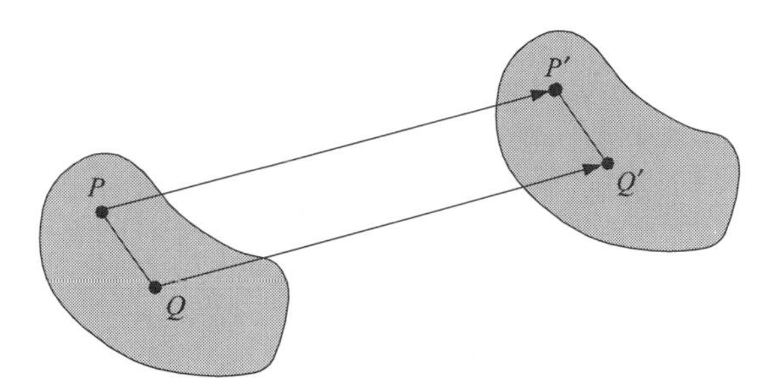

Figure 4.3 Translating an object.

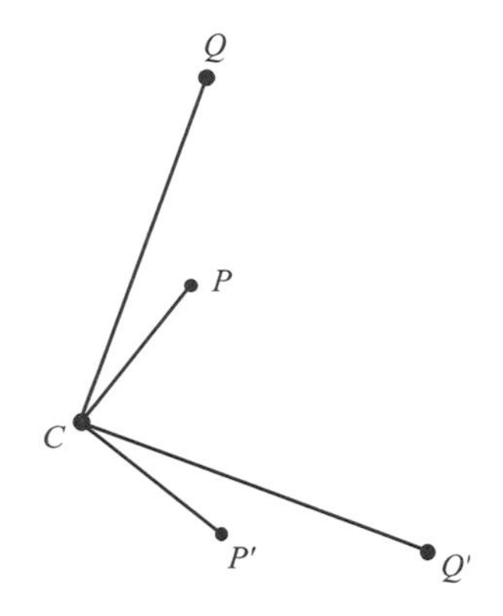

Figure 4.4 Rotating P and Q around C by $-90°$.

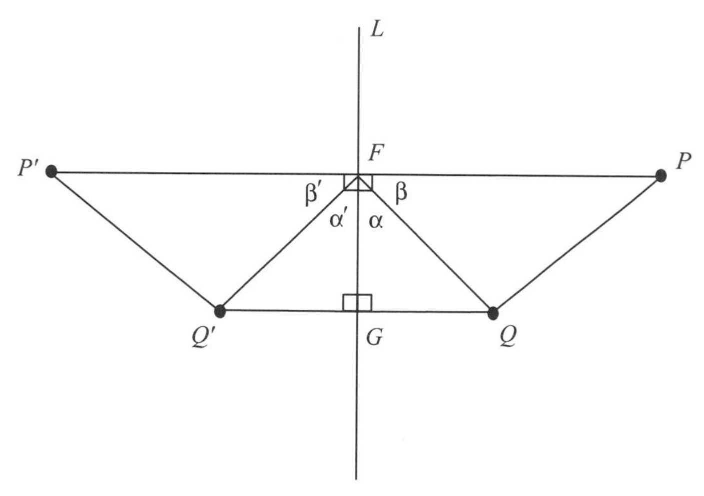

Figure 4.5 Reflection in line L is an isometry.

our terminology another way. We say that P' is the *image* of P. In fact, we use this terminology for functions that are not reflections. If S is any function, $S(P)$ is said to be the *image* of P.

There is another visual way of thinking about reflection. Open a spiral notebook flat and slide a pencil or stiff rod down through the spirals. Think of the notebook as the plane. Reflection is like spinning the notebook 180° on the rod (while keeping the two halves as one flat whole). The two halves of the notebook interchange. Imagine two points drawn in the notebook. After the reflection they are in new positions, but clearly their distance apart has not changed. This way of thinking causes us sometimes to call a reflection a *flip*.

4. *The identity.*

This function, denoted I, associates to each point P that very same point: $I(P) = P$ for all points P. Not a very interesting isometry, but one which makes the theory neat. For example, if we want to express the fact that rotation by 360° brings each point back to its original location, we can write $R(360°) = I$. Notice that if we translate with a distance of 0, this is just I. Likewise, $R(0°) = I$. ■

It is natural to think of translation and rotation as "continuous motions" in which a point moves along through some path to get to its final position. For example, under rotation about C we think of a point moving through a circular arc centered at C. But this is not an essential aspect of the definition, because we are concerned only with where the point starts and where it ends. If the point moved through some other path, we would still say the point was rotated provided it wound up in the same position as if it had taken a circular arc. For another example, consider applying $R(180°)$ and $R(-180°)$ to the same point P. If we think of these as continuous motions, they seem different since one is clockwise and the other is counterclockwise. But for any point P, whichever of these we apply to P gives the same image: $R(180°)(P) = R(-180°)(P)$. For this reason we say $R(180°) = R(-180°)$. This is an example of the following definition:

DEFINITION

If S_1 and S_2 are transformations, we say $S_1 = S_2$ provided, for *every* point P, $S_1(P) = S_2(P)$.

EXAMPLE 4.2: TRANSFORMATIONS WHICH ARE NOT ISOMETRIES

Figure 4.6 illustrates some functions that might be used in a computer graphics program and that are not isometries. We show the effect of these functions on a jack-o-lantern figure centered at the origin. The large sideways jack-o-lantern is produced by the following series of changes:

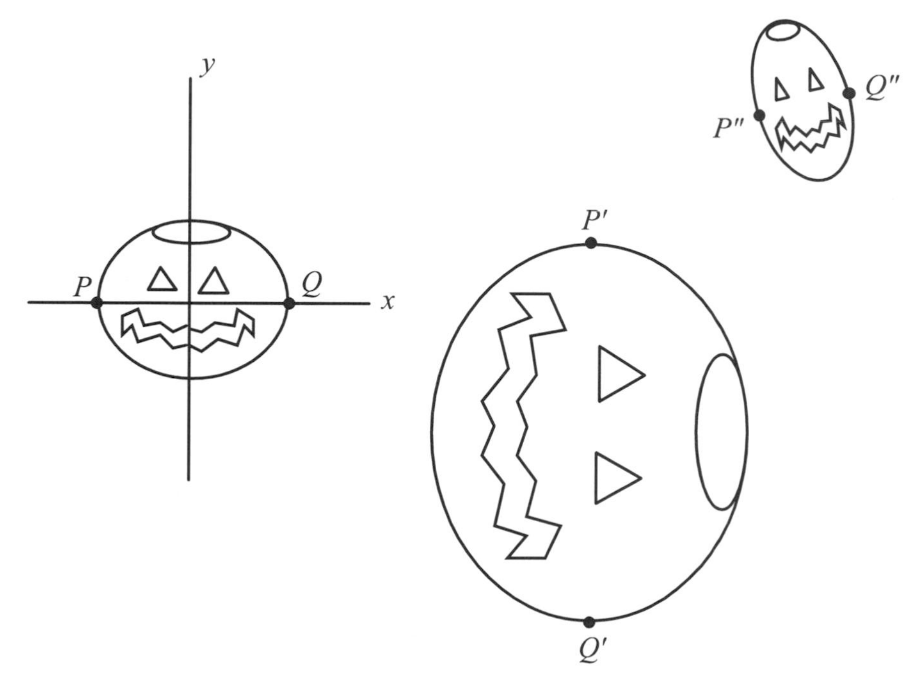

Figure 4.6 Transformations which are not isometries.

1. Double the size of the original. This is done by moving each point away from the origin O to a point twice as far. To be precise, (x, y) moves to $(2x, 2y)$.
2. Apply $R_O(-90^\circ)$.
3. Translate.

Notice that $P'Q' \neq PQ$.

To produce the skinny jack-o-lantern at the top right:

1. Shrink the original in the horizontal direction only. Each point is moved toward the y axis, halving its distance to that axis.
2. Rotate around O by a small positive angle.
3. Translate.

Once again, distance is not preserved: $P''Q'' \neq PQ$. ■

Although it may seem obvious that translation, rotation, and reflection are isometries, these facts need to be proved. In the case of translation, we need to show (Figure 4.3) that the fact that PP' is parallel and equal in length to QQ' guarantees that $P'Q' = PQ$. Can you prove this? In the case of rotation, we similarly need to show that $P'Q' = PQ$. We leave these as exercises, but here are the details for a reflection.

✯ THEOREM 4.1

A reflection M_L is an isometry (Figure 4.5).

PROOF

The proof strategy is to show that $\overline{PQ}$ and $\overline{P'Q'}$ are corresponding parts of congruent triangles PQF and $P'Q'F$. But first we have to work on triangles FQG and $FQ'G$. These last triangles are right triangles with a common side $\overline{FG}$. Furthermore, by the nature of reflection, $QG = Q'G$. Thus, triangles FQG and $FQ'G'$ are congruent by side-angle-side. Using the corresponding parts principle, we have $QF = Q'F$ and $\alpha = \alpha'$. Since $\angle GFP$ and $\angle GFP'$ are right angles by the definition of reflection, if we subtract the equal angles α and α' from the right angles, we get $\beta = \beta'$. Finally, we have $PF = P'F$ by the nature of reflection. Thus triangles PQF and $P'Q'F$ are congruent by side-angle-side. Then $PQ = P'Q'$ since these are corresponding parts in the two triangles.

Our proof is not yet complete. Our argument assumed that P and Q are on the same side of the mirror line. We leave the following cases for you to provide proofs for:

(a) P and Q are on opposite sides of the mirror line.

(b) One of P and Q is on the mirror line.

(c) Both P and Q are on the mirror line. ■

Invariants of Isometries

Our next goal in this section is to begin to find out more about the nature of isometries. The definition of isometry is remarkably short—a transformation that preserves distance—and it turns out there is much we can deduce from this definition to put flesh on its bare bones. Of course, it is true that for some examples of isometries—the translations, rotations, and reflections—we have a pretty good idea of what they are like. But what about isometries which are not of these types? For that matter, are there any isometries which are not of these types? We do not know the answer to the latter question yet (this will be revisited in later sections.) So, in what

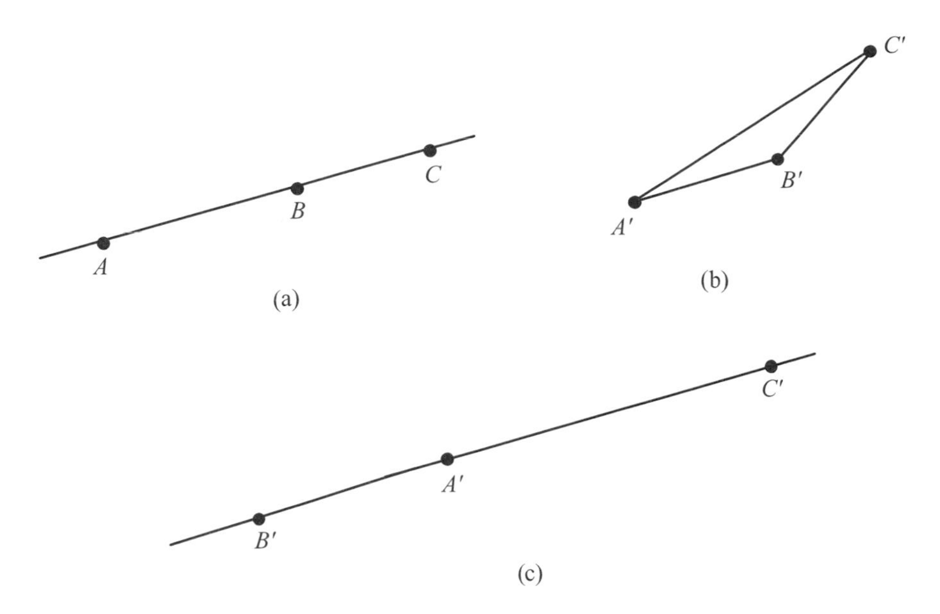

Figure 4.7 Three collinear points, as in configuration (a), cannot transform to (b) or (c) under an isometry because the lengths would not add up right.

follows we take the view that we are dealing with an arbitrary isometry. This means that all we know about it to start with is that it preserves distance. However, we will see that this is a powerful restriction on a transformation and from it we can deduce other *invariants*—properties of a figure that remain the same when the figure is subjected to the isometry. Our first theorem states that collinearity and "betweenness" are invariants of every isometry.

☆ THEOREM 4.2: LINE INVARIANCE

If B is between A and C on a line, then the images of these points under any isometry, A', B', C', are also collinear and B' is between A' and C' (Figure 4.7).

Proof

Since A, B, C are collinear and B is between A and C we have their lengths adding up as follows: $AB + BC = AC$ (see Figure 4.7). Consequently, because an isometry preserves lengths, $A'B' + B'C' = A'C'$. This last equation would be impossible if A', B', C' formed a triangle because of the triangle inequality (any side of a triangle is shorter than the sum of the other two sides). Therefore, A', B', C' are collinear. Furthermore, this equation also shows B' must be between A' and C'. ■

Technically, the theorem we have just proved does not say that the image of a line is an entire line—it leaves open the possibility that the image is only part of a line. Suppose D' is any point on the line determined by A' and B'. How do we know that there is a point D on line AB whose image is D'? It turns out there is such a point, but we have not proved it. We leave this as an exercise.

The style of proof given in the previous theorem can be adapted to show that noncollinearity is invariant. Here is a more detailed statement. The proof is left as an exercise.

✯ THEOREM 4.3

If A, B, C are not collinear, then the images A', B', C' are not collinear either. ■

✯ THEOREM 4.4

For any triangle ABC, the image triangle $A'B'C'$ is congruent to it.

Proof

Because distances are preserved, $A'B' = AB$, $A'C' = AC$, and $B'C' = BC$. Therefore, the SSS criterion can be applied to show the triangles congruent. ■

✯ THEOREM 4.5: ANGLE INVARIANCE

Angles are invariant under any isometry (Figure 4.8).

Proof

Let $\overline{AB}$ and $\overline{BC}$ be two segments making an angle of θ at B. We first consider the case where $\theta \neq 180°$. In this case, A, B, C make a triangle. By the previous theorem triangle $A'B'C'$ is congruent to triangle ABC. Therefore, the angle at B' has the same measure as the angle at B since these are corresponding parts in the congruence. This completes the proof for the case $\theta \neq 180°$.

Now suppose $\theta = 180°$. Then A, B, C are collinear with B between A and C, and the previous theorem implies that A', B', C' are collinear also with B' between A' and C'. Thus, $m\angle A'B'C' = 180°$. ■

From the previous theorem, we can see that the image of any polygon, whether or not it is a triangle, is another polygon with the same side lengths and angles, that is, a congruent polygon. If we have a curved shape, we can think of it as being almost the same as a polygon with very many very short sides. Therefore, it appears as if the image of a curved shape under an isometry is another congruent copy of itself.

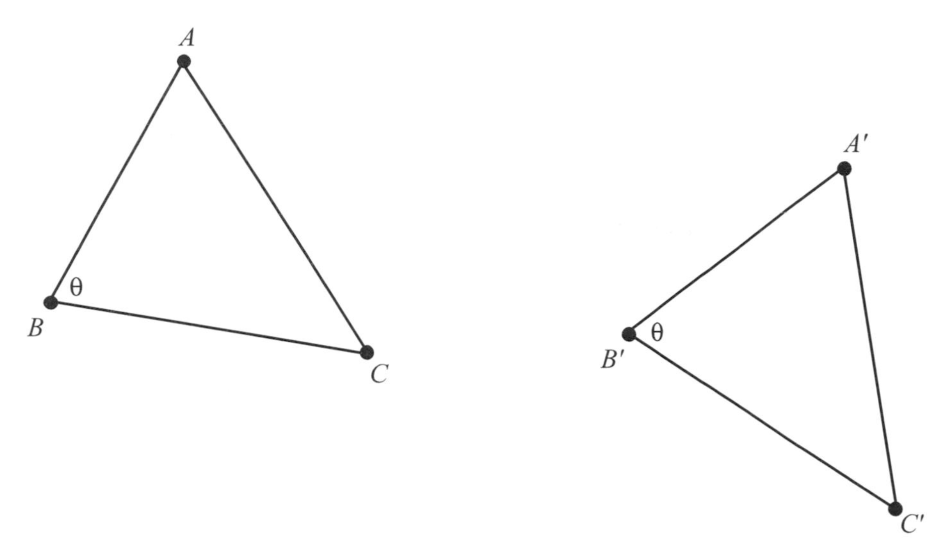

Figure 4.8 An isometry moves a triangle to a congruent copy.

EXAMPLE 4.3: SOMETHING WHICH IS NOT INVARIANT UNDER ISOMETRIES

If P and Q are two points and O denotes the origin, we call $m\angle POQ$ the *angular deviation* between P and Q. This is not an invariant of each isometry. For example, say P and Q lie on the positive x and y axes so that the angular deviation is $90°$. We could translate so that P' and Q' wind up collinear with O and the angular deviation between P' and Q' is $0°$. (Draw a figure to show this.)

The noninvariance of angular deviation should be compared with Theorem 4.5. The difference between the two situations is that in measuring the angular deviation we always measure the angle at the origin, both before and after the isometry is carried out. We do not switch to the image of the origin.

This example shows that the angular deviation is not invariant under translation and so we cannot say it is invariant under an arbitrarily given isometry. However, it is invariant under any rotation around the origin. Are there any reflections under which it is invariant? ■

✯ THEOREM 4.6

Parallelism is invariant under any isometry.

PROOF

Suppose L_1 and L_2 are parallel lines and L_1' and L_2' are their images. We need to show L_1' and L_2' are parallel. Select any points P_1 and P_2 on L_1 and L_2, respectively, and let M be the line joining P_1 and P_2. Since L_1 and L_2 are parallel and M is a transversal, alternate interior angles are equal. Now let M' be the image of the line M and consider the way M' crosses L_1' and L_2'. The alternate interior angles are equal since they are equal where M crosses L_1 and L_2 and since angles are invariant by the previous theorem. This implies that L_1' and L_2' are parallel. ■

EXERCISES

**Marks challenging exercises*

WHAT IS AN ISOMETRY?

1. Suppose T is an unknown translation, but you know that $T(P) = Q$ for known points P and Q. How many possibilities are there for T?

2. Answer Exercise 1 again but under the assumption that you do not know the exact location of $T(P)$, but you do know the direction of the translation (but not its length) and you know that $T(P)$ lies on a certain given line.

3. Suppose R is a rotation around some unknown center, by some unknown angle. Suppose there are known points P and Q where $Q = R(P)$.

   (a) Explain why you cannot determine where the center is.

   (b) There is a construction you could carry out that narrows down where the center might be. What is it?

4. Suppose M is an unknown reflection, but you know that $M(P) = Q$ for known points P and Q. How many possibilities are there for M?

5. Prove translation is an isometry. (In Figure 4.3 this means showing that $P'Q' = PQ$, based on the fact that segments $\overline{PP'}$ and $\overline{QQ'}$ are equal in length and parallel.)

6. Prove rotation is an isometry.

7. Suppose L is a line marked on a piece of paper and P and Q are marked points and P', the reflection of P in L, is marked. We want to construct Q', the reflection of Q in L. We have a ruler available, with which we can connect points with a straight line and can measure distances. But you have no compass. Describe how Q' can be found.

8. (a) Let R be a rectangle that is not a square. List as many isometries S as you can which keep R fixed. This means that if P is any point on the boundary of R, $S(P)$ is also on the boundary of R.
   (b) Answer the analogous question if R is a square.

9. Devise proofs to handle the uncovered cases mentioned at the end of the proof of Theorem 4.1.

*10. Suppose S is a transformation and P is one particular point and for every point Q, $PQ = PS(Q)$. Must S be an isometry? If so, prove it. If not, describe a transformation that is a counterexample. (*Hint:* It need not be a familiar transformation.)

Invariants of Isometries

11. Answer true or false for the following statements. Explain your answer.
    (a) "Circlehood" is an invariant of any isometry.
    (b) The slope of a line is an invariant of any isometry.
    (c) If D is the center of the circle inscribed in triangle ABC (i.e., that circle which is contained in the triangle and tangent to the sides), then D' is the center of the circle inscribed in triangle $A'B'C'$.

12. Give an example of an isometry that transforms the x axis into the y axis and has the origin as a fixed point—that is, the origin corresponds to itself under the isometry. Is there just one such isometry or can you find another?

13. Give an example of an isometry that preserves verticality, that is, the image of each vertical line is a vertical line. Can you find a second example? A third?

14. Is verticality an invariant of all isometries?

15. Suppose a certain isometry transforms the y axis to itself (but individual points on the axis are not necessarily invariant). What can you say about the image of the x axis?

16. Suppose S is a transformation in which every distance is doubled, that is, $P'Q' = 2PQ$. Prove that angles are invariant. (*Hint:* try to adapt the proof of Theorem 4.5.)

17. Prove Theorem 4.3. (*Hint*: Try a proof by contradiction.)

18. Let S be an isometry and let A and B be two points whose images under S are A' and B'. Let V be any third point on line $A'B'$. Show that there is a point U on line AB so that $U' = V$. (In this way we prove that the image of a line is a whole line.)

*19. (a) Give an example of a transformation that has the property mentioned in Exercise 16. (You need to give a rule for finding the image of each point P.)

(b) Show that your transformation preserves collinearity.

(c) Show that your transformation preserves parallelism.

20. Are there any reflections under which angular deviation is invariant? If so, describe the mirror lines for as many as you can.

21. If S is a transformation under which parallelism is invariant, does it follow that S is an isometry?

4.2 Composing Isometries

The Idea of Composition

We have made our acquaintance with isometries and studied three special types of them, the rotations, the translations, and the reflections. Are there any others? What if we followed one of these isometries with another isometry? It seems reasonable that the combination of the two isometries would be an isometry. In that case, maybe we could get some more isometries by combining the three isometries we know so far. To study these questions, we need to understand the theory of combining isometries. The following definition formalizes something you have already encountered in Example 4.2 of Section 4.1.

DEFINITION

If transformation S_1 is to be followed by S_2, we denote the combined transformation by $S_2 \circ S_1$. We call this the *composition* of S_1 followed by S_2. (Notice that the order in which we list the transformations in the composition notation is the reverse of the order in which we perform them.) If P is any point, then $(S_2 \circ S_1)(P)$ is, by this definition, $S_2[S_1(P)]$.

An example of a composition is the fact that two rotations around the same center "add up" to another rotation. For example, $R_C(30°) \circ R_C(40°) = R_C(70°)$. Other compositions involve rotations, translations, and reflections. Before seeing examples of these, let's pause for our first theorem, which shows why composition is interesting in relation to isometries.

✯ THEOREM 4.7

The composition of two isometries is an isometry.

Proof

Imagine two points P and Q and imagine their images after the first isometry. They are in new positions, P' and Q', but, because of the definition of isometry, their distance apart is the same as the previous distance:

$$P'Q' = PQ.$$

Now apply the second isometry. P' moves to P'' and Q' moves to P'' and Q'', but again the distances have not changed:

$$P''Q'' = P'Q'.$$

From these two equations, we deduce that $P''Q'' = PQ$, which tells us that there has been no change in distance resulting from applying one isometry after the other. ■

✯ THEOREM 4.8

If triangle EFG is congruent to triangle $E^*F^*G^*$ then there is an isometry which maps E to E^*, F to F^*, and G to G^* (Figure 4.9).

Proof

The idea of the proof is quite simple and is illustrated in Figure 4.9. We want to move the shaded triangle onto the congruent striped one. Start with a translation T to move EFG onto $E'F'G'$ where $E' = E^*$. Now we want to move G' onto G^* (without disturbing E') and we do this by a rotation R around E'. Let $E''F''G''$ result from applying R to $E'F'G'$.

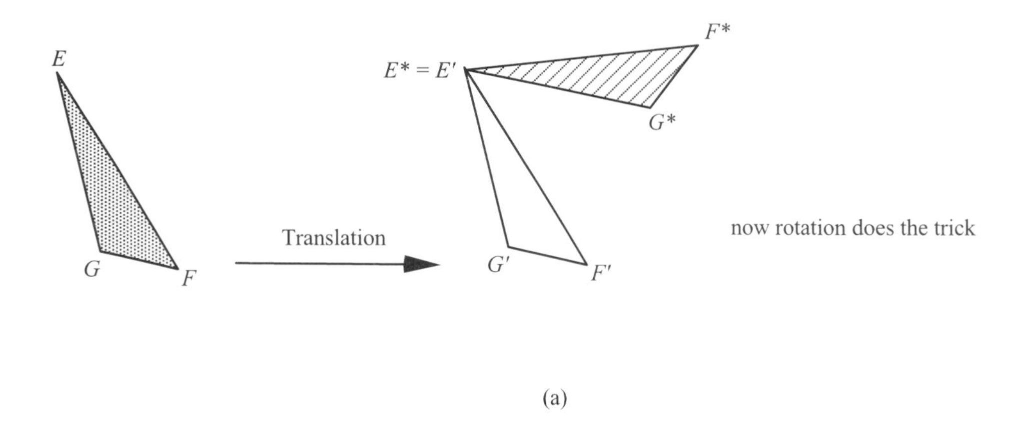

(a)

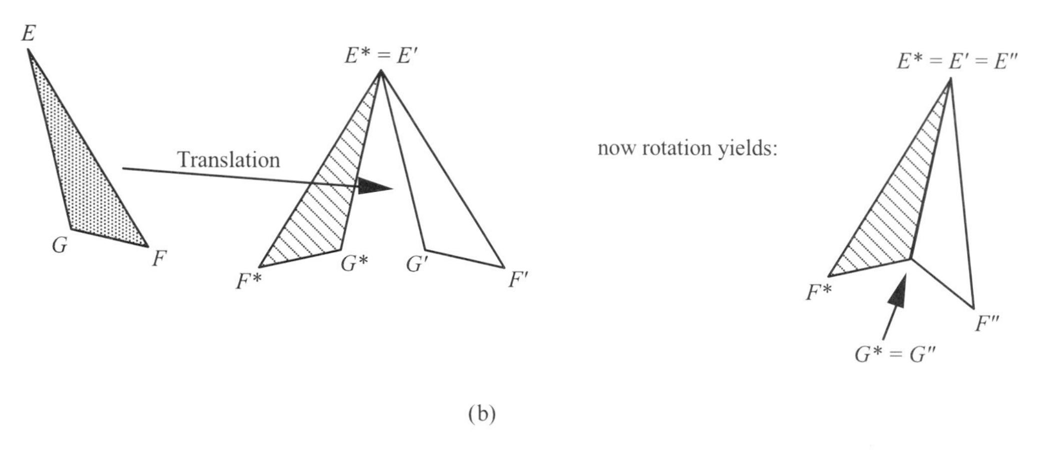

(b)

Figure 4.9 Mapping the shaded triangle to the congruent striped one by isometry (two cases).

At this point, we might hope that F' has rotated onto $F^\star$, that is, that $F'' = F^\star$. Figure 4.9a shows that this might happen, in which case we are done and $R \circ T$ is the isometry we want. But Figure 4.9b suggests that maybe $F'' \neq F^\star$. In that case we claim that a reflection M in $E''G''$ will move F'' onto $F^\star$ so that $M \circ R \circ T$ is the isometry we want. Can you verify this claim? ■

Special Compositions

In this subsection, we show that if we know the ingredients of a composition, we can often tell a good deal about the composite isometry.

EXAMPLE 4.4: COMPOSING TRANSLATIONS

A translation is completely described by giving the vector (directed line segment) from some arbitrarily chosen point to its image—since every other point will move in the same direction by the same amount. For example, in Figure 4.3 of the previous section, if we know the vector from P to its image P', we can readily find the image of Q by placing an instance of that vector with tail at Q. Suppose T_1 and T_2 are two translations whose vectors are $\mathbf{g}_1$ and $\mathbf{g}_2$, respectively (Figure 4.10). This example suggests that $T_2 \circ T_1$ is another translation. Imagine a troop of soldiers in formation who march 1 mile north followed by 1 mile east. Each soldier would wind up in the same place if, instead, he or she marched $\sqrt{2}$ miles northeast. (Draw a figure to illustrate this.) Later we give a more formal proof of this composition principle.

Although it is not important for our work in this chapter, in Chapter 6 we make use of the fact that the vector for the composite translation is $\mathbf{g}_1 + \mathbf{g}_2$, a fact that follows directly from the concept of vector addition. ■

☆ THEOREM 4.9

If T_1 and T_2 are two translations, $T_2 \circ T_1$ is also a translation (Figure 4.10).

Proof

Figure 4.10 shows P moving to $T_1(P)$ and then to $T_2[T_1(P)]$ by two separate translations. We could achieve the same effect if we move P to $T_2[T_1(P)]$ directly along the segment connecting these points.

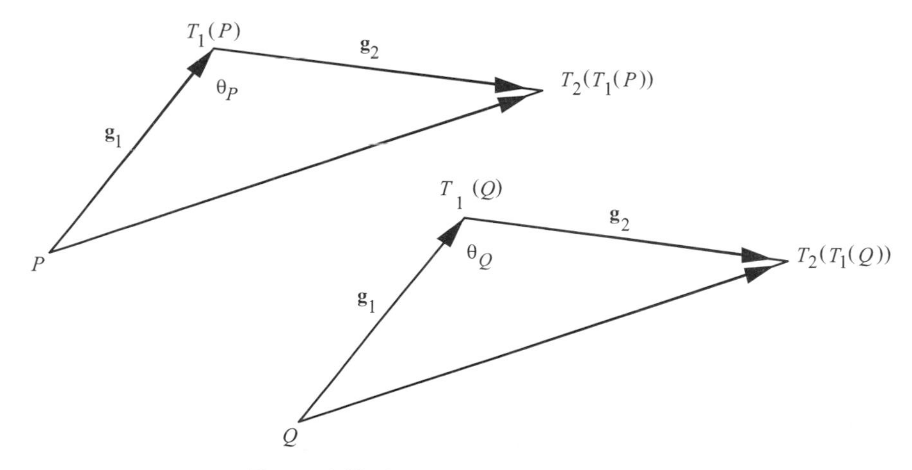

Figure 4.10 Composing two translations.

But we need to show that this would work for any point Q. More precisely, that the "one-step" move from Q to $T_2[T_1(Q)]$ would be by the same amount and direction. The key idea is to show that the two triangles are congruent. The two segments labeled $\mathbf{g}_1$ are congruent and parallel and so are the two labeled $\mathbf{g}_2$. The two parallelisms allow us to deduce that $m\angle\theta_P = m\angle\theta_Q$. This, together with the congruences of the segments and SAS, gives us the congruence of the triangles.

From the congruence of the triangles we get the congruence of the desired sides. But we need them to be parallel too. Can you explain why this follows? ■

EXAMPLE 4.5: COMPOSING REFLECTIONS IN TWO PARALLEL LINES

Suppose we take a reflection in a line L and compose it with reflection in a line L' parallel to L, $M_{L'} \circ M_L$. Figure 4.11 shows the effect on one particular point P. Is the composition equivalent to a transformation we are already familiar with or is it something new? We'll see that for points P that are on the other side of L from L' this composition has the same effect as a translation.

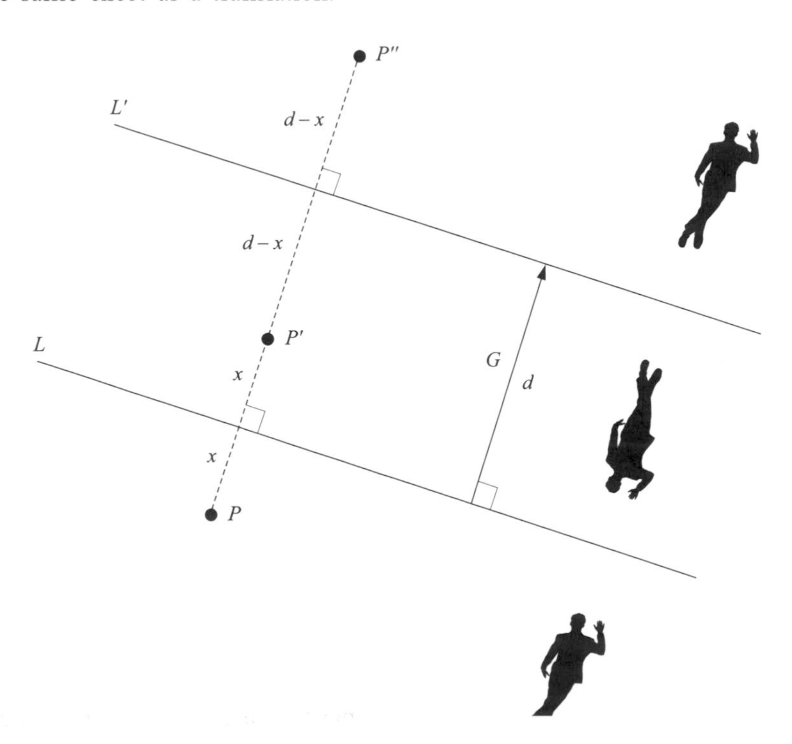

Figure 4.11 Reflections in two parallel lines.

SOLUTION

The two reflections move P perpendicular to the mirror lines. But by how much? Let d be the perpendicular distance between the lines and let x be the distance from P to line L. The number x will be different for different points P. Let $P' = M_L(P)$ and $P'' = M_{L'} \circ M_L(P)$. We can easily see from the figure that $PP'' = x + x + (d - x) + (d - x) = 2d$, an amount which is independent of x. Thus, as long as the position of P is as shown in Figure 4.11, its image after the two reflections is $2d$ units away in the direction from line L to line L'.

What would happen if P were in some other position, for example, between L and L'? To conclude that $M_{L'} \circ M_L$ is a translation by $2d$ in the direction from L to L', we need to show that no matter how we choose the point P, $M_{L'} \circ M_L$ has the same effect on P as that translation. It is a worthwhile exercise to work out all the possibilities, make a figure for each, and adapt our proof to each one. By checking all the possible cases, we can conclude that $M_{L'} \circ M_L$ is a translation.

In studying this example, notice that if we replaced L and L' by any other lines with the same distance apart and the same direction, we get the same translation. ■

We summarize the conclusions of the previous example in the following theorem:

✯ THEOREM 4.10

(a) If L and L' are two parallel lines with distance d between them, then $M_{L'} \circ M_L$ is a translation by an amount $2d$ in the direction from L to L', perpendicular to those lines.

(b) The translation is independent of the particular lines L and L' and depends only on their direction and their distance apart.

Example 4.5 showed that we do not get an isometry different from those we are already familiar with if we compose two reflections in parallel lines. What if the lines are not parallel? Here's the story.

✯ THEOREM 4.11

(a) If L and L' are two lines crossing at point C and making an angle of θ, then $M_{L'} \circ M_L = R_C(2\theta)$ (Figure 4.12).

(b) If L and L' are replaced by any other pair of lines meeting at C, making an angle of θ, we still get $M_{L'} \circ M_L = R_C(2\theta)$.

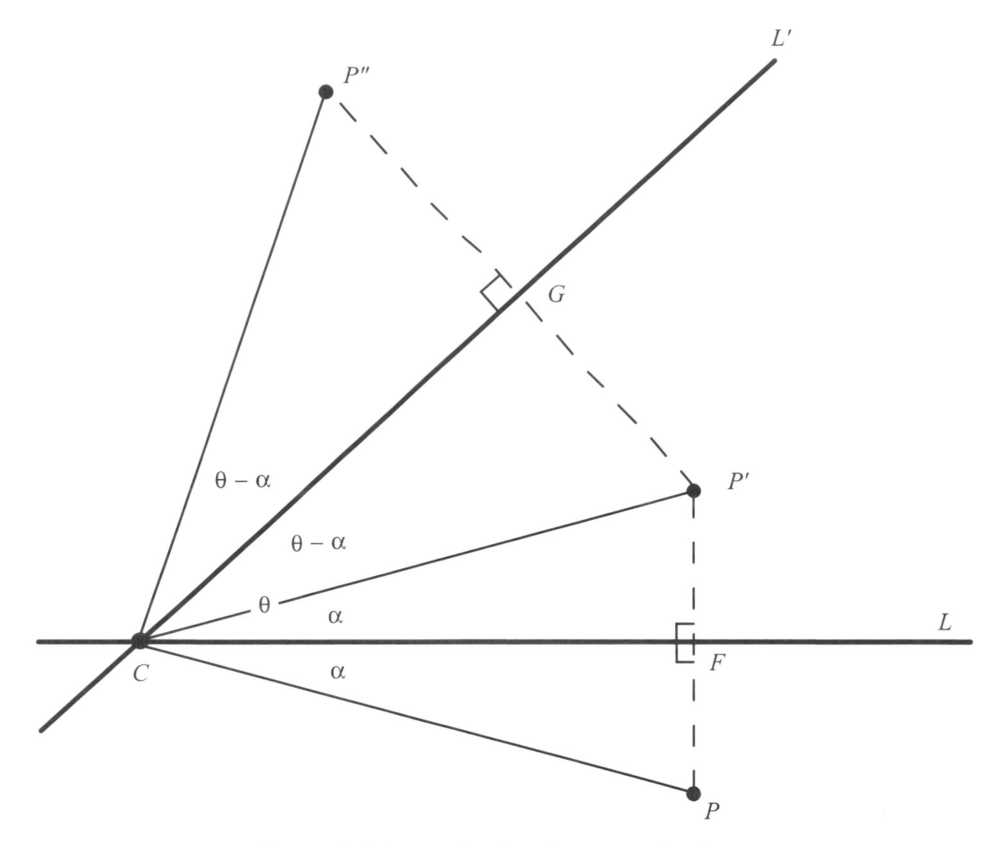

Figure 4.12 Two reflections in nonparallel lines.

Proof

Let's take any point P and follow it as it gets reflected. Drop a perpendicular from P to L, meeting it at F and extend to the image $P' = M_L(P)$. Let $P'' = M_{L'}(P') = [M_{L'} \circ M_L](P)$. Let α be the angle $\overline{CP}$ makes with the first line L. First observe that CPF and $CP'F$ are congruent triangles. (Use SAS and the fact that $PF = P'F$, $\mathrm{m}\angle PFC = \mathrm{m}\angle P'FC = 90^\circ$, and the fact that $\overline{CF}$ is common to both triangles.) Therefore, $\mathrm{m}\angle P'CF = \mathrm{m}\angle PCF = \alpha$ since these are corresponding parts of the congruent triangles. In a similar fashion, we show that $\mathrm{m}\angle P''CG = \mathrm{m}\angle P'CG = \theta - \alpha$. Adding the angles gives

$$\mathrm{m}\angle P''CP = \alpha + \alpha + (\theta - \alpha) + (\theta - \alpha) = 2\theta. \tag{4.2}$$

Using the congruent triangles again, we can see that $CP'' = CP' = CP$. This, together with Eq. (4.2) shows that P has been rotated through 2θ around C in order to reach P''. Therefore, $M_{L'} \circ M_L = R_C(2\theta)$.

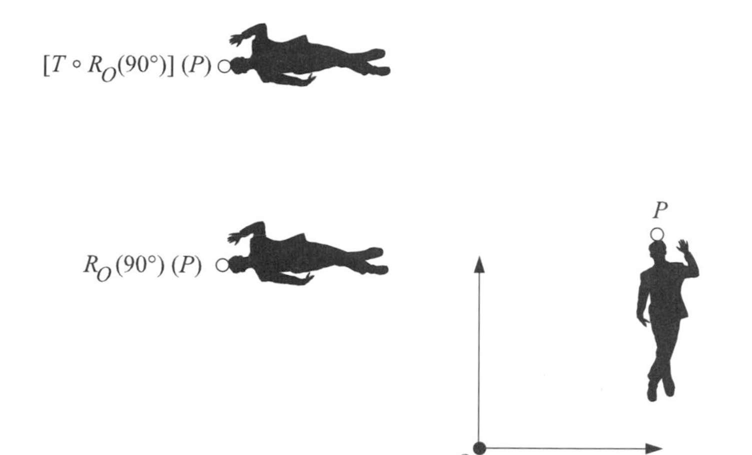

Figure 4.13 Rotation by 90° followed by translation by one unit vertically.

To prove part (b), notice that the only facts about L and L' we made use of were that they made an angle of θ and they met at C. ■

As with other theorems in this section, we have drawn a special case and given a proof for that case. You might like to think of other positions for P and adapt our proof.

EXAMPLE 4.6

In Figure 4.13, we illustrate $T \circ R_O(90^\circ)$ where T is translation by one unit in the vertical direction, and O denotes the origin as usual. Could this be a new type of isometry?

It is far from obvious, but this composition is equivalent to a rotation about some point other than the origin. The sections following this one provide the tools to prove this. Can you guess where that center of rotation is? ■

By now, perhaps you are thinking that any composition of rotation, reflection, and translation is another rotation, reflection, or translation. This is not true, as the next example shows.

EXAMPLE 4.7: GLIDE REFLECTION

Let T denote translation by two units to the right (parallel to the x axis) and let M_x denote reflection in the x axis (Figure 4.14). $M_x \circ T$ has the feature that it keeps no point *fixed* (unchanged). To see this, note that the x coordinate of any point increases by two units. By contrast, a rotation keeps the center point fixed and a reflection keeps the entire

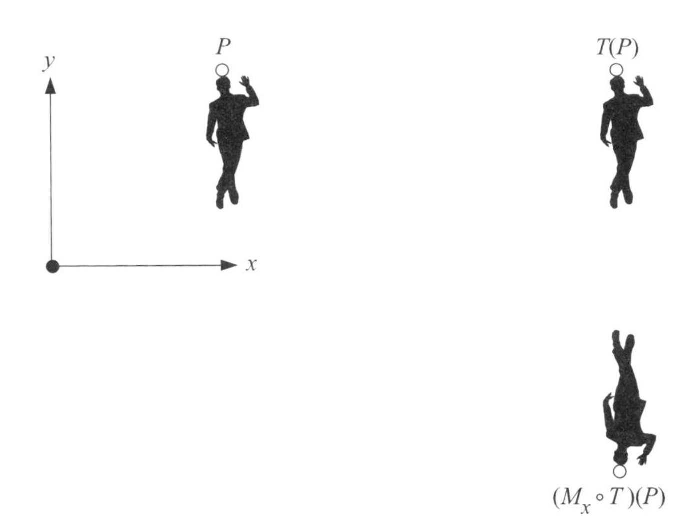

Figure 4.14 Glide reflection.

mirror line fixed. Consequently, $M_x \circ T$ cannot be a rotation or reflection in disguise. It also cannot be a translation in disguise. In a translation, all points move in the same direction. But in Figure 4.14, points under the x axis wind up "northeast" of their original position, while points over the line wind up "southeast." $M_x \circ T$ is an example of a *glide reflection*. By definition, a glide reflection is any translation followed by a reflection in any line parallel to the direction of translation. ■

The Algebra of Composition

Can you see that if we do the two isometries in Example 4.6 in the opposite order, we get a different image point: $[R_O(90°) \circ T](P) \neq [T \circ R_O(90°)](P)$ when P is the particular point shown in Figure 4.13? We summarize this by saying that composition of those two isometries is not *commutative*. For some pairs of isometries, it does not matter in which order we do them—the effect is the same on any point P. In this case, we say they *commute*. Here are some examples.

EXAMPLE 4.8: SOME COMMUTATIVE COMPOSITIONS

1. $R_C(\theta_2) \circ R_C(\theta_1) = R_C(\theta_1) \circ R_C(\theta_2)$ because either way we do the composition we get $R_C(\theta_1 + \theta_2)$. But note that our example has the same center for both rotations. If the centers were different, would the rotations still commute?

2. The translation and reflection that make up a glide reflection commute: For all points P, $[T \circ M_x](P) = [M_x \circ T](P)$.
3. Translations commute. Figure 4.15 shows how we can try to prove this. Say the translation T_1 moves A to B and T_2 moves B to C. If we start with T_2 and apply it to A, obtaining D, we have $\overline{AD}$ parallel and equal in length to $\overline{BC}$. Now if we apply T_1 to D do we get to C or to some other point E? See the exercises for a hint about how to resolve this question. ■

Whereas commutativity of isometries depends on which isometries one is considering, composition of three isometries is always associative, regardless of the three isometries involved:

☆ THEOREM 4.12: THE ASSOCIATIVE LAW

For any transformations S_1, S_2, S_3,

$$S_3 \circ (S_2 \circ S_1) = (S_3 \circ S_2) \circ S_1. \tag{4.3}$$

Proof

Let P be any point. Applying the left side of Eq. (4.3) to P gives:

$$\begin{aligned}[S_3 \circ (S_2 \circ S_1)](P) &= S_3([S_2 \circ S_1](P)) \\ &= S_3(S_2[S_1(P)]).\end{aligned}$$

Apply the right side of Eq. (4.3) and we get the same result:

$$\begin{aligned}[(S_3 \circ S_2) \circ S_1](P) &= [S_3 \circ S_2][S_1(P)] \\ &= S_3(S_2[S_1(P)]). \quad ■\end{aligned}$$

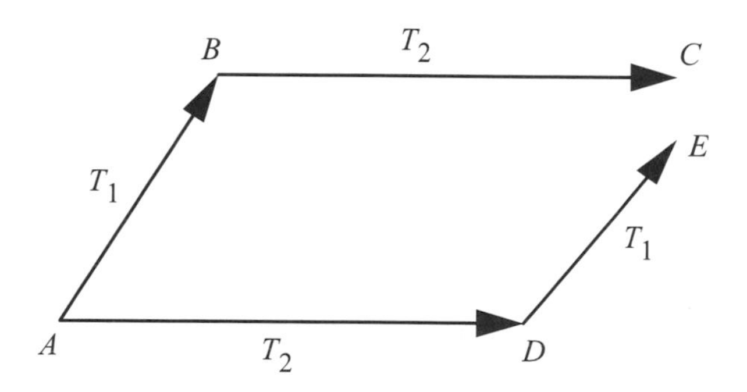

Figure 4.15 Do translations commute? (Is $E = C$?)

When we have two isometries in which the second cancels the effect of the first, and vice versa, we call either one the *inverse* of the other. For example, M_L is its own inverse: $M_L \circ M_L = I$.

DEFINITION

Let U and V be isometries where $U \circ V = V \circ U = I$. Then U is said to be the *inverse* of V and we write $U = V^{-1}$. Likewise, V is the inverse of U and we can write $V = U^{-1}$.

EXAMPLE 4.9: SOME INVERSES

1. For any line L, $M_L = M_L^{-1}$.
2. Let T be a translation by an amount d, and let S be a translation by the same amount d but in the opposite direction. Then, $T = S^{-1}$ and $S = T^{-1}$.
3. $R_C(-\theta) = [R_C(\theta)]^{-1}$ regardless of what C and θ are.
4. Let H be a glide reflection in line L where the distance to be translated is d. H^{-1} is the glide reflection in the same line L where the translation amount is d, but the direction of translation is opposite to that of H. ■

Does every isometry have an inverse? If so, is the inverse transformation an isometry? Since we don't have a catalog of all the isometries, we can't answer this question by examining all the individual types of isometries. Can you think of a way to get the answers?

In our final example, look for the use of inverses, associativity, and a clever use of part (b) of Theorem 4.10.

EXAMPLE 4.10: GLIDE REFLECTION IN DISGUISE

Let L be a line and let T be a translation that is not in the direction of L. Although $M_L \circ T$ is not a glide reflection involving L, we show that there is a second line N so that $M_L \circ T$ is a glide reflection involving line N.

SOLUTION

As Figure 4.16 shows, we can factor T as a composition of two translations,

$$T = T_\perp \circ T_L,$$

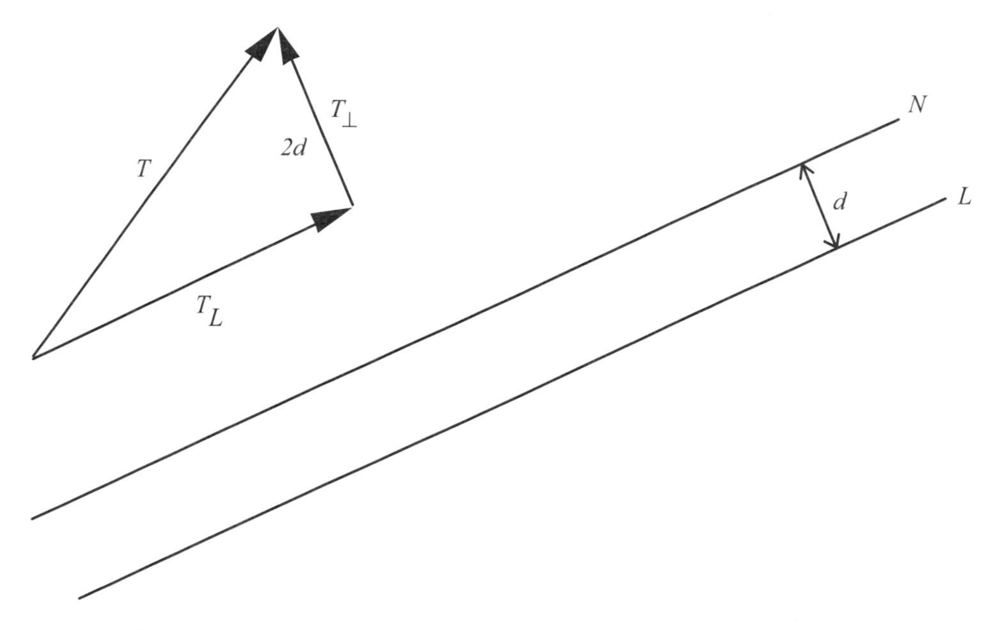

Figure 4.16 Factoring a translation into two perpendicular translations.

where T_L is in the direction of L and $T_\perp$ is perpendicular to L. Let $2d$ be the distance of the translation $T_\perp$.

According to Theorem 4.10, $T_\perp$ can, in turn, be factored into

$$T_\perp = M_N \circ M_L,$$

where N is a line parallel to L and a distance d from L in the direction of the translation $T_\perp$. Then

$$\begin{aligned} M_L \circ T &= M_L \circ (T_\perp \circ T_L) \\ &= (M_L \circ T_\perp) \circ T_L \\ &= [M_L \circ (M_L \circ M_N)] \circ T_L \\ &= [(M_L \circ M_L) \circ M_N] \circ T_L \\ &= (I \circ M_N) \circ T_L \\ &= M_N \circ T_L. \end{aligned}$$

But this is a glide reflection since T_L is in the direction of L, which is the same direction as N. ■

EXERCISES

Marks challenging exercises.

The Idea of Composition

1. Can you describe an isometry S (other than I) where $S \circ S \circ S = I$? Can you think of more than one?

2. Suppose you know that a rotation R about the origin followed by a translation is a rotation R' around some point other than the origin. Given the rotation angle and the translation distance and direction, how could you find the center for R'? (*Hint:* The center would be a point whose motion under the rotation is reversed by the translation.)

3. In the proof of Theorem 4.8, in order for the rotation to move G' onto $G^\star$ it would be necessary for $E'G' = E'G^\star$. Why are these lengths equal?

4. In the proof of Theorem 4.8, in order for the reflection to move F'' onto $F^\star$, it would be necessary for $E''G''$ to be the perpendicular bisector of $F''F^\star$. Why is this true?

Special Compositions

5. Answer true or false for the following statements. Explain your answer. (Use stick figures if you wish to illustrate.)
   (a) The composition of two reflections could be a reflection.
   (b) If S and S' are isometries then $S \circ S = S' \circ S'$ implies that $S = S'$.
   (c) If S is an isometry, and $S \circ S = I$, then S must be a reflection.
   (d) If A, B, C are the vertices of a triangle in clockwise order, and S is an isometry, then $S(A)$, $S(B)$, $S(C)$ are also in clockwise order.

6. (a) In Figure 4.11, show by a realistic picture that P could be below L and P' could be above L' and P'' could be between L and L'.
   (b) Can you prove Theorem 4.10 in this case?
   (c) How do d and x have to be related to make this happen?

7. In Figure 4.11, what if the point P we are reflecting is not under line L, but is above L but under L'. Now where are P' and P''? Can you do the proof of Theorem 4.10 in this case?

8. A pair of intersecting lines, like L and L' in Figure 4.12, make two different angles (which add to 180°). Which one is the θ referred to in Theorem 4.11? Of course, the correct angle is labeled in the figure. What you should supply is a description of how to tell that it is that labeled angle and not the other.

9. Use part (b) of Theorem 4.10 to show that the composition of three reflections in parallel lines is the same as a single reflection in some fourth line.

10. If L, L', and L'' are lines with point C in common, prove that $M_{L''} \circ M_{L'} \circ M_L$ is a single reflection in some line through C. (*Hint:* L and L' can be replaced by a more convenient pair of lines.)

The Algebra of Composition

11. Answer true or false for the following statements. Explain your answer.
    (a) Two translations always commute. (*Hint:* In Figure 4.15, join C to D. What can you prove about $ABCD$?)
    (b) Two reflections always commute.
    (c) Two rotations about the origin always commute.

12. Suppose H is a glide reflection. What is the nature of $H \circ H$?

13. In each of the following cases, do U and V commute? (*Hint:* Look for what happens to particular points and lines.)
    (a) $U = R_O(90°)$, $V =$ reflection in the y axis.
    (b) $U =$ reflection in the y axis, $V =$ reflection in the x axis.
    (c) $U = R_O(30°)$, $V = R_O(-30°)$.
    (d) $U = R_O(30°)$, $V = R_{(1,0)}(-30°)$.
    (e) $U =$ translation by one unit to the right, $V =$ glide reflection in the x axis and by two units to the right.

14. Show that $(M_{L'} \circ M_L)^{-1} = M_L \circ M_{L'}$ if the lines L and L' are parallel.

15. Is the equation in the previous exercise true if L and L' are not parallel? If not, give an example. If it is, can you prove it?

16. Let M_x denote reflection in the x axis and let M_y denote reflection in the y axis. Show that $R_O(180°) \circ M_x = M_y$. (*Hint:* Factor $R_O(180°)$ into two reflections, one involving the x axis.)

17. Let Q be the midpoint of $\overline{PR}$. Let T_{PR} denote the translation that sends P to R. Prove that $R_Q(180°) \circ R_P(180°) = T_{PR}$. (*Hint:* Factor the rotations into convenient reflections.)

18. Let T_{PO} be the translation that moves P to the origin O. What kind of isometry do you think $T_{PO}^{-1} \circ R_O(\theta) \circ T_{PO}$ is?

19. (a) $R_O(\theta) \circ M_x \circ R_O(-\theta) = M_L$ for a certain line L. Which line?
    (b) Prove the equation in part (a).

20. Prove that if S_1 and S_2 are isometries, then $(S_2 \circ S_1)^{-1} = S_1^{-1} \circ S_2^{-1}$.

*21. Could a glide reflection followed by rotation be a glide reflection? If not, prove it. If so, give an example and show that the composition really is a glide reflection.

*22. Could a glide reflection followed by rotation be a reflection? If not, prove it. If so, give an example and show that the composition really is a reflection.

*23. (a) Prove that if S is an isometry, then S is one-to-one, and therefore has an inverse transformation.

    (b) Show that the inverse transformation S^{-1} is an isometry.

24. Make a list of as many isometries as you can which are their own inverses.

4.3 There Are Only Four Kinds of Isometries

The main objective of this section is to show that the isometries which we have met so far are the only kinds there are: identity, translation, rotation, reflection, and glide reflection. This list of five can be reduced to four by leaving off the identity, since it is really a special type of rotation (through 0°) and a special type of translation. In particular, we prove here that no matter how examples of these are composed, you always get one or another of these four types as the result.

Circles and Isometries

On the face of it, an isometry is a complex object; to know it is to know what the image is of each point in the plane. But, as we look more closely, we will discover that the nature of an isometry is completely determined by its effect on three arbitrarily chosen noncollinear points. To see what we mean by this, let F_1, F_2, F_3 be any three noncollinear points of the plane. Can two *different* isometries have the same effect on these three points, both sending F_1 to F_1', both sending F_2 to F_2', and both sending F_3 to F_3'? If the isometries are different, then there would be some point X which would map to two different points under the two isometries. Our next objective is to prove that

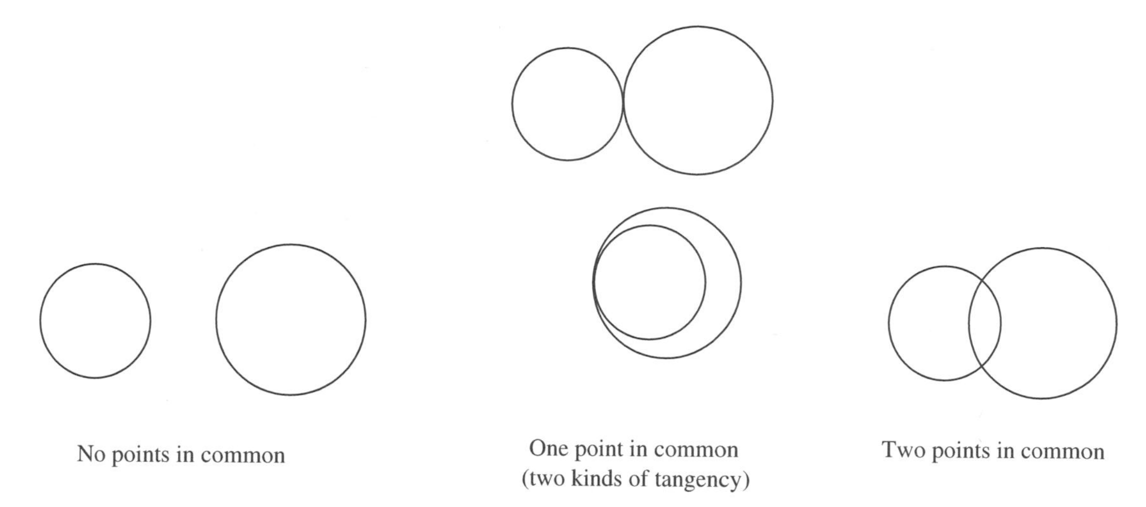

Figure 4.17 Four examples of how two circles can intersect.

this cannot happen: The image of a point X is completely determined by the images of F_1, F_2, F_3. The fate of these points determines the fate of all points. For some purposes, one can think of the isometry just in terms of these arbitrarily chosen three noncollinear points.

We shall see that this property arises out of facts about how circles can intersect. Two circles can intersect either not at all (Figure 4.17), can intersect once at a point of tangency, or can intersect twice. More than two intersections is impossible. (In the exercises at the end of this section, you will be led through proofs of some of the assertions made here about circles.)

Now take a case where two circles intersect twice at A and B and add a third circle (Figure 4.18). It is possible that it can pass through both A and B, but only if its center is on a line with the centers of the first two circles. Phrased another way, this gives us the principle we need.

★ THEOREM 4.13

If three circles have centers which are not collinear, then they intersect at most once (Figure 4.18).

Proof

See the exercises at the end of this section. ■

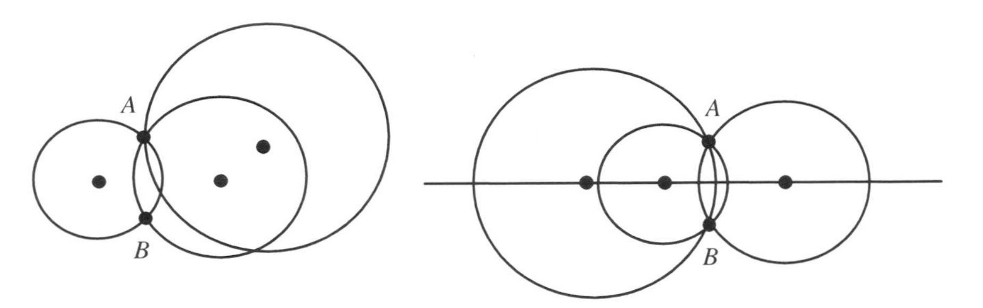

Figure 4.18 Three circles that intersect do so in two points only If the centers are collinear.

✯ THEOREM 4.14

An isometry is completely determined by its effect on any three noncollinear points F_1, F_2, F_3 (Figure 4.19).

PROOF

Let X be an arbitrary point and let it have distances r_1, r_2, and r_3 from F_1, F_2, F_3, respectively. Our objective is to show that there is only one place where X', the image of X, could be. Because the isometry preserves distances, X' must have these same

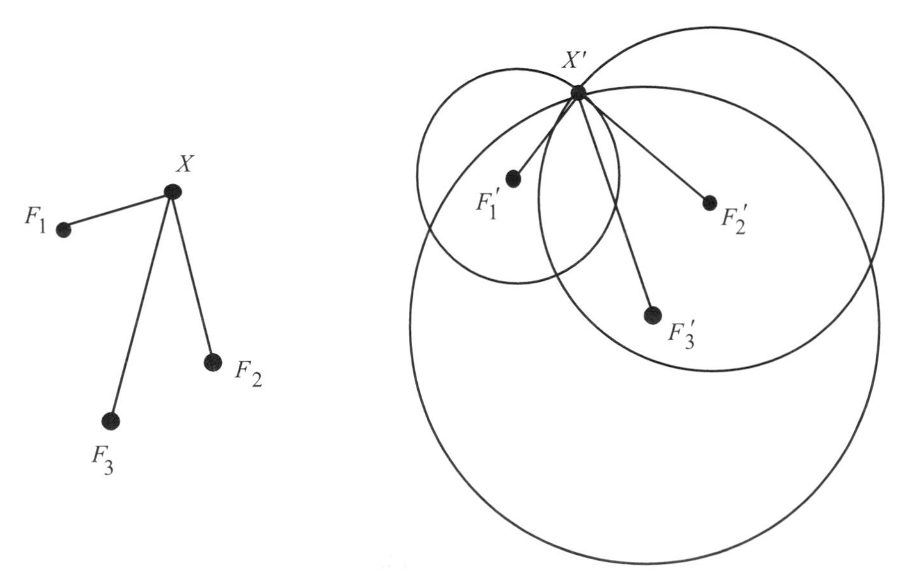

Figure 4.19 An isometry is determined by its effect on three noncollinear points.

distances r_1, r_2, r_3 to the images F_1', F_2', F_3'. Since X' has distance r_1 from F_1', it must be on the circle of radius r_1 drawn around F_1'. Likewise, it is on the circle of radius r_2 drawn around F_2' and on the circle of radius r_3 drawn around F_3'. So X' must be a point where these circles cross. Now since F_1, F_2, F_3 are not collinear, Theorem 4.3 of Section 4.1 in this chapter implies that F_1', F_2', F_3' are noncollinear. Therefore, Theorem 4.13 assures us that the three circles intersect just once. This crossing is the only place X' can be. ■

Isometries and Their Fixed Points

The previous theorem implies that if an unknown isometry S has the same effect on three noncollinear points as some known isometry S', then $S = S'$. This form of the theorem, with S' being I, is used to prove the following corollary. In this corollary we use the terminology S *fixes* P to mean $S(P) = P$. (Alternatively, we might say P is a *fixed point* of S.)

COROLLARY 4.15

An isometry S which fixes at least three noncollinear points F_1, F_2, F_3 fixes all points, that is, S is the identity.

PROOF

This follows from the previous theorem since the identity fixes F_1, F_2, F_3. ■

Consideration of fixed points leads to the following remarkable result. By thinking of composition as being a bit like multiplication, one can think of this next theorem as analogous to factoring a given number.

✯ THEOREM 4.16: FACTORING AN ISOMETRY

Every isometry is either the identity or the composition of at most three reflections.

We shall prove this factoring theorem as a series of three theorems, each dealing with a different assumption about how many fixed points S has.

✯ THEOREM 4.17

If S is an isometry that fixes at least the two points F_1 and F_2, then either $S = I$ or S is a reflection M in the line joining F_1 and F_2 (Figure 4.20).

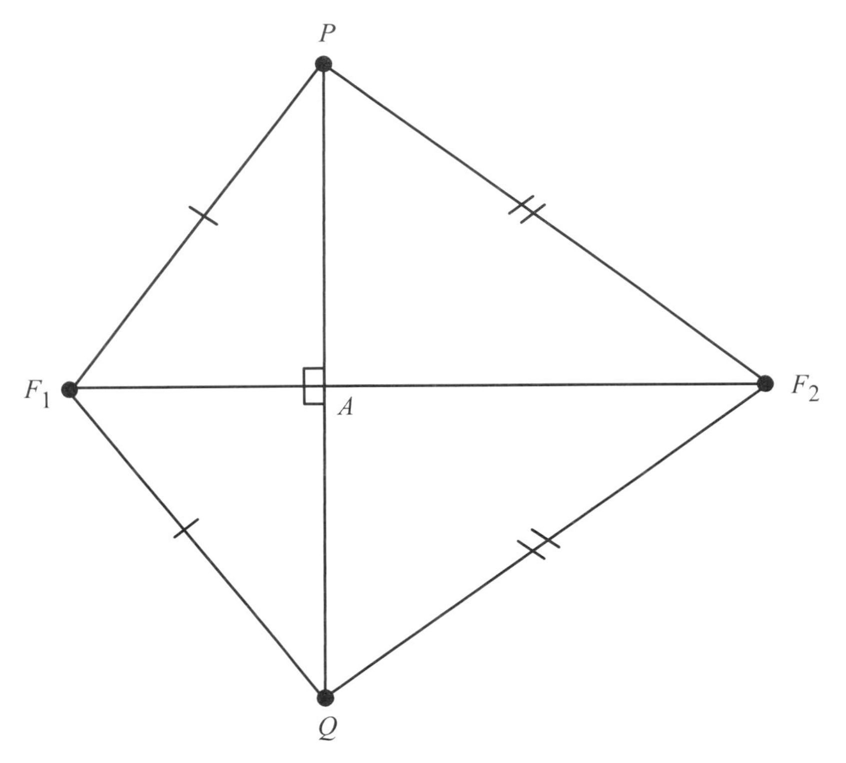

Figure 4.20 Two points equidistant from the ends of a segment.

Proof

Let P be any point not collinear with F_1 and F_2. If S fixes P, then S has the same behavior as the identity on F_1, F_2, and P. By Corollary 4.15, $S = I$.

If, on the other hand, S moves P to Q, say, then it can be shown (see the exercises at the end of this section) that $\overleftrightarrow{PQ}$ and $\overleftrightarrow{F_1F_2}$ are perpendicular. Furthermore, $\overleftrightarrow{F_1F_2}$ cuts $\overleftrightarrow{PQ}$ at a point A, which is the midpoint of $\overline{PQ}$. This means that the reflection M in line $\overleftrightarrow{F_1F_2}$ sends Q to P. Thus, $M \circ S$ fixes P, F_1, and F_2. By the first part of our proof,

$$\begin{aligned} M \circ S &= I, \\ M^{-1} \circ (M \circ S) &= M^{-1} \circ I \\ (M^{-1} \circ M) \circ S &= M^{-1}, \\ S &= M^{-1}. \end{aligned}$$

But $M^{-1} = M$, the reflection in line $\overleftrightarrow{F_1F_2}$, so the proof is completed. ■

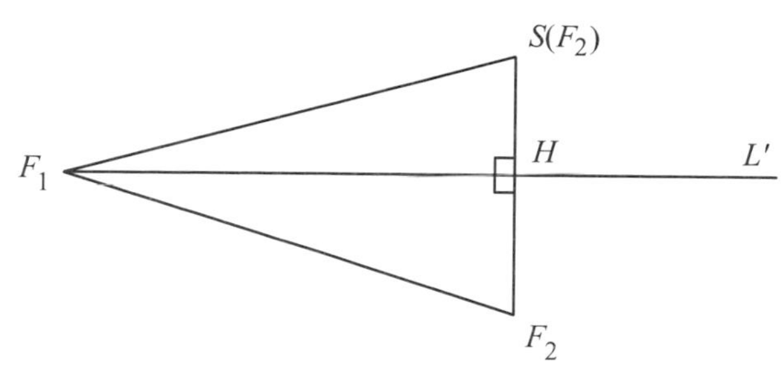

Figure 4.21 An isometry which fixes just one point.

★ THEOREM 4.18

If S is an isometry that fixes exactly one point F_1, then S is a composition of two reflections $S = M_{L'} \circ M_L$ (Figure 4.21).

PROOF

Let F_2 be any point other than F_1. Because S preserves distance and $F_1 = S(F_1)$, we have the following equal distances: $F_1S(F_2) = S(F_1)S(F_2) = F_1F_2$. Thus, the triangle $F_1F_2S(F_2)$ is isosceles. Let L' be the line connecting F_1 to the midpoint H of the base segment joining F_2 and $S(F_2)$. Since the triangle is isosceles, L' will be perpendicular to the base. This means that $M_{L'}[S(F_2)] = F_2$, or $[M_{L'} \circ S]\,(F_2) = F_2$. But $M_{L'} \circ S$ also fixes F_1. By the previous theorem, $M_{L'} \circ S = I$ or $M_{L'} \circ S = M_L$ for some line L. The case $M_{L'} \circ S = I$ can be ruled out. (Why?) Thus,

$$\begin{aligned} M_{L'} \circ S &= M_L, \\ M_{L'}^{-1} \circ (M_{L'} \circ S) &= M_{L'}^{-1} \circ M_L, \\ S &= M_{L'}^{-1} \circ M_L \\ &= M_{L'} \circ M_L. \quad \blacksquare \end{aligned}$$

★ THEOREM 4.19

If S is an isometry with no fixed points, it is the composition of at most three reflections (Figure 4.22).

PROOF

Let F_1 be any point and let L'' be the line that is the perpendicular bisector of segment $\overline{F_1S(F_1)}$. $M_{L''}[S(F_1)] = F_1$ so $M_{L''} \circ S$ has at least F_1 as a fixed point. Depending on how many other fixed points $M_{L''} \circ S$ has, one of the previous theorems applies

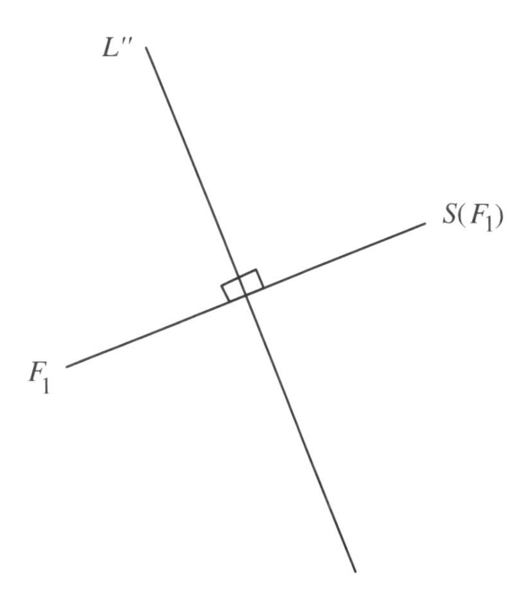

Figure 4.22 Reflecting the image of F_1 back to F_1.

and we have $M_{L''} \circ S$ being either the identity or a reflection or a composition of two reflections. Depending on which case we have, we get S to be either one reflection, or a composition of two or a composition of three. For example, say $M_{L''} \circ S$ is a composition of two reflections: $M_{L''} \circ S = M_{L'} \circ M_L$. Then

$$\begin{aligned} M_{L''}^{-1} \circ M_{L''} \circ S &= M_{L''}^{-1} \circ M_{L'} \circ M_L, \\ S &= M_{L''}^{-1} \circ M_{L'} \circ M_L \\ &= M_{L''} \circ M_{L'} \circ M_L. \quad \blacksquare \end{aligned}$$

✯ THEOREM 4.20: CLASSIFICATION OF ISOMETRIES

Every isometry S is either the identity, a reflection, a rotation, a translation, or a glide reflection.

Proof

We proceed according to how many reflections are in the composition guaranteed by Theorem 4.16. If there are two reflections, then Theorems 4.10 and 4.11 of Section 4.2 in this chapter show that S must be either a rotation or translation.

If there are three reflections, $S = M_{L''} \circ M_{L'} \circ M_L$, we proceed by two cases:

Case 1: Lines L and L' cross at a point P. Replace L and L' by lines K and K' obtained by rotating L and L' around P so that K' is parallel to L''. By part (b) of Theorem 4.11 of Section 4.2 in this chapter, $M_{L''} \circ M_{L'} = M_{K'} \circ M_K$ so $S = M_{L''} \circ M_{K'} \circ M_K = (M_{L''} \circ M_{K'}) \circ M_K$. Now the point of rotating the lines becomes clear as we recognize that the part in parentheses is a translation T, by Theorem 4.10 of Section 4.2 in this chapter. Thus, $S = T \circ M_K$. If T is the trivial translation by 0 distance (draw a picture of the three lines to show this could happen) we have just $S = M_L$. But if T is not trivial, then Example 4.10 of Section 4.2 in this chapter shows that S is a glide reflection.

Case 2: L and L' are parallel. Then $M_{L'} \circ M_L = T$, a translation, using Theorem 4.10 again. Then $S = M_{L''} \circ T$, which can be shown to be a glide reflection by an argument just like the one in Example 4.10 of Section 4.2. ■

Isometries and Sense

Suppose we are told that an isometry S has moved the left triangle in Figure 4.23 to the one on the right. Which type of isometry is it? If we knew how many reflections were in the factorization of Theorem 4.16 we might have a clue, but examining the triangles gives us no such information. If we knew how many fixed points S had, then Theorems 4.17 and 4.18 might help. But there is no such information available from studying the two triangles. A concept that does give a clue is the *sense* of a sequence of three noncollinear points.

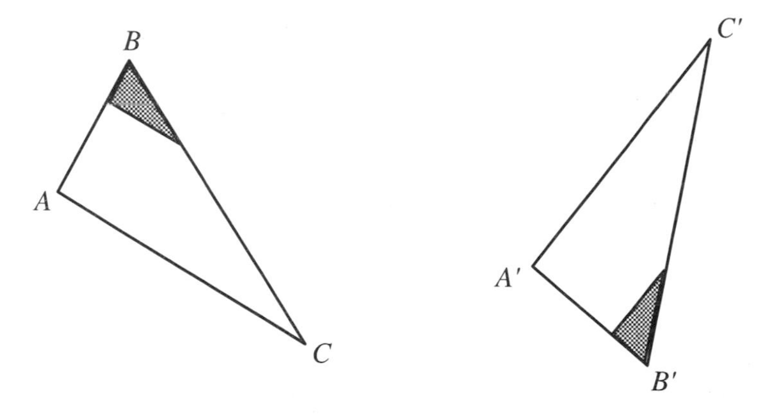

Figure 4.23 Triangles with opposite sense (shaded to identify larger acute angle).

DEFINITION

Given three points A, B, C not lying on the same line, if we travel in a counterclockwise direction as we go in order from A to B and then to C then we say $[A, B, C]$ has *counterclockwise sense*. If we travel clockwise, then we say the sequence $[A, B, C]$ has *clockwise sense*.

For example, in Figure 4.23, $[A, B, C]$ has clockwise sense while $[A', B', C']$ has counterclockwise sense. Notice that the idea of sense depends on the order in which we list the points, not just on the points themselves. If we take the same set of points but put them into a sequence in another way, say, $[C, B, A]$, we get a sequence with the opposite sense to the sense of $[A, B, C]$.

Suppose $A'B'C'$ is a triangle that is congruent to ABC with the correspondence $A \to A', B \to B', C \to C'$. Will the two ordered triangles $[A, B, C]$ and $[A', B', C']$ have the same sense? Not necessarily. Imagine that we apply a reflection to ABC in order to get $A'B'C'$. Reflection switches the sense of any sequence of three points. (Draw a picture to illustrate this.)

By contrast, translation and rotation clearly preserve the sense of any sequence of three points. Glide reflection, since it is the composition of a reflection that switches sense and a translation that does not, must also switch sense. This is an instance of the following fairly obvious theorem, whose proof we skip.

✯ THEOREM 4.21: SENSE AND COMPOSITION

If $S_3 = S_2 \circ S_1$ and S_1 and S_2 are isometries, then

(a) If S_1 and S_2 both preserve sense then so does S_3, and

(b) if one of S_1 and S_2 preserves sense while the other reverses it, then S_3 reverses sense, and

(c) if both S_1 and S_2 reverse sense, then S_3 preserves sense.

The upshot of our discussion of sense is that if an isometry has switched the sense of even a single triangle, then it must be a reflection or a glide reflection. If it has preserved the sense of even a single triangle, it must be either a rotation or translation.

EXAMPLE 4.11: USING SENSE

What kind of isometry is $S = T \circ R_O(\theta)$?

SOLUTION

CASE 1. If θ is a multiple of 360°, then $S = T$, not the most interesting case, so let's assume $R_O(\theta)$ is not the identity in disguise.

CASE 2. $R_O(\theta) \neq I$. Both rotation and translation preserve sense, so their composition S does also by Theorem 4.21. Therefore, $T \circ R_O(\theta)$ is either a translation or rotation. We can rule out translation. If it were a translation T', we would have

$$T \circ R_O(\theta) = T',$$
$$R_O(\theta) = T^{-1} \circ T'.$$

But T^{-1} is also a translation, so the right side is the composition of two translations, which is a translation. Thus, we have that $R_O(\theta)$ is a translation. But this is a contradiction because no rotation can also be a translation unless it is the identity. This shows that $T \circ R_O(\theta)$ is a rotation (around some point different from the origin unless T is the trivial translation). ■

EXERCISES

Marks challenging exercises.

CIRCLES AND ISOMETRIES

1. Prove that if two points F_1 and F_2 are both equidistant from the end of a segment $\overline{PQ}$, then lines $\overleftrightarrow{PQ}$ and $\overleftrightarrow{F_1F_2}$ are perpendicular. Furthermore, $\overleftrightarrow{F_1F_2}$ cuts $\overleftrightarrow{PQ}$ at a point A, which is the midpoint of $\overline{PQ}$ (Figure 4.20).

2. Prove that two circles cannot intersect more than twice.

3. Prove that if three points F_1, F_2, F_3 are all equidistant from the ends of the segment PQ, then F_1, F_2, F_3 all lie on the same line. (*Hint:* Use the previous exercise.)

4. Suppose there are three circles, and two points A and B with the property that each of the three circles passes through both A and B. Prove that the three centers are collinear. (*Hint:* Use a previous exercise.)

ISOMETRIES AND THEIR FIXED POINTS

5. Answer true or false for the following statements. Explain your answer.

(a) An isometry that fixes infinitely many points must be I.

(b) Every rotation has precisely one fixed point.

(c) Suppose M_1, M_2, M_3, M_4 are reflections, and consider $M_4 \circ M_3 \circ M_2 \circ M_1$. Because of Theorem 4.16, we could get the exact same isometry as this composition by removing one of the M_i from the formula.

6. Let F_1, F_2, F_3 be three points on the same line. Describe two isometries which are not the identity and which fix these three points.

7. Suppose S is an isometry with a fixed line L [for each P on L, $S(P)$ is also on L]. Name as many specific types of isometries as you can that have this property.

8. In the proof of Theorem 4.18, we ruled out $M_{L'} \circ S = I$. Why?

9. If S is a translation by two units along the x axis and you create $M_{L''} \circ S$ as in the proof of Theorem 4.19, what is the nature of $M_{L''} \circ S$? How many fixed points does it have?

10. Suppose S is a glide relection with the x axis as the reflection axis and the translation amount being two units in the positive x direction.

    (a) Let F_1 be any point. Show that the point where $\overline{F_1S(F_1)}$ crosses the x axis is the midpoint of $\overline{F_1S(F_1)}$.

    (b) If you create L'' as in the proof of Theorem 4.19, can you identify any particular point that lies on this line?

Isometries and Sense

11. (a) If P is rotated around the origin O by 30° to P', what is the sense of $[POP']$? $[OPP']$? $[P'OP]$?

    (b) Answer part (a) if the angle is 210°.

    (c) Answer part (a) if the angle is -90°.

12. If $S = U \circ V \circ W$ and S preserves sense, what can you say about how U, V, and W behave with respect to sense? What if S reverses sense?

13. Suppose S is an isometry and A and B are points. Suppose we know the positions of $S(A)$ and $S(B)$ but we have no information about other points or about whether S preserves sense. How many possibilities are there for S?

14. If U and V are isometries and $V^{-1} \circ U$ fixes two points A and B but is not the identity, what can you say about U and V?

15. Let A be a point on line L, and S an isometry that preserves sense. If $S(A)$ and the image of L are both known, does this determine S? If so, prove it. If not, how many possibilities are there for S?

16. Is there a sense-reversing isometry that has no fixed points? If so, give an example. If not, explain why not.

17. Is there a sense-reversing isometry that fixes more than one point? If so, give an example. If not, explain why not.

18. Is there a sense-preserving isometry, other than the identity, that fixes exactly two points? If so, give an example. If not, explain why not.

19. Prove that a sense-preserving isometry can always be written as a product of two reflections.

20. Prove that a sense-reversing isometry is either a reflection or can be written as a product of three reflections.

21. (a) If R designates a rotation and T is a translation, and $S = R \circ T$, explain why S is a rotation.
    (b) Let C be the center of this rotation and let A, B be any two points and A', B' the images. If the positions of A, B, A', and B' are known, how can we find the position of C? (*Hint:* See Exercise 1.)

Miscellaneous Exercises

*22. Let L, L', L'' be the angle bisectors of triangle $AA'A''$, with L bisecting $\angle A$, L' bisecting $\angle A'$, and L'' bisecting $\angle A''$.
    (a) Show that $S = M_{L''} \circ M_{L'} \circ M_L$ is a reflection. (You need to remember a remarkable fact about the three angle bisectors.)
    (b) Show that the line $\overleftrightarrow{AA''}$ is fixed under S. (This means each point of that line maps to a point of the same line, but not necessarily to the same point.)
    (c) Show that S is a reflection in a line perpendicular to a side of the triangle.

23. Suppose S is an isometry and A, B, C are noncollinear points. Suppose we know the positions of $S(A)$ and $S(B)$ but we have no information about the position of $S(C)$.
    (a) How many possibilities are there for $S(C)$?
    (b) If we also know whether S preserves or switches sense, how many possibilities are there for $S(C)$?

24. Prove that there is no series of translations and rotations, in any order or any number, which composes to a reflection.

25. Suppose A', B', C' result from A, B, C by an isometry. It is possible that the isometry is *not* a translation but we have that
    (a) the line $\overleftrightarrow{A'B'}$ is parallel to the line $\overleftrightarrow{AB}$,
    (b) the line $\overleftrightarrow{A'C'}$ is parallel to the line $\overleftrightarrow{AC}$, and
    (c) the line $\overleftrightarrow{B'C'}$ is parallel to the line $\overleftrightarrow{BC}$.

    Show an example of this and specify the isometry involved.

26. In the exercises of Section 4.2 of this chapter, you were asked to find one or more isometries where $S \circ S \circ S = I$. Can you find all possible isometries S that satisfy this equation?

27. Suppose S is a glide relection with the x axis as the reflection axis and the translation amount being two units in the positive direction. If you create $M_{L''} \circ S$ as in the proof of Theorem 4.19, can you say what kind of isometry this is? (*Hint:* Do Exercise 21 first.)

4.4 Symmetries of Patterns

Our goal in this section is to define and illustrate a method for characterizing the kind of symmetry a pattern has. To make things easy, we will stick to patterns like the one in Figure 4.24a (see also Figure 4.1), which are called *strip patterns*. A strip pattern lies between two parallel horizontal lines and repeats itself if we move along just the right amount from left to right (a more precise definition is coming up). Strip patterns are used for many kinds of ornamentation: borders for wallpaper, on pottery, etc. A strip pattern could have many colors, but we restrict ourselves to patterns in black and white. An actual strip pattern has a left and right end, or it might be a circle going around a jug or bowl, but we find it convenient for mathematical purposes to imagine that our strip pattern goes on indefinitely to the right and left.

Strip patterns are not the only kinds of patterns in which we can study symmetry. Figure 4.25 shows *wallpaper patterns*, patterns which extend indefinitely up and down as well as right and left. It is also possible to look at the repeating arrangement of molecules in a crystal and think of it as a three-dimensional pattern (Figure 4.26). The study of the symmetries of such three-dimensional patterns is of major concern to crystallographers. The material in this chapter is a good warmup for the mathematical study of crystals.

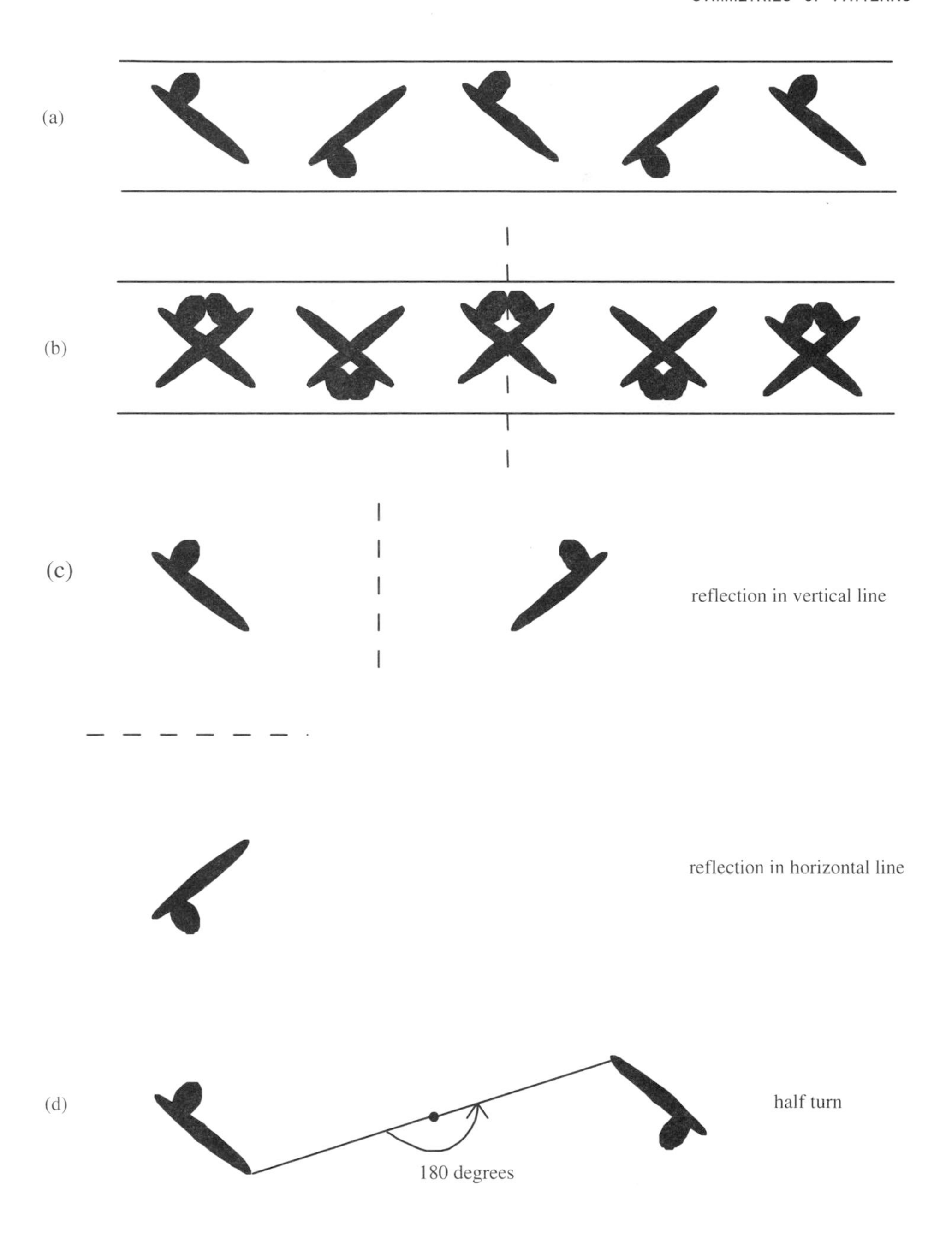

Figure 4.24 (a) A strip pattern. (b) Vertical reflection of strip pattern overlaid on original. (c) Vertical and horizontal reflections of a single motif. (d) Half-turn applied to a single motif.

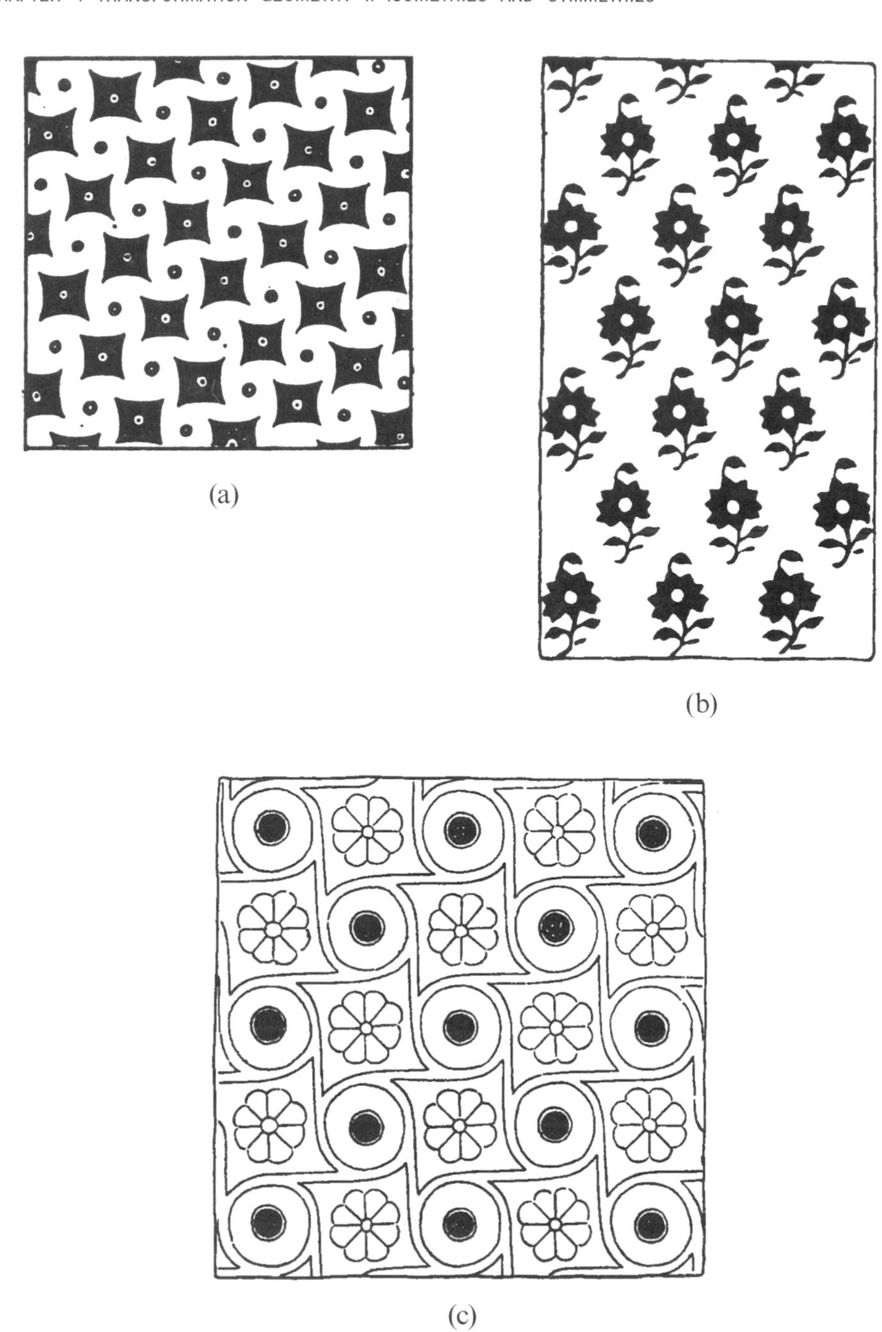

(a)

(b)

(c)

Figure 4.25 Three wallpaper patterns. Courtesy of Dover Publications.

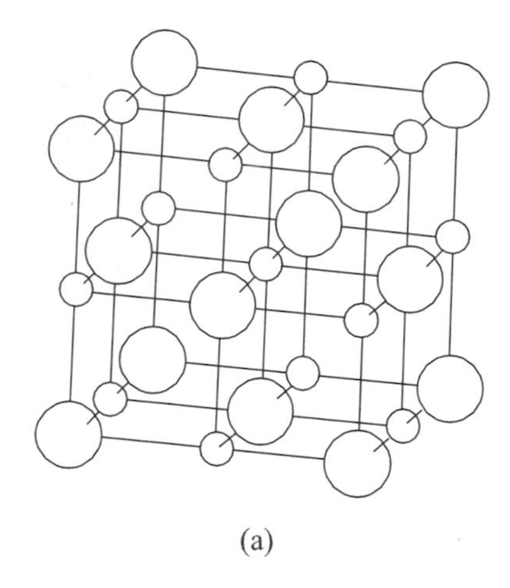

(a)

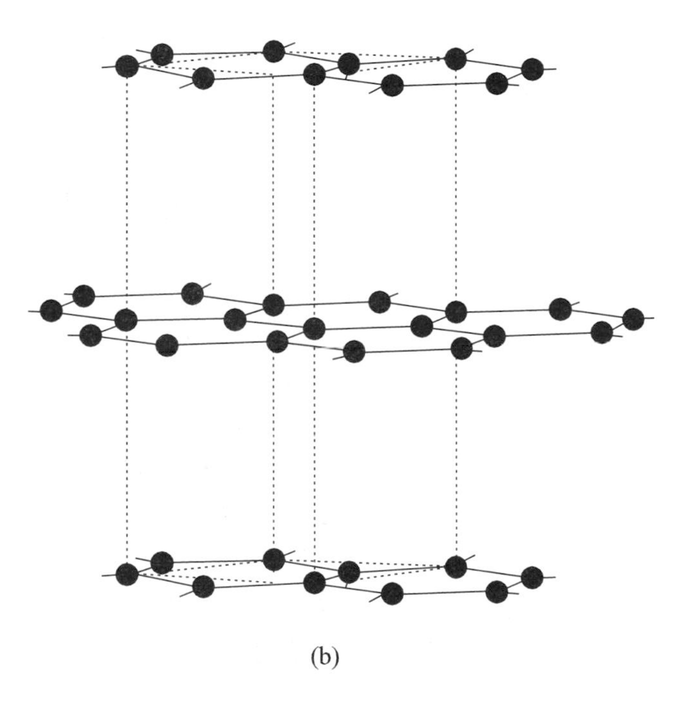

(b)

Figure 4.26 (a) A salt crystal (large spheres are chlorine and small ones are sodium). (b) A graphite crystal — carbon atoms arranged hexagonally in layers.

DEFINITION

A *symmetry* of a pattern is an isometry of the plane which carries the pattern onto itself. More precisely, S is a symmetry of a given pattern if S is an isometry and if, for every black point B in the pattern, $S(B)$ is also a black point, while for every white point W, $S(W)$ is a white point.

In the top strip pattern in Figure 4.1 in Section 4.1 of this chapter, if we translate the pattern either to the left or right, by the distance from one right-side-up bird's eye to the next, the pattern falls on top of itself exactly. Compare this to the operation of moving the pattern over by half that distance. Now the position that used to be occupied by a right-side-up bird is occupied by an upside-down bird. Thus, this is not a symmetry of the strip pattern. The middle and bottom strip patterns in Figure 4.1 also have translations as symmetries. In fact, this is precisely what we meant earlier when we said that a strip pattern "repeats itself as we move along just the right amount from left to right." Our more precise way of thinking allows us to give a proper mathematical definition of a strip pattern.

DEFINITION

A *strip pattern* is a black-and-white pattern that lies between two parallel lines and which has translational symmetries in the direction of the parallel lines. We always draw the parallel border lines of a strip pattern horizontally, although that is not strictly required by the definition.

Notice that, in our definition of strip pattern, we use the plural *symmetries*. If a strip pattern can be translated by a certain amount to fall on top of itself, then we can do the same by twice the amount, three times the amount, etc. Likewise, we can translate by these amounts in the other direction. A strip pattern may have symmetries other than the translational ones (but it is not required to in order to qualify as a strip pattern). For example, the middle pattern of Figure 4.1 has reflection in a variety of vertical axes as symmetries. For the top pattern of Figure 4.1, glide reflection in a horizontal axis running through the middle of the strip is a symmetry.

With this definition of symmetry in hand, we can make a symmetry checklist for a pattern. The checklist should include each possible type of isometry that a strip pattern could have. (See the end of the next example and Figure 4.28, shown later.) For each isometry, we check off whether the pattern we are studying has it as a symmetry or not. The main classification theorem of the previous section assures us that the only things we need to put on our checklist in order to get a complete profile of a pattern are translation, rotation, reflection, and glide reflection.

This is illustrated for Figure 4.24a in the next example. When you examine the checklist in the example, you will notice that some kinds of isometries are not listed. For example, reflections in diagonal lines are not present because performing a diagonal reflection results in a pattern that is no longer horizontal and couldn't possibly coincide with the original horizontal version of the strip pattern. The need to keep the strip pattern horizontal also accounts for the fact that the only kind of rotation is one through 180° (the so-called "half-turn"). Furthermore, the only possible location for the center of a half-turn is on the midline (the horizontal line halfway between the borders), because if the center were under the midline, the horizontal border lines of the pattern would not turn onto one another. Finally, the only glide reflections we need to look for are ones involving the midline.

EXAMPLE 4.12: Searching for Symmetries

Figure 4.24a shows a strip pattern for which we want the symmetry checklist. In doing this, it is helpful to concentrate on the basic unit that repeats itself and which we call a *motif*. In this case, the motif consists of two elliptical blobs stuck together, making a figure that looks a bit like a bat hitting a ball.

TRANSLATION: The fact that we are told that this is a strip pattern means that there are translational symmetries, and these are easy enough to see. If you move each motif to the right by two motifs, the pattern coincides with itself. So we enter "yes" in the checklist under translation.

REFLECTION IN A VERTICAL LINE: Figure 4.24b shows that if we reflect in the dotted line the image pattern does not coincide with the original. But what if we picked a different vertical line? Drawing other pictures like Figure 4.24b — or even imagining them — is more trouble than we need to go to. Just look at a single copy of the repeated two-ellipse motif (upper left of Figure 4.24c). No matter what vertical line we reflect in, that motif winds up looking like the one at the upper right of Figure 4.24c. There is no version of the motif like this (with the small ellipse pointing off to the upper left) in Figure 4.24a, so no reflection in a vertical line leaves the pattern invariant. In the checklist, we leave a blank under reflection in a vertical line (Vertical Flip).

REFLECTION IN THE HORIZONTAL MIDLINE: If we apply a reflection in the horizontal midline to the leftmost motif in Figure 4.24a, we get a motif that overlaps the original, which makes for a confusing picture. There is no harm in making viewing easier by using a lower mirror line. This is shown in Figure 4.24c. The reflected motif does appear in the original pattern, but farther along to the right. We conclude that reflection in the horizontal midline will not leave the strip pattern invariant, but reflection followed by translation, namely, a glide reflection, will do so.

GLIDE REFLECTION: We have already observed, during our discussion of horizontal reflection, that there is a glide reflection that keeps the pattern invariant.

HALF-TURN: Applying a half-turn to the basic motif is shown in Figure 4.24d. Regardless of where we choose the center of rotation, the resulting turned motif, with its small ellipse pointing to the lower left, does not appear anywhere in the original strip pattern of Figure 4.24a. Thus, a half-turn is not one of the symmetries of the original strip pattern.

The checklist that results is:

Pattern	Translation	Vertical Flip	Hor. Flip	Glide Reflection	Half-Turn
Figure 4.24a	Yes			Yes	

■

DEFINITION

Two patterns have the same symmetry if they have exactly the same set of symmetries. In practice, you need to make a checklist for each pattern to determine if they have the same symmetry.

For example, the pattern of Figure 4.24a has the same symmetry as the second pattern of Figure 4.28, shown later.

EXAMPLE 4.13: Searching for Symmetries Again

We carry out a search for symmetries for Figure 4.27, this time finding some symmetries we did not find in our previous example.

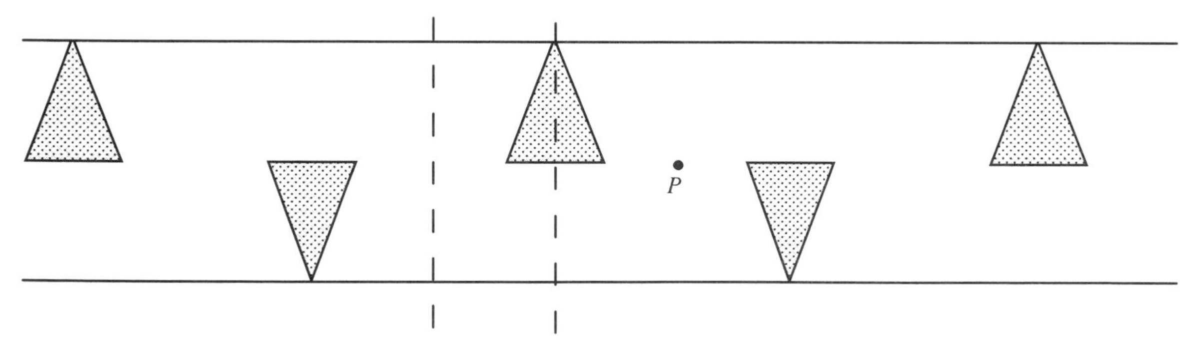

Figure 4.27 A strip pattern symmetric under half-turn and reflection in a vertical axis.

1. As usual, we see translational symmetry.

2. In searching for a vertical line of reflectional symmetry, suppose we tried to get the first triangle on the left to reflect onto the rightmost triangle. This would require the dotted reflection line through the top vertex of the middle triangle. This would also make the second triangle fall on top of the fourth. We can presume that the triangles not shown because they are "off the diagram" to the right or left will also reflect properly onto one another. Thus, we have found a vertical reflection line of symmetry. Are there other such vertical lines of symmetry? (How about the dotted line on the left?)

3. There is no horizontal mirror line for a reflectional symmetry.

4. There is a glide reflection symmetry.

5. In looking to see if there is a half-turn symmetry, we need to come up with some logical candidates for the center of rotation. As we have earlier observed, such a center (if it exists) would have to be on the horizontal center line of the strip, but that still leaves a lot of possible places. We know that a rotation by 180° turns a triangle upside down, so we might hope to move the center triangle onto the fourth one. To accomplish this, we could rotate around the point P midway between the triangles. We notice that under this rotation, the second triangle moves onto the fifth. The first triangle can be presumed to fall on top of a triangle off the diagram to the right. Likewise, triangles not shown are assumed to rotate onto other triangles not shown. Thus, we have found a center for a half-turn that keeps the pattern invariant. Are there any other points we could use to rotate around?

Pattern	Translation	Vertical Flip	Hor. Flip	Glide Reflection	Half-Turn
Figure 4.27	Yes	Yes		Yes	Yes

This symmetry checklist is the same as the one for the sixth pattern in Figure 4.28. Consequently, we regard these patterns as having the same symmetry. ■

Figure 4.28 shows checklists in which we record the symmetries for a number of patterns found in the San Ildefonso pueblo, an Indian settlement in the southwestern United States.

Anthropologists find it interesting to compare the pottery and weaving patterns of vanished peoples as well as modern people such as those of the San Ildefonso pueblo. In the case of a vanished culture, interesting conclusions can be drawn from symmetry

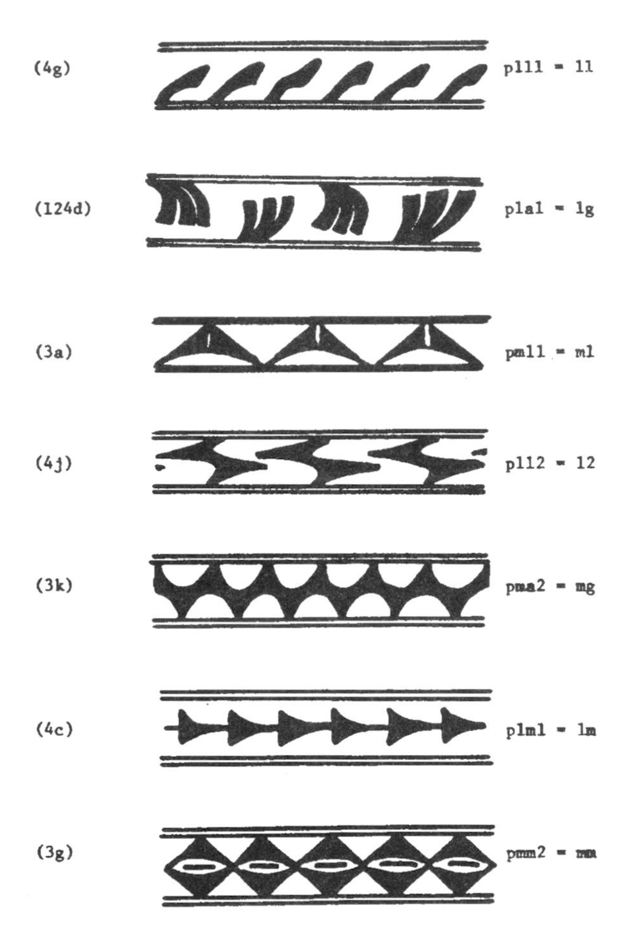

Table 1. Examples of the 7 monochromatic strip patterns. The numbers in parentheses are the Plate numbers from Chapman [1]. Both the standard crystallographic notation and the abbreviation of Senechal and Schattschneider are shown.

Figure 4.28 Strip patterns from the San Ildefonso pueblo. Courtesy of Hadronic Press.

analysis. For example, if the styles and symmetries found in two nearby settlements are similar, this is a clue that the people who occupied these different sites might have had a lot of communication with one another.

At this point we have an important question that will be answered in the next section. There are sequences of "yeses" and blanks (which mean "no") which are not shown in Figure 4.28; for example yes, blank, yes, blank, yes. Why is that? Is it because the San Ildefonso potters are ignorant of some kinds of strip patterns? Or do they find them distasteful? Or is there a mathematical reason why no strip pattern with that sequence of "yeses" and blanks can possibly exist?

EXERCISES

**Marks challenging exercises*

1. Make a checklist like the one in Figure 4.28 for the seven patterns shown in Figure 4.29.

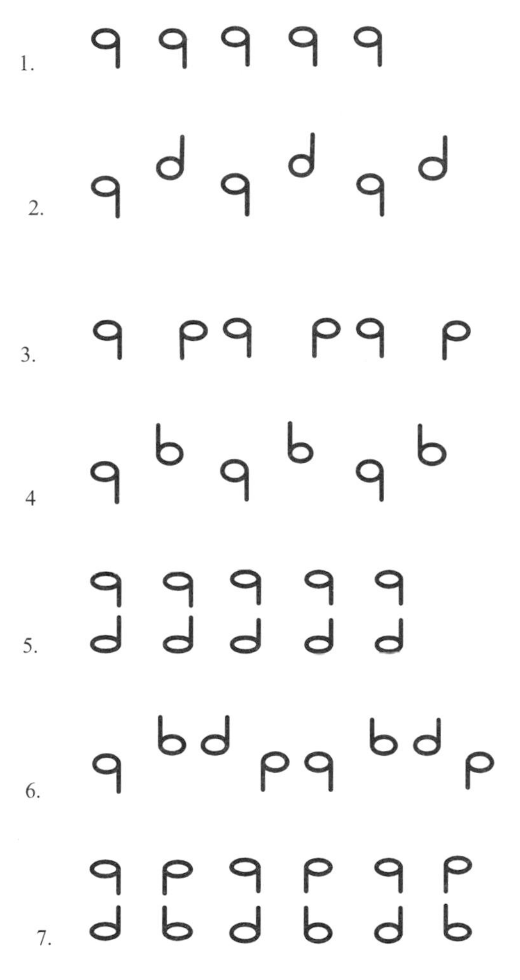

Figure 4.29 What are the symmetries of these seven strip patterns?

2. Verify the checklist entries for Figure 4.28. Each time an isometry is not a symmetry (blank in the table) explain why that isometry does not leave the pattern invariant.

*3. In our definition of symmetry, we required that the image of every white point be white. What if we leave off that requirement, but keep the requirement that every black point has a black image. Is the shorter definition equivalent? Can you find a pattern that qualifies under the shorter definition but not under the longer one?

DEFINITION

A *wallpaper pattern* is one for which there are two directions (making an angle different from 180°) in which the pattern can be translated by certain amounts and coincide with itself.

4. (a) Does Figure 4.25a have reflection in a diagonal line as one of its symmetries? (Keep in mind that the pattern goes on "to infinity.") If so, give an example of such a diagonal line.
   (b) Answer part (a) for Figure 4.25b.
   (c) Answer part (a) for Figure 4.25c.

5. Draw a wallpaper pattern that has among its symmetries rotation by 60°. To do this exercise, you must show a point about which you can rotate by 60° and have the new pattern be right on top of the old one. It can be tricky to verify whether an answer is correct — the eye is sometimes fooled. Try making two exact copies of your pattern, one of them on see-through paper or a transparency. Put one on top of the other, stick a pin through the center of both, and turn the top sheet by 60°.

6. Which of the wallpaper patterns in Figure 4.25 is symmetric by rotation through 90° around some point? Identify a point of rotation when such a symmetry exists.

4.5 What Combinations of Symmetries Can Strip Patterns Have?

From Section 4.3 we know that there are only four kinds of isometries (ignoring the identity): translations, rotations, reflections, and glide reflections. Each of these isometries can be found as the symmetry of some strip pattern in Section 4.4. Our interest in this section, however, is to see what *combinations* of these symmetries various

TABLE 4.1 HYPOTHETICAL AND ACTUAL SYMMETRY COMBINATIONS FOR STRIP PATTERNS

Symmetry Combination	Reflection in Vertical Line (M_V)	Reflection in Horizontal Line (M_H)	Glide Reflection G	Half-Turn R	Pattern Found in San Ildefonso?
$\mathcal{A}$	yes	yes	yes	yes	#7
$\mathcal{B}$	yes	yes	yes		
$\mathcal{C}$	yes	yes		yes	
$\mathcal{D}$	yes	yes			
$\mathcal{E}$	yes		yes	yes	#6
$\mathcal{F}$	yes		yes		
$\mathcal{G}$	yes			yes	
$\mathcal{H}$	yes				#3
$\mathcal{I}$		yes	yes	yes	
$\mathcal{J}$		yes	yes		#5
$\mathcal{K}$		yes		yes	
$\mathcal{L}$		yes			
$\mathcal{M}$			yes	yes	
$\mathcal{N}$			yes		#2
$\mathcal{O}$				yes	#4
$\mathcal{P}$					#1

strip patterns possess. For example, none of the strip patterns from the San Ildefonso set (Figure 4.28) has this combination of symmetries: a translation, both kinds of reflections (flips), but neither glide reflection nor a half-turn. Table 4.1 tabulates what combinations are present and which are missing from the San Ildefonso collection. (There is no column for translation. Every strip pattern has translational symmetry, so we leave it out to save space.) The rows of this table show every conceivable combination of "yes" and blank "no". There are 16 rows (possible combinations of symmetries) and only 7 are displayed by the San Ildefonso pottery.

One might suppose that the reason for some combinations being missing is that the San Ildefonso potters are unable to create them, or because they don't like what they look like, or perhaps they left them out by accident. But if we were to examine strip patterns from other cultures, we would discover that no culture has ever produced a combination of symmetries that has not been produced by the San Ildefonso potters. In fact, there is a mathematical reason why the missing combinations are never found: They are mathematically inconceivable. In other words, the San Ildefonso potters are as sophisticated as they can be as far as creating symmetric patterns go.

In the remainder of this section, we demonstrate why some of the symmetry combinations not found in the San Ildefonso collection are impossible. Those impossible combinations that we do not discuss are left for you as exercises. In reading over the proofs, notice that they all follow the same plan and are based partly on the following theorem.

✯ THEOREM 4.22: COMPOSITION OF SYMMETRIES

If U and V are symmetries of a strip pattern, then so is $V \circ U$.

PROOF

We just need to show that $V \circ U$ is an isometry and that it moves the pattern onto itself. Theorem 4.7 of Section 4.2 assures us it is an isometry. Because U and V separately move the pattern onto itself, doing U then V must do so also. ■

A typical impossibility proof will proceed by contradiction, assuming there is a strip pattern with the given combination of symmetries. Then we compose some of the symmetries present in that hypothetical combination and obtain a symmetry that is not present in the combination. The heart of the proof, in each case, is the composition of two isometries to produce a third. You might like to do some exploratory investigations of these compositions by filling in the blanks in Eqs. (4.4)–(4.8). In each case, the answer is either a translation, rotation, reflection, or glide reflection. In these equations we use the following notations:

M_H = reflection in the horizontal midline of the pattern
M_V = reflection in some vertical axis
R = rotation through 180° (half-turn)
G = glide reflection in the horizontal midline
T = translation in the horizontal direction

$$M_V \circ M_H = ?. \tag{4.4}$$

$$G \circ M_V = ?. \tag{4.5}$$

$$T \circ M_V = ?. \tag{4.6}$$

$$M_H \circ R = ?. \tag{4.7}$$

$$G \circ R = ?. \tag{4.8}$$

In working out the answers, you might find it helpful to take some asymmetrical motif, like the letter Q, and subject it to the given composition, as illustrated in the next example.

EXAMPLE 4.14

What is $T \circ M_V$ applied to the motif consisting of the letter Q? What single isometry would have the same effect? See Figure 4.30.

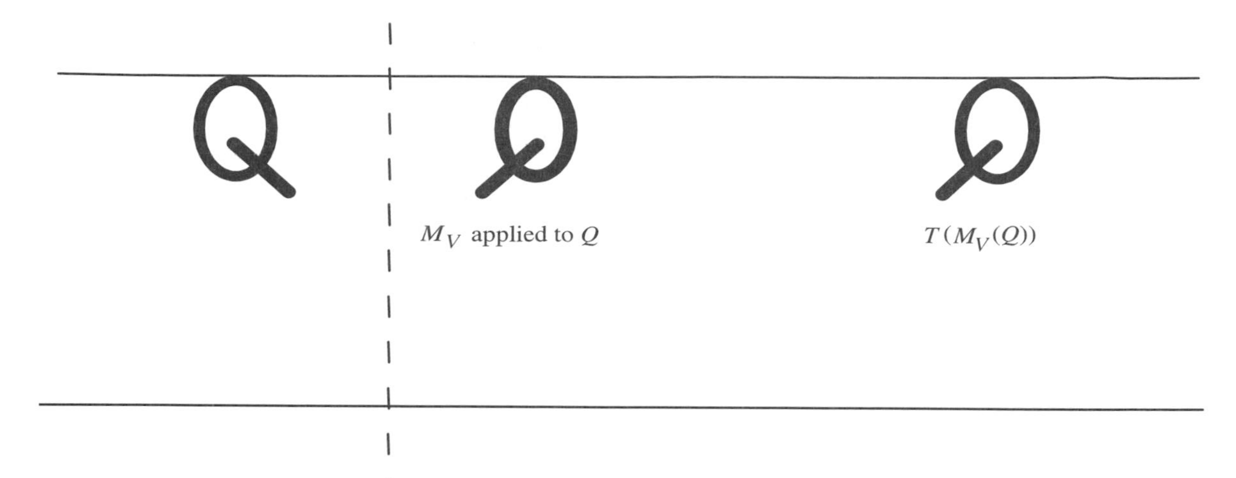

Figure 4.30 What single isometry moves the first motif to the third?

SOLUTION

What moves the first motif onto the third? Translation wouldn't since translation would keep the tail of the Q on the right. A half-turn wouldn't, because a half-turn turns the Q upside down. Glide reflection wouldn't work either. (Do you see why?) However, a reflection in a certain vertical line would move the first motif onto the third. Can you sketch the approximate position of the mirror line?

This solution "by picture" is fairly convincing, but purists are often suspicious of proofs that rely heavily on pictures because it can happen that the particular picture drawn is misleading. Can you see how a person might be misled if we had used a letter O instead of Q? When we give formal proofs below we will work out equations like $T \circ M_V = ?$ without pictures simply so you can see both approaches. ■

✯ THEOREM 4.23

No strip pattern has symmetry combination $\mathcal{B}$.

PROOF

A pattern of type $\mathcal{B}$ has reflections in horizontal and vertical lines M_H and M_V. We now compose them, expecting to obtain a symmetry of the pattern because of Theorem 4.22. $M_V \circ M_H = R(180^\circ)$ and so the strip pattern would have to have a "yes" in the column for half-turns. It does not, so this is a contradiction and symmetry combination $\mathcal{B}$ cannot exist. ■

✯ THEOREM 4.24

No conceivable strip pattern has symmetry combination $\mathcal{F}$.

PROOF

A pattern of type $\mathcal{F}$ has a reflection in a vertical line M_V and a glide reflection G. We now compose them, expecting to obtain a symmetry because of Theorem 4.22. In doing this we write the glide reflection as $G = T \circ M_H$ (the parts of a glide reflection commute):

$$\begin{aligned} G \circ M_V &= (T \circ M_H) \circ M_V \\ &= T \circ (M_H \circ M_V) \\ &= T \circ R. \end{aligned}$$

We have seen in Example 4.11 of Section 4.3 that a rotation around the origin followed by a translation is a rotation around some new center of rotation. The only rotations that keep the strip horizontal are half-turns about a point on the midline of the strip. Thus $G \circ M_V$ is a half-turn and there should be a "yes" in the half-turn column for symmetry combination $\mathcal{F}$. Since there isn't, we have our contradiction and symmetry combination $\mathcal{F}$ cannot exist. ■

✯ THEOREM 4.25

No strip pattern has symmetry combination $\mathcal{K}$.

PROOF

A pattern of type $\mathcal{K}$ has a reflection in a horizontal line M_H and a half-turn R. We now compose them, expecting to obtain a symmetry because of Theorem 4.22.

$$\begin{aligned} M_H \circ R &= M_H \circ (M_H \circ M_V) \\ &= (M_H \circ M_H) \circ M_V \\ &= I \circ M_V \\ &= M_V. \end{aligned}$$

Thus the strip pattern would have to have a "yes" in the column for reflection in a vertical line. It does not, so this is a contradiction and symmetry combination $\mathcal{K}$ cannot exist. ■

To analyze symmetry combination $\mathcal{M}$ we find the next result helpful. This lemma is just a rigorous version of Example 4.14, so consult Figure 4.30 as you read the proof.

LEMMA 4.26

The composition of reflection in vertical line L and a horizontal translation is a reflection in another vertical line L'.

PROOF

Let M_L be the reflection and T the translation. Let d be the distance of the translation. Let L' be a line obtained by moving L by $d/2$ to the right. By Theorem 4.10 of Section 4.2, $M_{L'} \circ M_L$ is a translation by $2(d/2) = d$, which means $M_{L'} \circ M_L = T$. Thus,

$$\begin{aligned} T \circ M_L &= (M_{L'} \circ M_L) \circ M_L \\ &= M_{L'} \circ (M_L \circ M_L) \\ &= M_{L'} \circ I \\ &= M_{L'}. \quad \blacksquare \end{aligned}$$

✯ THEOREM 4.27

No strip pattern has symmetry combination $\mathcal{M}$.

PROOF

If a strip pattern has symmetry pattern $\mathcal{M}$, then it has a half-turn R, which we can write as $R = M_H \circ M_V$. It also has a glide reflection G, which can be written as $G = T \circ M_H$ where T is a horizontal translation. We now compose R and G, expecting to obtain a symmetry because of Theorem 4.22.

$$\begin{aligned} G \circ R &= (T \circ M_H) \circ (M_H \circ M_V) \\ &= T \circ (M_H \circ M_H) \circ M_V \\ &= T \circ I \circ M_V \\ &= T \circ M_V. \end{aligned}$$

By the previous lemma, this is a reflection in a vertical line. Thus, the strip pattern would have to have a "yes" in the column for reflection in a vertical line. It does not, so this is a contradiction and symmetry combination $\mathcal{M}$ cannot exist. ■

EXERCISES

**Marks challenging exercises*

1. Create a "proof by picture," as in Example 4.14, to fill in the blank of Eq. (4.4).
2. Create a "proof by picture," as in Example 4.14, to fill in the blank of Eq. (4.5).
3. Create a "proof by picture," as in Example 4.14, to fill in the blank of Eq. (4.7).
4. Create a "proof by picture," as in Example 4.14, to fill in the blank of Eq. (4.8).
5. Why might a person be misled if the letter O were used in Example 4.14?
6. (a) Prove that no strip pattern has symmetry combination $\mathcal{C}$ from Table 4.1. (*Hint*: Show that the symmetries making up combination $\mathcal{C}$ imply that there is a glide reflection symmetry, contrary to the blank in that column.)
   (b) There is one more symmetry combination you can rule out by your proof in part (a). What is it?
7. Prove that no strip pattern has symmetry combination $\mathcal{G}$ from Table 4.1. (*Hint*: Express R in terms of reflections and determine the composition of M_V followed by R.)
8. Prove that no strip pattern has symmetry combination $\mathcal{I}$ from Table 4.1.
9. Prove that no strip pattern has symmetry combination $\mathcal{L}$ from Table 4.1.

Chapter 5

Vectors in Geometry

Prerequisites: We assume the reader is familiar with vectors to the extent commonly covered in a semester of multivariable calculus: addition, scalar multiplication, scalar products, cross-products and norms.

Ever since the development of coordinate geometry by Descartes (1596–1650), algebra and geometry have been inseparable companions. In this chapter we demonstrate the power of algebraic methods to solve geometric problems. In our use of algebra, we go Descartes one step better by making use of vectors, which hadn't been invented in his time. The essential idea here is to pack as much meaning as possible into as few symbols as possible.

Applications include topics in computer graphics, industrial design, robotics, and the global positioning system (GPS). The geometric ideas underlying these topics are diverse: lines, planes, spheres, curves. The "story line" connecting everything is the fact that our concise form of algebra, using vectors, is so widely useful.

Vectors are used again to a very mild extent in Chapter 6 and then intensively again in Sections 7.3, 7.4, and 7.5 of Chapter 7. Those sections of Chapter 7 can be regarded as a natural continuation of Sections 5.1, 5.2, and 5.3 of this chapter.

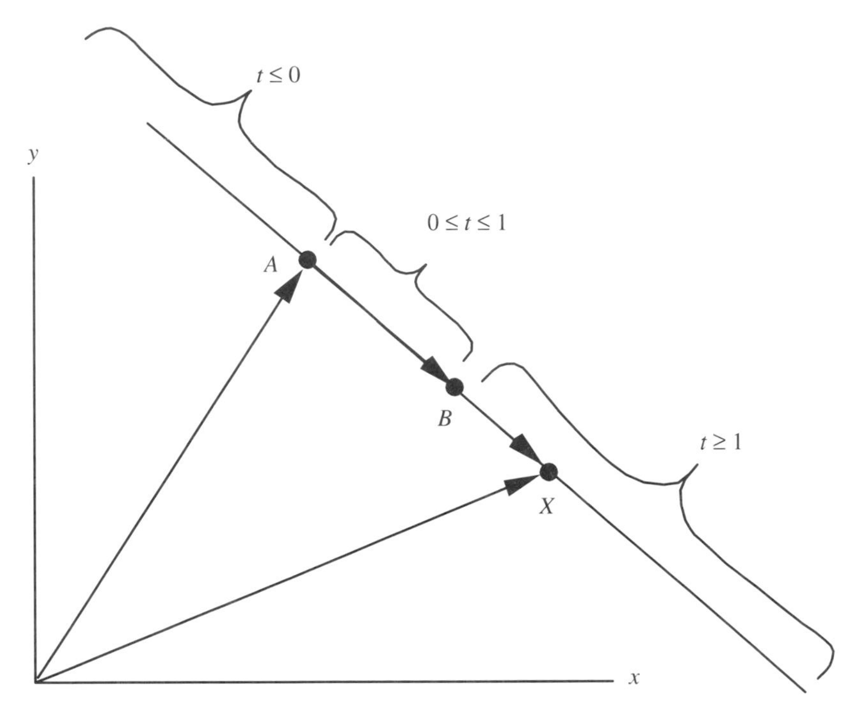

Figure 5.1 A parametric view of a straight line.

5.1 Parametric Equations of Lines

The simplest algebraic description of a straight line uses the equation $y = mx + b$ where b is the y intercept and m is the slope. This is fine for lines in the plane, but no such equation can be developed for lines in three-dimensional space. In this section, we overcome this shortcoming by developing *parametric equations*, which can be used for lines in three dimensions as well as two. The parametric equations have the further advantage that they allow a simple description of the line *segment* connecting points A and B. These ideas are applied to problems in robotics, computer graphics, and physics.

Basic Ideas

Our line equation is based on vectors (Figure 5.1). Recall that we draw a vector as a directed line segment (using an arrow) stretching from a point called the *tail* to a point called the *head*. Two such segments are considered to represent the same vector if they have the same direction and the same length. They need not have the same tail. When we look at a directed line segment we are actually looking at just one representative of a vector, although this distinction often gets lost when we refer to the directed line segment loosely as a vector.

If a vector has its tail at $A=(x_A, y_A)$ and its head at $B=(x_B, y_B)$ then the vector from A to B can be represented with numerical components as $\langle x_B - x_A, y_B - y_A \rangle$. If A and B are three-dimensional points, then we would have $\langle x_B - x_A, y_B - y_A, z_B - z_A \rangle$. Vectors will be denoted by lowercase, boldface letters such as **a**, **p**, **u**, **x**; for example, $\mathbf{p} = \langle 2, -3, 4 \rangle$. Boldfacing is awkward for work on a chalkboard or on paper, so you might use $\overrightarrow{p}$ in place of **p**, for example.

Vectors can be added, subtracted, and multiplied by scalars. For example, if $\mathbf{u} = \langle 1, 5, -3 \rangle$ and $\mathbf{v} = \langle -6, 2, -1 \rangle$, $\mathbf{u}+\mathbf{v} = \langle -5, 7, -4 \rangle$ and $\mathbf{u}-\mathbf{v} = \langle 7, 3, -2 \rangle$, $4\mathbf{u} = \langle 4, 20, -12 \rangle$ and $-2\mathbf{u} = \langle -2, -10, 6 \rangle$.

We rely here on a variety of facts about vectors that are usually introduced in the study of multivariable calculus. Three of these facts concern the geometric meaning of equality, addition, and scalar multiplication of vectors:

1. If **u** is the vector from B to A and **v** is the vector from D to C, then **u** and **v** have the same components if and only if the directed segment from B to A has the same length and direction as the directed segment from D to C.

2. If **u** and **v** are vectors, we can find $\mathbf{u}+\mathbf{v}$ as follows: Select a directed line segment representing **u** and then place a representative of **v** so its tail is at the head of **u**. The vector from the tail of **u** to the head of **v** is $\mathbf{u}+\mathbf{v}$.

3. If **u** is a vector and r a scalar (an ordinary number), then (a) if $r>0$, $r\mathbf{u}$ points in the same direction as **u** and its length is r times as long, and (b) if $r<0$, $r\mathbf{u}$ points in the opposite direction to **u** but its length is $|r|$ times as long,

The vector from the origin to a point P is called the *position vector* for P and we denote it **p**. In general, we denote a position vector by the lowercase, bold version of the capital letter used for the point. If P has coordinates (x_P, y_P, z_P) then the position vector $\mathbf{p} = \langle x_P, y_P, z_P \rangle$. As you can see, there is very little difference between points and position vectors, so you might wonder why we don't use the same notation P to stand for both, or why we introduce vectors at all. One reason is that a vector can be re-positioned to have its tail at any point we please. This is important in establishing the theory of vector calculations and in some applications. However, in our work, to avoid confusion, always visualize a position vector as having its tail at the origin.

Suppose we are given the coordinates of two points A and B and a third point X is given and we want to know if it is on the line determined by A and B. The points could be in two-dimensional space or three-dimensional space — our solution method will be the same. For definiteness, let's say we are in three dimensions so $A=(x_A, y_A, z_A)$ and $B=(x_B, y_B, z_B)$, $X=(x, y, z)$. X is on line $\overleftrightarrow{AB}$ (Figure 5.1) if and only if vector $\mathbf{x}-\mathbf{a}$ points in the same or opposite direction as $\mathbf{b}-\mathbf{a}$. This means

$$\mathbf{x}-\mathbf{a}=t(\mathbf{b}-\mathbf{a}) \quad \text{for some scalar } t. \tag{5.1}$$

(In Figure 5.1, $t=1.5$.) Equation (5.1) yields

$$\mathbf{x}=\mathbf{a}+t(\mathbf{b}-\mathbf{a}). \tag{5.2}$$

This can be thought of as a set of equations, one for each coordinate of the space we are in. For example, in three dimensions we have

$$\begin{aligned} x &= x_A + t(x_B - x_A), \\ y &= y_A + t(y_B - y_A), \\ z &= z_A + t(z_B - z_A). \end{aligned} \tag{5.3}$$

In two-dimensional space we just have the first two of these equations.

Equation (5.2) is perhaps the easiest way to remember the parametric equations, but when we use them we often need (Eq. 5.3) or the two-dimensional version of Eq. (5.3).

EXAMPLE 5.1

(a) Find the parametric equations of the line connecting the points $A(2, 1, -4)$ and $B(5, 0, 2)$ in three-dimensional space.

(b) Do the same for the points $A(0, a)$ and $B(b, 0)$ in two-dimensional space.

SOLUTION

(a) $\mathbf{b}-\mathbf{a}=\langle 5, 0, 2\rangle-\langle 2, 1, -4\rangle=\langle 3, -1, 6\rangle$ so we have

$$\begin{aligned} x &= 2+3t, \\ y &= 1-t, \\ z &= -4+6t. \end{aligned}$$

(b) $\mathbf{b}-\mathbf{a}=\langle b, 0\rangle-\langle 0, a\rangle=\langle b, -a\rangle$ so

$$\begin{aligned} x &= 0+bt \\ y &= a-at. \quad \blacksquare \end{aligned}$$

The vectors **a** and **b** do not enter Eq. (5.2) symmetrically: **a** appears twice and **b** just once. Given two points such as (2, 1) and (4, −1) how do we know which to use as **a** and which as **b**? The parametric equations will indeed be different depending on which choice you make, but when we use the parametric equations to solve problems, the final results will be the same.

EXAMPLE 5.2: POINT ON A LINE

Is $X = (-3, 1)$ on the line determined by $(5, -2)$ and $(1, 2)$?

SOLUTION

Let's take $\mathbf{a} = \langle 1, 2 \rangle$ and $\mathbf{b} = \langle 5, -2 \rangle$. If X is on the line, then there must be a t value such that Eq. (5.2) holds:

$$-3 = 1 + t(5 - 1) = 1 + 4t,$$
$$1 = 2 + t(-2 - 2) = 2 - 4t.$$

Each of these equations can easily be solved for t. If we get the same t value then the point is on the line. In this case, we get $t = -1$ and $= 1/4$, and so the point does not lie on the line. By contrast, you can verify that $(-3, 6)$ does lie on the line since the two equations would both have the solution $t = -1$. ■

How could we tell if $(-3, 6)$ is on the line *segment* $\overline{AB}$ connecting points A and B of Example 5.2? To answer this question, we analyze the three parts into which A and B divide the line they determine. The parts correspond naturally to different intervals of t values. Instead of thinking about the special point $(-3, 6)$, let's deal with an arbitrary position vector $\mathbf{x}$ whose head X lies on the line $\overleftrightarrow{AB}$ and the parameter value t that corresponds to $\mathbf{x}$.

1. If $t \geq 1$ then, according to Eq. (5.1), $\mathbf{x} - \mathbf{a}$ points in the same direction as $\mathbf{b} - \mathbf{a}$, and the length of $\mathbf{x} - \mathbf{a}$ is at least as great as the length of $\mathbf{b} - \mathbf{a}$. This means that X lies on the infinite part of the line starting at B and heading away from A.

2. If $t \leq 0$ then $\mathbf{x} - \mathbf{a}$ is either $\mathbf{0}$ (the zero vector), or points in the direction opposite to $\mathbf{b} - \mathbf{a}$. This means that X lies on the infinite part of the line starting at A and heading away from B.

3. If $0 \leq t \leq 1$ then $\mathbf{x} - \mathbf{a}$ is either $\mathbf{0}$ or points in the same direction as $\mathbf{b} - \mathbf{a}$, and the length of $\mathbf{x} - \mathbf{a}$ is no greater than the length of $\mathbf{b} - \mathbf{a}$. This means that X lies on $\overline{AB}$, the line segment determined by A and B.

For a point on the segment $\overline{AB}$, the corresponding t value tells us how far along the segment from A to B the point lies. For example, if X is the midpoint of $\overline{AB}$, then Eq. (5.1) shows that we must have $t = 1/2$. Likewise, for a point 1/3 of the way from A to B, $t = 1/3$, and so on.

EXAMPLE 5.3: MEDIANS OF A TRIANGLE

A *median* of a triangle is a segment connecting a vertex to the midpoint of the opposite side. In Figure 5.2 $\overline{AM_1}$, $\overline{BM_2}$, $\overline{CM_3}$ are medians. On median $\overline{AM_1}$ find a point P_1 that is 2/3 of the way from A to M_1. Do the same on each of the other two medians, finding the "two-thirds points" P_2 and P_3. Show that these points are all the same: $P_1 = P_2 = P_3$. This shows that the three medians meet at this single point called the *centroid* of the triangle.

SOLUTION

Because M_1 is the midpoint of $\overline{BC}$,

$$\begin{aligned} \mathbf{m}_1 &= \mathbf{b} + \frac{1}{2}(\mathbf{c} - \mathbf{b}) \\ &= \frac{1}{2}\mathbf{b} + \frac{1}{2}\mathbf{c}. \end{aligned}$$

Since P_1 is the "two-thirds point" of $\overline{AM_1}$,

$$\begin{aligned} \mathbf{p}_1 &= \mathbf{a} + \frac{2}{3}(\mathbf{m}_1 - \mathbf{a}) \\ &= \mathbf{a} + \frac{2}{3}\left(\frac{1}{2}\mathbf{b} + \frac{1}{2}\mathbf{c} - \mathbf{a}\right) \\ &= \mathbf{a} + \frac{1}{3}\mathbf{b} + \frac{1}{3}\mathbf{c} - \frac{2}{3}\mathbf{a} \\ &= \frac{1}{3}(\mathbf{a} + \mathbf{b} + \mathbf{c}). \end{aligned}$$

When we calculate $\mathbf{p}_2$ and $\mathbf{p}_3$ using the same method, we get the same formula. Try it as an exercise. ■

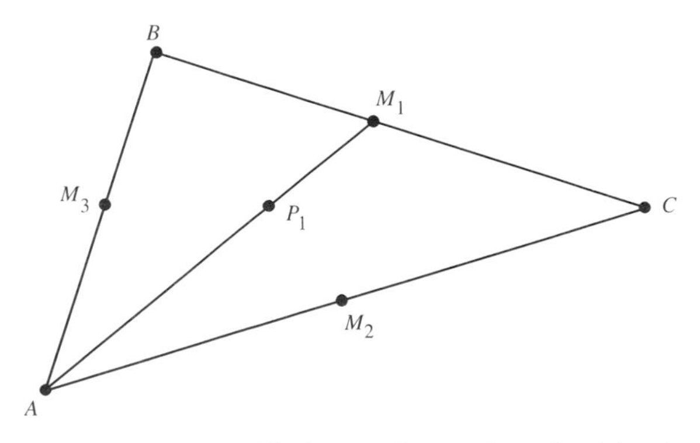

Figure 5.2 Three midpoints and a median of a triangle.

Applications of Lines in Two Dimensions

APPLICATION: Checking a Robot's Path

EXAMPLE 5.4: POINT ON A SEGMENT

Rita the robot is located at $S=(2, -1)$ and intends to move to $G=(4, 5)$ in a straight line. There is a pothole at $P=(3, 2)$. Does Rita's path contain the pothole?

SOLUTION

The path contains the pothole if and only if there is a t value so that $\mathbf{p}=\mathbf{s}+t(\mathbf{g}-\mathbf{s})$, that is,

$$3 = 2 + t(4 - 2),$$
$$2 = -1 + t[5 - (-1)].$$

The first equation has the solution $t = 1/2$, which also is the solution of the second. Since we have found a t that satisfies both equations, the pothole is on the *line* determined by Rita's start and end positions. But we can tell that it is on the *segment* $\overline{SG}$ determined by those positions by the fact that the t value is in [0, 1]. Actually, P is exactly halfway from S to G. ■

EXAMPLE 5.5: SEGMENT–SEGMENT INTERSECTION

Rita the robot is located at (2, 5) and intends to move to (4, 6). There is a doorway stretching from (5, 7) to (7, 5) as shown in Figure 5.3. Does Rita's path take her through the doorway?

SOLUTION

If we have a careful drawing, such as Figure 5.3, then we don't really need to do any computation to get the answer. But it is a lot simpler for a robot to do computations than to draw and examine diagrams.

$$\text{Rita's path:}\quad x = 2 + 2t,$$
$$y = 5 + t.$$

$$\text{Doorway segment:}\quad x = 5 + 2s,$$
$$y = 7 - 2s.$$

Notice that we use different parameters, s and t, for the path and the doorway. The reason is that if there is a point of intersection, it would probably have different parameter values on the two different segments.

If there is a point of intersection, the two equations for x have to give the same value at that point, so we equate the two expressions for x. Likewise, we can equate the two y expressions.

$$2 + 2t = 5 + 2s,$$
$$5 + t = 7 - 2s.$$

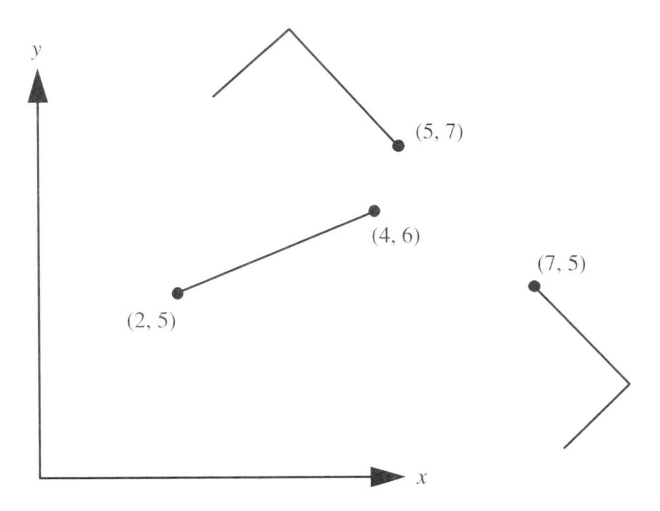

Figure 5.3 Does the robot path go through the doorway?

Rewriting in the more customary form for two linear equations in two unknowns gives:

$$2t - 2s = 3,$$
$$t + 2s = 2.$$

Add the two equations to obtain $t = 5/3$. Substitute this t value to get $s = 1/6$. Because t is not in [0, 1], we see that there is no intersection of the segments. ■

APPLICATION: Computer Graphics

EXAMPLE 5.6: POINT INSIDE A POLYGON

The chief of police wants a computer graphics system to display crime locations by lighting up dots on a city map on his computer screen. The color of the dot will be determined by the police precinct in which the dot is located. A police precinct is a polygon whose corners are known (by their coordinates) to the graphics program. For simplicity we show just one triangular precinct in Figure 5.4. Crime locations in this triangle are to be shown in blue. When a police officer enters the x and y coordinates of a place where a crime occurred, for example, $P(16, 4)$, how can the program compute whether the point should be shown in blue, that is, whether the point is inside the blue precinct?

SOLUTION

Here is an analysis that is almost correct. (In the exercises we'll ask you to supply a correction for some rare exceptional cases.) If you draw a horizontal ray from the crime location P, going left to infinity, this ray will cross exactly one triangle side if the location is inside the triangle. On the other hand, if P is outside the

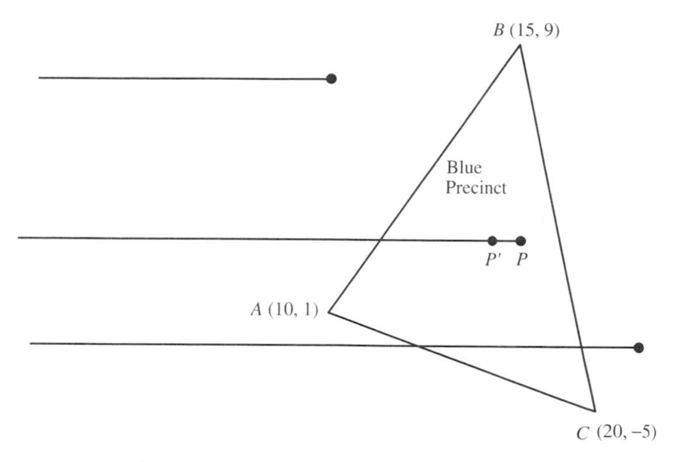

Figure 5.4 Is (16, 4) inside this triangle?

triangle it will cross no sides or two sides. So the program should calculate how many crossings there are to see if the point is inside or not. We put this in the form of a rule:

The Number of Crossings Rule (Imperfect Version)

1. Determine the parametric equations of the left-going ray starting at the given point P.
2. See how many triangle sides are crossed by this ray.
3. If the number of crossings is 0 or 2, the point is ouside the triangle. If it is 1, the point is inside.

To find a ray going left from $P(16, 4)$ begin by stepping one unit to the left to point P' and finding the coordinates of this point, $P'(15, 4)$. The ray starting at P passing through P' has equation $\mathbf{x} = \mathbf{p} + t(\mathbf{p}' - \mathbf{p})$ where $t \geq 0$. In nonvector format,

$$x = 16 + t \qquad \text{and} \qquad y = 4$$

with the restriction $t \geq 0$. We need to intersect this against the three line segments.

$$\text{Segment } \overline{AB}: \quad \mathbf{x} = \mathbf{a} + s(\mathbf{b} - \mathbf{a}),$$
$$x = 10 + 5s \qquad \text{and} \qquad y = 1 + 8s$$

with the restriction that s is in [0, 1]. Intersecting this segment with the ray gives: $s = 3/8$, $t = 33/8$. Because 3/8 does lie in [0, 1] and $33/8 \geq 0$, the intersection is on the segment $\overline{AB}$ and on the ray going left from (16, 4). So we have found one crossing. We go on to check the next two polygon sides:

$$\text{Segment } \overline{BC}: \quad \mathbf{x} = \mathbf{b} + s(\mathbf{c} - \mathbf{b}),$$
$$x = 15 + 5s \qquad \text{and} \qquad y = 9 - 14s$$

with the restriction that s is in [0, 1]. Intersecting this segment with the ray gives: $s = 5/14$, $t = -11/14$. Because $-11/14 < 0$, the point is not on the ray. No new crossing has been found.

Segment $\overline{CA}$: Here we can be clever and avoid some work by noticing that the ray has height of $y=4$, but the highest endpoint of the segment is at $y=1$. There can't be an intersection.

Altogether we have found one crossing of the left-pointing ray with triangle sides, so the point P is inside the triangle. ■

APPLICATION: Robot Collision Avoidance

EXAMPLE 5.7: POLYGON–SEGMENT INTERSECTION

Rita the robot intends to move in a straight line from $S(-6, 0)$ to $G(6, 8)$ as shown in Figure 5.5. There is a desk in the room and its footprint on the floor is a rectangle. Will the robot crash into the desk if it uses the proposed path?

SOLUTION

This problem can be solved by a series of segment–segment intersection problems of the sort illustrated in Example 5.5. The key point is that if Rita's path misses each of the edges of the desk footprint, the path is safe. But if we find an intersection with even one side, then Rita's path is not safe.

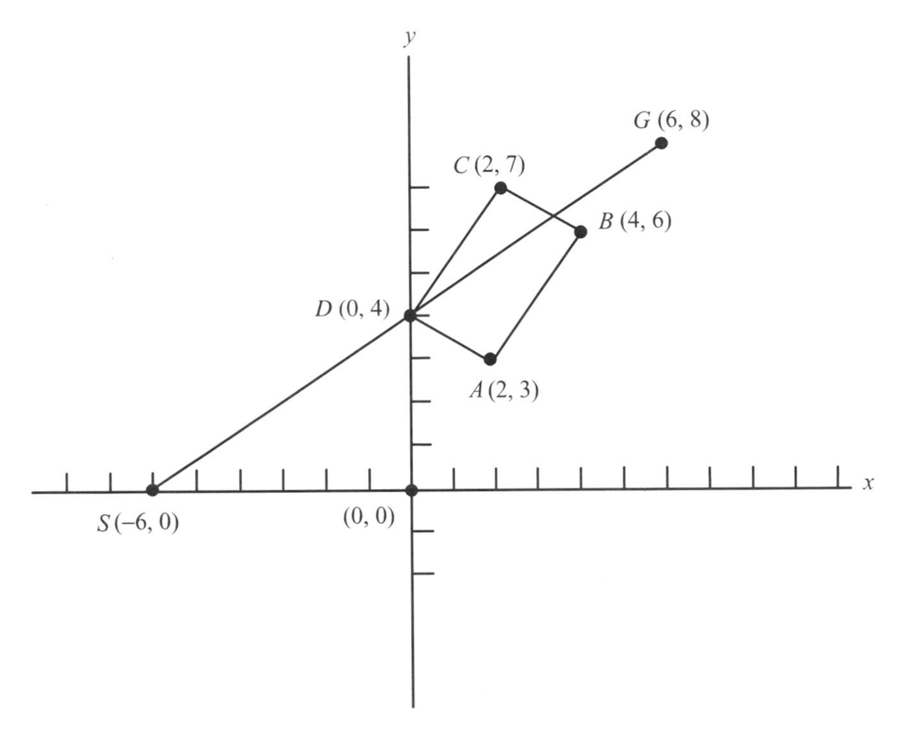

Figure 5.5 Will the robot hit the desk in its straight path?

Which desk segment should Rita test first? From the picture we can see that if Rita tested $\overline{AB}$ first, she would then have to test a second segment before she was done. On the other hand, if we tested $\overline{CD}$ she would get her answer with one test. But Rita does not have the picture to guide the computations so she will choose at random. Here is what the computation is like if Rita starts with $\overline{DA}$.

$$\text{Rita's path } \overline{SG}: \quad \mathbf{x} = \mathbf{s} + t(\mathbf{g} - \mathbf{s}),$$
$$x = -6 + 12t \quad \text{and} \quad y = 8t.$$

$$\text{Desk segment } \overline{DA}: \quad \mathbf{x} = \mathbf{d} + s(\mathbf{a} - \mathbf{d})$$
$$x = 2s \quad \text{and} \quad y = 4 - s;$$
$$-6 + 12t = 2s$$
$$8t = 4 - s;$$
$$12t - 2s = 6$$
$$8t + s = 4.$$

Multiplying the second equation by 2 and adding:

$$28t = 14, \text{ so } t = \frac{1}{2}.$$

Substituting $t = 1/2$ allows us to find that $s = 0$. Both parameters s and t are in $[0, 1]$, so the segments intersect. Rita has no need to test any other desk segments.

It may happen that the known obstacle has a curved shape. In this case, we can still check if a proposed straight-line path will collide with the object. There are at least two ways in which this might be done. First, we could find a piecewise linear approximation to the curved shape (see Figure 5.6). Then

Figure 5.6 Curved object with piecewise-linear approximation.

we could examine each segment in our approximation and test it for intersection with the robot's path. There are some shortcomings to this approach (can you see any?), but with a little care it can be useful in many cases.

If the curved object has a convenient equation, we can avoid piecewise linear approximations. Instead, substitute the parametric equations for x and y into the equation of the curved object and solve the resulting equation for t as in the next example. ■

Applications of Lines in Three Dimensions

EXAMPLE 5.8: SEGMENT–SPHERE INTERSECTION

1. An X-ray beam is aimed in a straight path from $S(2, 4, 3)$ toward $G(5, -2, 0)$. There is an atom centered at the origin and part of the beam will become absorbed if it comes within $\sqrt{17}$ units of the center of the atom. Will there be an absorbtion?

2. An X-ray beam is aimed in a straight path from $S(2, 4, 3)$ toward $G(5, -2, 0)$. There is a cancerous tumor roughly shaped like a sphere centered at the origin with radius $\sqrt{17}$. If the X ray hits the tumor, it will help kill it. Will this X ray hit the tumor?

SOLUTION

These two problems are obviously the same, so let's forget the story and do the math. We need to see if the segment intersects the sphere $x^2 + y^2 + z^2 = 17$. The segment has equations

$$x = 2 + 3t,$$
$$y = 4 - 6t,$$
$$z = 3 - 3t.$$

Substituting into the sphere equation:

$$(2 + 3t)^2 + (4 - 6t)^2 + (3 - 3t)^2 = 17$$
$$(4 + 12t + 9t^2) + (16 - 48t + 36t^2) + (9 - 18t + 9t^2) = 17$$
$$54t^2 - 54t + 12 = 0$$
$$9t^2 - 9t + 2 = 0$$
$$(3t - 2)(3t - 1) = 0$$
$$t = \frac{2}{3}, \quad t = \frac{1}{3}.$$

Since both t values are in $[0, 1]$ the line segment crosses the sphere twice. Therefore, the beam does come within $\sqrt{17}$ of the center of the atom and part of the beam will be absorbed. Likewise, the X ray does hit the tumor. ■

APPLICATION: Stereo Vision

How can we tell how far away something is by looking at it? We do this all the time, so it might seem there is nothing to explain. But if you close one eye your "depth perception" will be impaired. The reason we can tell how far nearby objects are is, in part, because we have two eyes. Each eye records a slightly different direction to the object. The brain combines this directional information[1] to give us the perception of how far an object is. When scientists build computer vision systems they often find that two cameras help in the same way that two eyes are better than one.

Before explaining the value of two cameras—an arrangement called *stereo vision*—let's examine just the first of the cameras in the simple model of image formation shown in Figure 5.7. (You can think of the box as an eyeball instead, although the example we work out in detail is the computer vision one. Either way, our pinhole camera model is only an approximation—but a useful one.) Light emanates from point P in all directions, and one ray will pass through the lens, modeled as a pinhole the size of a single point C_1 in the box. The light continues on to the imaging surface at the back of the box (the film or the retina), which, in our model, is a part of a plane. The light ray creates an image at P'.

In the case of a computer vision system, the three-dimensional coordinates of the point P' can be measured. In addition, the three-dimensional coordinates of C_1 are known. So what does this tell us about P? Knowing P' and C_1 tells us only that P lies on the line connecting them; that is, we know the direction from the first camera pinhole to P:

$$\mathbf{p} = \mathbf{p}' + t_1(\mathbf{c}_1 - \mathbf{p}') \text{ for some value of } t_1. \tag{5.4}$$

Now let's bring an additional camera into the picture. It has its own pinhole (lens) C_2, and it measures the location of the image point at P''. This second camera locates the point P on the line connecting P'' and C_2:

$$\mathbf{p} = \mathbf{p}'' + t_2(\mathbf{c}_2 - \mathbf{p}'') \text{ for some value of } t_2. \tag{5.5}$$

The point P creating these images is the same in both cases so we can equate the right-hand sides of Eqs. (5.4) and (5.5):

$$\mathbf{p}' + t_1(\mathbf{c}_1 - \mathbf{p}') = \mathbf{p}'' + t_2(\mathbf{c}_2 - \mathbf{p}'').$$

This is actually three equations, one for the x component, one for the y, and one for the z. By choosing two of these, one can solve for t and t' as illustrated in the next example.

---

[1] Along with other clues. For example if an object is known to be large but looks small, the mind concludes it is far away.

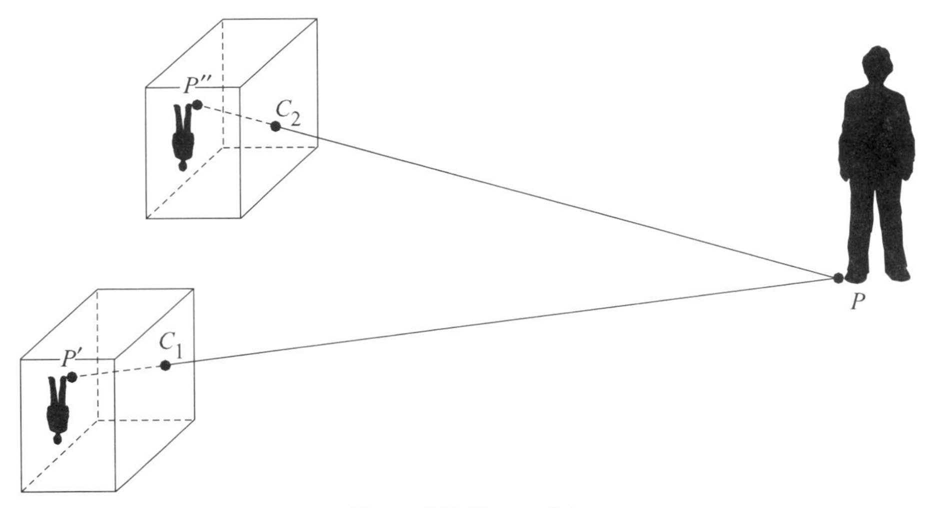

Figure 5.7 Stereo vision.

EXAMPLE 5.9: INTERSECTION OF TWO LINES IN THREE DIMENSIONS

Let $P' = (1, 2, 0), C_1 = (3, 3, 1)$, and $P'' = (-2, 8, 6), C_2 = (1, 7, 5)$. Find the coordinates of the intersection P (see Figure 5.7).

$$\mathbf{p} = \langle 1,\ 2,\ 0\rangle + t_1\langle 2,\ 1,\ 1\rangle$$
$$\mathbf{p} = \langle -2,\ 8,\ 6\rangle + t_2\langle 3,\ -1,\ -1\rangle.$$

Setting x components equal: $1 + 2t_1 = -2 + 3t_2$.

Setting y components equal: $2 + t_1 = 8 - t_2$.

Setting z components equal: $t_1 = 6 - t_2$.

We obtain the following system of equations:

$$2t_1 - 3t_2 = -3$$
$$t_1 + t_2 = 6$$
$$t_1 + t_2 = 6.$$

The last equation is redundant so we choose the first two to solve simultaneously. The solution is $t_1 = 3,\ t_2 = 3$. Substituting $t_1 = 3$ into the first equation for $\mathbf{p}$ gives

$$\mathbf{p} = \langle 1,\ 2,\ 0\rangle + 3\langle 2,\ 1,\ 1\rangle = \langle 7,\ 5,\ 3\rangle.$$

Although it is, strictly speaking, unnecessary, we could substitute $t_2 = 3$ into the second equation to verify that we get the same position vector $\mathbf{p}$. ■

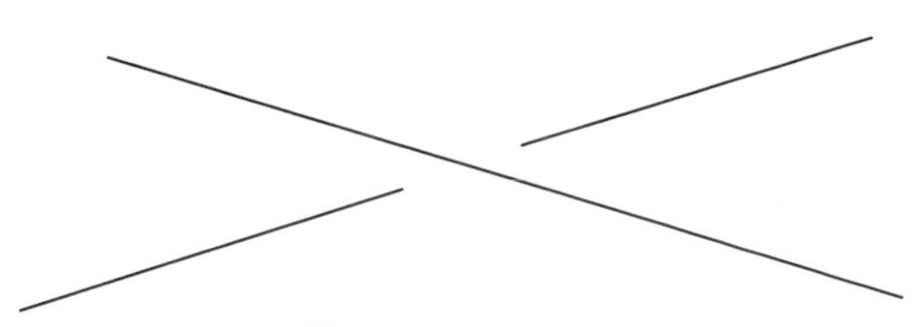

Figure 5.8 Skew lines.

The method used in the previous example is a general method for finding the intersection of two lines in three-dimensional space and can be used for other applications. However, it is essential to understand that two arbitrarily given lines in three-dimensional space may not have a point of intersection. (In our stereo vision application, we knew there was a point P on both lines.) One way this can occur is when the lines are parallel — that is, point in the same direction. But there is another possibility, which has no counterpart in our study of two-dimensional space: The lines may be *skew* (Figure 5.8). For example, the line where the front wall of a room meets the ceiling is skew to the line where the right wall meets the floor. These lines are not parallel, but they don't meet either. You can tell you are dealing with skew or parallel lines if you can't solve the equations simultaneously because they are inconsistent.

EXAMPLE 5.10: SKEW LINES

Show that the line connecting $P'(0, 1, 1)$ with $C_1(1, 2, 2)$ and the line connecting $P''(-1, -2, 0)$ and $C_2(2, 3, -4)$ are either parallel or skew.

SOLUTION

$$\text{First line:} \quad \mathbf{x} = \mathbf{p}' + t_1(\mathbf{c}_1 - \mathbf{p}') = \langle 0, 1, 1\rangle + t_1\langle 1, 1, 1\rangle.$$
$$\text{Second line:} \quad \mathbf{x} = \mathbf{p}'' + t_2(\mathbf{c}_2 - \mathbf{p}'') = \langle -1, -2, 0\rangle + t_2\langle 3, 5, -4\rangle.$$

$$\text{Setting } x \text{ components equal:} \quad t_1 = -1 + 3t_2.$$
$$\text{Setting } y \text{ components equal:} \quad 1 + t_1 = -2 + 5t_2.$$
$$\text{Setting } z \text{ components equal:} \quad 1 + t_1 = -4t_2.$$

We obtain the following system of equations:

$$t_1 - 3t_2 = -1,$$
$$t_1 - 5t_2 = -3,$$
$$t_1 + 4t_2 = -1.$$

Subtracting the second from the first gives $2t_2 = 2$, which gives $t_2 = 1$ and $t_1 = 2$. But these values of t_1 and t_2 don't satisfy the third equation so the equations have no solution. Thus the lines don't intersect. To distinguish whether the lines are parallel or skew requires some additional theory, which can be found in Section 7.4 of Chapter 7. ■

EXERCISES

**Marks challenging exercises.*

Basic Ideas

1. Find a set of parametric equations for the line connecting (2, −3, 4) and (3, −6, 2).

2. Find a set of parametric equations for the line connecting (0, 1, 2) and (2, 1, 0).

3. Rework Example 1.2, switching A and B; that is, let $A = (5, -2)$ and $B = (1, 2)$ and determine whether (−3, 1) and (−3, 6) lie on the line.

4. Is (2, 3) on the line connecting (−2, 4) and (6, 5)? Is it on the segment connecting these points?

5. Is (−3, 5) on the segment connecting $S(-5, 3)$ and $G(1, 9)$? If so, what fraction of the way is it from S to G?

6. Find the coordinates of the centroid of the triangle whose coordinates are (2, 4), (−1, 7), and (−4, 3) (see Example 5.3).

7. Given that one vertex of a triangle is at (0, 4) and the centroid is at (6, 2), find the coordinates of the midpoint of the side opposite (0, 4).

8. In Example 5.3, show that $\mathbf{p}_2 = (1/3)\,(\mathbf{a} + \mathbf{b} + \mathbf{c})$. Next, derive a formula for $\mathbf{p}_3$.

9. Let $ABCD$ be a quadrilateral. Let M_1 be the midpoint of $\overline{AB}$, let M_2 be the midpoint of $\overline{BC}$, and so on. Show that these four midpoints form a parallelogram. (*Hint:* Compare vectors formed by opposite sides of the quadrilateral formed by the M_i.)

Applications of Lines in Two Dimensions

10. Determine whether the point (14, −3) is inside the triangle of Example 5.6. Use the number of crossings rule (imperfect version).

11. Determine whether the point (15, 3) is inside the triangle of Example 5.6. Use the number of crossings rule (imperfect version).

12. In Example 5.6 we used rays going to the left. We could instead have used rays going to the right. Determine whether (15, 3) is inside the polygon by using rays going to the right. Use the number of crossings rule (imperfect version).

13. In Example 5.6 we used rays going to the left. We could instead have used rays going up. Determine whether (14, 3) is inside the polygon by using rays going up. Use the number of crossings rule (imperfect version).

14. The number of crossings rule (imperfect version) in Example 5.6 for checking whether a point is inside a polygon may fail if the horizontal ray passes through an endpoint of a triangle side.

    (a) Illustrate this with a drawing and explain exactly how the problem comes about.

    (b) There is a simple way to revise the rule so it still works. Find the revised rule.

15. The number of crossings rule in Example 5.6 for checking whether a point is inside a polygon fails if the polygon is nonconvex as in the example of Figure 5.9 where there are "dents." There is a way to revise the rule so it still works. Find the revised rule. Is your rule vulnerable to the kinds of problems mentioned in the previous exercise?

16. In Example 5.7, show by calculation that $\overline{SG}$ crosses $\overline{BC}$.

17. In Example 5.7, show by calculation that $\overline{SG}$ does not cross $\overline{AB}$.

18. A robot proposes to move in a straight line from $(-11, 2)$ to $(10, 5)$. Will it cross the circle of radius 5 centered at $(0, 0)$?

19. Does the line connecting $(0, 0)$ to $(1, \sqrt{3})$ cross the circle of radius $\sqrt{2}$, centered at $(\sqrt{3}, -1)$? If so, in how many places?

20. Suppose the pothole of Example 5.4 is too large to be represented by a point. Suppose it is a circle of radius 3, centered at $(-3, 1)$. Does the robot's path cross this pothole?

21. What shortcomings do you see to using a piecewise linear approximation for a curved obstacle when doing collision detection problems for a mobile robot?

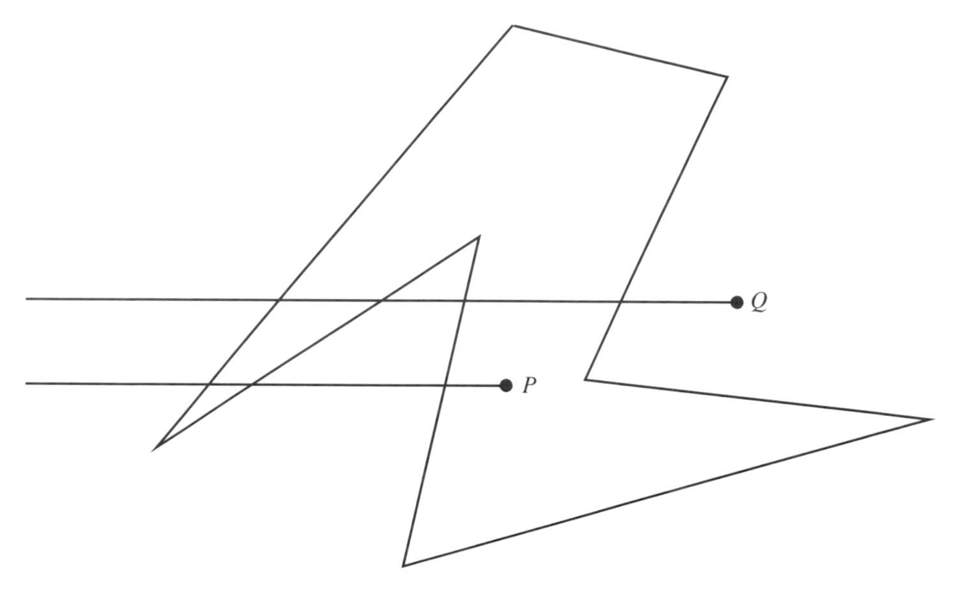

Figure 5.9 A nonconvex polygon.

Applications of Lines in Three Dimensions

22. Determine whether the line segment connecting $S(1, 7, -2)$ to $G(6, -3, 3)$ crosses the sphere of radius $\sqrt{354}$ centered at $(0, 0, 0)$.

23. Find the coordinates of the points where the line segment crosses the sphere in Example 5.8.

24. Explain a method you could use to test whether the line connecting two given points was tangent to a circle whose center and radius are given. (You do not have to give formulas; just describe in general terms what calculations need to be done.)

*25. In our examples so far, a mobile robot has been assumed to be so small it could be modeled as a point. Suppose instead that its footprint on the floor is a circle of radius 1. It intends to move from (3, 6) to (10, 8). There is a circular obstacle of radius 2 centered at (5, 5). Will there be a collision? (*Hint:* There is a way to "grow" an obstacle so that it compensates for our desire to imagine the robot as just its center. Add to the obstacle a danger zone of locations for the robot center where an overlap of robot and obstacle circles would occur. What is the shape of this danger zone? What is the shape of the grown obstacle = obstacle $\cup$ danger zone?)

26. Find the point of intersection of the line connecting $P'(2, 4, 5)$ to $C_1(3, 3, 2)$ with the line connecting $P''(-2, 6, -3)$ to $C_2(1, 4, -2)$.

27. Find the point of intersection of the line connecting $P'(3, 0, 1)$ to $C_1(5, 1, 2)$ with the line connecting $P''(1, 9, 0)$, $C_2(3, 8, 1)$.

28. Which of the following pairs of lines intersect and which don't intersect?

    (a) Line 1 connects $A(-4, 3, 1)$ and $B(-2, 5, 3)$; line 2 connects $A'(2, -1, 0)$ and $B'(3, 0, 1)$.

    (b) Line 1 connects $A(0, 4, 2)$ and $B(3, 4, -1)$; line 2 connects $A'(4, -3, 2)$ and $B'(5, -3, 1)$.

    (c) Line 1 connects $A(4, 4, 2)$ and $B(10, 1, 8)$; line 2 connects $A'(3, 3, 2)$ and $B'(8, 1, 8)$.

    (d) Line 1 connects $A(1, 1, 1)$ and $B(2, 3, 4)$; line 2 connects $A'(4, 1, -1)$ and $B'(2, 5, -1)$.

5.2 Scalar Products, Planes, and the Hidden Surface Problem

Planes and Scalar Products

In the last section we dealt with lines in three-dimensional space using vector concepts. In this section we deal with planes. The central algebraic idea is the scalar product of two vectors, a concept often introduced in multivariable calculus courses. Readers who have this fresh in their minds can proceed quickly through the beginning of this section. We end with a discussion of the hidden surface problem of computer graphics, which forms one of the most important applications of geometry in use today.

Recall that if $\mathbf{u} = \langle x_u, y_u, z_u\rangle$ and $\mathbf{v} = \langle x_v, y_v, z_v\rangle$ are two vectors, their *scalar product* is

$$\mathbf{u} \cdot \mathbf{v} = x_u x_v + y_u y_v + z_u z_v.$$

For example $\langle 1, -2, 4\rangle \cdot \langle -5, 2, 0\rangle = (1)(-5) + (-2)(2) + (4)(0) = -9$.

Here are some important properties of the scalar product that arise directly from the definition.

$$\text{Commutativity: for any vectors } \mathbf{u} \text{ and } \mathbf{v},\ \mathbf{u}\cdot\mathbf{v} = \mathbf{v}\cdot\mathbf{u}. \tag{5.6}$$

Linearity: for any vectors $\mathbf{u}$, $\mathbf{v}$, $\mathbf{u}_1$, $\mathbf{u}_2$, $\mathbf{v}_1$, $\mathbf{v}_2$ and any scalar r,

$$(r\mathbf{u}) \cdot \mathbf{v} = r(\mathbf{u} \cdot \mathbf{v}) \quad \text{and} \quad \mathbf{u} \cdot (r\mathbf{v}) = r(\mathbf{u} \cdot \mathbf{v}), \tag{5.7a}$$

$$(\mathbf{u}_1 + \mathbf{u}_2) \cdot \mathbf{v} = \mathbf{u}_1 \cdot \mathbf{v} + \mathbf{u}_2 \cdot \mathbf{v} \quad \text{and} \quad \mathbf{u} \cdot (\mathbf{v}_1 + \mathbf{v}_2) = \mathbf{u} \cdot \mathbf{v}_1 + \mathbf{u} \cdot \mathbf{v}_2, \tag{5.7b}$$

$$(\mathbf{u}_1 - \mathbf{u}_2) \cdot \mathbf{v} = \mathbf{u}_1 \cdot \mathbf{v} - \mathbf{u}_2 \cdot \mathbf{v} \quad \text{and} \quad \mathbf{u} \cdot (\mathbf{v}_1 - \mathbf{v}_2) = \mathbf{u} \cdot \mathbf{v}_1 - \mathbf{u} \cdot \mathbf{v}_2. \tag{5.7c}$$

EXAMPLE 5.11: SCALAR PRODUCT ALGEBRA

Expand $(3\mathbf{u} - \mathbf{v}) \cdot (\mathbf{u} + \mathbf{v})$ and evaluate it numerically based on the assumptions that $\mathbf{u} \cdot \mathbf{u} = 2$, $\mathbf{u} \cdot \mathbf{v} = -1$, and $\mathbf{v} \cdot \mathbf{v} = 3$.

SOLUTION

$$\begin{aligned}
(3\mathbf{u} - \mathbf{v}) \cdot (\mathbf{u} + \mathbf{v}) &= (3\mathbf{u}) \cdot (\mathbf{u} + \mathbf{v}) - \mathbf{v} \cdot (\mathbf{u} + \mathbf{v}) && \text{(Rule 5.7c)} \\
&= (3\mathbf{u}) \cdot \mathbf{u} + (3\mathbf{u}) \cdot \mathbf{v} - (\mathbf{v} \cdot \mathbf{u} + \mathbf{v} \cdot \mathbf{v}) && \text{(Rule 5.7b)} \\
&= 3\mathbf{u} \cdot \mathbf{u} + 3\mathbf{u} \cdot \mathbf{v} - \mathbf{v} \cdot \mathbf{u} - \mathbf{v} \cdot \mathbf{v} && \text{(Rule 5.7a)} \\
&= 3\mathbf{u} \cdot \mathbf{u} + 3\mathbf{u} \cdot \mathbf{v} - \mathbf{u} \cdot \mathbf{v} - \mathbf{v} \cdot \mathbf{v} && \text{(Rule 5.6)} \\
&= 3\mathbf{u} \cdot \mathbf{u} + 2\mathbf{u} \cdot \mathbf{v} - \mathbf{v} \cdot \mathbf{v} && \text{(Combining terms)} \\
&= 3(2) + 2(-1) - 3 && \text{(Substituting)} \\
&= 1.
\end{aligned}$$

■

The *norm* of a vector $\mathbf{u}$ is defined

$$\|\mathbf{u}\| = \sqrt{\mathbf{u} \cdot \mathbf{u}}$$

and represents the length of the segment from the tail of **u**'s arrow to the head of **u**'s arrow. If P is the head of a vector and Q is the tail of a vector, then the vector[2] is $\mathbf{p} - \mathbf{q}$ and substituting this for $\mathbf{u}$ in the previous equation gives

$$\|\mathbf{p} - \mathbf{q}\| = \sqrt{(\mathbf{p} - \mathbf{q}) \cdot (\mathbf{p} - \mathbf{q})} = \sqrt{(x_P - x_Q)^2 + (y_P - y_Q)^2 + (z_P - z_Q)^2}.$$

If Q is the origin, so that we are finding the length of the position vector from the origin to P, this simplifies to

$$\|\mathbf{p}\| = \sqrt{x_P^2 + y_P^2 + z_P^2},$$

the distance of point P from the origin. A very important property of the norm is

$$\|r\,\mathbf{p}\| = |r|\,\|\mathbf{p}\|. \tag{5.8}$$

For example, we can easily calculate that $\|\langle 1, \ 1, \ 1 \rangle\| = \sqrt{3}$. Consequently, by Eq. (5.8), $\|\langle -18, \ -18, \ -18 \rangle\| = 18\sqrt{3}$.

---

[2] As described in the previous section, we use the same letter for a point and its position vector.

EXAMPLE 5.12: NORMS AND SPHERES

$P(0, 3, 4)$ lies on a sphere of radius 5 centered at the origin. On what sphere centered at the origin does $Q(0, 30, 40)$ lie? What about $R(0, -21, -28)$?

SOLUTION

The fact that P lies on the sphere of radius 5 centered at the origin can be expressed $\|\langle 0, 3, 4\rangle\| = 5$. By Eq. (5.8), $\|\langle 0, 30, 40\rangle\| = \|10\langle 0, 3, 4\rangle\| = 10\|\langle 0, 3, 4\rangle\| = 50$. Thus P lies on the sphere of radius 50 centered at the origin. In the same way, R lies on a sphere of radius 35 centered at the origin. ■

EXAMPLE 5.13: NORMS AND DISTANCE

Let P and Q be points whose position vectors are $\mathbf{p}$ and $\mathbf{q}$. Use norms and their properties to show that $(1/2)\mathbf{p}+(1/2)\mathbf{q}$ is the position vector of a point M with the property that $PM=(1/2)PQ$.

SOLUTION

$$PM = \left\|\mathbf{p} - \left(\frac{1}{2}\mathbf{p}+\frac{1}{2}\mathbf{q}\right)\right\| = \left\|\frac{1}{2}\mathbf{p}-\frac{1}{2}\mathbf{q}\right\| = \left\|\frac{1}{2}(\mathbf{p}-\mathbf{q})\right\| = \frac{1}{2}\|\mathbf{p}-\mathbf{q}\| = \frac{1}{2}PQ. \quad ■$$

It is possible to show that, given a plane and a point P on it, there is a line through P which is perpendicular (orthogonal) to every line through P that lies in the plane. Translating into vector language, this tell us that for every plane there is a *normal vector* $\mathbf{n}$ so that if P and X are any two points in the plane, the angle between $\mathbf{n}$ and $\mathbf{x}-\mathbf{p}$ is 90°. This is illustrated in Figure 5.10 where we take advantage of the fact that we can place the tail of a vector anywhere we like. We choose a version of $\mathbf{n}$ whose tail is at point P so as to display the orthogonality most clearly.

To express the orthogonality of $\mathbf{n}$ and $\mathbf{x}-\mathbf{p}$ algebraically, we use the following fundamental equation, which holds for any two vectors $\mathbf{u}$ and $\mathbf{v}$:

$$\mathbf{u}\cdot\mathbf{v} = \|\mathbf{u}\|\,\|\mathbf{v}\|\cos\theta, \tag{5.9}$$

where θ is the angle between the two vectors. For $\mathbf{u}$ and $\mathbf{v}$ to be orthogonal, that is, $\theta=90°$, it is necessary and sufficient for $\cos\theta=0$. Therefore, from Eq. (5.9) we get the following condition:

ORTHOGONALITY CONDITION

$\mathbf{u}$ and $\mathbf{v}$ are orthogonal if and only if $\mathbf{u}\cdot\mathbf{v}=0$.

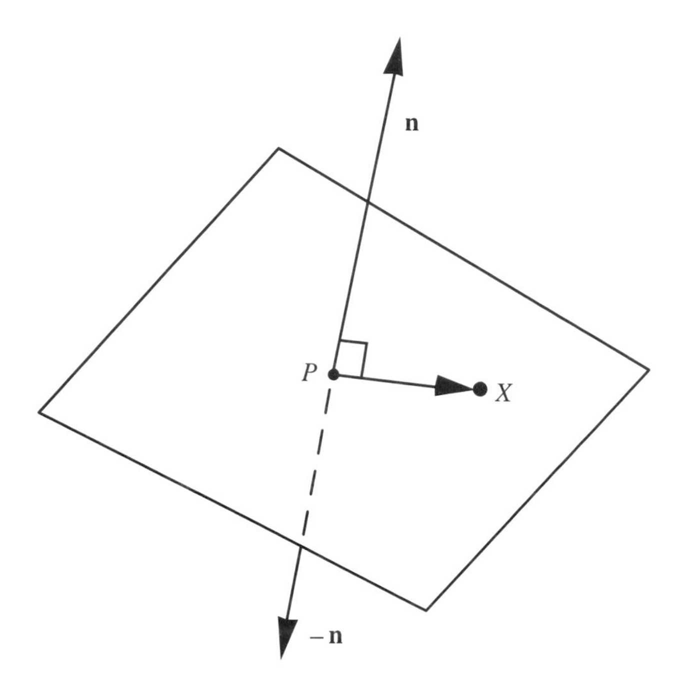

Figure 5.10 A plane and two normal vectors.

Returning to our plane, this means that X lies on the plane if and only if $(\mathbf{x}-\mathbf{p})\cdot\mathbf{n}=0$. Expanding gives $\mathbf{x}\cdot\mathbf{n}-\mathbf{p}\cdot\mathbf{n}=0$ or $\mathbf{x}\cdot\mathbf{n}=\mathbf{p}\cdot\mathbf{n}$. The vectors $\mathbf{p}$ and $\mathbf{n}$ are typically known, so the right side of the last equation is often a known constant k:

VECTOR EQUATIONS OF A PLANE

$$(\mathbf{x}-\mathbf{p})\cdot\mathbf{n}=0, \tag{5.10}$$

$$\mathbf{x}\cdot\mathbf{n}=k \tag{5.11}$$

In these equations, $\mathbf{p}$ corresponds to a fixed point of the plane, $\mathbf{n}$ is a constant vector, $\mathbf{x}$ is a position vector corresponding to a variable point X on the plane, and k is constant.

Not only does every plane have an equation in the form Eq. (5.10) and (5.11), but every equation fitting one of those forms, as long as $\mathbf{n}$ is not the zero vector, has a plane as its solution set. Sometimes it is useful to replace the short vector forms of the equation of a plane with a form in which the vectors are replaced with their components: $\mathbf{x} = \langle x, y, z\rangle$ and $\mathbf{n} = \langle x_n, y_n, z_n\rangle$. Thus, expanding Eq. (5.11) gives the following equation:

SCALAR EQUATION OF A PLANE

$$xx_n + yy_n + zz_n = k.$$

EXAMPLE 5.14: EQUATIONS OF PLANES

(a) Find a vector and scalar form of the equation of a plane with normal $\langle 3, -5, 2\rangle$ and passing through $(1, 1, 2)$.

(b) Show that planes with the same normal, $\mathbf{x}\cdot\mathbf{n} = k_1$ and $\mathbf{x}\cdot\mathbf{n} = k_2$, do not meet if $k_1 \neq k_2$. Equivalently, show that if there is a point X_0 on both planes then $k_l = k_2$.

SOLUTION

(a)
$$\begin{aligned}
&(\mathbf{x} - \langle 1, 1, 2\rangle)\cdot\langle 3, -5, 2\rangle = 0,\\
&\mathbf{x}\cdot\langle 3, -5, 2\rangle - \langle 1, 1, 2\rangle\cdot\langle 3, -5, 2\rangle = 0,\\
&\mathbf{x}\cdot\langle 3, -5, 2\rangle = 2 \qquad \text{(vector form)},\\
&3x - 5y + 2z = 2 \qquad \text{(scalar form)}.
\end{aligned}$$

(b) If X_0 lies on both planes, $\mathbf{x}_0$ satisfies both equations: $k_1 = \mathbf{x}_0\cdot\mathbf{n} = k_2$. Consequently, if $k_1 \neq k_2$ then the planes are parallel. ■

Given three points P, Q, and R, which are not collinear, Axiom 10 (Section 1.3 of Chapter 1) guarantees that there is exactly one plane containing them. How can we find the normal $\mathbf{n}$ of that plane? The cross-product of two vectors comes to the rescue. Recall that $\mathbf{u}\times\mathbf{v}$ is orthogonal to both $\mathbf{u}$ and $\mathbf{v}$. We know that the normal to the plane will be orthogonal to $\mathbf{q}-\mathbf{p}$ and $\mathbf{r}-\mathbf{p}$. Thus, we can calculate the normal by $\mathbf{n} = (\mathbf{q}-\mathbf{p})\times(\mathbf{r}-\mathbf{p})$.

EXAMPLE 5.15: FINDING NORMALS

(a) Find a normal vector to the plane containing $P(1, 1, 2)$ and $Q(2, -3, 1)$ and $R(-5, 2, 0)$.

(b) Find the equation of the plane passing through the three points in part (a).

SOLUTION

(a) $\mathbf{q}-\mathbf{p}=\langle 1,\ -4,\ -1\rangle$ and $\mathbf{r}-\mathbf{p}=\langle -6,\ 1,\ -2\rangle$ so

$$(\mathbf{q}-\mathbf{p})\times(\mathbf{r}-\mathbf{p}) = \begin{vmatrix} \mathbf{i} & \mathbf{j} & \mathbf{k} \\ 1 & -4 & -1 \\ -6 & 1 & -2 \end{vmatrix}$$
$$= 9\mathbf{i} + 8\mathbf{j} - 23\mathbf{k}$$
$$= \langle 9,\ 8,\ -23\rangle.$$

(b)
$$(\mathbf{x}-\langle 1,\ 1,\ 2\rangle)\cdot\langle 9,\ 8,\ -23\rangle = 0,$$
$$\mathbf{x}\cdot\langle 9,\ 8,\ -23\rangle - \langle 1,\ 1,\ 2\rangle\cdot\langle 9,\ 8,\ -23\rangle = 0,$$
$$\mathbf{x}\cdot\langle 9,\ 8,\ -23\rangle = -29,$$
$$9x + 8y - 23z = -29.$$
■

Planes, Polyhedra, and Hidden Surfaces

A polyhedron is any solid with flat sides, such as the box shown later in Figure 5.12. This is a rough definition — in Section 8.3 of Chapter 8 we'll go into more detail — but good enough for the purposes of this section. Polyhedra have always been of interest to mathematicians and lately they have been used a good deal in computer graphics. In graphics, a curved three-dimensional object, like the human head shown in Figure 5.11, is generally approximated by polyhedral surfaces as one step along the way to creating an image on the screen. The pictures in Figure 5.11 come from the Oscar-winning short film *Geri's Game* by Pixar. Figure 5.11a shows an early stage in the creation of a picture — a polyhedral model of Geri's head. But this will not appear on the screen. A variety of processes are carried out on this polyhedral model so that the final view completely lacks the "edgy" appearance of a polyhedron. Figure 5.11b is typical of what might appear on the screen.

We now consider how scalar products are used in dealing with polyhedra. If $\mathbf{u}$ and $\mathbf{v}$ are orthogonal, then how about $\mathbf{u}$ and $r\mathbf{v}$? They are also orthogonal because $\mathbf{u}\cdot(r\mathbf{v}) = r(\mathbf{u}\cdot\mathbf{v}) = (r)(0) = 0$. It follows that if $\mathbf{x}$ satisfies the equation of a plane $(\mathbf{x}-\mathbf{p})\cdot\mathbf{n} = 0$, then it also satisfies $(\mathbf{x}-\mathbf{p})\cdot(r\mathbf{n}) = 0$. In simple language: If $\mathbf{n}$ is a normal to a plane then so is $r\mathbf{n}$; in particular, taking $r = -1$, we see that $-\mathbf{n}$ is a normal. (This should be obvious visually, but it is nice to know that the algebra confirms our intuition.) Now $\mathbf{n}$ and $-\mathbf{n}$ point in opposite directions (see Figure 5.12) and we can take advantage of this when describing polyhedra because it is conventional to use

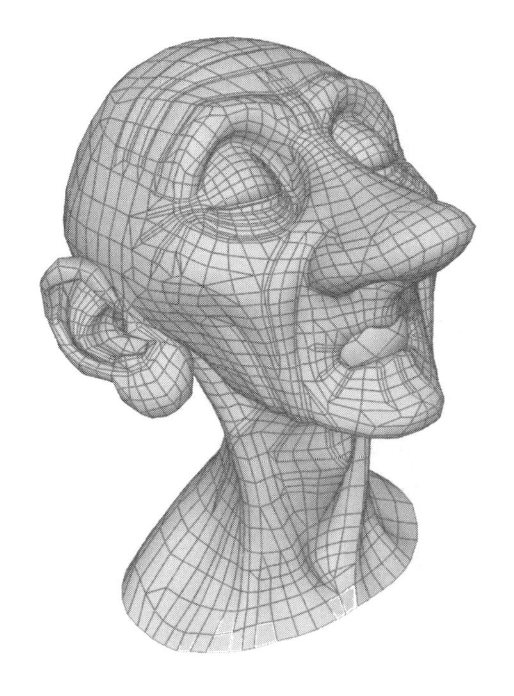

(a)

(b)

Figure 5.11 (a) A polyhedral model. (b) A completely rendered image from *Geri's Game* (courtesy of Pixar).

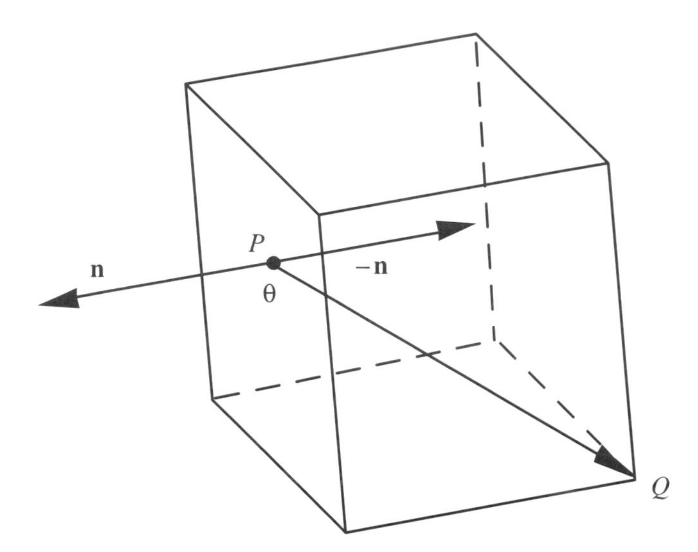

Figure 5.12 One of these is an outer normal.

a normal that points away from the interior of the polyhedron, an *outer normal.* If the normal $\mathbf{n}$ that we have for a face doesn't point to the outside of the polyhedron, we use $-\mathbf{n}$ in its place.

One large class of interesting polyhedra is the convex polyhedra (see Chapter 8, Section 8.3 for a detailed discussion), which have the following property: For each plane determined by a face of the polyhedron, all the corners of the polyhedron are on the same side of that face plane. The box of Figure 5.12 has this property. For examples that don't have this property, see the head in Figure 5.11a and see the box with a tunnel drilled through it in Figure 8.25 of Chapter 8. Because computations with polyhedra often start with the corners being given, convexity makes it relatively easy to find an outer normal for a face plane. Begin by computing any normal $\mathbf{n}$ as in Example 5.15. This may or may not be an outer normal. If we attach a copy of this vector to a corner P of the face in question and also consider the vector leading from P to some corner Q not on the face, then the angle θ between these vectors $\mathbf{n}$ and $\mathbf{q}-\mathbf{p}$ is greater than $90°$ if and only if our normal is an outer normal. If, however, $\theta < 90°$ then we need to replace our normal with its negative. We can avoid actually computing the angle because $\theta > 90°$ if and only if $\cos\theta < 0$. Applying Eq. (5.9) with $\mathbf{u}=\mathbf{q}-\mathbf{p}$ and $\mathbf{v}=\mathbf{n}$ we see that, because $\|\mathbf{q}-\mathbf{p}\| > 0$ and $\|\mathbf{n}\| > 0$, $\cos\theta < 0$ if and only if $(\mathbf{q}-\mathbf{p})\cdot\mathbf{n} < 0$. Thus we have the following algorithm.

Computing an Outer Normal for a Face of a Polyhedron

1. Find any normal **n** (for example, using the cross-product), a corner on the face *P*, and a corner off the face *Q*.
2. If $(\mathbf{q}-\mathbf{p})\cdot\mathbf{n}<0$, then **n** is an outer normal. If $(\mathbf{q}-\mathbf{p})\cdot\mathbf{n}>0$, replace **n** by $-\mathbf{n}$.

EXAMPLE 5.16: AN OUTER NORMAL

A box has the following corners:

$A(3.2,\ 2.6,\ -2)$
$B(3.2,\ 2.6,\ -3)$
$C(3.8,\ 3.4,\ -3)$
$D(3.8,\ 3.4,\ -2)$
$E(4,\ 2,\ -2)$
$F(4,\ 2,\ -3)$
$G(4.6,\ 2.8,\ -3)$
$H(4.6,\ 2.8,\ -2)$

One of its faces is *ABCD*. Find an outer normal for that face.

SOLUTION

To get a normal we calculate

$$\begin{aligned}\mathbf{n}=(\mathbf{b}-\mathbf{a})\times(\mathbf{c}-\mathbf{a}) &= \begin{vmatrix}\mathbf{i} & \mathbf{j} & \mathbf{k}\\ 0 & 0 & -1\\ 0.6 & 0.8 & -1\end{vmatrix}\\ &= \langle 0.8,\ -0.6,\ 0\rangle.\end{aligned}$$

To see if this normal is an outer normal, we pick *A* on the face and *F* off it and apply our algorithm:

$$\begin{aligned}\mathbf{f}-\mathbf{a} &= \langle 4,\ 2,\ -3\rangle - \langle 3.2,\ 2.6,\ -2\rangle\\ &= \langle 0.8,\ -0.6,\ -1\rangle,\\ (\mathbf{f}-\mathbf{a})\cdot\mathbf{n} &= \langle 0.8,\ -0.6,\ -1\rangle\cdot\langle 0.8,\ -0.6,\ 0\rangle\\ &= 1.\end{aligned}$$

Since $1>0$, we replace **n** by $\langle -0.8,\ 0.6,\ 0\rangle$ and this is an outer normal. ■

APPLICATION: The Hidden Surface Problem of Computer Graphics

Imagine painting the six sides of a cereal box with six different colors. Now hold the box out in front of you at any angle you like. How many colors will you see? Surely not all of them because some faces of the box are hidden by others. Now suppose a computer graphics program has been asked to display this box. It needs to decide which faces are visible and which are hidden so that only the visible ones are drawn and colored on the screen. Of course, the graphics program needs to be able to solve the same problem for any polyhedron.

Let's be more specific about the computer's hidden surface problem (Figure 5.13). The user of the program has specified the box in enough detail that the program can find the coordinates of the corners. The user has specified a color for each face. The computer has worked out outer normals for each face, just as in Example 5.16. In addition, the user has specified the coordinates of the point S, which is the "seeing position"—where the observer's eye or the lens of a camera is located. The computer's task is to show what the observer would see. An important part of this task is to determine which faces are visible and which are hidden by others.

Here is a method we can use to test a face for visibility in the case of the box or any other convex polyhedron (Figure 5.13). Imagine that you are standing at some point P on the face of interest and looking in the direction of the outer normal. If you are looking more or less away from the eye point S, as in the case of point P_2, you are on a hidden face. If you are looking more or less toward S, as in the case of point P_1, you are visible from the eye point. Assuming the outer normal of the face is denoted $\mathbf{n}$, here is how this would be calculated.

1. Choose a point P in the face to be tested—if you know a corner, that is a good choice.
2. Work out the vector $\mathbf{v} = \mathbf{s} - \mathbf{p}$ from P to S.
3. Let θ be the angle between $\mathbf{v}$ and $\mathbf{n}$. If $0° \leq \theta < 90°$, then the face is visible. If $\theta = 90°$ then the line of sight lies in the plane of the face in question. You are seeing the face "edge-on." If $90° < \theta \leq 180°$, then the face is hidden.

We can tell the size of θ from its cosine. For θ between $0°$ and $90°$, $\cos\theta \geq 0$, while for θ between $90°$ and $180°$, $\cos\theta \leq 0$. In view of Eq. (5.9) and the fact that the norm of a vector is always nonnegative, we have the following:

Hidden Surface Computation for Convex Polyhedra

If $(\mathbf{s} - \mathbf{p}) \cdot \mathbf{n} > 0$, the face is visible. If $(\mathbf{s} - \mathbf{p}) \cdot \mathbf{n} < 0$, the face is hidden. If $(\mathbf{s} - \mathbf{p}) \cdot \mathbf{n} = 0$, the line of sight runs along the face (just one or more edges of the face are visible.) Remember, $\mathbf{n}$ must be an outer normal.

EXAMPLE 5.17

Suppose the eye is at $S(1, -2, 4)$ and a face of a box has an outer normal $\mathbf{n} = \langle 2, -6, 5\rangle$. $P(10, 12, 20)$ is a corner of the face. Is the face visible?

SOLUTION

$(\mathbf{s} - \mathbf{p}) \cdot \mathbf{n} = (\langle 1, -2, 4\rangle - \langle 10, 12, 20\rangle) \cdot \langle 2, -6, 5\rangle = \langle -9, -14, -16\rangle \cdot \langle 2, -6, 5\rangle = -14$ so the face is hidden. ■

The algorithm given earlier does not work for nonconvex polyhedra, nor does it help if there are a number of polyhedra, one in front of another. You can read about methods for such cases in almost any computer graphics textbook.

EXERCISES

**Marks challenging exercises.*

Planes and Scalar Products

1. In each case below, calcuate the scalar product $\mathbf{u} \cdot \mathbf{v}$.

   (a) $\mathbf{u} = \langle 1, 0, 0\rangle$, $\mathbf{v} = \langle 0, 1, 0\rangle$

   (b) $\mathbf{u} = \langle 3, -8, 2\rangle$, $\mathbf{v} = \langle 1, -2, 3\rangle$

   (c) $\mathbf{u} = \langle 1, 2, -1\rangle + t\langle 0, 1, 2\rangle$, $\mathbf{v} = \langle 1, 2, -1\rangle - t\langle 0, 1, 2\rangle$ (Give the answer in terms of t.)

2. Find the norm of the following vectors.

   (a) $\mathbf{a} = \langle 1, -3, 5\rangle$

   (b) $\mathbf{b} = \langle 3, 4, 0\rangle$

   (c) $\mathbf{c} = \langle 12, 0, 5\rangle$

3. (a) Let A have position vector $\langle 1, 1, 2\rangle$ and let B have position vector $\langle 2, -3, 1\rangle$. Verify that A and B lie on the plane whose equation is $\mathbf{x} \cdot \langle 7, 3, -5\rangle = 0$. Then verify that any point on the line connecting A and B lies on the plane.

   (b) Carry out part (a) more generally. Let the plane have equation $\mathbf{x} \cdot \mathbf{n} = k$ and suppose $\mathbf{a}$ and $\mathbf{b}$ are some unspecified vectors that satisfy this equation. Show that $\mathbf{a} + t(\mathbf{b} - \mathbf{a})$ does also.

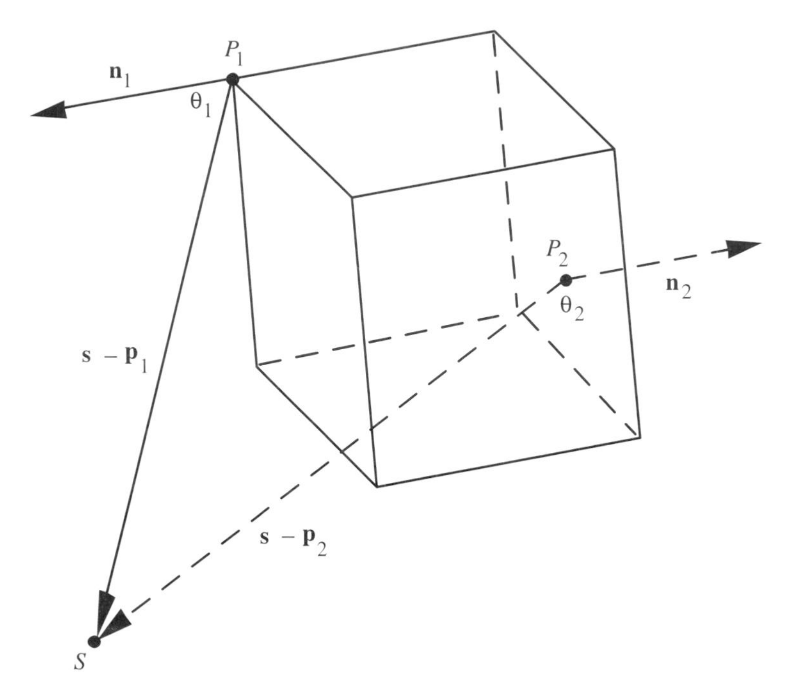

Figure 5.13 The hidden surface problem for a box.

4. (a) Expand and simplify $(a\mathbf{u}-b\mathbf{v}) \cdot (a\mathbf{u}+b\mathbf{v})$.

   (b) Simplify this further in the case for which $\mathbf{u}$ and $\mathbf{v}$ have unit length.

5. Use the algebraic properties of scalar products and vectors to prove $(\mathbf{x}-\mathbf{y}) \cdot (\mathbf{x}+\mathbf{y}) = \|\mathbf{x}\|^2 - \|\mathbf{y}\|^2$.

6. For each of the following assertions, decide whether it is true or false. If it is false, explain what misconception about vector algebra is used. Wherever possible, illustrate with an example.

   (a) $\mathbf{x} \cdot \mathbf{y} = 0$ means one of $\mathbf{x}$ and $\mathbf{y}$ must be the zero vector.

   (b) If $\mathbf{x} \cdot \mathbf{y} = 1$, we can multiply both sides by $\mathbf{x}^{-1}$ to get $\mathbf{y} = \mathbf{x}^{-1}$.

   (c) Let $\mathbf{x}$ be some unit vector so $\mathbf{x} \cdot \mathbf{x} = 1$. Let $\mathbf{y}$ be any vector at all. We'll try to prove $\mathbf{y}$ is a multiple of $\mathbf{x}$: Denote $\mathbf{x} \cdot \mathbf{y} = k$. We have $\mathbf{y} = (1)\mathbf{y} = (\mathbf{x} \cdot \mathbf{x})\mathbf{y} = \mathbf{x} \cdot (\mathbf{x} \cdot \mathbf{y}) = \mathbf{x}k = k\mathbf{x}$ as we wished to prove.

   (d) $\|\mathbf{x} - \mathbf{s}\|^2 = \|\mathbf{x}\|^2 - 2\|\mathbf{x}\|\|\mathbf{s}\| + \|\mathbf{s}\|^2$.

*7. Let $ABCD$ be a parallelogram. Define the vectors $\mathbf{u}$ and $\mathbf{v}$ to be $\mathbf{u}=\mathbf{b}-\mathbf{a}$, $\mathbf{v}=\mathbf{d}-\mathbf{a}$.

(a) Express the diagonal vectors $\mathbf{c}-\mathbf{a}$ and $\mathbf{b}-\mathbf{d}$ in terms of $\mathbf{u}$ and $\mathbf{v}$.

(b) Use the algebraic properties of vector norms and scalar products to show that if we add the squares of the lengths of the diagonals we get twice the sum of the squares of the sides.

8. Determine whether the following are true or false.

(a) $\langle 1, -3, 6\rangle$ is on the sphere of radius 46 centered at $(0, 0, 0)$.

(b) $\langle 2, 2, 2\rangle$ is on the sphere of radius $2\sqrt{3}$ centered at $(0, 0, 0)$.

(c) $\langle 8, 5, -3\rangle$ is on the sphere of radius $\sqrt{164}$ centered at $(2, \text{-}3, 5)$.

(d) $\langle 0, 3, 1\rangle$ is on the sphere of radius 8 centered at $(2, -3, 5)$.

9. Find the cosine of the angle θ between the following vectors. Then, without computing θ, determine whether $0 \le \theta < 90$, or $\theta = 90$, or $90° < \theta \le 180°$.

(a) $\mathbf{a}$ and $\mathbf{b}$ of Exercise 2.

(b) $\mathbf{a}$ and $\mathbf{c}$ of Exercise 2.

(c) $\mathbf{c}$ and $\mathbf{b}$ of Exercise 2.

10. The sides of a box make a rectangle. If the points in Example 5.16 really make a box, where $ABCD$ is a face, then $ABCD$ should have four right angles. Check this.

11. Do the previous exercise for face $EFGH$ of Example 5.16.

12. Find the equation of the plane with the given normal $\mathbf{n}$, passing through the given point P. Give the equation in both vector form and scalar form.

(a) $\mathbf{n}=\langle 2, -4, 7\rangle$, $P=(0, 0, 3)$

(b) $\mathbf{n}=\langle 1, 1, 1\rangle$, $P=(0, 0, 0)$

(c) $\mathbf{n}=\langle 0.8, -0.6, 0\rangle$, $P=(1, 2, -3)$.

13. Find the normal to the plane determined by the three given points. Then find the equation of the plane.

(a) $A(1, -2, 1)$, $B(3, 0, 2)$, $C(8, -1, -1)$

(b) $A(2, -4, 5)$, $B(1, -3, 4)$, $C(7, -5, 3)$

(c) $A(6, -2, 1)$, $B(0, 0, 1)$, $C(1, 2, 0)$.

Planes, Polyhedra, and Hidden Surfaces

14. In the previous exercise, suppose the three points are corners of one face of a convex polyhedron and one additional corner F is given that does not lie on the face. In each of the following cases, determine an outer normal to the polyhedron.

    (a) A, B, and C as in part (a) of the previous exercise and $F = (1, 1, 1)$.

    (b) A, B, and C as in part (b) of the previous exercise and $F = (-2, -1, 1)$.

    (c) A, B, and C as in part (c) of the previous exercise and $F = (5, 0, 3)$.

15. Here are the faces of the box in Example 5.16:

    face #1 is $ABCD$,

    face #2 is $EFGH$,

    face #3 is $ABFE$,

    face #4 is $DCGH$,

    face #5 is $BCGF$,

    face #6 is $ADHE$.

    Determine which faces of the box in Example 5.16 are visible if $S = (2, 6, 7)$.

16. Do the previous exercise if $S = (4, 2, 5)$.

17. How would a typical hidden surface calculation come out if the eye point S is taken inside the box?

*18. Let A_1, A_2, A_3, A_4 be the four corners of a four-sided polyhedron called a tetrahedron (see Figure 8.23 in Chapter 8). Let F_i denote the face that does not contain A_i (e.g., F_1 is determined by A_2, A_3, A_4). Let $\mathbf{n}_1$, $\mathbf{n}_2$, $\mathbf{n}_3$, $\mathbf{n}_4$ be outer normals whose norms are 1 for F_1, F_2, F_3, F_4 respectively. Show that $\text{Area}(F_1)\mathbf{n}_1 + \text{Area}(F_2)\mathbf{n}_2 + \text{Area}(F_3)\mathbf{n}_3 + \text{Area}(F_4)\mathbf{n}_4 = \mathbf{0}$. (*Hint:* Think of connections between areas, cross-products, and normals.)

*19. Suppose P is a a pyramid with a four-sided base. If we slice a plane through the apex and a diagonal of the base, we can cut it into two tetrahedra that meet on a common face. Prove that if the faces of P are F_1, F_2, F_3, F_4, F_5, with the unit normal to face F_i being $\mathbf{n}_i$ then $\Sigma\, \text{Area}(F_i)\mathbf{n}_i = \mathbf{0}$. (*Hint:* Use the previous exercise.)

*20. Explain how a polyhedron can be divided into pyramids with various kinds of bases. What does this, along with the previous exercise, suggest about the most general form of the theorem we are dealing with in the last two exercises?

5.3 Norms, Spheres, and the Global Positioning System

Here is a plan for retrieving a stolen car: The car has a radio transmitter and sends a message to the police station saying exactly where it is located. Believable? This is far from a fantasy: Many cars are outfitted with such locating systems today. How is it done? Part of the story is a hidden radio transmitter that can send a message—an old technology. But how does the car know where it is? One answer uses a very new technology, the Global Positioning System (GPS). This is a system of space satellites, receivers on earth that communicate with the satellites, and some vector geometry algorithms. These algorithms are hardwired into the receivers, which are about the size of hand calculators and can be purchased in a store. If you are carrying one of these receivers, you can find your location on the face of the earth (latitude, longitude, and altitude above sea level). Our goal in this section is to tell part of the story using vector geometry.

If there is a GPS handheld device at X in Figure 5.14, it can receive signals from each of four satellites. These signals include the time they were sent, so the handheld device can compute, after noting the arrival time using its own clock, how long the signals were under way. The device can then divide by the speed of the signals (a well-known constant of nature) to find the distance to each satellite.

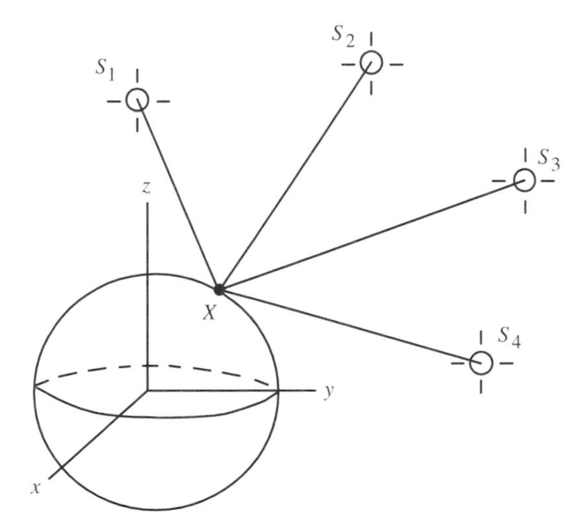

Figure 5.14 Satellites locating X using the GPS.

At this point we have a geometry problem: locating X if we know the distance to the four known satellite locations. But before we show how this can be solved using vector algebra, we should remark that the accuracy of this scheme depends on how well the times can be measured. Atomic clocks can be extremely accurate and are getting better and cheaper all the time. However, at present there are some inaccuracies in the GPS system that can cause errors of 100 yards or more in estimating the coordinates of X. This might be good enough to guide a pilot to a ship in distress, but some applications need better accuracy. For example, earthquake researchers have installed 250 receivers at "fixed" locations in southern California and each needs to be able to measure an inch or two of movement of one crustal plate relative to another. Figure 5.15 shows a GPS receiver at work in Hawaii where volcanic and tectonic changes alter the shape of the island by amounts that can actually be measured year by year. The highest quality handheld GPS units have ways of reducing the errors to a level where very demanding applications are possible.

One source of inaccuracy is that the time signals sent by the satellites are deliberately dithered by the U.S. military, for whom the GPS was constructed, in a

Figure 5.15 A GPS receiver at work on a Hawaiian beach. (Courtesy of Professor Gilbert Strang.)

way that can be decoded and corrected only by the military. (For this reason, only the United States is able to use the sytem to guide missiles accurately to their destinations.) The U.S. president has proposed discontinuing the corruption of the time signals in the future so that more accurate nonmilitary applications of the GPS system will become easier.

Our presentation assumes that the time signals are accurate — even though it may be some years before this is the case. We conclude with some remarks about how present-day receivers can compensate for the inaccuracies in the system today.

Figure 5.14 shows an unknown position X on or above the surface of the earth whose latitude and longitude and height above sea level we want to find. Each of these three measurements can be determined if we know the (x, y, z) coordinates of X in the coordinate system with origin at the center of the earth; consequently, we restrict our attention to finding x, y, and z for point X. Each satellite is in a known position; that is, the coordinates of S_i, (x_i, y_i, z_i), are known. We now describe how it is possible to calculate the location of X from the points S_i and their distances to X.

GPS and Sphere Intersections

The mathematics starts with a geometric visualization of the problem in terms of spheres. If we know that the distance from X to S_1 is d_1, then we know that X is located on the sphere of radius d_1 around the known point S_1. Thus the location of the unknown point X has become a little less unknown. In the same way, we know that X lies on three other spheres. Figure 5.16 suggests that the intersection of these four spheres is normally a single point. Each part of the figure takes the intersection from the part above and adds one more sphere to the intersection. (There are some exceptional cases in which spheres don't intersect in this way. More detail about this later.) This leaves the question of how to calculate the coordinates of the point of intersection.

The scalar product and the norm are key concepts in calculating the coordinates of X. If we have a sphere of radius d_i, centered at S_i, and if X is a variable point on the sphere we can write

$$\|\mathbf{x} - \mathbf{s}_i\|^2 = (\mathbf{x} - \mathbf{s}_i) \cdot (\mathbf{x} - \mathbf{s}_i) = d_i^2.$$

Using the algebraic properties of the dot product, and applying them for $i = 1, 2, 3, 4$:

$$\begin{aligned} \mathbf{x} \cdot \mathbf{x} - 2\mathbf{x} \cdot \mathbf{s}_1 + \mathbf{s}_1 \cdot \mathbf{s}_1 &= d_1^2, \\ \mathbf{x} \cdot \mathbf{x} - 2\mathbf{x} \cdot \mathbf{s}_2 + \mathbf{s}_2 \cdot \mathbf{s}_2 &= d_2^2, \\ \mathbf{x} \cdot \mathbf{x} - 2\mathbf{x} \cdot \mathbf{s}_3 + \mathbf{s}_3 \cdot \mathbf{s}_3 &= d_3^2, \\ \mathbf{x} \cdot \mathbf{x} - 2\mathbf{x} \cdot \mathbf{s}_4 + \mathbf{s}_4 \cdot \mathbf{s}_4 &= d_4^2. \end{aligned} \tag{5.12}$$

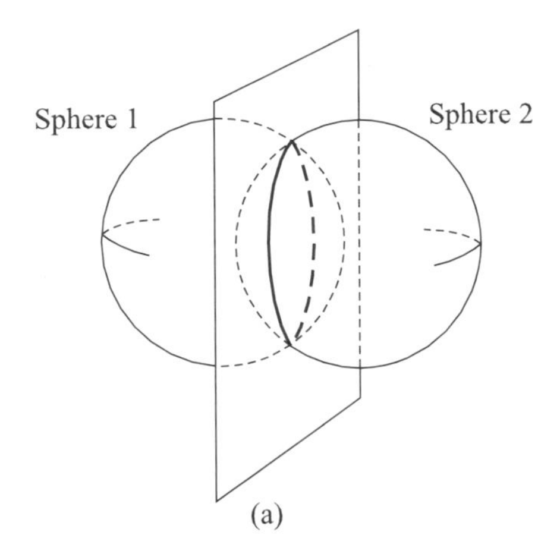

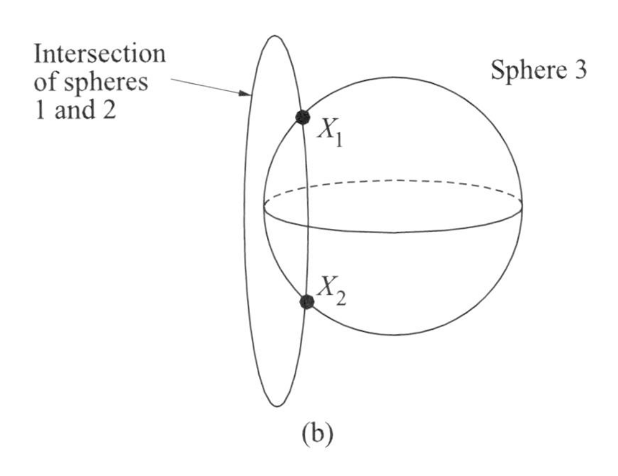

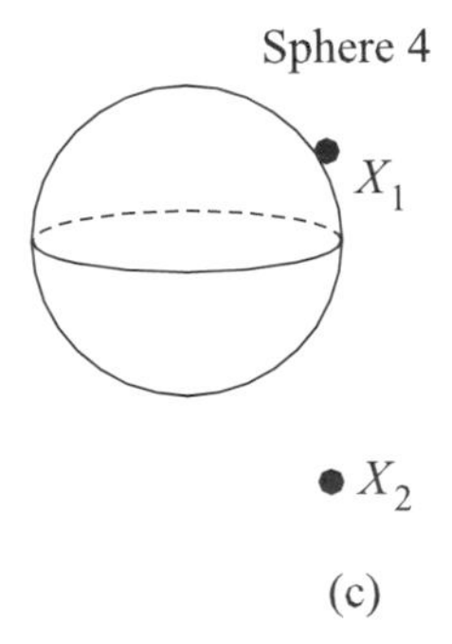

Figure 5.16 (a) Two spheres intersect in a circle lying in the radical plane. (b) A third sphere cuts intersection down to two points. (c) A fourth sphere cuts intersection down to one point.

Radical Planes

Can we solve this system of equations for the unknown vector $\mathbf{x}$? Let's examine the first and the fourth equations. If $\mathbf{x}_0$ satisfies both equations, then we have Eqs. (5.13) and (5.14) below. We could subtract one from the other, obtaining an equation of a plane called the *radical plane* of the two spheres.

$$\mathbf{x}_0 \cdot \mathbf{x}_0 - 2\mathbf{x}_0 \cdot \mathbf{s}_1 + \mathbf{s}_1 \cdot \mathbf{s}_1 = d_1^2 \qquad \text{(means } X_0 \text{ is on sphere 1)}, \tag{5.13}$$

$$\mathbf{x}_0 \cdot \mathbf{x}_0 - 2\mathbf{x}_0 \cdot \mathbf{s}_4 + \mathbf{s}_4 \cdot \mathbf{s}_4 = d_4^2 \qquad \text{(means } X_0 \text{ is on sphere 4)}, \tag{5.14}$$

$$2\mathbf{x}_0 \cdot (\mathbf{s}_4 - \mathbf{s}_1) = d_1^2 - d_4^2 + \mathbf{s}_4 \cdot \mathbf{s}_4 - \mathbf{s}_1 \cdot \mathbf{s}_1 \qquad \text{(means } X_0 \text{ is on radical plane)}. \tag{5.15}$$

What we have shown is that if a point satisfies the first and fourth sphere equations, then its position vector satisfies

$$2\mathbf{x} \cdot (\mathbf{s}_4 - \mathbf{s}_1) = d_1^2 - d_4^2 + \mathbf{s}_4 \cdot \mathbf{s}_4 - \mathbf{s}_1 \cdot \mathbf{s}_1. \tag{5.16}$$

We have *not* shown that every solution to Eq. (5.16) is a vector corresponding to a point that lies on both spheres. In fact, there will be points satisfying Eq. (5.16) that do not lie on both spheres. On the other hand, if Eqs. (5.15) and (5.14) hold for example, we could add them to show that Eq. (5.13) holds. We'll soon see the geometric interpretation of this.

Equations just like Eq. (5.16) are obtained with subscript 2 in place of 1, and with subscript 3 in place of 1. We summarize these three equations with a single one as follows:

$$2\mathbf{x} \cdot (\mathbf{s}_4 - \mathbf{s}_i) = d_i^2 - d_4^2 + \mathbf{s}_4 \cdot \mathbf{s}_4 - \mathbf{s}_i \cdot \mathbf{s}_i, \qquad \text{for } i = 1, 2, 3. \tag{5.17}$$

Since the $\mathbf{s}_i$ are known vectors, what we have in Eq. (5.17) is three different plane equations, one for $i=1$, one for $i=2$, and one for $i=3$. All that remains is to solve these equations simultaneously.

EXAMPLE 5.18

Find the intersection of the following four spheres:

	Center, S_i	Radius, d_i
Sphere 1	$(-2, 0, 0)$	$2\sqrt{2}$
Sphere 2	$(1, 0, 0)$	$\sqrt{5}$
Sphere 3	$(-2, 0, 2)$	2
Sphere 4	$(0, 1, 0)$	$\sqrt{5}$

SOLUTION

Equation (5.17) becomes

$$2\mathbf{x} \cdot (\langle 0, 1, 0\rangle - \mathbf{s}_i) = d_i^2 - 4 - \mathbf{s}_i \cdot \mathbf{s}_i, \qquad \text{for } i = 1, 2, 3.$$

$$\begin{array}{llcr} \text{For } i = 1: & 2\mathbf{x} \cdot \langle 2, 1, 0\rangle = 0 & \text{or} & 4x + 2y = 0, \\ \text{For } i = 2: & 2\mathbf{x} \cdot \langle -1, 1, 0\rangle = 0 & \text{or} & -2x + 2y = 0, \\ \text{For } i = 3: & 2\mathbf{x} \cdot \langle 2, 1, -2\rangle = -8 & \text{or} & 4x + 2y - 4z = -8. \end{array}$$

We gather these into a system of three equations in three unknowns:

$$\begin{aligned} 4x + 2y &= 0, \\ -2x + 2y &= 0, \\ 4x + 2y - 4z &= -8. \end{aligned}$$

The first two equations can be solved to give $x = 0, y = 0$. Substituting these values into the third gives $z = 2$, so $X = (0, 0, 2)$. You should check that this point lies on all the spheres. Calculate its distance to each center. ■

In the next example, we see that it may not be possible to solve the system of equations for a unique solution.

EXAMPLE 5.19

Find X for the same data as in the previous example, except that the third sphere is replaced by one with $S_3 = (0, -2, 0)$ and $d_3 = 2\sqrt{2}$.

SOLUTION

Since spheres 1, 2, and 4 are the same as in the previous example, the equations for $i = 1$ and $i = 2$ are the same. As before, these tell us that $x = 0, y = 0$. But for $i = 3$, using the replacement sphere, we get the equation

$$2\mathbf{x} \cdot \langle 0, 3, 0\rangle = 0 \qquad \text{or} \qquad 6y = 0.$$

This tells us $y = 0$, which we already knew, but gives us no information about the value(s) of z. In fact, a more extensive analysis involving parametric equations of the z axis shows that there are two possible z values, $z = 2$ and $z = -2$. You can verify by substitution into the sphere equations that the two corresponding points, $(0, 0, 2)$ and $(0, 0, -2)$, lie on all four spheres. ■

The geometric reason that we get two solutions in the previous example can be seen by reference to Figure 5.16. Just imagine that the last sphere (bottom figure) happens to go through both points X_1 and X_2. Although this would be an unlikely coincidence, it could happen and it calls our attention to the fact that Figure 5.16 only shows what the "normal" state of affairs is. In what other ways can four distinct spheres intersect? Here are the possibilities:

1. The intersection is empty. This is geometrically possible for randomly chosen spheres, but it can't happen in the GPS problem because we know that the place where the handheld device is located (X in Figure 5.14) lies on each of the spheres.
2. One point in the intersection, the usual case.
3. Two points in the intersection, as in the previous example.
4. The intersection is an entire circle of points, an infinite set.
5. The intersection is an entire sphere.

It turns out to be impossible to have a three-point intersection or any other finite number of points greater than two. Cases 3, 4, and 5 are called *degeneracies*, which means that some unusual and unlikely coincidences have occurred. Case 4 occurs only in the unlikely case when the four sphere centers are collinear. Case 3 happens only when the spheres have their centers all in the same plane, also unlikely. Can you explain Case 5?

In planning the earth orbits for the GPS satellites, it was obviously necessary to ensure that none of the degenerate situations would occur. But there are also other problems of a practical, nonmathematical nature that affect the satellites. Because it is necessary for the handheld device to be exchanging radio signals with the satellite, a satellite is useless if it is invisible because it is below the horizon, that is, if the straight line from X to S_i passes through the earth. If there are only 4 satellites altogether, it is certain that there will be locations on the earth that cannot see at least one of the satellites. In fact, there are 24 satellites now in orbit, and in most cases many more than 4 satellites are visible. This makes it possible to avoid degeneracies and the problem of invisible satellites. It also leads to the interesting question of whether a more accurate answer for X can be obtained by somehow using more than 4 satellites. Least squares estimation methods have been devised to get extra accuracy from each extra satellite.

The equations in (5.12) are clearly equations of planes. But what planes? What is their relationship to the spheres we started with? Consider the first plane equation, which we got from the first and fourth sphere equations. We have noted that each point on the intersection of the first and fourth spheres lies in this plane. This plane is therefore the plane determined by the circle of intersection, provided the spheres intersect in a circle. We call it the *radical plane* corresponding to the two spheres whose equations were used to get it (Figure 5.16a). As you can see from Eq. (5.17), the normal vector of the radical plane formed by spheres 4 and i is $\mathbf{s}_4 - \mathbf{s}_i$, so that radical plane is perpendicular to the vector connecting the sphere centers.

It is important to be aware that there are points on the radical plane — in fact, most of them — which are not in the intersection of the two spheres. However, it is true that if a point is in the radical plane, and is on one of the spheres, it is also on the other sphere.

☆ THEOREM 5.1

Suppose R is the radical plane of two spheres, Sphere 1, Sphere 2. If X_0 lies in the intersection of any two of the sets R, Sphere 1, and Sphere 2, it lies in the third also.

PROOF

Suppose we denote the centers of the spheres as S_1 and S_4 and the radii as d_1 and d_4 so that we can refer to Eqs. (5.12)–(5.15). If X_0 is on the two spheres, then Eqs. (5.13) and (5.14) hold. Subtracting Eq. (5.14) from (5.13), we get Eq. (5.15), which means X_0 is on R.

But now suppose X_0 is on R and the sphere is centered at S_4. This means Eqs. (5.15) and (5.14) hold. We can add these equations to get Eq. (5.13). Thus X_0 is on the sphere centered at S_1.

We leave to the reader the task of showing that if X_0 is on R and is also on the sphere centered at S_1 then it is on the sphere centered at S_4. ■

Dealing with Inaccuracy

The theory we have developed so far assumes that the distances d_i are accurate — a situation that is not really the case at present. However, at any given time, the inaccuracy in each distance is nearly the same. If d'_i denotes the measured distance from X to the ith satellite S_i, and d_i denotes, as usual, the true distance then:

$$d_1 - d'_1 = d_2 - d'_2 = d_3 - d'_3 = d_4 - d'_4. \tag{5.18}$$

From the first equation we can conclude that

$$d_1 - d_2 = d'_1 - d'_2.$$

The right side of this equation is a known quantity, since we compute it by subtracting one measured distance from another. Setting this to be the constant c_{12} and converting to vector and norm notation, we get

$$\|\mathbf{x} - \mathbf{s}_1\| - \|\mathbf{x} - \mathbf{s}_2\| = c_{12}.$$

The locus of points that satisfies this equation is called a *hyperboloid*, a three-dimensional relative of the hyperbola. By considering subscripts 1 and 3 in Eq. (5.18) and then subscripts 1 and 4 we obtain two more hyperboloids on which the unknown point $\mathbf{x}$ must lie. Thus we have the problem of finding the intersection of three hyperboloids, which is a bit more difficult than the sphere intersection problem we have when the distances are accurate.

EXERCISES

**Marks challenging exercises.*

GPS and Sphere Intersections

1. Find the equation of the sphere centered at (1, −2, 3) and having radius 2. First express it using norms. Then express your equation using scalar products. Finally, expand and simplify, as we did to get the equations in (5.12).

2. Find the equation of the sphere centered at (0, 2, −4) and having radius $\sqrt{5}$. First express it using norms. Then express your equation using scalar products. Finally, expand and simplify, as we did to get the equations in (5.12).

3. Find the intersection of the following four spheres:

   (a) $S_1 = (1, 0, 0)$, radius $\sqrt{3}$

   (b) $S_2 = (2, 0, 3)$, radius $\sqrt{5}$

   (c) $S_3 = (3, 2, 3)$, radius $\sqrt{6}$

   (d) $S_4 = (3, 0, 1)$, radius $\sqrt{2}$.

4. Find the intersection of the following four spheres:

   (a) $S_1 = (4, 5, 7)$, radius 1

   (b) $S_2 = (6, 5, 6)$, radius 2

   (c) $S_3 = (4, 8, 6)$, radius 3

   (d) $S_4 = (3, 4, 6)$, radius $\sqrt{2}$

Radical Planes

5. Find the radical plane of these two spheres: Sphere 1 is centered at $S_1 = (2, 3, -4)$ and has radius 5, while Sphere 2 is centered at $S_2 = (5, -6, 0)$ and has radius 7.

6. Find the radical plane of the following two spheres: $S_1 = (2, 4, -1)$ and radius 3; $S_2 = (0, 1, -2)$ and radius 3. Express it in vector form, $\mathbf{x} \cdot \mathbf{n} = k$, where k and $\mathbf{n}$ are specified numerically. Then express it in terms of the components of $\mathbf{x}$, namely, x, y, and z.

7. Find the radical plane of the following two spheres: $S_1 = (0, 0, 0)$ and radius $\sqrt{3}$; $S_2 = (3, 3, 3)$ and radius $2\sqrt{3}$. Express it in vector form, $\mathbf{x} \cdot \mathbf{n} = k$, where k and $\mathbf{n}$ are specified numerically. Then express it in terms of the components of $\mathbf{x}$, namely x, y, and z.

8. Does the radical plane of two spheres, centered at S_1 and S_2 respectively, always pass through the midpoint of the segment from S_1 to S_2? If this is true, prove it. Otherwise, give a persuasive argument or show a counterexample and do the calculations to prove that it really is a counterexample.

To understand spheres and radical planes and their vector algebra, it can be helpful to first deal with circles, lines in the plane, and their vector algebra. The following series of five exercises explores this. In these exercises all points have two coordinates and all vectors have two components.

9. In two-dimensional geometry, find a vector form of the equation of a line through P and orthogonal to $\mathbf{n}$. (Examine the theory for planes.) How does this differ from the equation of a plane?

10. Let C denote a point in the plane and let d be a positive number. Find the vector equation of the circle centered at C with radius d. First express it in terms of norms, then in terms of scalar products. How does the scalar product form differ from the equation of a sphere?

11. The radical axis of two circles in the plane is the line whose equation results by subtracting one circle equation from the other. Find the radical axis (in vector form) of two circles, each of radius 2, the first centered at (1, 2) and the second at (4, 3).

12. Show, using vector algebra, that if $\mathbf{x}_1$ and $\mathbf{x}_2$ are position vectors for points X_1 and X_2 that lie in the plane on a circle centered at the origin and with radius 1, then the midpoint of $\overline{X_1X_2}$, whose position vector is $(\mathbf{x}_1 + \mathbf{x}_2)/2$, does not lie on the circle. [*Hint:* use Eqs. (5.6)–(5.9) to estimate the square of the distance of the midpoint from the center.]

13. In two-dimensional geometry show that there is only one point, a point of tangency, on the circle of radius 1 around the origin and on the line $\mathbf{x} \cdot \mathbf{u} = 1$ where $\mathbf{u}$ is a unit length vector. [*Hint:* Use Eqs. (5.6) and (5.7) of the previous section to evaluate $\|\mathbf{x} - \mathbf{u}\|^2$.]

14. Show that there is only one point, a point of tangency, on the sphere of radius 1 around the origin and on the plane $\mathbf{x} \cdot \mathbf{u} = 1$ where $\mathbf{u}$ is a unit length vector. (*Hint:* What can you find out about $\mathbf{x} - \mathbf{u}$?)

15. Prove that if a plane and sphere of radius r centered at the origin have nonempty intersection, then the intersection is a point or a circle. Here is one approach using vectors. Let $\mathbf{x}$ be the position vector of a point in the intersection of the sphere with a certain plane whose equation is $\mathbf{x} \cdot \mathbf{n} = k$. Imagine a perpendicular dropped from the center (the origin) to the plane, meeting it at F.

    (a) Sketch this.

    (b) How is $\mathbf{f}$ related to $\mathbf{n}$?

    (c) Study $\|\mathbf{x} - \mathbf{f}\|^2$ and show that the distance from X to F is determined by $\mathbf{f}$, $\mathbf{n}$, k, and r; that is, it comes out the same for every X in the intersection.

16. Suppose a sphere is centered at a point S, which lies on the x-y plane, $z = 0$. Suppose (a, b, c) lies on the sphere. Show that $(a, b, -c)$ does also.

17. Suppose four spheres all have their centers S_i on the plane $z = 0$. Show that if they have a point in common above the x-y plane, then they also have a point below the x-y plane in their intersection. (*Hint*: See the previous exercise.)

18. Suppose a sphere of radius r is centered at a point S which lies on the plane $\mathbf{x} \cdot \mathbf{u} = k$ where $\mathbf{u}$ is a unit length vector. Suppose P lies on the sphere.

    (a) Show that $\mathbf{p}' = \mathbf{p} - 2(\mathbf{p} \cdot \mathbf{u} - k)\mathbf{u}$ is also the position vector of a point P' on the sphere.

    (b) Use vector algebra to show that the midpoint of $\overline{PP'}$ is on the plane.

    (c) Show that segment $\overline{PP'}$ is perpendicular to the plane.

19. Suppose four spheres all have their centers S_i on the plane $\mathbf{x} \cdot \mathbf{u} = k$ where $\mathbf{u}$ is a unit length vector. Show that if the spheres have a point in common that is not on this plane, then they also have another point in their intersection. (*Hint:* Consider the previous exercise.)

20. Consider the following four spheres:

    Sphere 1 has center at (0, 0, 0) and radius $\sqrt{3}$;

    Sphere 2 has center at (0, 1, 0), radius $\sqrt{2}$;

    Sphere 3 has center at (0, 2, 0), radius $\sqrt{3}$; and

    Sphere 4 has center at (0, 3, 0), radius $\sqrt{6}$.

    Show by calculation that the radical plane for Sphere 1 and Sphere 4 is the same as for Sphere 2 and Sphere 4, which is the same as for Sphere 3 and Sphere 4. What does this mean for the intersection of all four spheres?

*21. (a) Show that if three spheres have their centers S_1, S_2, and S_3 on a straight line and if the spheres have a point in common, then no matter which pair of spheres we pick, their radical plane is the same. (*Hint:* Think about the parametric equations of a line. Show that the radical plane equations are all equivalent.)

    (b) Suppose four spheres have their centers S_i on a straight line and have a point in common. What does this mean about the intersection of the four spheres?

22. It can be proved that a line intersects a sphere at most twice. Demonstrate this by finding the intersection of the sphere of radius 10 centered at the origin and the line joining the points (−1, 2, 0) and (1, 1, 1). Suppose the line joins points A and B whose coordinates are not known?

23. If the intersection of four spheres contains at least three noncollinear points P, Q, and R, then no matter which pair of spheres we pick, their radical plane is the same. Explain why this is so. What does this mean for the intersection of the four spheres? (*Hint:* Given three points which are not on a line, there is precisely one plane containing them. Also, see the previous exercise.)

5.4 Curve Fitting with Splines

Artists have always been interested in drawing graceful curves—sweeping through two-dimensional space in ways more appealing to the eye than the relatively predictable straight lines and circles of elementary geometry. In "practical" subjects too, such as the design of ships and aircraft, there is a need for curves with custom twists and turns. Geometry, with a little help from calculus, can completely satisfy all of these needs.

The problem of drawing custom curves can be thought of as the inverse of the familiar problem of plotting points to make a graph of a given equation. Suppose, for example, we have parametric equations $x = t^2 - 3$, $y = t - 1$. We might select values $t = 0, 1, 2, 3$ to substitute in order to get four points (x, y) to plot. Substituting $t = 0$ gives $P_0 = (0^2 - 3, 0 - 1) = (-3, -1)$. Likewise, we get $P_1 = (-2, 0)$, $P_2 = (1, 1)$, $P_3 = (6, 2)$. But now we want to stand this process on its head:

DEFINITION

Suppose we are given the numerical coordinates of points $P_0, P_1, \ldots, P_n$ and we are looking for equations that create a curve passing through all of them. This is called fitting a curve to the given points, or simply *curve fitting*.

APPLICATION: Magnifying a Curve

EXAMPLE 5.20

For a scenario in which the need for curve fitting comes about, imagine an artist drawing a curve on a computer screen by moving a mouse on a mouse pad. Every hundredth of a second or so the computer records the current mouse position in terms of x and y screen coordinates. In this way the computer might obtain a few hundred closely spaced (x, y) pairs. If the computer puts a little dot of color at each of these screen positions, we will perceive the result as a continuous curve, with no visible gaps between the dots. But what if the drawing is to be magnified to 100 times its original size? To do this, the computer would first multiply each of the recorded x and y coordinates by 100 in order to space out the points. If the computer's next step were to simply put a dot of color at each of these new points, now there would be visible spaces between the dots (compare Figures 5.17a and 5.17b). It would be handy to have formulas that allow the computer to calculate as many new "in-between" dots as we need so the eye will be fooled into seeing a curve with no gaps.

Say, for example, that the first two points before magnification are (0.01, 0.01) and (0.02, 0.03). In the table of magnified values where we have multiplied by 100, the first two points are $P_0(1, 1)$ and $P_1(2, 3)$. Suppose we have found that the following parametric equations give a curve passing through P_0 at $t = 0$ and P_1 at $t = 1$:

$$x(t) = -t^3 + 2t^2 + 1, \tag{5.19}$$

$$y(t) = -3t^3 + 4t^2 + t + 1. \tag{5.20}$$

Now substitute closely spaced t values between 0 and 1, say, $t = 0.01, 0.02, \ldots, 0.99$, to get points to plot that will fill in the gap between P_0 and P_1.

In this example, we have not explained how the computer might obtain Eqs. (5.19) and (5.20). As Figures 5.17c and d suggest, some possibilities may be better than others. How shall we choose? Read on. ■

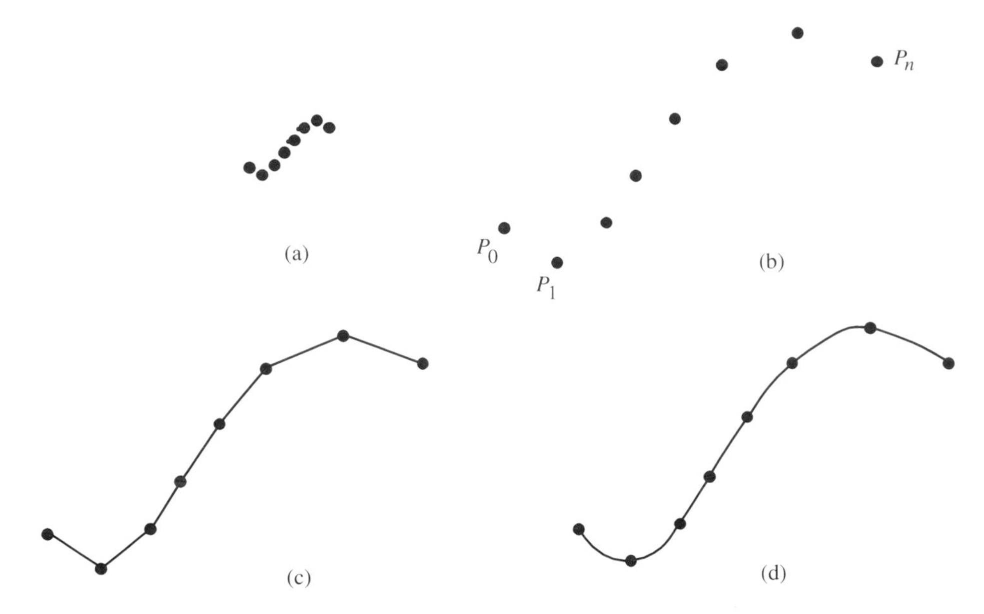

Figure 5.17 (a) and (b) Magnifying a curve makes gaps more visible. (c) and (d) How should we connect the dots?

In our work we will be dealing with *vector polynomials*, functions of the form

$$\mathbf{p}(t) = \mathbf{c}_0 t^m + \mathbf{c}_1 t^{n-1} + \ldots + \mathbf{c}_m .$$

Each $\mathbf{c}_i$ is a two-component vector of constants $\mathbf{c}_i = \langle a_i, b_i \rangle$. You may think of these either as column vectors or row vectors since we won't need to multiply them by matrices. For any t value, the right side evaluates to a two-component vector, so $\mathbf{p}(t)$ can be regarded as a position vector for a point in two-dimensional space varying along a curve parameterized by t. In the previous example, Eqs. (5.19) and (5.20) can be combined to form one cubic vector polynomial

$$\mathbf{p}(t) = \langle x(t), \ y(t) \rangle = \langle -1, \ -3 \rangle t^3 + \langle 2, \ 4 \rangle t^2 + \langle 0, \ 1 \rangle t + \langle 1, \ 1 \rangle .$$

One of our goals in this section is to see that, in some ways, we can deal with vector polynomials as if they were ordinary polynomials.

EXAMPLE 5.21

Suppose a moving point's x and y coordinates at time t are given by $x(t) = -2t^2 + 3t - 1$ and $y(t) = t^2 + t + 4$. Find the $\mathbf{c}_i$ for the vector polynomial that describes the curve along which the point moves.

SOLUTION

$\langle x(t),\ y(t)\rangle = \langle -2t^2 + 3t - 1,\ t^2 + t + 4\rangle$. We break this vector apart into a second degree term, a first degree term, and the constant term

$$\begin{aligned} \mathbf{p}(t) &= \langle x(t),\ y(t)\rangle \\ &= \langle -2t^2 + 3t - 1,\ t^2 + t + 4\rangle \\ &= \langle -2,\ 1\rangle t^2 + \langle 3,\ 1\rangle t + \langle -1,\ 4\rangle. \end{aligned}$$

$$\mathbf{c}_0 = \langle -2,\ 1\rangle, \qquad \mathbf{c}_1 = \langle 3,\ 1\rangle, \qquad \mathbf{c}_2 = \langle -1,\ 4\rangle. \ ■$$

We discuss two strategies for fitting the points $P_0, P_1, \ldots, P_n$ with a curve:

1. Finding a curve that is described by a single vector polynomial, and
2. Finding a curve which requires a number of vector polynomials, called *splines*, to describe it.

Curve Fitting with a Single Vector Polynomial

EXAMPLE 5.22: FITTING WITH A SINGLE QUADRATIC

Find a quadratic vector polynomial that passes through (1, 2) at $t=0$, through (2, 5) at $t=1$, and through (3, 10) at $t=2$. Symbolically these conditions become:

$$\mathbf{p}(t) = \mathbf{c}_0 t^2 + \mathbf{c}_1 t + \mathbf{c}_2, \tag{5.21}$$

$$\mathbf{p}(0) = \langle 1,\ 2\rangle, \tag{5.22}$$

$$\mathbf{p}(1) = \langle 2,\ 5\rangle, \tag{5.23}$$

$$\mathbf{p}(2) = \langle 3,\ 10\rangle. \tag{5.24}$$

SOLUTION

Substituting t values 0, 1, 2 into the quadratic vector polynomial:

$$\begin{aligned} \mathbf{c}_2 &= \langle 1,\ 2\rangle, \\ \mathbf{c}_0 + \mathbf{c}_1 + \mathbf{c}_2 &= \langle 2,\ 5\rangle, \\ 4\mathbf{c}_0 + 2\mathbf{c}_1 + \mathbf{c}_2 &= \langle 3,\ 10\rangle. \end{aligned}$$

Substitute from the first equation into the second and third:

$$\begin{aligned} \mathbf{c}_0 + \mathbf{c}_1 + \langle 1,\ 2\rangle &= \langle 2,\ 5\rangle, \\ 4\mathbf{c}_0 + 2\mathbf{c}_1 + \langle 1,\ 2\rangle &= \langle 3,\ 10\rangle. \end{aligned}$$

Carrying out vector subtractions of $\langle 1,\ 2\rangle$ from both sides:

$$\mathbf{c}_0 + \mathbf{c}_1 = \langle 1,\ 3\rangle,$$
$$4\mathbf{c}_0 + 2\mathbf{c}_1 = \langle 2,\ 8\rangle.$$

Now, in this new system of equations, subtract twice the first from the second to get

$$2\mathbf{c}_0 = \langle 0,\ 2\rangle$$
$$\mathbf{c}_0 = \langle 0,\ 1\rangle.$$

Substitute this into an earlier equation to obtain $\mathbf{c}_1 = \langle 1,\ 2\rangle$ and we get $\mathbf{p}(t) = \langle 0,\ 1\rangle t^2 + \langle 1,\ 2\rangle t + \langle 1,\ 2\rangle$. ■

Most curve fitting problems involve a large number of points for the curve to pass through. Normally, if there are $n + 1$ points $P_0, P_1, \ldots, P_n$ to be passed through at $t = 0, 1, 2, \ldots, n$, a vector polynomial will have to have degree at least n in order to be able to pass through all of them. In place of finding three vector coefficients $\mathbf{c}_0$, $\mathbf{c}_1$, $\mathbf{c}_2$, as we did in the previous example, now we will have to find all $n + 1$ coefficients of $\mathbf{p}(t) = \mathbf{c}_0 t^n + \mathbf{c}_1 t^{n-1} + \ldots + \mathbf{c}_n$. In place of the three requirements of Eqs. (5.22)–(5.24) we have $n + 1$. This system of requirements is a system of $n + 1$ linear equations in the $n + 1$ unknown vectors $\mathbf{c}_i$. In principle, it is possible to solve the system, but if n is large it is quite time-consuming. Our next method gives a quicker solution.

Curve Fitting with Cubic Splines

The method of splines involves breaking the task into pieces. We look for a vector polynomial to provide a short curve C_1 stretching from P_0 to P_1. Then we look for another vector polynomial to stretch from P_1 to P_2 and so on—a total of n vector polynomials. Geometrically, we are getting our curve by putting together a lot of individual shorter curves.

The simplest example of a vector polynomial linking P_0 to P_1 would be the first degree equation $\mathbf{p}(t) = \mathbf{p}_0 + t(\mathbf{p}_1 - \mathbf{p}_0)$. As you will recall from Section 5.1, this is the equation of a straight line. Unfortunately, connecting the dots with straight-line segments will give corners at the points where the lines meet (Figure 5.17c) and this is undesirable for most applications. Instead, we will take each piece to be a cubic vector polynomial—a function of the form

$$\mathbf{p}(t) = \mathbf{c}_0 t^3 + \mathbf{c}_1 t^2 + \mathbf{c}_2 t + \mathbf{c}_3\,. \tag{5.25}$$

We'll find a cubic vector polynomial to provide us with a short curve C_1 stretching from P_0 to P_1. We'll call this vector polynomial $\mathbf{p}_{01}(t)$. A second cubic vector polynomial $\mathbf{p}_{12}(t)$

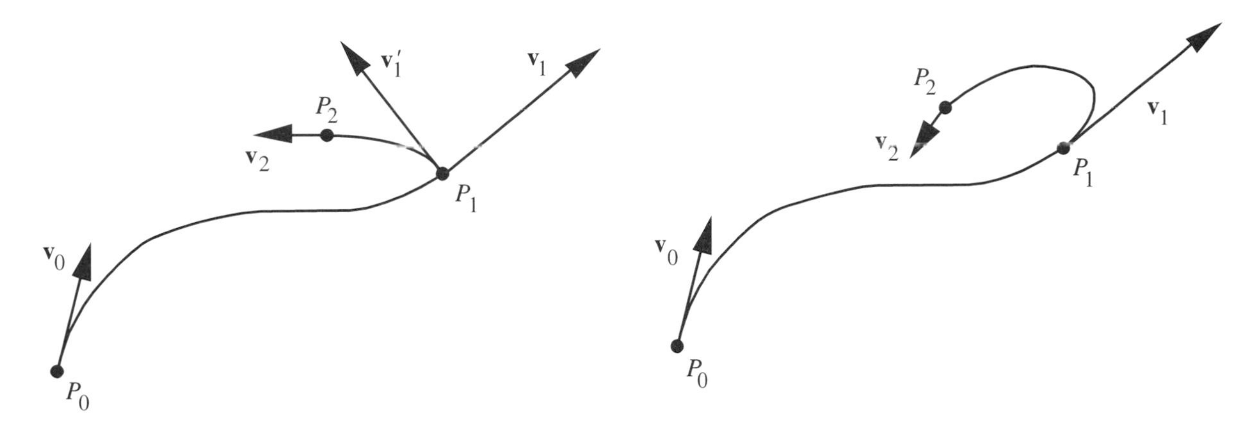

Figure 5.18 (a) C_1 and C_2 making a corner (different tangent vectors $\mathbf{v}_1$ and $\mathbf{v}_1'$). (b) C_1 and C_2 with the same tangent vector $\mathbf{v}_1$ (no corner).

will give a curve C_2 from P_1 to P_2. If we're not careful, C_1 and C_2 could make a corner where they meet at P_1 (Figure 5.18a) so we arrange things so our cubic curves have the same tangent vector at P_1. We continue in this way till we get to P_n. These short curves are called *cubic splines.* There will be no single polynomial formula for the overall curve formed by putting the splines together.

To see how we can find the individual cubic splines that fit together without corners, we need to recall some facts about parametric equations of curves. Suppose we have a curve given by $\mathbf{p}(t)$. Since we are only interested in short stretches of curve, we restrict t to $0 \leq t \leq 1$. If we pick a t value in this interval, t_0, we may ask: What is the tangent vector at $\mathbf{p}(t_0)$? You will recall from calculus that this is computed using derivatives as follows:

$$\text{tangent vector} = \mathbf{p}'(t_0) = \langle x'(t_0),\ y'(t_0)\rangle. \tag{5.26}$$

We are interested in tangent vectors in part because of the following fact:

CURVES THAT HAVE THE SAME TANGENT VECTOR AT THE POINT WHERE THEY MEET WILL APPEAR TO MEET SMOOTHLY WITHOUT CORNERS.

The next theorem gives us a familiar and convenient alternative way to express the derivative in the event $\mathbf{p}(t)$ has the form given in Eq. (5.25).

✯ THEOREM 5.2

If $\mathbf{p}(t)$ has the form in Eq. (5.25) where $\mathbf{c}_i = \langle a_i, b_i \rangle$, then

$$\mathbf{p}'(t) = 3\mathbf{c}_0 t^2 + 2\mathbf{c}_1 t + \mathbf{c}_2. \tag{5.27}$$

PROOF

The formula we need to prove looks just like the formula for the derivative of a nonvector polynomial, which we recall from calculus. This formula for ordinary polynomials is exactly what we rest our proof on.

Equation (5.25) can be written

$$\begin{aligned}\mathbf{p}(t) &= \langle a_0, b_0\rangle t^3 + \langle a_1, b_1\rangle t^2 + \langle a_2, b_2\rangle t + \langle a_3, b_3\rangle \\ &= \langle a_0 t^3 + a_1 t^2 + a_2 t + a_3,\ b_0 t^3 + b_1 t^2 + b_2 t + b_3\rangle.\end{aligned}$$

Now we apply Eq. (5.26):

$$\begin{aligned}\mathbf{p}'(t) &= \langle x'(t), y'(t)\rangle \\ &= \langle 3a_0 t^2 + 2a_1 t + a_2,\ 3b_0 t^2 + 2b_1 t + b_2\rangle \\ &= \langle 3a_0 t^2, 3b_0 t^2\rangle + \langle 2a_1 t, 2b_1 t\rangle + \langle a_2, b_2\rangle \\ &= 3\langle a_0, b_0\rangle t^2 + 2\langle a_1, b_1\rangle t + \langle a_2, b_2\rangle \\ &= 3\mathbf{c}_0 t^2 + 2\mathbf{c}_1 t + \mathbf{c}_2. \quad \blacksquare\end{aligned}$$

EXAMPLE 5.23: TANGENT VECTORS

(a) Find the tangent vector at $t=1$ if $\mathbf{p}(t)$ is given by Eqs. (5.19) and (5.20).

(b) Suppose two curves are given by

$$\begin{aligned}\mathbf{p}(t) &= \langle 1, 0\rangle t^2 + \langle 0, -1\rangle t + \langle 1, 3\rangle \\ \mathbf{q}(t) &= \langle 0, 1\rangle t^2 + \langle 3, 1\rangle t + \langle 2, 2\rangle.\end{aligned}$$

Show that the $\mathbf{p}$ curve at $t=1$ meets the $\mathbf{q}$ curve at $t=0$. Determine whether the curves meet smoothly in the sense that the tangent vectors are equal. Sketch the tangent vectors, with their tails at the meeting points.

SOLUTION

(a) Using Eq. (5.26),

$$\begin{aligned}\mathbf{p}(t) &= \langle x(t), y(t)\rangle = \langle -t^3 + 2t^2 + 1,\ -3t^3 + 4t^2 + t + 1\rangle, \\ \mathbf{p}'(t) &= \langle -3t^2 + 4t,\ -9t^2 + 8t + 1\rangle, \\ \mathbf{p}'(1) &= \langle -3 + 4,\ -9 + 8 + 1\rangle \\ &= \langle 1, 0\rangle.\end{aligned}$$

(b) Substituting $t = 0$ and $t = 1$,

$$\mathbf{p}(1) = \langle 1,\ 0\rangle + \langle 0,\ -1\rangle + \langle 1,\ 3\rangle = \langle 2,\ 2\rangle \text{ while}$$
$$\mathbf{q}(0) = \langle 2,\ 2\rangle,$$

so the curves meet at the point $\langle 2,\ 2\rangle$

Applying Eq. (5.27) to each curve, and noting that $\mathbf{c}_0 = \mathbf{0}$ in each case:

$$\mathbf{p}'(t) = \langle 2,\ 0\rangle t + \langle 0,\ -1\rangle \qquad \text{so } \mathbf{p}'(1) = \langle 2,\ -1\rangle$$
$$\mathbf{q}'(t) = \langle 0,\ 2\rangle t + \langle 3,\ 1\rangle \qquad \text{so } \mathbf{q}'(0) = \langle 3,\ 1\rangle.$$

The tangent vectors are not the same (nor do they even point in the same direction; see Figure 5.19). The meeting is not smooth as required. ■

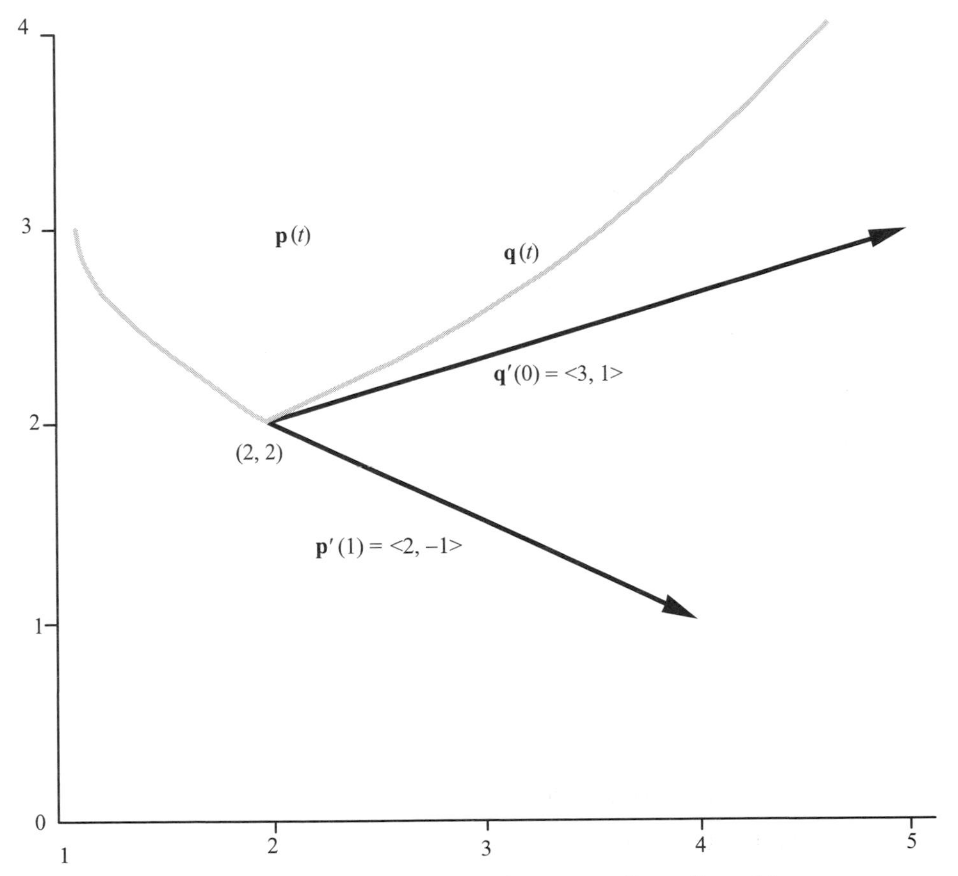

Figure 5.19 Unequal tangent vectors make nonsmooth meeting of two curves.

As you will see, to solve our curve fitting problem, we will be applying the thought process of Example 5.23 in reverse: Instead of having formulas and computing a derivative, we will have derivatives and points the curves must pass through and this will help us find the formulas for the individual splines $\mathbf{p}_{01}(t)$, $\mathbf{p}_{12}(t)$, etc. The splines we find will fit together smoothly.

In curve fitting with cubic splines we make the following requirements for the first spline: It should be a cubic vector polynomial that stretches from points P_0 to P_1 and has given tangent vector $\mathbf{v}_0$ at P_0 and given tangent vector $\mathbf{v}_1$ at P_1. In symbols, requirements for one cubic spline are as follows:

$$\mathbf{p}_{01}(t) = \mathbf{c}_0 t^3 + \mathbf{c}_1 t^2 + \mathbf{c}_2 t + \mathbf{c}_3 \qquad \text{(the } \mathbf{c}_i \text{ to be found)} \tag{5.28}$$

$$\mathbf{p}_{01}(0) = \mathbf{p}_0 \qquad \text{(given)} \tag{5.29}$$

$$\mathbf{p}_{01}(1) = \mathbf{p}_1 \qquad \text{(given)} \tag{5.30}$$

$$\mathbf{p}'_{01}(0) = \mathbf{v}_0 \qquad \text{(given)} \tag{5.31}$$

$$\mathbf{p}'_{01}(1) = \mathbf{v}_1 \qquad \text{(given).} \tag{5.32}$$

You should compare these to Eqs. (5.21) – (5.24).

We need to find the appropriate numerical values for the vectors $\mathbf{c}_i$. Using Eqs. (5.29) through (5.32) to make substitutions in the formulas for $\mathbf{p}_{01}(t)$ and $\mathbf{p}'_{01}(t)$ gives

$$\begin{aligned} \mathbf{c}_3 &= \mathbf{p}_0, \\ \mathbf{c}_0 + \mathbf{c}_1 + \mathbf{c}_2 + \mathbf{c}_3 &= \mathbf{p}_1, \\ \mathbf{c}_2 &= \mathbf{v}_0, \\ 3\mathbf{c}_0 + 2\mathbf{c}_1 + \mathbf{c}_2 &= \mathbf{v}_1. \end{aligned} \tag{5.33}$$

The system of Eqs. (5.33) is not hard to solve; $\mathbf{c}_3$ and $\mathbf{c}_2$ are immediately known. Making these substitutions into the second and fourth equations produces two equations with two unknowns:

$$\begin{aligned} \mathbf{c}_0 + \mathbf{c}_1 &= \mathbf{p}_1 - \mathbf{p}_0 - \mathbf{v}_0, \\ 3\mathbf{c}_0 + 2\mathbf{c}_1 &= \mathbf{v}_1 - \mathbf{v}_0. \end{aligned}$$

Since vectors can be added, subtracted, and multiplied by scalars using the same rules as if they were numbers, we can solve this system by the same techniques as are used for simultaneous linear equations. (For example, begin by subtracting twice the first equation from the second to eliminate the vector $\mathbf{c}_1$.) In this way we obtain:

$$\begin{aligned} \mathbf{c}_0 &= \mathbf{v}_1 - \mathbf{v}_0 - 2(\mathbf{p}_1 - \mathbf{p}_0 - \mathbf{v}_0) = 2\mathbf{p}_0 - 2\mathbf{p}_1 + \mathbf{v}_0 + \mathbf{v}_1, \\ \mathbf{c}_1 &= \mathbf{p}_1 - \mathbf{p}_0 - \mathbf{v}_0 - (2\mathbf{p}_0 - 2\mathbf{p}_1 + \mathbf{v}_0 + \mathbf{v}_1) = -3\mathbf{p}_0 + 3\mathbf{p}_1 - 2\mathbf{v}_0 - \mathbf{v}_1, \\ \mathbf{c}_2 &= \mathbf{v}_0, \\ \mathbf{c}_3 &= \mathbf{p}_0. \end{aligned} \tag{5.34}$$

Our next task is to find the next cubic spline connecting P_1 to P_2, having tangent vectors $\mathbf{v}_1$ and $\mathbf{v}_2$ at these points. It is important to note that we do not have free choice of the tangent vector at P_1—we must choose it to be $\mathbf{v}_1$, the tangent vector chosen (or imposed) for the previous spline at P_1. We get the same formulas as Eqs. (5.34), but with subscript 1 replaced by 2 and subscript 0 replaced by 1. In general, the ith cubic spline (connecting P_{i-1} with P_i) has these coefficients.

$$\begin{aligned} \mathbf{c}_0 &= 2\mathbf{p}_{i-1} - 2\mathbf{p}_i + \mathbf{v}_{i-1} + \mathbf{v}_i, \\ \mathbf{c}_1 &= -3\mathbf{p}_{i-1} + 3\mathbf{p}_i - 2\mathbf{v}_{i-1} - \mathbf{v}_i, \\ \mathbf{c}_2 &= \mathbf{v}_{i-1}, \\ \mathbf{c}_3 &= \mathbf{p}_{i-1}. \end{aligned} \tag{5.35}$$

EXAMPLE 5.24: FITTING THREE POINTS WITH TWO CUBIC SPLINES

Find the two cubic splines for the following data:

$$\begin{aligned} &P_0 = (2,\ 6), \quad &P_1 = (5,\ 10), \quad &P_2 = (4,\ 16), \\ &\mathbf{v}_0 = \langle 1,\ 2\rangle, \quad &\mathbf{v}_1 = \langle 4,\ 2\rangle, \quad &\mathbf{v}_2 = \langle -1,\ 1\rangle. \end{aligned}$$

SOLUTION

Applying Eqs. (5.34) we get the first spline:

$$\begin{aligned} \mathbf{c}_0 &= 2\langle 2,\ 6\rangle - 2\langle 5,\ 10\rangle + \langle 1,\ 2\rangle + \langle 4,\ 2\rangle = \langle -1,\ -4\rangle, \\ \mathbf{c}_1 &= -3\langle 2,\ 6\rangle + 3\langle 5,\ 10\rangle - 2\langle 1,\ 2\rangle - \langle 4,\ 2\rangle = \langle 3,\ 6\rangle, \\ \mathbf{c}_2 &= \langle 1,\ 2\rangle, \\ \mathbf{c}_3 &= \langle 2,\ 6\rangle, \end{aligned}$$

so $\mathbf{p}_{01}(t) = \langle -1,\ -4\rangle t^3 + \langle 3,\ 6\rangle t^2 + \langle 1,\ 2\rangle t + \langle 2,\ 6\rangle$.

Applying Eqs. (5.35) with $i = 2$ we get the second spline:

$$\begin{aligned} \mathbf{c}_0 &= 2\langle 5,\ 10\rangle - 2\langle 4,\ 16\rangle + \langle 4,\ 2\rangle + \langle -1,\ 1\rangle = \langle 5,\ -9\rangle, \\ \mathbf{c}_1 &= -3\langle 5,\ 10\rangle + 3\langle 4,\ 16\rangle - 2\langle 4,\ 2\rangle - \langle -1,\ 1\rangle = \langle -10,\ 13\rangle, \\ \mathbf{c}_2 &= \langle 4,\ 2\rangle, \\ \mathbf{c}_3 &= \langle 5,\ 10\rangle, \end{aligned}$$

so $\mathbf{p}_{12}(t) = \langle 5,\ -9\rangle t^3 + \langle -10,\ 13\rangle t^2 + \langle 4,\ 2\rangle t + \langle 5,\ 10\rangle$. ■

The extensive modern theory of splines has its roots in the shipbuilding and aircraft industries where the structures to be built were so large and the exact shapes so crucial that paper drawings of the shapes were only a first step. For really large structures, each shape was eventually laid out on a floor in a large room, usually a loft, with a long flexible strip of wood or plastic called a *loftsman's spline*. The process was called *lofting*. To keep the spline from wiggling or straightening out, weights called *ducks* were laid down as barriers (Figure 5.20). The effect of these ducks was to force the curve to pass through the "duck points." By taking into account the physics of how a loftsman's spline bends, it can be proved mathematically that the shape of the spline from one duck point to the next is that of a cubic polynomial as in Eq. (5.25). The mathematical process described in this section is a replacement for the time-consuming task of lofting.

Figure 5.20 Aeronautical engineer David Davis lays out a spline with ducks in 1941 (courtesy of General Dynamics).

The theory we have sketched here is for curves in two-dimensional space. It can easily be extended to curves that twist and spiral through given points in three dimensions as well.

EXERCISES

**Marks challenging exercises*

1. (a) Express the following as a vector polynomial: $x(t) = 2t^3 - 3t^2 + 1$, $y(t) = -t^3 + 4t + 3$.

   (b) Find the tangent vector of this vector polynomial at $t=0$, then at $t=1$.

2. (a) Express the following as a vector polynomial: $x(t) = 4t^2 + 1$, $y(t) = -t^4 + 4t + 3$.

   (b) Find the tangent vector of this vector polynomial at $t=0$, then at $t=1$.

Curve Fitting with a Single Vector Polynomial

3. Fit the following three points with a single quadratic vector polynomial: At $t=0$ it should pass through $P_0(2, 5)$, at $t=1$ through $P_1(3, 7)$, at $t=2$ through $P_2(4, 13)$.

4. Fit the following three points with a single quadratic vector polynomial: At $t=0$ it should pass through $P_0(2, -3)$, at $t=1$ through $P_1(3, -1)$, at $t=2$ through $P_2(4, -3)$.

5. The points $P_0(2, 5)$, $P_1(3, 9)$, $P_2(4, 13)$ lie on a line; that is, they could be fit with a single linear vector polynomial $\mathbf{p}(t) = \mathbf{c}_0 t + \mathbf{c}_1$ passing through P_i when $t=i$. But what happens if we try to fit them with a single quadratic vector polynomial? Work this out.

6. (a) Suppose you wanted to fit the following four points with a single quadratic vector polynomial: $P_0(1, 4)$, $P_1(2, 3)$, $P_2(3, 4)$, $P_3(4, 8)$, passing through P_i when $t=i$. If you can do so, do it. If you can't, explain what goes wrong when you try.

   (b) How would you answer part (a) if the last point were changed to (4, 7)?

7. Suppose you wanted to fit the four points in part (a) of Exercise 6 with a single cubic vector polynomial. Can you do it?

8. Find the single cubic spline for the following data: $P_0 = (2, -4)$, $\mathbf{v}_0 = \langle 1, -1 \rangle$, $P_1 = (5, 1)$, $\mathbf{v}_1 = \langle 2, 0 \rangle$.

9. Find the single cubic spline for the following data: $P_0 = (1, 1)$, $\mathbf{v}_0 = \langle 0, 1 \rangle$, $P_1 = (2, 3)$, $\mathbf{v}_1 = \langle 1, 0 \rangle$.

10. Suppose you want a fourth degree vector polynomial $\mathbf{p}(t)$ to

    (i) pass through given points P_0 and P_1,

    (ii) have given tangent vectors $\mathbf{v}_0$ and $\mathbf{v}_1$ at these points, and

    (iii) have a given vector $\mathbf{w}$ as $\mathbf{p}''(0)$.

    (a) What system of equations involving the polynomial coefficients do you obtain?

    (b) Can you solve this system and find the formula for $\mathbf{p}(t)$?

Curve Fitting with Cubic Splines

11. Fit the following data with two cubic splines: $P_0 = (2, 5)$, $P_1 = (3, 7)$, $P_2 = (4, 13)$, $\mathbf{v}_0 = \langle 1, 1 \rangle$, $\mathbf{v}_1 = \langle 1, 4 \rangle$, $\mathbf{v}_2 = \langle 1, 6 \rangle$.

12. Fit the following data with two cubic splines: $P_0 = (2, -3)$, $P_1 = (3, -1)$, $P_2 = (4, -3)$, $\mathbf{v}_0 = \langle 1, 4 \rangle$, $\mathbf{v}_1 = \langle 1, 0 \rangle$, $\mathbf{v}_2 = \langle 1, -4 \rangle$.

*13. (a) In Example 5.24, find the second derivative vector $\mathbf{p}''_{01}(t)$ of the first spline at P_1. [$\mathbf{p}''_{01}(t)$ gives information about the curvature at a point. In some applications it is considered useful for the splines to have the same second derivative vector where they meet.] Find $\mathbf{p}''_{12}(t)$ for the second spline at P_1.

    (b) Find formulas for the two second derivative vectors found in part (a). Your formulas should involve $\mathbf{p}_0$, $\mathbf{p}_1$, $\mathbf{p}_2$, $\mathbf{v}_0$, $\mathbf{v}_1$, $\mathbf{v}_2$.

    (c) Prove that if $3(\mathbf{p}_2 - \mathbf{p}_0) = \mathbf{v}_0 + 4\mathbf{v}_1 + \mathbf{v}_2$ then the second derivatives are equal at the point where the splines meet.

*14. Suppose we are fitting three points P_0, P_1, P_2 with two cubic splines. As usual $\mathbf{v}_0$, $\mathbf{v}_1$, $\mathbf{v}_2$ are given or have been somehow selected. Suppose we want the first cubic spline parameter to run from 0 to s (instead of 0 to 1) and the second to run from 0 to q. The idea is that s and q might be selected differently for the two splines.

    (a) Find formulas analogous to Eqs. (5.34) and (5.35), but now involving s and q as well.

(b) Rework parts (a) and (b) of Exercise 13 taking this change into account.

(c) Show how we can select the coefficients of the two splines and s and q so as to make the second derivative vectors equal where the two splines meet at P_1.

*15. An engineer decides to break up the task of fitting n points in a way different from our cubic spline method. The first curve will be required to pass through the first three points P_0, P_1, P_2 at $t=0, 1, 2$ and to have a given tangent vector $\mathbf{v}_2$ at P_2. The second curve will start at P_2 (at $t=0$) and pass through P_3 (at $t=1$) and end at P_4 (at $t=2$). It will be required to have the same tangent vector at P_2 as the first curve. There is no requirement on the tangent vectors at other points. He intends to continue in this way, each new spline passing through three points, and having the same tangent vector where it meets the previous spline and the same tangent vector where it meets the next spline. Assume there are an odd number of points so that this is feasible. What degree should the individual splines have? Describe a set of requirements for this problem. Derive formulas for the vector coefficients of the first spline.

Exercises with a Graphing Utility

16. In Exercise 8, what is the effect of changing the length of the given tangent vector $\mathbf{v}_0$? Use a graphing utility to make a series of plots for different scalar multiples of the original $\mathbf{v}_0$ [for example, try $2\mathbf{v}_0$ and $(-1)\mathbf{v}_0$].

17. Suppose an artist wants two cubic spline curves to fit the points P_0, P_1, P_2 as given in Example 5.24. Tangent vectors $\mathbf{v}_0$ and $\mathbf{v}_2$ are the same as in that example, but the artist has the freedom to choose $\mathbf{v}_1$. She wants to do it so that the curve made by the two splines passes through the points as directly as possible with no unnecessary curving or looping around. She begins with $\mathbf{v}_1 = \langle 4, 2\rangle$ and draws the graph, but then compares the following alternatives (which are each 90° around from the previous one): $\mathbf{v}_1 = \langle -2, 4\rangle$, $\mathbf{v}_1 = \langle -4, -2\rangle$, $\mathbf{v}_1 = \langle 2, -4\rangle$. Use a graphing utility to carry out these experiments. Which $\mathbf{v}_1$ would the artist choose?

18. The artist of Exercise 17 is interested in being able to make a choice of $\mathbf{v}_1$ according to some formula so that she doesn't have to conduct trial and error experiments—as in the last exercise—for every new problem. The formula may involve $\mathbf{p}_0$, $\mathbf{p}_1$, $\mathbf{p}_2$, $\mathbf{v}_0$, $\mathbf{v}_2$. Experiment with some formulas that you could use to calculate $\mathbf{v}_1$. Use a graphing utility to study whether or not they achieve the objective of "directness" in some specific cases you devise.

19. (a) Use a graphing utility to plot the splines in Exercise 11.

    (b) In Exercise 11, suppose you had mistakenly used $-\mathbf{v}_1$ as the first tangent vector for the second spline, while correctly using the given $\mathbf{v}_1$ for the second tangent vector of the first spline. Try to predict the difference it would make to the overall shape formed by the two splines. Make a rough sketch.

    (c) Use the graphing utility to make a precise picture of the overall shape that results from the error described in part (b).

Chapter 6

Transformation Geometry II: Isometries and Matrices

Prerequisites: We assume the following ideas about isometries in the plane (all of which can be found in the first three sections of Chapter 4): the definitions of transformation, isometry, and composition; the definitions of rotation (including the fact that clockwise rotations have negative angles), translation, and reflection; and the fact that an isometry is determined when the images of three given points are known. A little experience with vectors, for example, the ideas commonly found in third semester calculus, is needed. The previous chapter can serve to review this, but the specific topics of the previous chapter are not needed here. Our use of matrices is fairly elementary and the reader is only expected to be familiar with their multiplication and the associative law for multiplication.

It would be hard to name a more important aspect of the world around us than the fact that things move and wind up in new positions. Many of the applications of this chapter — in robotics, computer graphics, CAD/CAM — deal with objects in different positions. The underlying mathematics to deal with this is often the same as in Chapter 4. But now we want to compute locations numerically. This means we will think of points as position vectors, with numerical components, and we will work out formulas and matrices for isometries.

6.1 Equations and Matrices for Familiar Transformations

Changes in the position of an object can often be expressed as rotation, translation, or (less often) reflection, transformations we have encountered in Section 4.1 of Chapter 4, where they were called isometries. Our approach to these isometries is a little different in this chapter: We want to express points in terms of coordinates and to express isometries with formulas and matrices. We also want to know what matrix operation corresponds to doing one isometry after another, the composition of the isometries in the language of Chapter 4.

Transformations and Equations

Rotation

EXAMPLE 6.1: WAVING THE FLAG

Fred is creating a computer simulation of the Indianapolis 500 car race and he wants a figure of a man who will wave the starter's flag (Figure 6.1). The only part of the figure that needs to move is the arm holding the flag straight out. Furthermore, the flag will not change its shape or size through wrinkling, crumpling, or fluttering so what we have is a pure rotation around the point where the arm meets the body. Suppose the coordinates of a corner on the flag (before it is waved) are $P(x, y)$. If the rotation

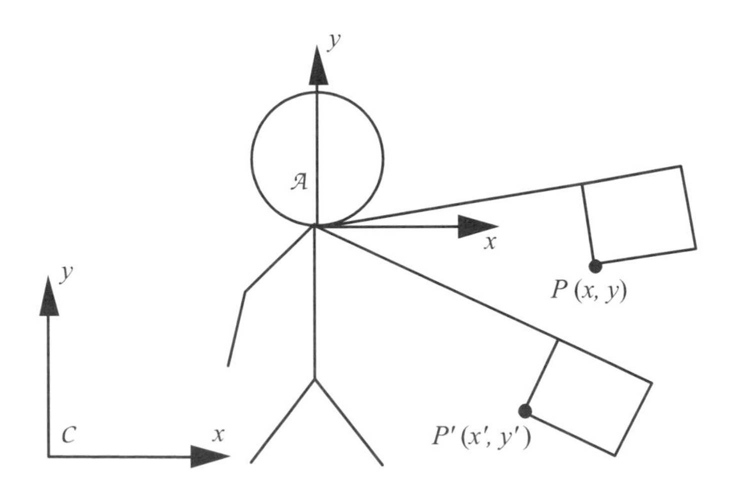

Figure 6.1 Rotating the point P.

of the arm is 30° clockwise, what are the new coordinates, x', y', of that corner of the flag? The graphics system needs to be able to compute this in order to display the new picture on the screen.

The problem actually comes in two versions. In the easier (but less realistic) version, the coordinates x, y are measured from the coordinate frame $\mathcal{A}$ whose origin is at the center of rotation, and the answers x', y' are wanted in terms of $\mathcal{A}$ as well. In that case Eqs. (6.1) given later provide the answer. But a graphics system is more likely to use a coordinate frame, $\mathcal{C}$, at the corner of the screen. What if x, y are measured from $\mathcal{C}$ and the answer is needed in terms of $\mathcal{C}$? We will solve this problem in the next section. ■

We have expressed this problem using the idea of a coordinate frame, which is defined as follows.

DEFINITION

A *coordinate frame* in two-dimensional space is a set of two vectors having unit length and which make a right angle with one another. The vectors are called the x-vector and the y-vector. (It is often just a matter of convenience which is x and which is y, as long as you don't switch in the middle of a problem.)

Frames come in two varieties, called *senses*: *counterclockwise* and *clockwise*. In a frame with counterclockwise sense, the y-vector is 90° around in the counterclockwise direction from the x-vector, while in a clockwise sense frame the y-vector is 90° around in the clockwise direction from the x-vector. (Both frames in Figure 6.1 have counterclockwise sense.)

From a coordinate frame one can extend the x- and y-vectors to get the infinite x and y axes. Using these axes, points can be given coordinates in the usual way. On first glance, a coordinate frame seems little different from a pair of infinite x and y coordinate axes since its purpose is the same: endowing points with coordinates. In a later section we will see an advantage of thinking about frames instead of infinite axes.

In much of mathematics, one coordinate system or frame is given. However, in many applications of geometry, it can be convenient or necessary to have more than one. This is partly because there may be various sensible possibilities for how to position the coordinate frames on a computer screen, on a floor, or wherever the application takes place. If there is more than one coordinate system in a problem and if a point's coordinates are given, make sure you understand which frame was used to obtain the coordinates. Coordinates that are correct for one frame are probably incorrect for a different frame.

If $P(x, y)$ is rotated through α around the origin, it moves to the point P' whose coordinates, (x', y'), are given by the familiar formulas (see the exercises at the end of this section for an outline of a proof):

$$\begin{aligned} x' &= x \cos \alpha - y \sin \alpha, \\ y' &= x \sin \alpha + y \cos \alpha. \end{aligned} \tag{6.1}$$

Note that α can be a positive or negative angle, depending on whether the direction of rotation is counterclockwise (+) or clockwise (−).

EXAMPLE 6.2: COMPUTING THE WAVE

Let's say that the coordinates of P with respect to frame $\mathcal{A}$ are $(7, -1)$ and we wish to subject P to a clockwise rotation of 30°. Then substituting into Eqs. (6.1) gives the coordinates of the new position P':

$$x' = 7 \cos(-30^\circ) - (-1) \sin(-30^\circ) = 7\frac{\sqrt{3}}{2} - (-1)\frac{-1}{2} = \frac{7\sqrt{3}-1}{2} = 5.562,$$

$$y' = 7 \sin(-30^\circ) + (-1) \cos(-30^\circ) = 7\left(\frac{-1}{2}\right) + (-1)\frac{\sqrt{3}}{2} = \frac{-7-\sqrt{3}}{2} = -4.366.$$

If we did not have the coordinates of P with respect to $\mathcal{A}$, but had them with respect to $\mathcal{C}$, say, (13, 4), then we could not work out the coordinates of P' using Eqs. (6.1) because that equation assumes that x and y are coordinates with respect to a coordinate frame whose origin is at the point of rotation. We will see one way to overcome this obstacle in the next section. ■

TRANSLATION

Another common kind of motion is one in which there is no turning and no distortion of the object's shape by stretching or crumpling. Such a motion is called *translation* and we can find equations for it as follows.

Let Q be some point in the body (Figure 4.3 of Chapter 4) and let Q' be the new position after the translation. Recall that we use a boldface, lowercase version of the letter for a point to represent the position vector of that point. Then the *glide vector* or *translation vector* $\mathbf{g} = \mathbf{q}' - \mathbf{q}$ summarizes the motion of the chosen point.[1] We can use the glide vector to find the new location of *any other* given point $P = (x, y)$. Let $P'(x', y')$

[1] See Section 5.1 of Chapter 5 for the notational conventions we use for vectors.

be the new position of P after translation. Because there has been no turning or distortion of the object's size or shape, the line segment $\overline{P'Q'}$ is parallel and equal in length to $\overline{PQ}$. It can be proved that in this case $QQ'P'P$ is a parallelogram. This means that the directed segment PP' is parallel and equal to the segment QQ'. In vector terms, $\mathbf{p}' - \mathbf{p} = \mathbf{q}' - \mathbf{q} = \mathbf{g}$ Thus,

$$\mathbf{p}' = \mathbf{p} + \mathbf{g} \tag{6.2}$$

holds for any point P. We call $\mathbf{g}$ the *translation vector* or *glide vector.*

If we know the components of $\mathbf{g}$, say (x_g, y_g), then we can express Eq. (6.2) in terms of individual components as

$$\begin{aligned} x' &= x + x_g\,, \\ y' &= y + y_g\,. \end{aligned} \tag{6.3}$$

EXAMPLE 6.3: TRANSLATION

A Zamboni glides along the ice of a skating rink without turning. Its center of gravity P moves from $(-1, 3)$ to $(5, 7)$. What is the new location of the corner of the vehicle, which starts out at $(2, -2)$?

SOLUTION

We first find the translation vector by substituting the known values of P and P' into Eq. (6.2): $\mathbf{g} = \mathbf{p}' - \mathbf{p} = \langle 5, 7\rangle - \langle -1, 3\rangle = \langle 6, 4\rangle$. Using Eq. (6.2) again, we then add $\mathbf{g}$ to $\langle 2, -2\rangle$ obtaining $\mathbf{q}' = \mathbf{q} + \mathbf{g} = \langle 2, -2\rangle + \langle 6, 4\rangle = \langle 8, 2\rangle$. ■

In the previous problem, we presented the vectors as row vectors so that the addition could be expressed in a single line. Had we used column vectors we would have written the addition as

$$\begin{pmatrix} 2 \\ -2 \end{pmatrix} + \begin{pmatrix} 6 \\ 4 \end{pmatrix} = \begin{pmatrix} 8 \\ 2 \end{pmatrix}.$$

We will usually express vectors as columns just when we want to multiply them on the left by matrices.

Reflection

For a reflection in the x axis, we have

$$\begin{aligned} x' &= x\,, \\ y' &= -y\,. \end{aligned} \tag{6.4}$$

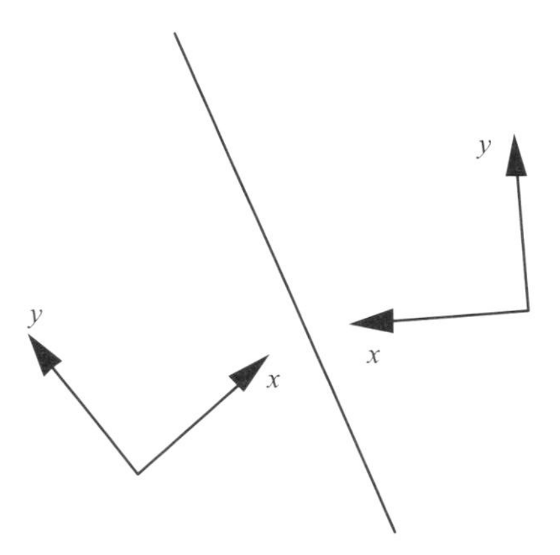

Figure 6.2 Reflection reverses the sense of a frame.

For a reflection in the y axis, we have

$$\begin{aligned} x' &= -x\,, \\ y' &= y\,. \end{aligned} \tag{6.5}$$

For the equations of reflection in an arbitrary line, see the exercises at the end of this section.

It is useful to realize that reflection turns a counterclockwise sense coordinate frame into one with clockwise sense and vice versa (Figure 6.2). By contrast, rotation and translation preserve the sense of a coordinate frame.

Transformations and Matrices

2×2 Matrices

Transformations of the sort we are discussing here are commonly represented by matrices. For example the following matrix equation is another way of writing the identity transformation equations $x'=x$, $y'=y$.

$$\begin{pmatrix} x' \\ y' \end{pmatrix} = \begin{pmatrix} 1 & 0 \\ 0 & 1 \end{pmatrix} \begin{pmatrix} x \\ y \end{pmatrix}.$$

It is not hard to see how to represent the equations for rotation and for reflection in the axes in matrix terms.

Rotation by α — see Eqs. (6.1):

$$\begin{pmatrix} x' \\ y' \end{pmatrix} = \begin{pmatrix} \cos\alpha & -\sin\alpha \\ \sin\alpha & \cos\alpha \end{pmatrix} \begin{pmatrix} x \\ y \end{pmatrix}.$$

We often abbreviate the matrix for rotation around the origin through α by $\mathbf{R}_O(\alpha)$, or more simply by $\mathbf{R}(\alpha)$ or even just $\mathbf{R}$, when it is clear which center and/or angle is involved. (We generally use the same letter for an isometry and its matrix, except that we use the boldface version for the matrix.) Thus, an abbreviated way to write the preceding matrix–vector equation is

$$\mathbf{p}' = \mathbf{R}_O(\alpha)\mathbf{p}.$$

In this equation, it is essential to regard $\mathbf{p}$ and $\mathbf{p}'$ as column vectors. We maintain this convention for all equations involving matrix–vector products in this book.

Reflection in the x axis — see Eqs. (6.4):

$$\begin{pmatrix} x' \\ y' \end{pmatrix} = \begin{pmatrix} 1 & 0 \\ 0 & -1 \end{pmatrix} \begin{pmatrix} x \\ y \end{pmatrix}.$$

Reflection in the y axis — see Eqs. (6.5):

$$\begin{pmatrix} x' \\ y' \end{pmatrix} = \begin{pmatrix} -1 & 0 \\ 0 & 1 \end{pmatrix} \begin{pmatrix} x \\ y \end{pmatrix}.$$

These reflection matrices are abbreviated $\mathbf{M}_x$ and $\mathbf{M}_y$, respectively, so the previous two matrix–vector equations are written $\mathbf{p}' = \mathbf{M}_x\mathbf{p}$ and $\mathbf{p}' = \mathbf{M}_y\mathbf{p}$.

EXAMPLE 6.4: REFLECTION AND ROTATION BY 2×2 MATRICES

Reflect the point $P(3, 4)$ in the y axis, to get P'. Reflect P' in the x axis to get P''. Compare P'' to the result of rotating P through 180° to obtain Q. Do all calculations with matrices and vectors.

SOLUTION

First we do the reflections

$$\mathbf{p}' = \mathbf{M}_y\mathbf{p} = \begin{pmatrix} -1 & 0 \\ 0 & 1 \end{pmatrix} \begin{pmatrix} 3 \\ 4 \end{pmatrix} = \begin{pmatrix} (-1)(3) + (0)(4) \\ (0)(3) + (1)(4) \end{pmatrix} = \begin{pmatrix} -3 \\ 4 \end{pmatrix},$$

$$\mathbf{p}'' = \mathbf{M}_x\mathbf{p}' = \begin{pmatrix} 1 & 0 \\ 0 & -1 \end{pmatrix} \begin{pmatrix} -3 \\ 4 \end{pmatrix} = \begin{pmatrix} -3 \\ -4 \end{pmatrix}.$$

Now for the rotation, substitute 180° into the rotation matrix and multiply by $\mathbf{p}$:

$$\mathbf{q} = \mathbf{R}_O(180^\circ)\mathbf{p} = \begin{pmatrix} -1 & 0 \\ 0 & -1 \end{pmatrix} \begin{pmatrix} 3 \\ 4 \end{pmatrix} = \begin{pmatrix} -3 \\ -4 \end{pmatrix}.$$

These calculations show that $P'' = Q$. We have simply confirmed part (a) of Theorem 4.11 of Section 4.2 in Chapter 4. Draw a picture to illustrate this. ■

Homogeneous Coordinates and 3×3 Matrices

There is a fly in our ointment: We can't express translation using 2×2 matrices. This may not seem much of a loss, since the translation equation, $\mathbf{p}' = \mathbf{p} + \mathbf{g}$, is simple enough. But there are reasons, which we will explain later, why it would be nice if translation and other isometries were "homogeneous" in the sense that all could be expressed as multiplication of a column vector $\mathbf{p}$ by a matrix. To see how this could be done we first present the concept of homogeneous coordinates and homogeneous matrices.

DEFINITION

The *standard homogeneous form* for a point in the plane $P(x, y)$ is the ordered triple of numbers $(x, y, 1)$.

For our current purposes , the idea of adding a "1" on the end of a 2-tuple is best thought of as a simple accounting trick to make our formulas work. However, it is worth noting that there are some extra wrinkles to homogeneous coordinates — especially in the study of projective transformations[2] — that make them much more than an accounting trick.

Suppose we have a point in homogeneous form, $P(x, y, 1)$, and another one $P'(x', y', 1)$. Suppose P and P' are related by the following matrix equation:

$$\begin{pmatrix} x' \\ y' \\ 1 \end{pmatrix} = \begin{pmatrix} 1 & 0 & x_g \\ 0 & 1 & y_g \\ 0 & 0 & 1 \end{pmatrix} \begin{pmatrix} x \\ y \\ 1 \end{pmatrix}. \tag{6.6}$$

If we carry out the matrix multiplication we get

$$\begin{pmatrix} x' \\ y' \\ 1 \end{pmatrix} = \begin{pmatrix} x + x_g \\ y + y_g \\ 1 \end{pmatrix}.$$

Equating the corresponding entries of these two column vectors gives three equations. The last is $1 = 1$, which is uninteresting and we ignore it. The first two give the translation equations, Eqs. (6.3). For this reason, the matrix $\mathbf{T}_g$ in Eq. (6.6) is called the homogeneous translation matrix corresponding to the glide vector $\mathbf{g} = \langle x_g, y_g \rangle$.

---

[2] See Chapter 2 of Wylie (1970).

Now we can also create homogeneous forms for the rotation and reflection matrices. For example, rotation can be represented by the following homogeneous equation:

$$\begin{pmatrix} x' \\ y' \\ 1 \end{pmatrix} = \begin{pmatrix} \cos\alpha & -\sin\alpha & 0 \\ \sin\alpha & \cos\alpha & 0 \\ 0 & 0 & 1 \end{pmatrix} \begin{pmatrix} x \\ y \\ 1 \end{pmatrix}.$$

Notice how the homogeneous rotation matrix is closely related to the 2×2 rotation matrix: One simply puts a 1 in the third row and third column and fills out the rest of the third row and third column with 0's. You should check that the same method can be applied to the reflection matrices involving the x and y axes to get the 3×3 homogeneous versions. The general situation can be described by saying that the 3×3 homogeneous matrix $\mathbf{S}_3$ corresponding to a 2×2 matrix $\mathbf{S}_2$ is

$$\mathbf{S}_3 = \begin{pmatrix} \mathbf{S}_2 & \mathbf{0} \\ \mathbf{0} & 1 \end{pmatrix}.$$

In this matrix, you must interpret the upper right boldface $\mathbf{0}$ as a column of two 0's; likewise the lower left boldface $\mathbf{0}$ stands for a row of two 0's. Notice also that we use the same letter $\mathbf{S}$ for the upper left 2×2 matrix as for the entire 3×3 matrix, using subscripts to distinguish them. This use of subscripts to distinguish the size of matrices is sometimes dropped, as you will see in the next example.

EXAMPLE 6.5: 3×3 MATRICES FOR ISOMETRIES

The 3×3 rotation matrix for 30° is

$$\mathbf{R}_O(30^\circ) = \begin{pmatrix} \frac{\sqrt{3}}{2} & \frac{-1}{2} & 0 \\ \frac{1}{2} & \frac{\sqrt{3}}{2} & 0 \\ 0 & 0 & 1 \end{pmatrix}.$$

The 3×3 translation matrix corresponding to a glide vector of $\mathbf{g} = \langle 4, -2 \rangle$ is

$$\mathbf{T}_{\langle 4,-2 \rangle} = \begin{pmatrix} 1 & 0 & 4 \\ 0 & 1 & -2 \\ 0 & 0 & 1 \end{pmatrix}.$$

The 3×3 matrix for reflection in the x axis is

$$\mathbf{M}_x = \begin{pmatrix} 1 & 0 & 0 \\ 0 & -1 & 0 \\ 0 & 0 & 1 \end{pmatrix}. \quad ■$$

EXERCISES

**Marks challenging exercises.*

Transformations and Equations

1. Use the appropriate equations to do the following.

   (a) Find the images of the following points after rotation around the origin by 90°: $P(1, -2)$, $Q(1, 1)$.

   (b) Find the images of P and Q after rotation by 30°.

   (c) Find the images of P and Q after reflection in the x axis.

   (d) Find the images of P and Q after reflection in the y axis.

2. Find the rotation angle required to turn the unit vector $\mathbf{p} = \langle -3/5, 4/5 \rangle$ into $\mathbf{q} = \langle -\sqrt{2}/2, \sqrt{2}/2 \rangle$.

3. Suppose a point P has coordinates (1, 1) in coordinate frame $\mathcal{W}$. Describe or sketch another coordinate frame in which P has the same coordinates. How many such other coordinate frames are there? What do they have in common, if anything?

4. The point $(5, -2)$ is rotated around the origin through 90° in a continuous motion taking 1 second. The motion takes place so that the point moves at a constant velocity. Where has the point moved to by $\frac{1}{2}$ second? $\frac{1}{3}$ second? Express the location of the point after the time elapsed is t, using a formula involving t.

5. Prove that the equations for reflection in the line $y = x$ are $x' = y$, $y' = x$.

The following two exercises result in a derivation of the rotation equations, Eqs. (6.1).

6. Show that if the point $P(x, y)$ is rotated by 90° around the origin, the image is $(-y, x)$ (see Figure 6.3). Do this without using rotation equations (6.1). (Try congruent triangles. For simplicity, deal with the case where P is in the first quadrant.)

*7. Let $P'(x', y')$ be the image of $P(x, y)$ after rotation through α (see Figure 6.4). Let Q be the foot of the perpendicular dropped from P' to $\overline{OP}$.

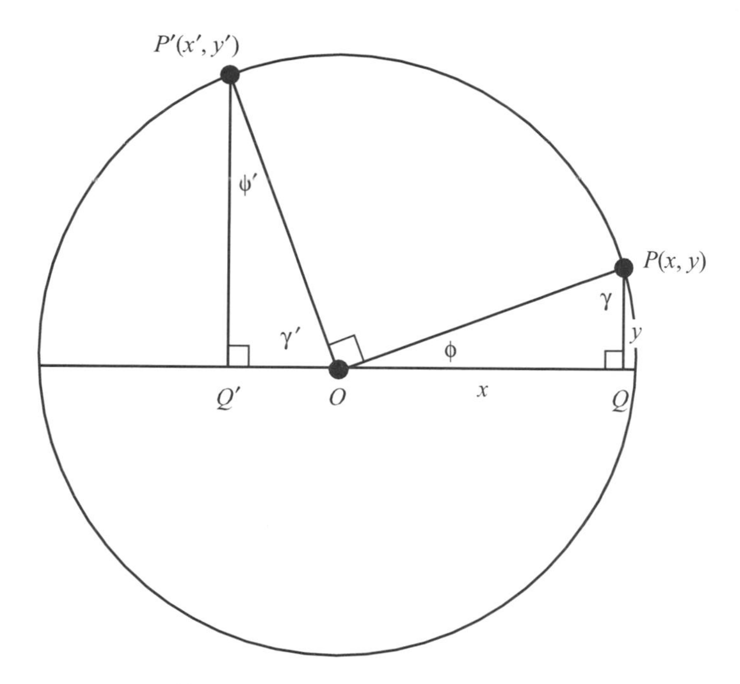

Figure 6.3 Rotating through 90°.

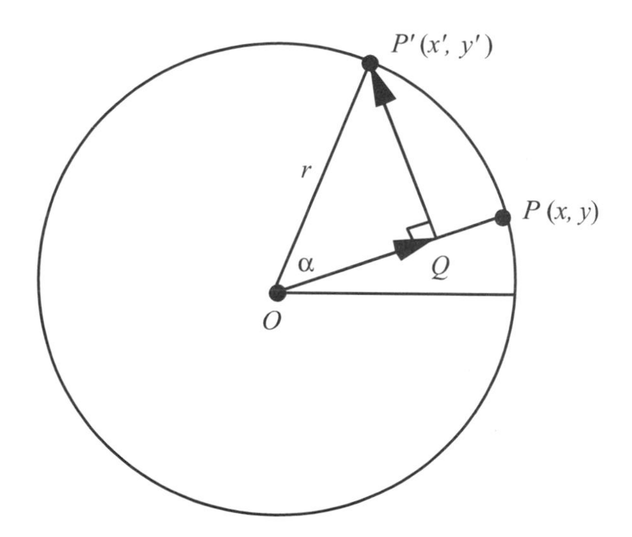

Figure 6.4 Rotating through α.

(a) Express the ratio OQ/OP in terms of α.

(b) Vector $\mathbf{q}$ is a scalar multiple of vector $\mathbf{p}$. What is it?

*(c) Show that vector $\mathbf{p}' - \mathbf{q}$ is $\langle -y \sin \alpha, x \sin \alpha \rangle$. (*Hint*: See the previous exercise.)

(d) Express vector $\mathbf{p}'$ as a sum and thereby derive the rotation equations.

8. Are the rotation equations correct if we are using a frame with clockwise sense? Explain and illustrate your answer.

9. Fill in the gap in our discussion of translation (See Figure 4.3 of Section 4.1 in Chapter 4.): In quadrilateral $QQ'P'P$, suppose $\overline{P'Q'}$ is parallel and equal in length to $\overline{PQ}$. Prove that in this case $QQ'P'P$ is a parallelogram.

Transformations and Matrices

10. Answer Exercise 1 by using the appropriate 2×2 matrices. Then answer Exercise 1 by using the appropriate 3×3 matrices.

11. What is the 2×2 matrix that represents the transformation whose equations are

$$x' = 3x + 7y,$$
$$y' = 4x + 8y.$$

12. What is the 3×3 homogeneous matrix representing the identity transformation?

13. What is the 2×2 matrix for rotation through 45° about the origin? What is the homogeneous 3×3 matrix? What is the image of $(4, -3)$?

14. Suppose a certain translation T sends $P(x_P, y_P)$ to $Q(x_Q, y_Q)$. Find the 3×3 matrix $\mathbf{T}$ that represents this (its entries expressed in terms of the coordinates of P and Q).

15. Find the equations for reflection in the line $x = a$. What is the 3×3 matrix for these equations?

16. Find the equations for reflection in the line $y = b$. What is the 3×3 matrix for these equations?

17. A classmate claims that

$$\begin{pmatrix} \frac{\sqrt{2}}{2} & -\frac{\sqrt{2}}{2} & 2 \\ \frac{\sqrt{2}}{2} & \frac{\sqrt{2}}{2} & 2-2\sqrt{2} \\ 0 & 0 & 1 \end{pmatrix}$$

is the matrix of some reflection. She doesn't tell you the mirror line but claims that it is not one of the ones studied in this section, so it is not entirely routine to check whether this is the matrix of a reflection or not. As a matter of fact, your classmate is wrong. Your job is to find some calculation that will reveal that this matrix does not represent a reflection.

18. Central inversion is the mapping that associates to each point $P(x, y)$ the point $-P = (-x, -y)$. Make a careful figure to illustrate the effect of this mapping on two different points. (You may pick any two points.) Do you think this is an isometry? Give reasons for your answer. Can you find a matrix to represent this? Is this related to any transformation we have studied in this section?

6.2 Composition and Matrix Multiplication

We often want to follow one planar isometry by another. For example, we may wish to apply a rotation R to a point and then to apply translation T to the resulting point. This combining of transformations is called *composition* and the resulting transformation is written $T \circ R$. If $\mathbf{R}$ and $\mathbf{T}$ are the corresponding 3×3 homogeneous matrices, we find the final image of a point P whose position vector is $\mathbf{p}$ by two matrix multiplications:

$$\mathbf{p} \rightarrow \mathbf{Rp} \rightarrow \mathbf{T(Rp)}.$$

Thus, the new position of P, P', is related to P by two matrix–vector multiplications:

$$\mathbf{p}' = \mathbf{T(Rp)}. \tag{6.7}$$

Recall that the multiplication of matrices and vectors is *associative*. This means we also have

$$\mathbf{p}' = \mathbf{(TR)p}. \tag{6.8}$$

EXAMPLE 6.6: ILLUSTRATING THE ASSOCIATIVE LAW

Let $\mathbf{p}$ be the position vector $\langle 3, -4\rangle$; let $\mathbf{R}$ be the matrix for rotation by 30° around the origin; let $\mathbf{T}$ be the matrix for translation by $\langle 3, 4\rangle$. These are the matrices of Example 6.5 of the previous section. Work out $\mathbf{p}'$ using Eq. (6.7) and then using Eq. (6.8). Do you get the same result?

SOLUTION

Working out Eq. (6.7):

$$\mathbf{Rp} = \begin{pmatrix} \frac{\sqrt{3}}{2} & \frac{-1}{2} & 0 \\ \frac{1}{2} & \frac{\sqrt{3}}{2} & 0 \\ 0 & 0 & 1 \end{pmatrix} \begin{pmatrix} 3 \\ -4 \\ 1 \end{pmatrix} = \begin{pmatrix} \frac{3\sqrt{3}}{2} + 2 \\ \frac{3}{2} - 2\sqrt{3} \\ 1 \end{pmatrix},$$

$$\mathbf{T}(\mathbf{Rp}) = \begin{pmatrix} 1 & 0 & 4 \\ 0 & 1 & -2 \\ 0 & 0 & 1 \end{pmatrix} \begin{pmatrix} \frac{3\sqrt{3}}{2} + 2 \\ \frac{3}{2} - 2\sqrt{3} \\ 1 \end{pmatrix} = \begin{pmatrix} \frac{3\sqrt{3}}{2} + 6 \\ -\frac{1}{2} - 2\sqrt{3} \\ 1 \end{pmatrix}.$$

Working out Eq. (6.8):

$$\mathbf{TR} = \begin{pmatrix} 1 & 0 & 4 \\ 0 & 1 & -2 \\ 0 & 0 & 1 \end{pmatrix} \begin{pmatrix} \frac{\sqrt{3}}{2} & \frac{-1}{2} & 0 \\ \frac{1}{2} & \frac{\sqrt{3}}{2} & 0 \\ 0 & 0 & 1 \end{pmatrix} = \begin{pmatrix} \frac{\sqrt{3}}{2} & \frac{-1}{2} & 4 \\ \frac{1}{2} & \frac{\sqrt{3}}{2} & -2 \\ 0 & 0 & 1 \end{pmatrix},$$

$$(\mathbf{TR})\mathbf{p} = \begin{pmatrix} \frac{\sqrt{3}}{2} & \frac{-1}{2} & 4 \\ \frac{1}{2} & \frac{\sqrt{3}}{2} & -2 \\ 0 & 0 & 1 \end{pmatrix} \begin{pmatrix} 3 \\ -4 \\ 1 \end{pmatrix} = \begin{pmatrix} \frac{3\sqrt{3}}{2} + 6 \\ -\frac{1}{2} - 2\sqrt{3} \\ 1 \end{pmatrix}.$$

We see that $\mathbf{T}(\mathbf{Rp}) = (\mathbf{TR})\mathbf{p}$ for this particular point *P*, confirming one instance of the associative law. ■

EXAMPLE 6.7: THE ASSOCIATIVE LAW IS A MIGHTY GOOD LAW

We can take tremendous advantage of the associativity of matrix and vector multiplication to speed calculations. For example, in a graphics program we might have 10,000 points needing to have 3×3 matrices $\mathbf{R}_3$ then $\mathbf{T}_3$ applied. Working according to Eq. (6.7) we first compute $\mathbf{R}_3\mathbf{p}$ for the first point *P*. Multiplying the homogeneous vector $\mathbf{p}$ by the 3×3 matrix $\mathbf{R}_3$ requires 9 multiplications. We'll ignore additions for the time being. Next we need 9 more multiplications to get $\mathbf{T}_3(\mathbf{R}_3\mathbf{p})$, a total of 18 for the single point P. Now do this repeatedly for the remaining 9999 points, making a total of 10,000(18) multiplications. Now compare the cost of using the plan of calculations summarized by Eq. (6.8). First we multiply the matrices $\mathbf{T}_3$ and $\mathbf{R}_3$ to get $\mathbf{T}_3\mathbf{R}_3$, requiring 27 multiplications. This is done just once. For each of the 10,000 points, we need to take $\mathbf{T}_3\mathbf{R}_3$ and multiply it by that point's homogeneous vector, requiring 9 multiplications each, a total of $27 + 10{,}000(9)$ multiplications—many fewer than the number 10,000(18). The disparity between the two methods increases as the number of points gets even larger. ■

Now let us consider how we can express the fate of a point that is subjected to a series of transformations involving some translations. Our objective is to display the advantage of the homogeneous coordinates and the related 3×3 matrices over ordinary xy coordinates and 2×2 matrices. For example, suppose the following transformations are applied in order:

1. Translation by vector $\mathbf{g}$
2. Rotation whose 2×2 matrix is $\mathbf{R}_2$
3. Translation by $\mathbf{g}'$
4. Reflection by 2×2 matrix $\mathbf{M}_2$.

If the points are in nonhomogeneous form, we would have the final point $\mathbf{p}' = \begin{pmatrix} x' \\ y' \end{pmatrix}$ related to $\mathbf{p} = \begin{pmatrix} x \\ y \end{pmatrix}$ by:

$$\mathbf{p}' = \mathbf{M}_2[\mathbf{R}_2(\mathbf{p} + \mathbf{g}) + \mathbf{g}'].$$

It is a complicated formula that is hard to manipulate or remember because of the alternation of vector addition with matrix multiplication. We could multiply through, but there is not much improvement:

$$\mathbf{p}' = \mathbf{M}_2\mathbf{R}_2\mathbf{p} + \mathbf{M}_2\mathbf{R}_2\mathbf{g} + \mathbf{M}_2\mathbf{g}'.$$

Now suppose we have the homogeneous forms

$$\mathbf{p} = \begin{pmatrix} x \\ y \\ 1 \end{pmatrix} \quad \text{and} \quad \mathbf{p}' = \begin{pmatrix} x' \\ y' \\ 1 \end{pmatrix}$$

and we switch to 3×3 matrices. The big advantage here is that the translations can be carried out by multiplying by 3×3 matrices $\mathbf{T}_3$ and $\mathbf{T}_3'$. Then we have

$$\mathbf{p}' = \mathbf{M}_3\mathbf{T}_3'\mathbf{R}_3\mathbf{T}_3\mathbf{p}.$$

There is a conceptual and notational clarity in having one tool — matrix multiplication — do everything we need.

When an isometry S is followed by a translation T, it is particularly easy to find the matrix for $T \circ S$ if we know $\mathbf{S}$, the 3×3 matrix for S, and the components for the glide vector $\mathbf{g}$. The next theorem tells us that we simply take the x and y components of the glide vector and add them to the first and second entries of the last column of $\mathbf{S}$.

✯ THEOREM 6.1

Let

$$\mathbf{S} = \begin{pmatrix} a & b & c \\ d & e & f \\ 0 & 0 & 1 \end{pmatrix}$$

be the matrix of an isometry S and let T denote translation by the glide vector $\mathbf{g} = \langle x, y \rangle$. The matrix for $T \circ S$ is

$$\begin{pmatrix} a & b & c+x \\ d & e & f+y \\ 0 & 0 & 1 \end{pmatrix}.$$

PROOF

We work out the matrix for $T \circ S$ by multiplying the matrices for S and T:

$$\mathbf{TS} = \begin{pmatrix} 1 & 0 & x \\ 0 & 1 & y \\ 0 & 0 & 1 \end{pmatrix} \begin{pmatrix} a & b & c \\ d & e & f \\ 0 & 0 & 1 \end{pmatrix} = \begin{pmatrix} a & b & c+x \\ d & e & f+y \\ 0 & 0 & 1 \end{pmatrix}.$$

■

Notice that the previous theorem says nothing about the case where we take the translation first, $S \circ T$. As an exercise try deducing from this theorem that $T_{\langle c, d \rangle} \circ T_{\langle a, b \rangle} = T_{\langle a+c,\ b+d \rangle}$. This, in turn, implies that two translations always commute: $T_{\langle c, d \rangle} \circ T_{\langle a, b \rangle} = T_{\langle a, b \rangle} \circ T_{\langle c, d \rangle}$.

EXAMPLE 6.8

Find the matrix for rotation by 30° followed by translation by ⟨-5, 3⟩.

SOLUTION

The rotation matrix we need is the one in Example 6.6. Then apply Theorem 6.1 to obtain

$$\begin{pmatrix} \frac{\sqrt{3}}{2} & \frac{-1}{2} & -5 \\ \frac{1}{2} & \frac{\sqrt{3}}{2} & 3 \\ 0 & 0 & 1 \end{pmatrix}.$$

■

Rotation around a Point Other Than the Origin

We now obtain a 3×3 matrix for rotation by angle α around a point C that is not necessarily the origin. Notice how composition of isometries plays a central role. Let P be the point to be rotated and P' the new point we want to compute, and let $\mathbf{p}$ and $\mathbf{p}'$ be their homogeneous column vectors. What we will do is to get to P' in stages (Figure 6.5). First imagine that we translate P, C, and P' by vector $-\mathbf{c}$ so that C moves to the origin, P moves to a new point Q, and P' moves to a new point

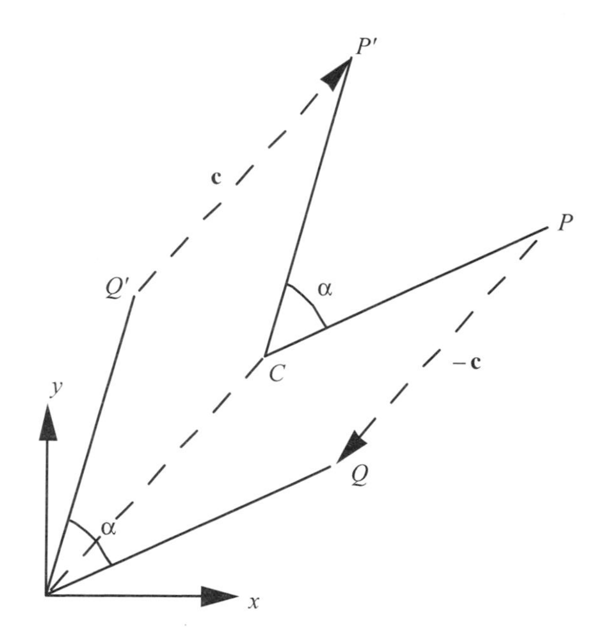

Figure 6.5 Rotation around C.

Q'. Here, **c** is a two-component vector $\langle x_C, y_C \rangle$. Notice that Q' can be obtained from Q by rotation of α around the origin. Thus we rotate Q to Q' by the matrix $\mathbf{R}_O(\alpha)$. Finally we translate back by vector **c** so that Q' moves back to P'.

$$P \underset{\text{translate by } -\mathbf{c}}{\rightarrow} Q \underset{\text{rotate}}{\rightarrow} Q' \underset{\text{translate by } \mathbf{c}}{\rightarrow} P'.$$

Expressing this by matrix multiplication:

$$\mathbf{p}' = [\mathbf{T}_\mathbf{c}\mathbf{R}_O(\alpha)\mathbf{T}_{-\mathbf{c}}]\mathbf{p}.$$

What we have actually done here is to derive the following theorem.

✯ THEOREM 6.2

$$\mathbf{R}_C(\alpha) = \mathbf{T}_\mathbf{c}\mathbf{R}_O(\alpha)\mathbf{T}_{-\mathbf{c}}.$$

■

In thinking about this theorem, keep in mind that matrix multiplication is not normally commutative. For if commutativity held in this case, we could rewrite the right hand side as $\mathbf{T}_\mathbf{c}\mathbf{T}_\mathbf{c}\mathbf{R}_O(\alpha) = \mathbf{I}\mathbf{R}_O(\alpha) = \mathbf{R}_O(\alpha)$. But this would imply $\mathbf{R}_C(\alpha) = \mathbf{R}_O(\alpha)$ which is false unless $C = O$.

To work out the entries of $\mathbf{R}_C(\alpha)$ using this theorem, first note that Theorem 6.1 gives us a matrix for the product $\mathbf{T}_\mathbf{c}\mathbf{R}_O(\alpha)$, so we only have to do one more multiplication:

$$\begin{aligned}[\mathbf{T}_\mathbf{c}\mathbf{R}_O(\alpha)]\mathbf{T}_{-\mathbf{c}} &= \begin{pmatrix} \cos\alpha & -\sin\alpha & x_C \\ \sin\alpha & \cos\alpha & y_C \\ 0 & 0 & 1 \end{pmatrix}\begin{pmatrix} 1 & 0 & -x_C \\ 0 & 1 & -y_C \\ 0 & 0 & 1 \end{pmatrix} \\ &= \begin{pmatrix} \cos\alpha & -\sin\alpha & -x_C\cos\alpha + y_C\sin\alpha + x_C \\ \sin\alpha & \cos\alpha & -x_C\sin\alpha - y_C\cos\alpha + y_C \\ 0 & 0 & 1 \end{pmatrix}.\end{aligned} \tag{6.9}$$

The method we used to get this matrix is worth studying and is perhaps more important than the formula itself. This method of moving the configuration so it is situated more conveniently can also be used to deal with reflection in an arbitrary line (see the exercises at the end of this section).

EXAMPLE 6.9: ROBBIE THE ROBOT TURNS

Figure 6.6 shows a "footprint" of a robot vacuum cleaner which scoots around on the floor of a room. Of course, it is a three-dimensional object, but most of the mathematics

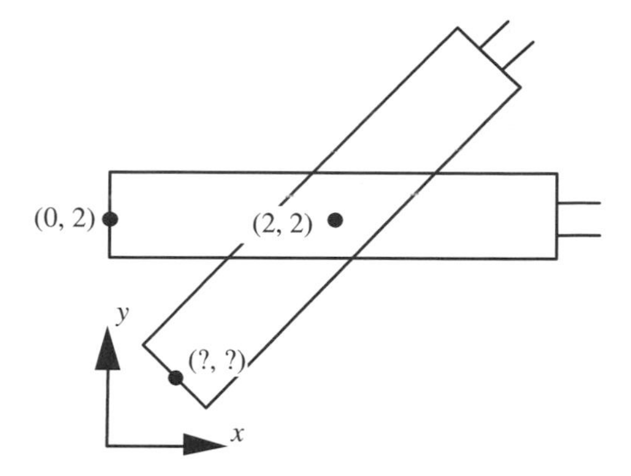

Figure 6.6 Robbie the robot pivots 45° on its center.

needed to deal with it will be two-dimensional geometry applied to the footprint of the robot. The robot moves by turning and by straight-line motion. The turning is a rotation around its center and is depicted in Figure 6.6. Suppose the center is (2, 2), the turn angle is 45°, and we want to know the new location of the center of Robbie's tail light, which starts out at (0, 2).

SOLUTION

Using Eq. (6.9),

$$\begin{pmatrix} \frac{\sqrt{2}}{2} & -\frac{\sqrt{2}}{2} & 2 \\ \frac{\sqrt{2}}{2} & \frac{\sqrt{2}}{2} & 2-\sqrt{2} \\ 0 & 0 & 1 \end{pmatrix} \begin{pmatrix} 0 \\ 2 \\ 1 \end{pmatrix} = \begin{pmatrix} 2-\sqrt{2} \\ 2-\sqrt{2} \\ 1 \end{pmatrix}.$$

So far the only reflections that we know how to represent with matrices are reflections in the x and y axes (and any other horizontal or vertical lines if you did Exercises 15 and 16 of the previous section). Could it be that if we had a mirror line that was not vertical or horizontal, perhaps did not even go through the origin, that there was no matrix that represented it?

✭ THEOREM 6.3

Any reflection can be represented by a 3×3 matrix.

Proof

Exercise 14 at the end of this section shows how to derive the appropriate matrix from data about the mirror line. ■

EXERCISES

**Marks challenging exercises.*

1. Work out the 3×3 matrix for the transformation that results if you perform a reflection in the y axis followed by a reflection in the x axis. Compare this to $\mathbf{R}_O(180°)$. What do you conclude? (Example 6.4 of the previous section is relevant here.)

2. Suppose T is a translation with vector $\mathbf{g} = \langle -3, 0 \rangle$. Define $H = M_x \circ T$. (Readers of Chapter 4 will recall that this is called a glide reflection.) What is the 3×3 matrix of H? What is the 3×3 matrix of $H \circ H$?

3. Let T_1 be translation by the vector $\langle a, 0 \rangle$ and let T_2 be translation by $\langle 0, b \rangle$. What is the 3×3 matrix of $T_2 \circ T_1$? Take $a = 1$ and $b = 2$ and compute and draw the effect of $T_2 \circ T_1$ on the points (1, 0), (0, 1), (2, 3). Do T_1 and T_2 commute?

4. In each of the following cases, determine, using matrix multiplication, whether U and V commute.

   (a) U = rotation around (0,0) by 90°, V = reflection in the y axis.

   (b) U = reflection in the y axis, V = reflection in the x axis.

   (c) U = rotation around (0, 0) by 30°, V = rotation around (0, 0) by $-30°$.

   (d) U = rotation around (0, 0) by 30°, V = rotation around (1, 0) by $-30°$.

5. Suppose you know that a rotation about the origin followed by a translation by $\mathbf{g}$ is a rotation around some point other than the origin. Given the rotation angle θ and the translation vector $\mathbf{g}$, how could you find a formula for the center?

6. Find the 3×3 matrix **S** that represents the transformation obtained by first rotating around the origin by $45°$ and then translating by $\langle -2, 6 \rangle$. Then find the 3×3 matrix that represents the result of doing these in the reverse order. Are they the same? What do you conclude?

7. Work out how many multiplications are required for determining the images of 100 points, first according to Eq. (6.7) and then according to Eq. (6.8). In the case of Eq. (6.8), what fraction of the work is the computation of the 3×3 matrix **TR**? How does this fraction change if we have 1000 points? 10,000?

8. Work out formulas for how many additions and how many multiplications are required for determining the images of N points, first according to Eq. (6.7), and then according to Eq. (6.8) with once-only precomputation of **TR**. Express your answers as functions of N.

9. Do Exercise 7 but now count the number of additions plus the number of multiplications.

10. Suppose **S** is a 3×3 matrix and you need to compute $\mathbf{S}^n$.

    (a) If you mutiply the matrices one at a time (first compute $\mathbf{S}^2$; then take this square and get $\mathbf{S}^3$ by computing $\mathbf{S}(\mathbf{S}^2)$; next compute $\mathbf{S}^4$ by $\mathbf{S}(\mathbf{S}^3)$; etc.), how many multiplications of individual entries are there? Work out the answer for $n=2$, 3, 4. Then give the formula for general n.

    (b) Suppose $n=4$ and, motivated by $\mathbf{S}^4 = (\mathbf{S}^2)(\mathbf{S}^2)$, you proceed by calculating $\mathbf{A} = \mathbf{S}^2$ and then calculating $\mathbf{A}^2$. Compare the number of multiplications of individual entries with the plan in part (a). What does this suggest about the least number of multiplications for $n=8$, 16, 32?

    (c) If n is a power of 2, $n=2^k$, say how many (in terms of k) multiplications are involved if you use the plan of calculation in part (b).

11. Use Theorem 6.1 to prove that $T_{\langle c,\ d\rangle} \circ T_{\langle a,\ b\rangle} = T_{\langle a+c,\ b+d\rangle}$.

12. If an isometry S fixes the origin and is represented by the matrix
$$\mathbf{S} = \begin{pmatrix} a & b & c \\ d & e & f \\ 0 & 0 & 1 \end{pmatrix},$$
what can you say about the last column of its matrix?

*13. Let $T_{\mathbf{g}}$ denote translation with glide vector $\mathbf{g} = \langle u,\ v \rangle$. Is it true that $S \circ T_{\mathbf{g}} = T_{S(\mathbf{g})} \circ S$, where S is any isometry?

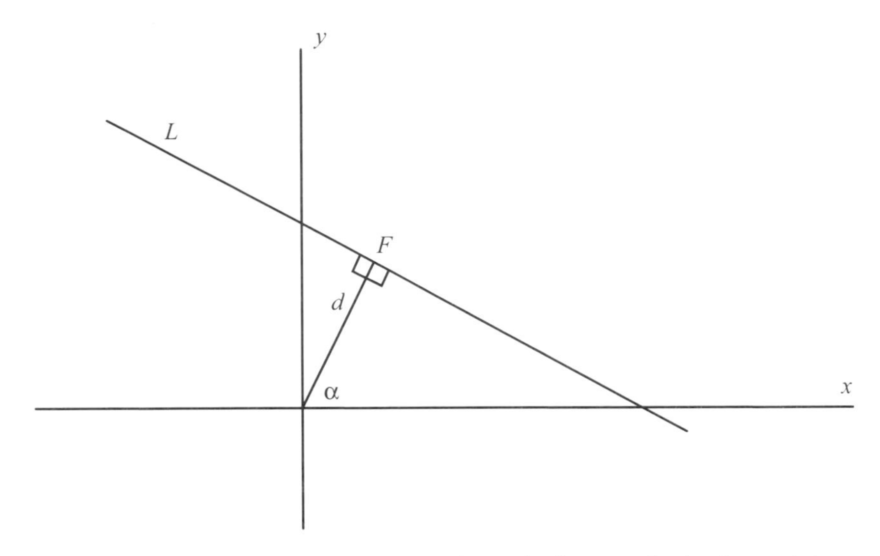

Figure 6.7 Line specified by angle and length of perpendicular from origin.

*14. *Deriving the general reflection matix.* Let L be a line whose perpendicular distance from the origin is d and where the perpendicular segment from the origin to the line makes an angle of α with the x axis (Figure 6.7).

(a) If this line were to be rotated around the origin by $-\alpha$, what would the equation of the new line be? What would the reflection matrix in the rotated line be?

(b) Show that reflection in the original line has matrix

$$\begin{pmatrix} \sin^2\alpha - \cos^2\alpha & -2\sin\alpha\cos\alpha & 2d\cos\alpha \\ -2\sin\alpha\cos\alpha & \cos^2\alpha - \sin^2\alpha & 2d\sin\alpha \\ 0 & 0 & 1 \end{pmatrix}.$$

(*Hint*: Follow the general strategy used to derive Theorem 6.2.)

(c) Suppose the original line has equation $y = mx + b$. Can you express the reflection matrix of part (b) in terms of m and b?

15. Since the composition of two isometries $S_2 \circ S_1$ is represented by the product $\mathbf{S}_2\mathbf{S}_1$ where $\mathbf{S}_1$ and $\mathbf{S}_2$ are 3×3 matrices, one might suppose that the matrix sum $\mathbf{S}_2 + \mathbf{S}_1$ would have some significance. Does it represent an isometry? Illustrate your answer with a numerical example.

6.3 Frames and How to Represent Them

Frames and Matrices

The question we wish to ask in this section is "How can we represent a frame in a convenient numerical way, suitable for computation by hand or computer?" Recall that a frame is a pair of perpendicular unit vectors emanating from any chosen point called the origin of the frame. Example 6.10 shows a typical use of frames. A great many kinds of computer software that deal with geometric ideas rest on describing and manipulating frames: robot control software, CAD/CAM systems, various computer graphics applications, guidance systems for aircraft, etc. In addition to helping us with these practical matters, our study of frames will lead to the important theoretical result that every isometry can be represented by a matrix.

EXAMPLE 6.10: WHERE IS THIS ROBOT?

Robbie the security robot moves through a department store at night and every few seconds radios its position back to the central computer. If it were a human being sending back its location, the message might be something like: "I'm in sporting goods, near the soccer balls." But for a robot it may be more natural and precise to specify position through numerical coordinates. Therefore, let's suppose there is a coordinate frame $\mathcal{W}$ on the floor, fixed once and for all.

A frame that stays fixed throughout a problem — while other frames may move — is typically called a *world frame* and denoted $\mathcal{W}$. In this book, and most applications of geometry, world frames are taken to have counterclockwise sense.

The robot could report the coordinates of its center with respect to $\mathcal{W}$. But this would give no information about how the robot is pointing, that is, its *orientation*. A number of schemes could be devised for reporting the robot's orientation. We could imagine a unit vector running from the center to the front of the robot, and we could measure the angle this makes with the direction of the x-vector of $\mathcal{W}$ (θ in Figure 6.8). The method we will actually use is to imagine an entire frame $\mathcal{F}$ embedded in the robot, with origin at the center of the robot. Its x-vector will be the unit vector just mentioned, pointing toward the front of the robot. The y-vector will be 90° around in the counterclockwise direction (this way we get a frame of counterclockwise sense). This brings us back to the question with which we opened this section: How do we describe a frame numerically? ■

Various methods have been proposed for describing a frame numerically. For example, in the next section we describe a method based on specifying the coordinates

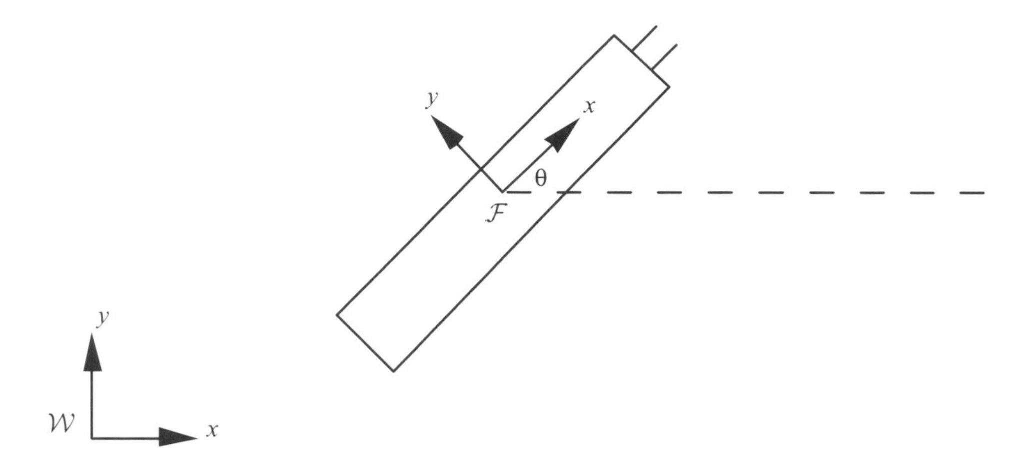

Figure 6.8 Where is this robot in relation to $\mathcal{W}$?

of the frame's origin and the components of its *x*- and *y*-vectors. But, in this section we describe the representation we prefer. It hinges on the fact that if we establish a world frame $\mathcal{W}$, used to give coordinates to the points of our space, then any other frame $\mathcal{F}$ can be regarded as the result of applying an isometry to the world frame $\mathcal{W}$. We can use the matrix of this isometry as a numerical representation of the frame. For example, in Figure 6.10 (shown later) we can say that $\mathcal{F}$ is represented by rotation by 45° followed by translation by the vector $\langle 1, 1\rangle$.

Of course, to propose this form of representation for a frame, we must first prove that any frame can be obtained as the result of applying some isometry to the world frame $\mathcal{W}$. After proving this, there will be a bonus: We'll see that any isometry can always be described by a matrix. (In Sections 6.1 and 6.2 we discovered that some familiar isometries can be represented by matrices. Now we will see that "some" can be changed to "all.")

✯ THEOREM 6.4

If two frames have the same origin, the same *x*-vector, and the same sense, then they have the same *y*-vector also, that is, the frames are identical.

Proof

This is evident if you draw a picture or two so we skip a formal proof. ■

☆ THEOREM 6.5

1. If $\mathcal{W}$ is a world frame and $\mathcal{F}$ is any other frame (Figure 6.9), there is an isometry which moves $\mathcal{W}$ to $\mathcal{F}$.
2. There is only one such isometry.
3. (a) If $\mathcal{F}$ has clockwise sense, this isometry can be carried out as $T \circ R \circ M$ where M is a reflection, R is a rotation, and T is a translation.
   (b) If $\mathcal{F}$ has counterclockwise sense this isometry can be carried out as $T \circ R$.

Proof

We'll do the proof in two cases, according to whether $\mathcal{F}$ has clockwise or counterclockwise sense. Recall that $\mathcal{W}$, being a world frame, has counterclockwise sense.

Case 1. $\mathcal{F}$ has counterclockwise sense (like $\mathcal{F}_1$ in Figure 6.9). Now apply a rotation R that moves the x-vector of $\mathcal{W}$ to point in the same direction as the x-vector of the frame $\mathcal{F}$. This produces a frame $\mathcal{R}$. Next apply a translation T that moves the origin onto the origin of $\mathcal{F}$. Since translation preserves direction, the x-vector of $\mathcal{R}$ has been moved onto the x-vector of $\mathcal{F}$. Both R and T preserve the sense of a frame, so the new position of $\mathcal{W}$ has the same origin, the same x-vector, and the same sense as $\mathcal{F}$. By Theorem 6.4, $T \circ R$ has moved $\mathcal{W}$ onto $\mathcal{F}$.

Case 2. $\mathcal{F}$ has clockwise sense (like $\mathcal{F}_2$ in Figure 6.9). We begin by reflecting $\mathcal{W}$ in the x axis, producing a clockwise frame. Call this reflection M. Now apply a rotation R that moves the new x-vector to point in the same direction as the x-vector of the frame

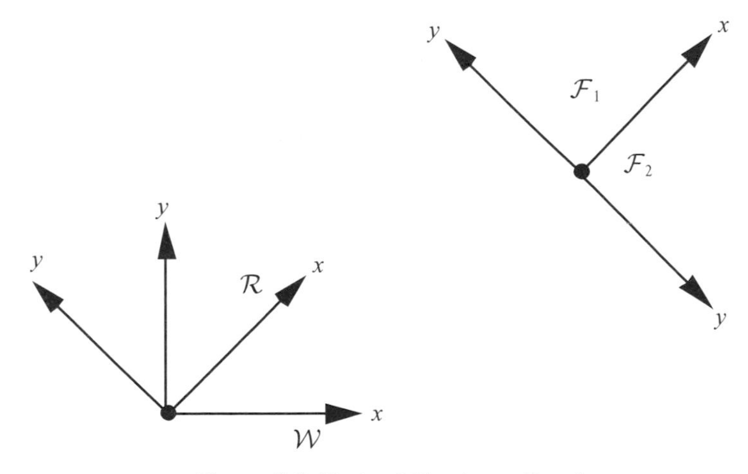

Figure 6.9 Moving $\mathcal{W}$ onto another frame.

$\mathcal{F}$. Next apply a translation T that moves the world origin onto the origin of $\mathcal{F}$. The new position of $\mathcal{W}$ has the same origin, the same x-vector, and the same sense as $\mathcal{F}$. By Theorem 6.4, $T \circ R \circ M$ has moved $\mathcal{W}$ onto $\mathcal{F}$. This completes parts 1 and 3 of the proof.

Now for part 2: Suppose there were two isometries that move frame $\mathcal{W}$ to $\mathcal{F}$. These isometries would have the same effect on the origin of $\mathcal{W}$ (sending it to the origin of $\mathcal{F}$) and the tips of the x- and y-vectors (sending them to the tips of the x- and y-vectors of $\mathcal{F}$). But this contradicts Theorem 4.14 of Chapter 4, Section 4.3. ■

In thinking about Theorem 6.5, keep in mind that R could be the trivial rotation through 0° and T could be the trivial translation by a zero vector. (There is no such thing as the "trivial reflection.")

EXAMPLE 6.11

In Figure 6.10, $\mathcal{F}$ is obtained from $\mathcal{W}$ by the composition of a 45° rotation around the world origin, followed by translation by $\langle 1, 1 \rangle$— that is, by $T_{\langle 1,\ 1 \rangle} \circ R_O(45^\circ)$. $\mathcal{F}$ can also be obtained by the composition $R_{(1,\ 1)}(45^\circ) \circ T_{\langle 1,\ 1 \rangle}$. Note that the rotation in this composition is not around the origin. This is no contradiction to part 2 of Theorem 6.5— we simply have two compositions yielding the same isometry.

$\mathcal{G}$ is obtained from $\mathcal{W}$ by reflection in the x axis followed by translation by $\langle 2, 0 \rangle$. Can you find another composition that moves $\mathcal{W}$ to $\mathcal{G}$? ■

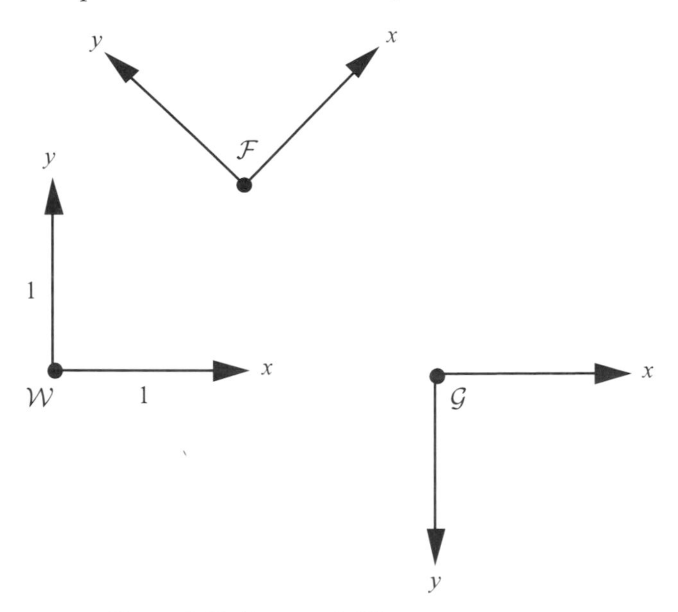

Figure 6.10 Frames of different senses.

We have established that it is possible to represent a frame $\mathcal{F}$ by the isometry which moves $\mathcal{W}$ to $\mathcal{F}$. But we wanted an idea that was numerical and convenient. As far as being numerical goes, recall that rotations, reflection in the x axis, and translations can be described by matrices, and the composition of a number of isometries can be described by the matrix product of the corresponding matrices (Section 6.2). In the present case, if R, M, T are represented by 3×3 homogeneous matrices $\mathbf{R}_3$, $\mathbf{M}_3$, $\mathbf{T}_3$, then we can assert that:

Any frame can be represented by either a matrix product of the form $\mathbf{T}_3\mathbf{R}_3\mathbf{M}_3$ *or one of the form* $\mathbf{T}_3\mathbf{R}_3$ *— the correct formula depends on the sense of the frame* $\mathcal{F}$. *(In these formulas, one or both of the translation or rotation matrices may be the identity.)*

EXAMPLE 6.12

In Figure 6.10, $\mathcal{F}$ would be represented by

$$\mathbf{T}_3\mathbf{R}_3 = \begin{pmatrix} 1 & 0 & 1 \\ 0 & 1 & 1 \\ 0 & 0 & 1 \end{pmatrix} \begin{pmatrix} \sqrt{2}/2 & -\sqrt{2}/2 & 0 \\ \sqrt{2}/2 & \sqrt{2}/2 & 0 \\ 0 & 0 & 1 \end{pmatrix}$$
$$= \begin{pmatrix} \sqrt{2}/2 & -\sqrt{2}/2 & 1 \\ \sqrt{2}/2 & \sqrt{2}/2 & 1 \\ 0 & 0 & 1 \end{pmatrix}.$$

The product we have just computed could be obtained without multiplication from Theorem 6.1 of Section 6.2. Using that theorem, we see that $\mathcal{G}$ of Figure 6.10 would be represented by

$$\mathbf{T}_3\mathbf{M}_3 = \begin{pmatrix} 1 & 0 & 2 \\ 0 & -1 & 0 \\ 0 & 0 & 1 \end{pmatrix}. \quad \blacksquare$$

COROLLARY 6.6

Any isometry S can be written in the form $T \circ R \circ M$ or in the form $T \circ R$.

PROOF

Let $\mathcal{F}$ be the frame that results from applying S to $\mathcal{W}$. We know that we can find R and T and M (if needed) so that either $T \circ R \circ M$ or $T \circ R$ moves $\mathcal{W}$ to $\mathcal{F}$. But by Theorem 6.5, there can be only one isometry that moves $\mathcal{W}$ to $\mathcal{F}$. Thus, either $S = T \circ R \circ M$ or $S = T \circ R$ according to the sense of $\mathcal{F}$. ■

COROLLARY 6.7

Any isometry can be represented by a matrix.

PROOF

Apply the previous corollary together with the fact that the matrix of a composition of isometries is the product of the matrices representing the individual isometries. ■

Changing Coordinates

While it is true that a frame gives a good specification of the position and orientation of a robot — or any other object — there are important questions related to where the object is which the frame doesn't tell us without extra work. We conclude this section with two examples in which it is necessary to change coordinates. If there are two frames in a problem then we can use either of them to find the coordinates of a point. In our examples, the coordinates are known in the $\mathcal{F}$ frame but we want to know what they are in the $\mathcal{W}$ frame.

APPLICATION: Mobile Robots

EXAMPLE 6.13: LOCAL TO WORLD COORDINATES AND COLLISION DETECTION

Robbie the robot is contemplating a move that would put its frame at

$$\mathbf{F} = \begin{pmatrix} 2\frac{\sqrt{5}}{5} & -\frac{\sqrt{5}}{5} & 5 \\ \frac{\sqrt{5}}{5} & 2\frac{\sqrt{5}}{5} & 3 \\ 0 & 0 & 1 \end{pmatrix}.$$

Robbie is operating in a room (Figure 6.11) with one wall having equation $y = 3.8$. This equation is in terms of the world frame $\mathcal{W}$. Would the move under consideration involve a collision with that wall? An accurate drawing (ours is only moderately accurate) would show that the desired position would require Robbie's front-left corner to be beyond the wall. How could this be computed so that accurate pictures would not be needed?

SOLUTION

To solve this problem we would have to check the corners of the robot and see if any have y coordinate >3.8. Here's how we could find the coordinates of a typical corner such as the front-left corner P'. We do a "Gedanken experiment" (German for an experiment carried out purely in the mind) in which the robot's

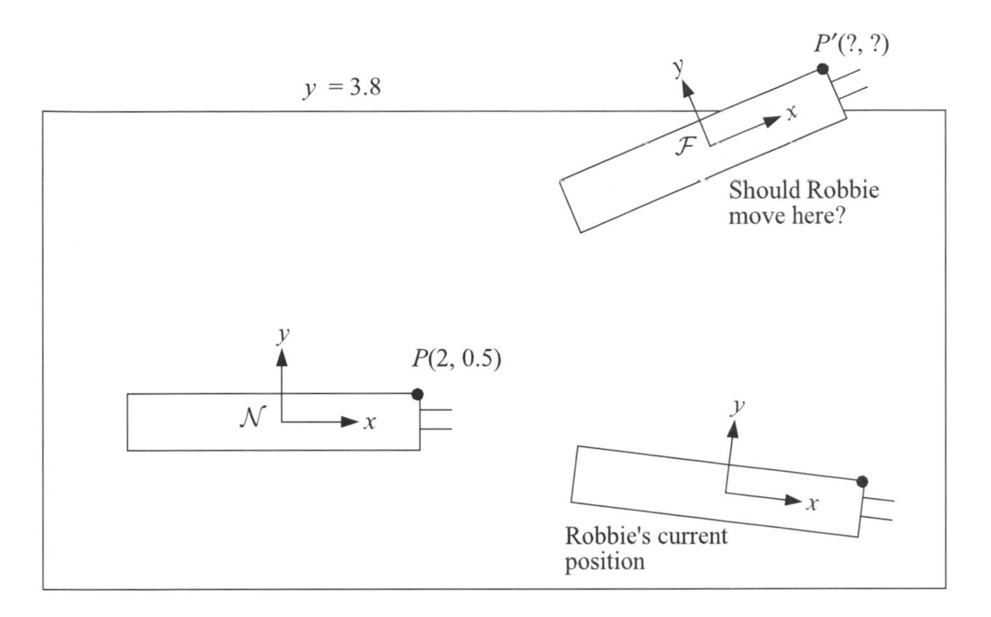

Figure 6.11 Robbie pokes through a wall.

frame is brought into coincidence with the world frame. Imagine that we know the coordinates of the front-left corner P when the robot is in its home position at $\mathcal{W}$ are (2, 0.5). Now the isometry that moves $\mathcal{W}$ to $\mathcal{F}$ will also move P to P'. Thus we compute the position vector of P' by

$$\mathbf{p}' = \mathbf{F}\mathbf{p} = \begin{pmatrix} 2\frac{\sqrt{5}}{5} & -\frac{\sqrt{5}}{5} & 5 \\ \frac{\sqrt{5}}{5} & 2\frac{\sqrt{5}}{5} & 3 \\ 0 & 0 & 1 \end{pmatrix} \begin{pmatrix} 2 \\ 0.5 \\ 1 \end{pmatrix} = \begin{pmatrix} 6.57 \\ 4.34 \\ 1 \end{pmatrix}.$$

Since $4.34 > 3.8$, we see that the front-left corner really is beyond the wall.

You might wonder how we would know the coordinates of P. The answer is that when the robot is designed, nowadays with the help of a computer screen and CAD/CAM software, an engineer begins by making a drawing of the robot, including the frame at its center, and specifies the coordinates of each corner with respect to that frame. Although this is done on a computer screen, it is really no different from doing the drawing on the floor, using $\mathcal{W}$ as the frame. The coordinates the designer gets for the front-left corner while designing on the screen are the same as they would be on the floor. Thus, the coordinates of P (and every other corner) in terms of the frame $\mathcal{W}$ are obtained from the design and programmed into the computer's database as part of the robot's self-knowledge. ■

EXAMPLE 6.14: LOCAL TO WORLD COORDINATES IN COMPUTER VISION

Robbie the robot has a vision system that it carries around. This system locates a point of interest having coordinates at $Q'(3, 2)$ measured in the robot's coordinate frame (see Figure 6.11). These local coordinates are the "natural," that is, convenient, way for the vision system to locate a point. What are the coordinates of this point in the world frame coordinate system?

SOLUTION

Solving this problem requires a way of thinking somewhat like that in the previous example. But first, imagine moving Robbie's frame $\mathcal{F}$ back to $\mathcal{W}$ and let the point of interest Q' move along with it, ending up at Q. Since Q' has moved the same way $\mathcal{F}$ has, the coordinates of Q in $\mathcal{W}$ are the same as Q' in $\mathcal{F}$, namely, (3, 2). Now we are in a situation just like the previous example. To find the world coordinates of Q', apply to Q the isometry that moves $\mathcal{W}$ to $\mathcal{F}$. Numerically, this means

$$\mathbf{q}' = \mathbf{F}\mathbf{q} = \begin{pmatrix} 2\frac{\sqrt{5}}{5} & -\frac{\sqrt{5}}{5} & 5 \\ \frac{\sqrt{5}}{5} & 2\frac{\sqrt{5}}{5} & 3 \\ 0 & 0 & 1 \end{pmatrix} \begin{pmatrix} 3 \\ 2 \\ 1 \end{pmatrix} = \begin{pmatrix} 6.79 \\ 6.13 \\ 1 \end{pmatrix}.$$

■

We close this section with an example of frames in use for the construction of a bicycle track. Bicycle tracks used to be designed by hunches and hopes, without much science. The track shown in Figure 6.12, built especially for the 1996 Atlanta Olympics,

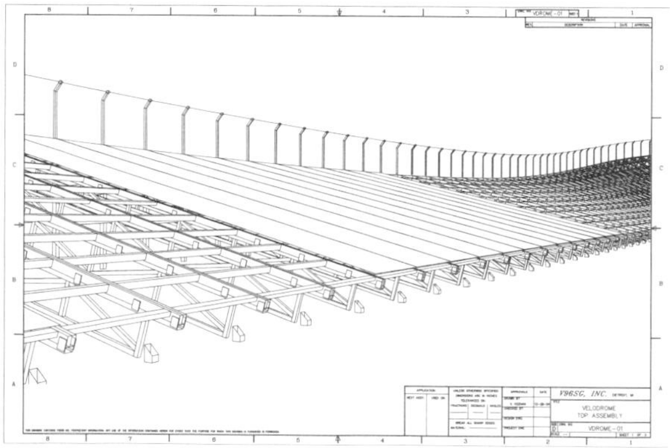

Figure 6.12 The Atlanta olympic velodrome. Courtesy of Chris Nadovitch and Mike Gladu.

was one of the first to be designed with the help of mathematical software. There were plenty of naysayers, but they have been silenced by the many Olympic and world records that have been set on this track, attesting to how fast it is. Because of the curving shape of the track, each steel "plank" is a different size and shape. In addition, each one is tilted a bit differently in order to create the perfect curve and banking of the track. For each plank, its frame had to be calculated to show how that plank needed to be positioned.

EXERCISES

**Marks challenging exercises.*

Frames and Matrices

1. Suppose a frame results from the world frame by reflection in the world x axis, then rotation through 30° followed by translation by $\langle 2, 1\rangle$. What is the matrix for the new frame?

2. Suppose a frame results from the world frame by rotation through 90°, followed by translation by $\langle -4, 2\rangle$. What is the matrix for the new frame?

3. For the frame $\mathcal{E}$ in Figure 6.15 of the next section:

   (a) Describe in words an isometry that moves $\mathcal{W}$ to $\mathcal{E}$. Be as specific as you can about quantities.

   (b) Is this situation described by case (a) or case (b) of part 3 of Theorem 6.5?

   (c) Write down matrices for R, M, and T (in case a) or R and T (in case b) as described in Theorem 6.5.

4. In Example 6.11, can you find another composition that moves $\mathcal{W}$ to $\mathcal{G}$? (You don't need to give the matrix for it.)

5. If $\theta = 45°$ in Figure 6.8, and if the origin of $\mathcal{F}$ is at (5, 4), what is the matrix of the isometry that moves $\mathcal{W}$ onto $\mathcal{F}$?

6. In Figure 6.9, show how to move $\mathcal{W}$ onto $\mathcal{F}_2$ by first rotating around some point and then reflecting in some line and then translating. A verbal description without equations or matrices is acceptable. Does this contradict part 2 of Theorem 6.5?

7. In Section 4.3 of Chapter 4 we defined the sense of a sequence of three noncollinear points. In this section we defined the sense of a frame. What is the connection between these concepts?

*8. Can two different matrices represent the same isometry? If so, give an example. If not, prove that it is not possible.

*9. Corollary 6.6 could be given an alternate proof by relying on a result from Section 4.3 of Chapter 4. Can you find this alternate proof?

Changing Coordinates

10. Suppose, in Example 6.13, the frame $\mathcal{F}$ were obtained from $\mathcal{W}$ by a turn of 120°, followed by a translation by $\langle -3, 2 \rangle$.

    (a) What is the matrix of $\mathcal{F}$?

    (b) What is the new position of the left-front corner?

11. In Example 6.13, we asserted that the coordinates of P in the rooms world frame could be obtained from the original design of the robot, which is usually done on a computer screen. Is this still true if the coordinate frame on the screen has clockwise sense? Illustrate your answer.

12. Robbie the robot (see Figure 6.11) wants to move so that its frame matrix is

$$\begin{pmatrix} \sqrt{2}/2 & -\sqrt{2}/2 & 4 \\ \sqrt{2}/2 & \sqrt{2}/2 & 2 \\ 0 & 0 & 1 \end{pmatrix}.$$

The right-hand wall has equation $y=6$. Will point P poke through the wall if this motion is undertaken?

13. Robbie's vision system locates a point Q' of interest at $(-2, 7)$ measured in Robbie's current frame, whose matrix is the one given in Exercise 12. What are the world frame coordinates of this point?

14. Robbie is currently located with its frame coinciding with the world frame. Robbie rotates by 30° around the world origin and then translates by $\langle 4, -3 \rangle$. In this location, its vision system detects a point of interest whose coordinates in terms of Robbie's frame are $(2, 1)$. What are the world frame coordinates of this point?

*15. Suppose you know the matrix **F** of a frame $\mathcal{F}$. Suppose $(a, b, 1)$ are the homogeneous coordinates of a point in terms of the $\mathcal{W}$ frame. How would you find the coordinates in terms of $\mathcal{F}$? (This requires an idea about matrices that has not been covered in this book.)

*16. Suppose a frame $\mathcal{F}$ has 3×3 matrix **F** and this frame is subjected to an isometry that moves it to $\mathcal{G}$. Suppose **G** is the 3×3 matrix of this isometry when we pretend that $\mathcal{F}$ is the world frame [e.g., if the isometry is rotation around the origin of $\mathcal{F}$, the matrix would be $\mathbf{R}_O(\theta)$, not $\mathbf{R}_P(\theta)$ where P gives the coordinates of $\mathcal{F}$'s origin in terms of the $\mathcal{W}$ frame]. Suppose $P = (a, b, 1)$ are the homogeneous coordinates of a point measured in terms of $\mathcal{G}$. How would you get the coordinates in terms of $\mathcal{W}$?

*17. Suppose a frame has x-vector $\langle a, b \rangle$ and y-vector $\langle c, d \rangle$. Explain how the sense of the frame can be computed numerically from these vectors.

6.4 Properties of the Frame Matrix

Positional Representation of a Frame

In the previous section we put forward the advantages of the matrix representation of a frame. But perhaps you were not convinced. After all when we think of a frame, we think of a dot representing the origin and two perpendicular vectors leading from it. Wouldn't it be more natural to represent the frame by the x and y coordinates of the origin and the x and y components of the two vectors? After all, that information would actually allow us to sketch the frame. This kind of information is called the *positional representation* of the frame.

DEFINITION

The positional representation of a frame consists of three items of data:

1. The origin
2. The x-vector
3. The y-vector

Each of these is expressed in terms of its coordinates or components with respect to a world frame.

The key point of this section is that the positional representation is extremely easy to get from the frame matrix. Conversely, given the positional representation, one can write down the frame matrix in no time at all. To see the principles involved and why they work, let's look at an example.

EXAMPLE 6.15

Suppose a frame $\mathcal{F}$ has matrix

$$\begin{pmatrix} 0.6 & -0.8 & 4 \\ 0.8 & 0.6 & 1 \\ 0 & 0 & 1 \end{pmatrix}.$$

(a) Find the origin of $\mathcal{F}$.

(b) Find the coordinates of the head of the x-vector of $\mathcal{F}$. Then find the x and y components of the x-vector of $\mathcal{F}$.

(c) Do the same for the y-vector of $\mathcal{F}$.

SOLUTION

The key to all these questions is this: Since the matrix **F** represents the isometry that moves $\mathcal{W}$ to $\mathcal{F}$, we will simply multiply the key points in $\mathcal{W}$ by **F** to get the key points of $\mathcal{F}$.

(a) The origin of $\mathcal{F}$ is

$$\begin{pmatrix} 0.6 & -0.8 & 4 \\ 0.8 & 0.6 & 1 \\ 0 & 0 & 1 \end{pmatrix}\begin{pmatrix} 0 \\ 0 \\ 1 \end{pmatrix} = \begin{pmatrix} 4 \\ 1 \\ 1 \end{pmatrix}.$$

Notice that the 0's in the origin vector have the effect that all but the last column of the matrix is irrelevant for this particular calculation—all we are doing is picking out the last column. This gives the following principle:

The origin of a frame is given by the first two entries of the last column of the frame matrix.

(b) The head of the x-vector of $\mathcal{F}$ is

$$\begin{pmatrix} 0.6 & -0.8 & 4 \\ 0.8 & 0.6 & 1 \\ 0 & 0 & 1 \end{pmatrix}\begin{pmatrix} 1 \\ 0 \\ 1 \end{pmatrix} = \begin{pmatrix} 0.6+4 \\ 0.8+1 \\ 0+1 \end{pmatrix} = \begin{pmatrix} 0.6 \\ 0.8 \\ 0 \end{pmatrix} + \text{origin of } \mathcal{F}.$$

The pattern of 0's and 1's in the vector we are multiplying means that we are just adding the first and last columns to get the head of the x-vector of $\mathcal{F}$. To get the x-vector itself, subtract the x and y coordinates of the tail (the origin of $\mathcal{F}$) from those of the head, obtaining the vector $\begin{pmatrix} 0.6 \\ 0.8 \end{pmatrix}$. Here is the principle this calculation illustrates:

The x-vector of a frame is given by the first two entries of the first column of the frame matrix.

(c) The head of the y-vector of $\mathcal{F}$ is

$$\begin{pmatrix} 0.6 & -0.8 & 4 \\ 0.8 & 0.6 & 1 \\ 0 & 0 & 1 \end{pmatrix} \begin{pmatrix} 0 \\ 1 \\ 1 \end{pmatrix} = \begin{pmatrix} -0.8 + 4 \\ 0.6 + 1 \\ 0 + 1 \end{pmatrix} = \begin{pmatrix} -0.8 \\ 0.6 \\ 0 \end{pmatrix} + \text{origin of } \mathcal{F}.$$

The pattern of 0's and 1's in the vector we are multiplying means that we are just adding the middle and last columns to get the head of the y-vector of $\mathcal{F}$. To get the y-vector itself, subtract the x and y coordinates of the tail (the origin of $\mathcal{F}$) from those of the head, obtaining the vector $\begin{pmatrix} -0.8 \\ 0.6 \end{pmatrix}$. Here is the principle this calculation illustrates:

The y-vector of a frame is given by the first two entries of the second column of the frame matrix.

The results of these calculations are illustrated in Figure 6.13. ■

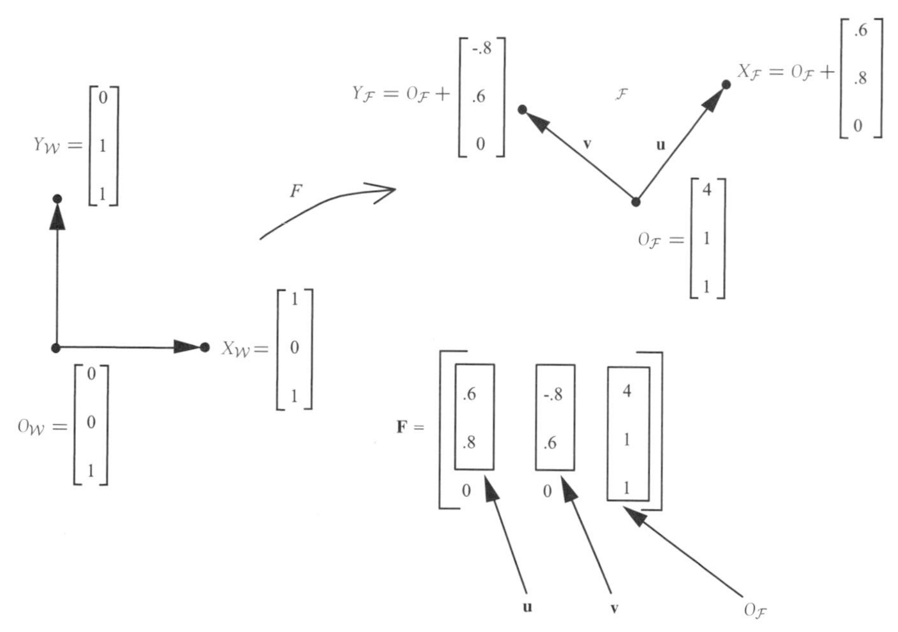

Figure 6.13 The anatomy of a frame matrix.

EXAMPLE 6.16

A robot is located so its frame coincides with the world frame $\mathcal{W}$. It undergoes an isometry whose matrix is

$$F = \begin{pmatrix} 5/13 & 12/13 & 3 \\ 12/13 & -5/13 & 2 \\ 0 & 0 & 1 \end{pmatrix}.$$

Sketch the new robot frame, showing the coordinates of the origin and the components of **u** and **v**, the x- and y-vectors.

SOLUTION

After studying Figure 6.14, do you see anything unusual about this robot's motion? ■

EXAMPLE 6.17

A security patrol robot starts its route with its frame at $\mathcal{W}$ (see Figure 6.15). Its first step tonight will be to get on the elevator. It has been instructed to wait for the elevator with its frame in the position shown by frame $\mathcal{E}$. Find the matrix that represents the isometry it needs to carry out in order to get from $\mathcal{W}$ to $\mathcal{E}$.

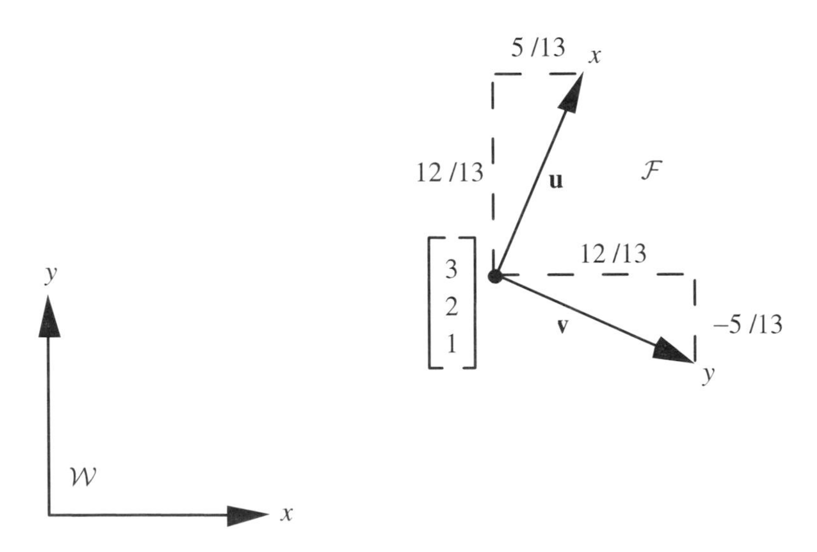

Figure 6.14 Finding a frame from its matrix.

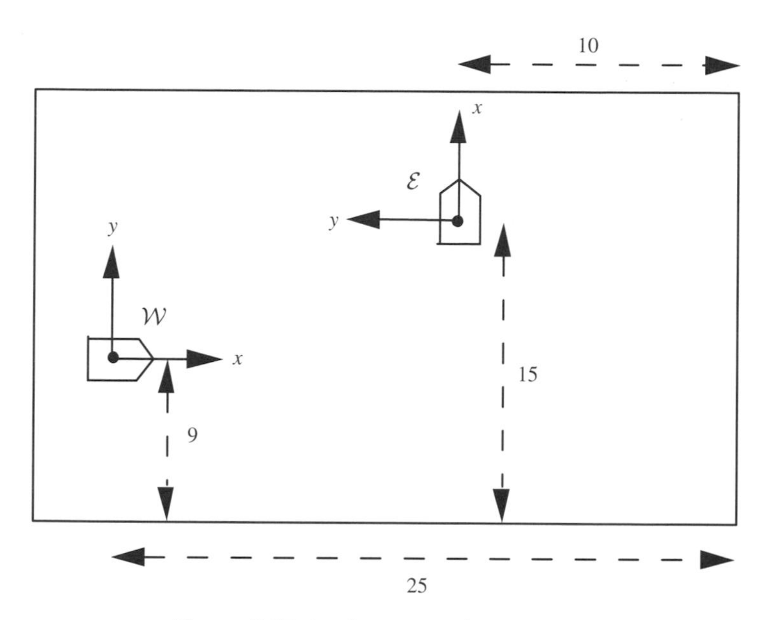

Figure 6.15 A robot moves from $\mathcal{W}$ to $\mathcal{E}$.

SOLUTION

From the lengths in the figure, we calculate that the origin of $\mathcal{E}$ is (15, 6). The x-vector of $\mathcal{E}$ is $\begin{pmatrix} 0 \\ 1 \end{pmatrix}$ and the y-vector is $\begin{pmatrix} -1 \\ 0 \end{pmatrix}$, so

$$\mathbf{E} = \begin{pmatrix} 0 & -1 & 15 \\ 1 & 0 & 6 \\ 0 & 0 & 1 \end{pmatrix}. \qquad \blacksquare$$

Moving a Frame

We have given one piece of evidence regarding the convenience of the isometry matrix representation of a frame (using the matrix of an isometry), namely, the ease with which we can switch back and forth to the positional representation (where we list the origin and the x- and y-vectors). The next theorem shows a second convenience, a principle that is used frequently in graphics and robotics software.

✭ THEOREM 6.8

Given a frame $\mathcal{F}$, described by an isometry matrix $\mathbf{F}$ (Figure 6.16), suppose this frame is moved to a new frame $\mathcal{G}$ by an isometry S, with matrix $\mathbf{S}$. The matrix of $\mathcal{G}$ is $\mathbf{SF}$.

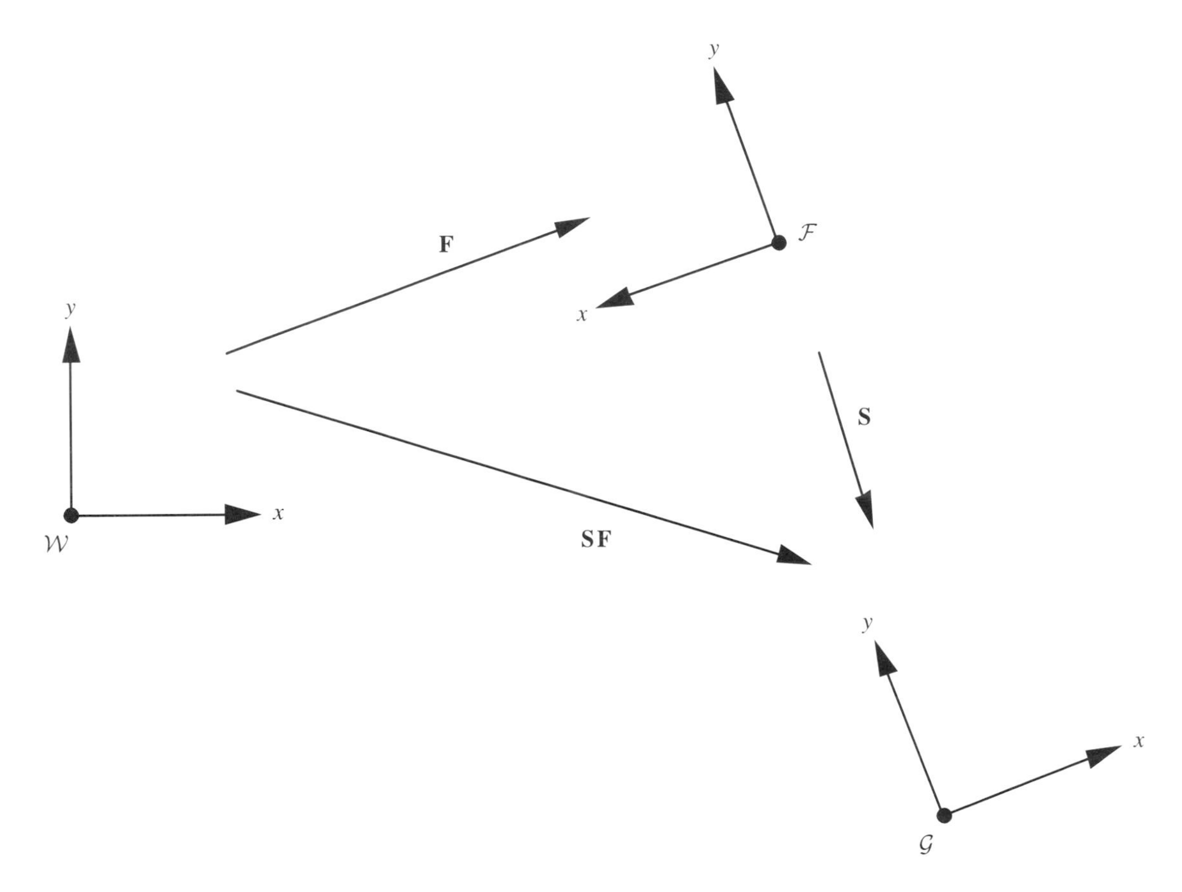

Figure 6.16 The matrix describing a moved frame.

Proof

We need a matrix that moves $\mathcal{W}$ onto $\mathcal{G}$. The isometry that moves $\mathcal{W}$ onto $\mathcal{G}$ is the composition of the isometry that moves $\mathcal{W}$ onto $\mathcal{F}$ with the one that moves $\mathcal{F}$ onto $\mathcal{G}$. Thus the matrix is **SF**. ■

EXAMPLE 6.18

The Skytop restaurant is at the top of a skyscraper and the dining area rotates around the center of the building so diners can see the whole city during the course of a meal. Skytop boasts the world's first robot waiter. To keep track of the robot, a world frame $\mathcal{W}$ is imagined with origin at the center of rotation of the floor and with x-vector pointing east and y-vector pointing north. This frame does not turn as the floor moves.

The robot's current location is specified by its internal frame, which is currently at

$$\mathbf{F} = \begin{pmatrix} \frac{5}{13} & \frac{-12}{13} & 2 \\ \frac{12}{13} & \frac{5}{13} & 4 \\ 0 & 0 & 1 \end{pmatrix}.$$

The robot does not move its wheels for a while but it moves anyhow because during this time the floor rotates around the origin of $\mathcal{W}$ by 90°. Find the matrix that represents the robot frame's new location.

SOLUTION

Rotation by 90° around the origin has matrix

$$\mathbf{R}_O(90^\circ) = \begin{pmatrix} 0 & -1 & 0 \\ 1 & 0 & 0 \\ 0 & 0 & 1 \end{pmatrix}.$$

Thus, the new frame has matrix

$$\mathbf{G} = \mathbf{R}_O(90^\circ)\mathbf{F} = \begin{pmatrix} 0 & -1 & 0 \\ 1 & 0 & 0 \\ 0 & 0 & 1 \end{pmatrix} \begin{pmatrix} \frac{5}{13} & \frac{-12}{13} & 2 \\ \frac{12}{13} & \frac{5}{13} & 4 \\ 0 & 0 & 1 \end{pmatrix}$$

$$= \begin{pmatrix} \frac{-12}{13} & \frac{-5}{13} & -4 \\ \frac{5}{13} & \frac{-12}{13} & 2 \\ 0 & 0 & 1 \end{pmatrix}.$$

■

EXERCISES

**Marks challenging exercises.*

Positional Representation of a Frame

1. Suppose a frame $\mathcal{F}$ has its origin at (4, -1) and the x- and y-vectors are

$$\begin{pmatrix} \frac{12}{13} \\ -\frac{5}{13} \end{pmatrix} \quad \text{and} \quad \begin{pmatrix} -\frac{5}{13} \\ -\frac{12}{13} \end{pmatrix}.$$

What is the matrix that represents the frame?

2. Suppose a frame $\mathcal{F}$ has its origin at (−2, 1) and the x- and y-vectors are

$$\begin{pmatrix} \frac{4}{5} \\ \frac{3}{5} \end{pmatrix} \quad \text{and} \quad \begin{pmatrix} -\frac{3}{5} \\ \frac{4}{5} \end{pmatrix}.$$

What is the matrix that represents the frame?

3. If a frame $\mathcal{F}$ is represented by the following matrix, what is the origin of the frame and what are the x- and y-vectors? Sketch the frame showing its relation to a world frame.

$$\mathbf{F} = \begin{pmatrix} \frac{\sqrt{3}}{2} & \frac{-1}{2} & \frac{3}{4} \\ \frac{1}{2} & \frac{\sqrt{3}}{2} & \frac{1}{4} \\ 0 & 0 & 1 \end{pmatrix}.$$

4. If a frame is represented by the following matrix, what is the origin of the frame and what are the x- and y-vectors? Sketch the frame showing its relation to a world frame.

$$\mathbf{F} = \begin{pmatrix} \frac{\sqrt{2}}{2} & -\frac{\sqrt{2}}{2} & \frac{\sqrt{3}}{2} \\ \frac{\sqrt{2}}{2} & \frac{\sqrt{2}}{2} & -\frac{1}{2} \\ 0 & 0 & 1 \end{pmatrix}.$$

5. What, if anything, can you say about a 3×3 matrix if all you know about it is that it represents an isometry that doesn't move the origin?

6. Suppose the x axis of a frame $\mathcal{F}$ with counterclockwise sense makes an angle of θ with the x axis of $\mathcal{W}$. Suppose the origin of $\mathcal{F}$ has polar coordinates ρ, α in the coordinate system of $\mathcal{W}$. The three numbers ρ, α, θ can serve to represent the frame numerically (and more briefly than the matrices we have been using).

   (a) If we didn't know that the sense of the frame was counterclockwise, would this representation be useful?

   (b) Show how the 3×3 matrix of $\mathcal{F}$ can be obtained from ρ, α, θ (give formulas for the entries of the matrix). If you think it is necessary, assume that $\mathcal{F}$ is a counterclockwise frame.

7. Suppose

$$\mathbf{F} = \begin{pmatrix} a & c & e \\ b & d & f \\ 0 & 0 & 1 \end{pmatrix}$$

   is a matrix of a frame. Prove that $a^2 + b^2 = c^2 + d^2 = 1$.

8. Answer the question posed at the end of Example 6.16.

Moving a Frame

9. A mobile robot has its frame at the world frame $\mathcal{W}$ and it rotates around the world origin by 60°. Then it moves 10 units in the direction of its rotated x-vector. What is the matrix of the new frame?

10. Suppose the frame of Exercise 3 is subjected to rotation by 60° around the origin followed by translation by $\left\langle \frac{1}{2}, \frac{\sqrt{3}}{2} \right\rangle$. What is the new frame?

11. Suppose the frame of Exercise 4 is subjected to the isometry in Example 6.16. What is the new frame?

12. Suppose the frame of Exercise 1 is subjected to rotation by 60° around the origin followed by translation by $\left\langle \frac{1}{2}, \frac{\sqrt{3}}{2} \right\rangle$. What is the new frame?

13. Suppose the frame of Exercise 2 is subjected to the isometry in Example 6.17. What is the new frame?

14. Suppose the robot waiter in Example 6.18 moves through a translation of $10\sqrt{2}$ units in the northeast direction at a time when the dining area has not yet begun to rotate. Then the robot stops and the dining area rotates through 30°. What is the translation vector $\langle a, b \rangle$? Find the new robot frame.

15. Suppose the frame $\mathcal{E}$ of Example 6.17 is moved by an isometry S represented by the matrix **F** of Exercise 4. Can you tell whether the origin of the new robot frame is still inside the room? If there is not enough information to tell, explain what is missing.

16. Verify the principles of Example 6.15 by using the frame matrix

$$\mathbf{F} = \begin{pmatrix} \frac{\sqrt{3}}{2} & -\frac{1}{2} & 2 \\ \frac{1}{2} & \frac{\sqrt{3}}{2} & 7 \\ 0 & 0 & 1 \end{pmatrix}.$$

Go through steps (a), (b), and (c) of Example 6.15 in order to find the origin and the x- and y-vectors of $\mathcal{F}$.

*17. Let $\mathcal{F}$ be a frame with counterclockwise sense whose matrix is given. We know $\mathbf{F} = \mathbf{T}_{\langle a,b \rangle}\mathbf{R}_O(\theta)$ for a suitable translation matrix $\mathbf{T}_{\langle a,b \rangle}$ and rotation matrix $\mathbf{R}_O(\theta)$.

(a) Explain how we can compute the entries of $\mathbf{T}_{\langle a,b \rangle}$ and the rotation angle θ from the entries of **F**.

(b) Suppose we want to find an isometry G that sends $\mathcal{F}$ back to $\mathcal{W}$. Show how this can be done.

6.5 Forward Kinematics for a Simple Robot Arm

End Effector Position

In this section we study a two-dimensional robot arm consisting of a number of solid links hinged together and operating entirely in the plane (Figure 6.17). Our two-dimensional robot is attached to a base that doesn't move. There is a rotation axis through point O in the base, perpendicular to the plane, so the first link can make any angle with the horizontal. The robot can also bend around the rotation axis that passes through point B and is perpendicular to the plane at B. This enables the second link to rotate with respect to the first. At the end of the second link, there is some kind of *end effector* to do useful work. In the present case, we depict a gripper type of end effector. The only motion the gripper is capable of is opening and closing its jaws. Other types of end effectors in use today are welders, drills, paint sprayers, and fingered hands.

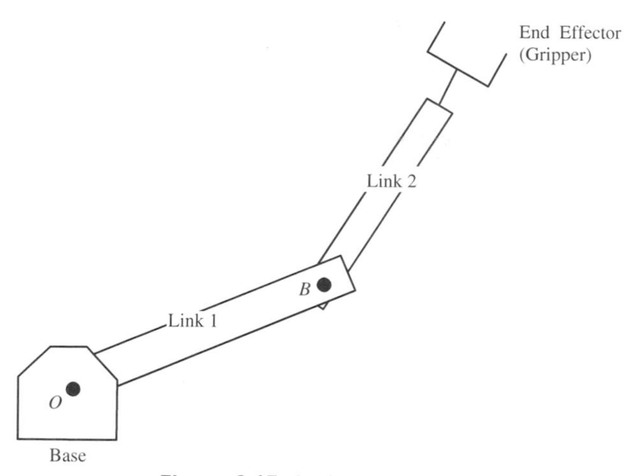

Figure 6.17 A planar robot arm.

Most useful robots differ from our two-dimensional robot because they have joints that enable them to move in all three dimensions (see Figure 6.18 showing the six rotation axes of the PUMA robot). However, some two-dimensional robots like ours have been built to test robot control software or for other experimental purposes. It is also interesting to note that the claw of a crab is essentially a two-joint two-dimensional arm like the one depicted in Figure 6.17. But the main justification for considering our two-dimensional robot arm is that the mathematics needed to deal with it is an excellent introduction to the mathematics needed for the kinds of three-dimensional robot arms at work in factories today.

Figure 6.19a shows a mathematical idealization of our planar arm in which some details, like the thicknesses of the links, have been removed because they are not essential. In addition, a world coordinate frame has been added with origin at the rotation point O. The rotations of the links are measured by θ_1 and θ_2. θ_1 is the angle through which the x vector of $\mathcal{W}$ has to turn to coincide with the first link. If the turn is counterclockwise, θ_1 is positive; otherwise, it is negative (see Figure 6.20). θ_2 is the angle through which the extension of the first link (dotted) must turn to coincide with the second link—a positive or negative angle according to whether the turn is counterclockwise or clockwise.

The end effector point G is an imaginary point centered between the tips of the jaws of the gripper. The first link extends from the origin of $\mathcal{W}$ to the elbow point $\mathcal{B}$, and its length is denoted l_1. The second link extends from the elbow $\mathcal{B}$ all the way to the end effector point G (even though some of this is thin air), and this length is l_2.

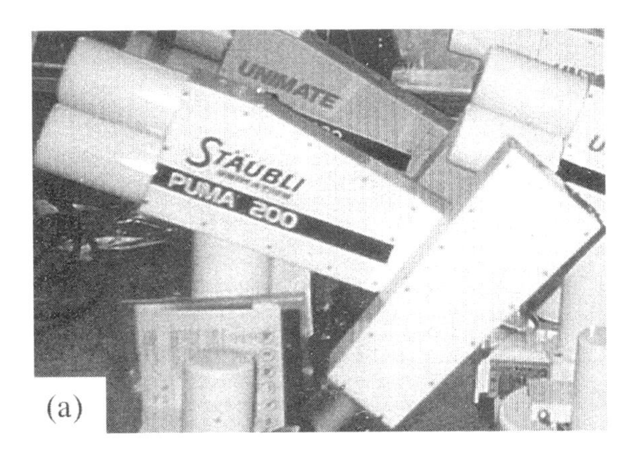

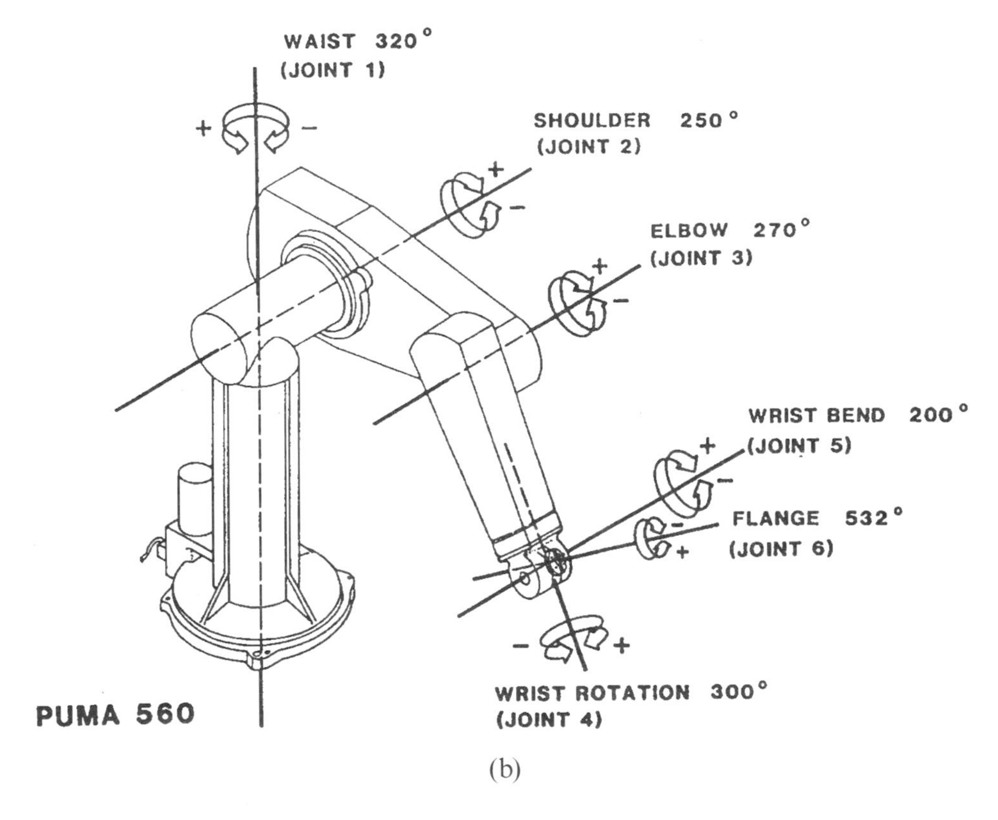

Figure 6.18 (a) A PUMA robot. (Courtesy of RP Automation.) (b) Drawing of the PUMA showing its rotation axes. (Courtesy of Schäubli.)

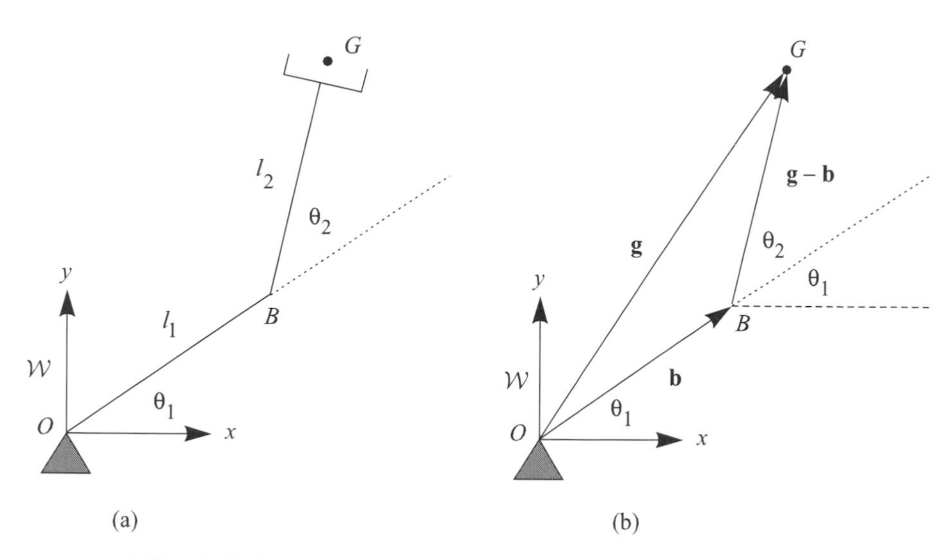

Figure 6.19 (a) Mathematical model of a planar robot arm. (b) The links as vectors.

By moving its joints, the robot moves the end effector point around. How can we calculate the coordinates of G (in the world coordinate system $\mathcal{W}$) in terms of the angles θ_1 and θ_2 and the link lengths l_1 and l_2? In robotics, this is the first part of what is called the *forward kinematic problem*, the most basic question in the theory of robot arm motion.

The first step in calculating G is to imagine that each link is a vector (Figure 6.19b) by putting a direction on the line segment representing the link. We choose the directions so that if we start at the origin O (base) and follow the directions, we will arrive at the end effector point. Recall that the coordinates of G and B are also the components of position vectors $\mathbf{g}$ and $\mathbf{b}$, respectively. Then the link vectors just described would be $\mathbf{b}$ and $\mathbf{g} - \mathbf{b}$. Naturally:

$$\mathbf{g} = \mathbf{b} + (\mathbf{g} - \mathbf{b}). \tag{6.10}$$

Now we need to find formulas for $\mathbf{b}$ and $\mathbf{g} - \mathbf{b}$ in terms of the angles and the link lengths. $\mathbf{b}$ makes an angle θ_1 with the horizontal and has length l_1. Thus, using the familiar polar coordinates formula,

$$\mathbf{b} = \langle l_1 \cos \theta_1, l_1 \sin \theta_1 \rangle. \tag{6.11}$$

To find the components of the vector $G - B$, we begin by noting that the angle the second link makes with the rightward horizontal is $\theta_1 + \theta_2$ (see Figure 6.19b). Thus,

$$\mathbf{g} - \mathbf{b} = \langle l_2 \cos(\theta_1 + \theta_2),\ l_2 \sin(\theta_1 + \theta_2) \rangle. \tag{6.12}$$

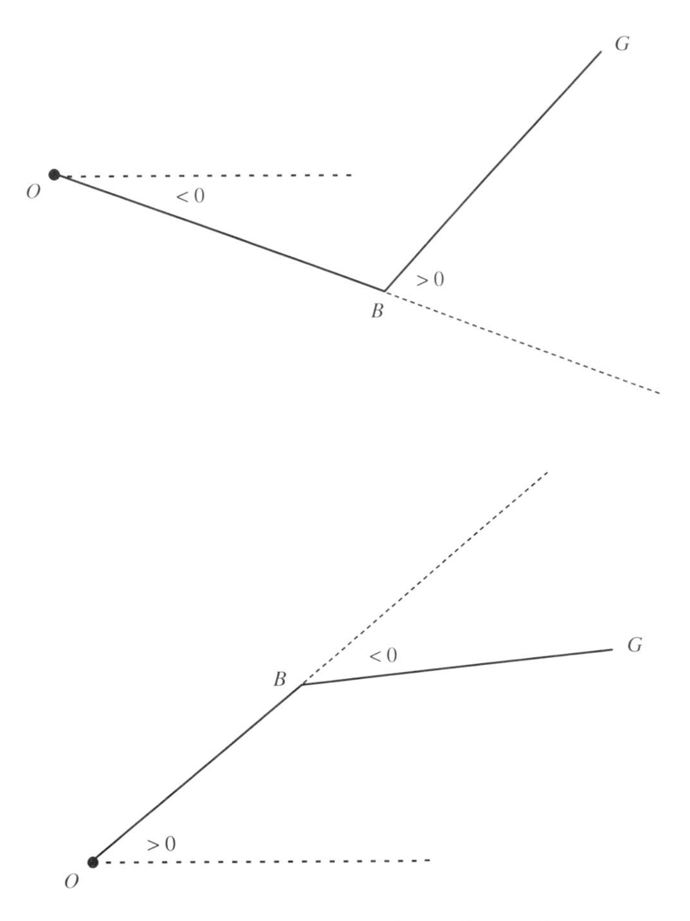

Figure 6.20 Positive and negative angles.

Finally, substituting from Eq. (6.11) and Eq. (6.12) into Eq. (6.10), we get

$$\mathbf{g} = \langle l_2 \cos \theta_1 + l_2 \cos(\theta_1 + \theta_2),\ \ l_1 \sin \theta_1 + l_2 \sin(\theta_1 + \theta_2)\rangle. \tag{6.13}$$

Therefore, the x and y components for $\mathbf{g}$ are:

$$\begin{aligned} x_g &= l_1 \cos \theta_1 + l_2 \cos(\theta_1 + \theta_2), \\ y_g &= l_1 \sin \theta_1 + l_2 \sin(\theta_1 + \theta_2). \end{aligned} \tag{6.14}$$

EXAMPLE 6.19: FINDING THE END EFFECTOR POINT

Suppose $l_1 = 10$, $l_2 = 15$, $\theta_1 = 30^\circ$, $\theta_2 = -45^\circ$. Find the coordinates of G.

SOLUTION

One approach is simply to substitute into either Eqs. (6.13) or (6.14). However, we shall work out Eqs. (6.11) and (6.12) and add these vectors as in Eq. (6.10).

$$\mathbf{b} = \langle 10\cos 30^\circ, 10\sin 30^\circ\rangle = \left\langle 10\frac{\sqrt{3}}{2},\ 10\left(\frac{1}{2}\right)\right\rangle$$
$$= \langle 8.66, 5\rangle.$$
$$\mathbf{g} - \mathbf{b} = \langle 15\cos(30^\circ - 45^\circ), 15\sin(30^\circ - 45^\circ)\rangle = \langle 15\cos(-15^\circ), 15\sin(-15^\circ)\rangle$$
$$= \langle 14.49, -3.88\rangle.$$
$$\mathbf{g} = \langle 8.66, 5\rangle + \langle 14.49, -3.88\rangle = \langle 23.15, 1.12\rangle. \quad \blacksquare$$

End Effector Frame

We have just learned how to find where the end effector point of a two-dimensional robot arm is located based on the joint angles and link lengths. But there is more that we might need to know about the end effector. Figure 6.21 shows two configurations of an arm holding a cup. In one case, the cup holds water, and in the other, it doesn't. The difference in the configurations, of course, is in "how the gripper is pointing"—what is normally called the *orientation* of the gripper. We need a numerical representation of this orientation, and then we need a way to calculate it from the angles and link lengths.

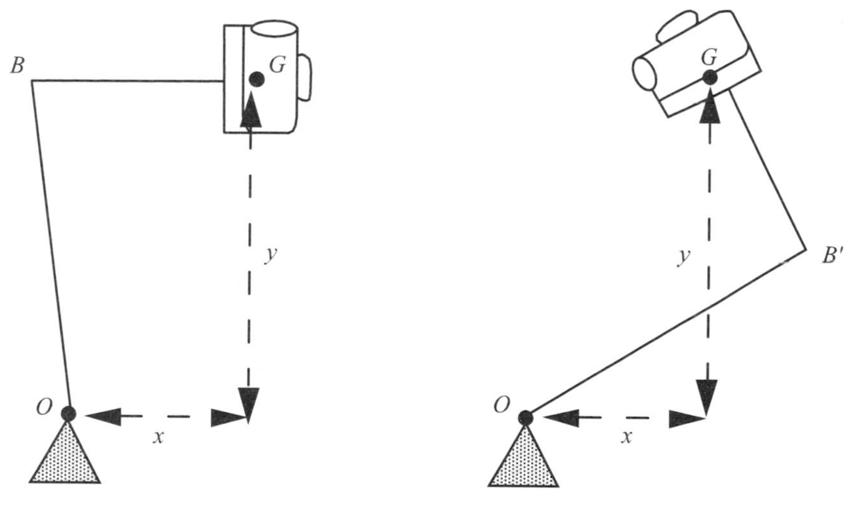

Figure 6.21 Two configurations with the same end effector point.

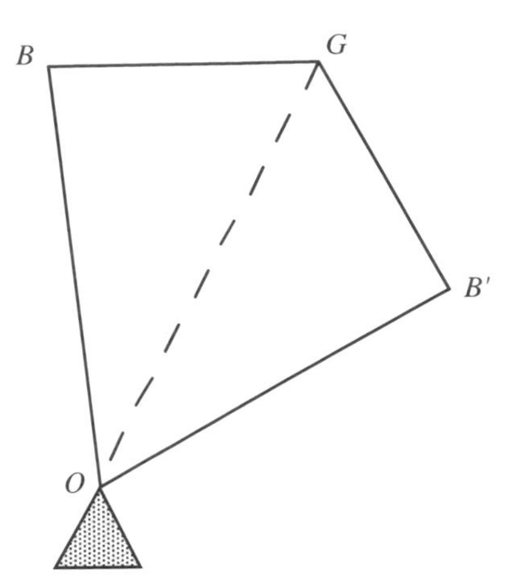

Figure 6.22 Triangles of the two configurations.

An interesting sidelight on Figure 6.21 is that the end effector point G is the same in both cases (the x and y values for G in the two cases are the same). Clearly, the end effector point does not determine how the gripper is pointing. Figure 6.22 shows a simplified picture that makes the geometry clearer. The upper left triangle OBG represents the left configuration of Figure 6.21. Two sides of the triangle are links, and the dotted third side is the connector from base O to end effector point G. Reflect the triangle on this dotted side and you get new positions for the links that keep the end effector point the same as before.

To describe the gripper's orientation, we install a frame $\mathcal{E}$ with origin at the end effector point G (Figure 6.23a). The x axis of this frame is in the direction of the last link and the y axis is perpendicular to it, making a frame of counterclockwise sense. If we know this frame, we know everything we need to know about the position and orientation of the gripper. The matrix $\mathbf{E}$ of this frame is what we take to be the numerical representation of the position and orientation of the gripper. Finding $\mathbf{E}$ in terms of θ_1, θ_2, l_1, and l_2 is the second part of the forward kinematic problem. To solve this problem we find the isometry matrix which moves $\mathcal{W}$ onto $\mathcal{E}$.

The following series of steps moves the world frame $\mathcal{W}$ onto $\mathcal{E}$ with a series of four intermediate frames: $\mathcal{M}_1$, $\mathcal{M}_2$, $\mathcal{M}_3$, $\mathcal{M}_4$, where $\mathcal{M}_4 = \mathcal{E}$.

1. Rotate $\mathcal{W}$ around its origin by θ_1 until the new x-vector points along the first link. The frame that results from this is called $\mathcal{M}_1$ (Figure 6.23a). This rotation is $R_O(\theta_1)$. Remember, we use the subscript "O" to emphasize that the rotation is around the origin (the next rotation will not be).

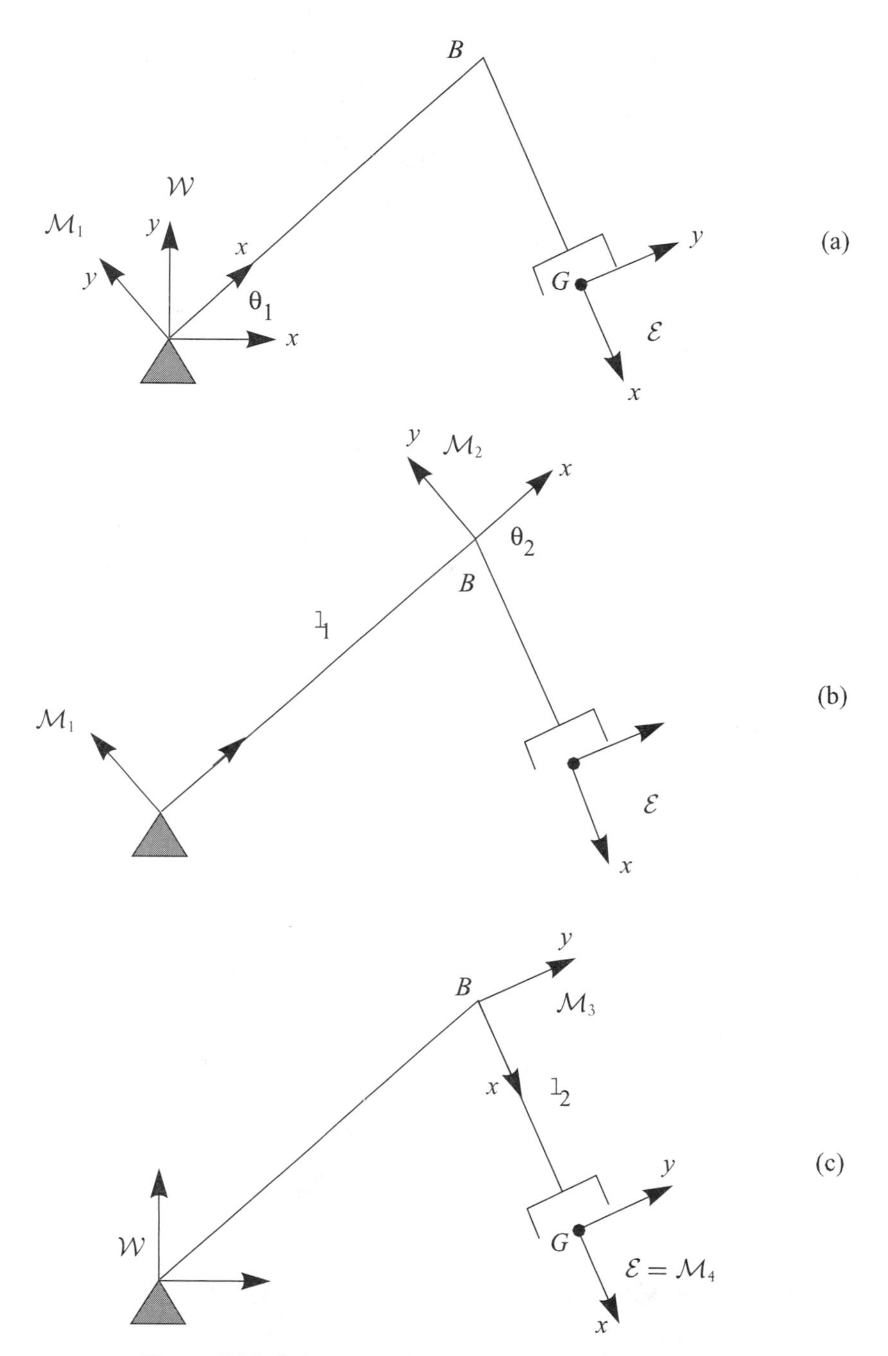

Figure 6.23 A frame moving to the end effector frame $\mathcal{E}$.

2. Translate $\mathcal{M}_1$ along the first link till the frame origin coincides with the elbow point B (Figure 6.23b). The frame resulting from this is $\mathcal{M}_2$. The translation vector for this move is $\mathbf{b}$, so we denote this translation by $T_{\mathbf{b}}$.

3. Rotate $\mathcal{M}_2$ around B by θ_2 so that the frame x-vector points in the direction of the second link. This brings us to $\mathcal{M}_3$ (compare Figures 6.23b and c). This rotation is $R_B(\theta_1)$, the subscript B reminding us that the center of rotation is not the origin, but the point B.

4. Translate $\mathcal{M}_3$ along the second link by a distance l_2, resulting in $\mathcal{M}_4$, which is identical with $\mathcal{E}$ (Figure 6.23c). The translation vector for this is $\mathbf{g}-\mathbf{b}$, so the translation is denoted $T_{\mathbf{g}-\mathbf{b}}$.

The isometry E which sends $\mathcal{W}$ to $\mathcal{E}$ is the composition of these isometries:

$$E = T_{\mathbf{g}-\mathbf{b}} \circ R_B(\theta_2) \circ T_{\mathbf{b}} \circ R_O(\theta_1). \tag{6.15}$$

The next logical step seems to be to write down the matrices for these four isometries and multiply them together. The matrix for $R_B(\theta_2)$ is unpleasantly complex (see Theorem 6.2 and Eq. (6.9) of Section 6.2) so we'll use a little theory to simplify Eq. (6.15).

☆ THEOREM 6.9

If B is any point at all and θ is any angle,

$$R_B(\theta) \circ T_{\mathbf{b}} = T_{\mathbf{b}} \circ R_O(\theta). \tag{6.16}$$

Proof

If you draw a picture of what this equation says, it should be fairly obvious. But we'll give two algebraic proofs as well. First, an algebraic manipulation from Theorem 6.2 of Section 6.2. Restating the formula of that theorem in terms of transformations instead of their matrices:

$$R_B(\theta) = T_{\mathbf{b}} \circ R_O(\theta) \circ T_{-\mathbf{b}}.$$

Now simply compose both sides with $T_{\mathbf{b}}$ on the right, and the result follows. ■

Our second proof follows by verifying that the left and right sides of the alleged Equation (6.16) have the same effect on the world frame. The instructions of the left side first move the world frame by translation so its origin is at B, and then rotate around B until the x axis is inclined at an angle of θ to the rightward horizontal. Now let's look at the right side. We turn the frame first, until the x axis is inclined at an angle of θ to the rightward horizontal. Next we translate until the origin is at B. This translation doesn't turn the x

axis, so that axis is still inclined at an angle of θ to the rightward horizontal. The y-vectors also wind up pointing the same way, whether we use the right or left side of Eq. (6.16). (Do you see why?) In other words, the frame we get by following the instructions on the right is the same as the frame we get from the instructions on the left. Since an isometry is determined by its effect on the origin and tips of the x- and y-vectors of a frame (see Theorem 4.14 in Section 4.3 of Chapter 4), the right and left sides of Eq. (6.16) are equal. ■

☆ THEOREM 6.10

The matrix **E** of the end effector frame $\mathcal{E}$ of the robot in Figure 6.23 is

$$\mathbf{E} = \begin{pmatrix} \cos(\theta_1+\theta_2) & -\sin(\theta_1+\theta_2) & l_1 \cos\theta_1 + l_2 \cos(\theta_1+\theta_2) \\ \sin(\theta_1+\theta_2) & \cos(\theta_1+\theta_2) & l_1 \sin\theta_1 + l_2 \sin(\theta_1+\theta_2) \\ 0 & 0 & 1 \end{pmatrix}. \tag{6.17}$$

PROOF

We return to Eq. (6.16) and simplify it by the previous theorem,

$$\begin{aligned} E &= T_{\mathbf{g}-\mathbf{b}} \circ R_B(\theta_2) \circ T_{\mathbf{b}} \circ R_O(\theta_1) \\ &= T_{\mathbf{g}-\mathbf{b}} \circ T_{\mathbf{b}} \circ R_O(\theta_2) \circ R_O(\theta_1). \end{aligned} \tag{6.18}$$

But $R_O(\theta_2) \circ R_O(\theta_1) = R_O(\theta_1+\theta_2)$ and $T_{\mathbf{g}-\mathbf{b}} \circ T_{\mathbf{b}} = T_{\mathbf{g}}$ (see Exercise 11 of Section 6.2 in this chapter.) Thus, the right side of Eq. (6.18) is $T_{\mathbf{g}} \circ R_O(\theta_1+\theta_2)$. Thus, by Theorem 6.1 in Section 6.2, the matrix for this isometry is

$$\mathbf{E} = \begin{pmatrix} \cos(\theta_1+\theta_2) & -\sin(\theta_1+\theta_2) & x_g \\ \sin(\theta_1+\theta_2) & \cos(\theta_1+\theta_2) & y_g \\ 0 & 0 & 1 \end{pmatrix}.$$

The components of **g** have been worked out in terms of the angles and link lengths in Eqs (6.14). Substituting these in the last formula gives Eq. (6.17), which is what we wished to prove. ■

EXAMPLE 6.20: THE END EFFECTOR FRAME

(a) What is the orientation of the end effector if $l_1 = 30$, $l_2 = 10$, $\theta_1 = 90°$, $\theta_2 = 45°$?

(b) What is the origin of the end effector frame? What are the x- and y-vectors of the frame?

SOLUTION

(a) Substituting into Eq. (6.17) gives

$$\begin{pmatrix} \cos(90^\circ + 45^\circ) & -\sin(90^\circ + 45^\circ) & 30 \cos 90^\circ + 10 \cos(90^\circ + 45^\circ) \\ \sin(90^\circ + 45^\circ) & \cos(90^\circ + 45^\circ) & 30 \sin 90^\circ + 10 \sin(90^\circ + 45^\circ) \\ 0 & 0 & 1 \end{pmatrix}$$

$$= \begin{pmatrix} -0.707 & -0.707 & 30(0) + 10(-0.707) \\ 0.707 & -0.707 & 30(1) + 10(0.707) \\ 0 & 0 & 1 \end{pmatrix}$$

$$= \begin{pmatrix} -0.707 & -0.707 & -7.07 \\ 0.707 & -0.707 & 37.07 \\ 0 & 0 & 1 \end{pmatrix}$$

(b) As described in the previous section, the origin is given by the last column and the x- and y-vectors are obtained from the first and second column. Origin =

$$(-7.07,\ 37.07),\ x\text{-vector} = \begin{pmatrix} -0.707 \\ 0.707 \end{pmatrix},\ y\text{-vector} = \begin{pmatrix} -0.707 \\ -0.707 \end{pmatrix}. \quad \blacksquare$$

So far, we have concerned ourselves with the two parts of the forward kinematic problem. It is also useful to be able to do the problem "in reverse." This problem, called the *inverse kinematic problem*, is: given the end effector matrix **E**, and given the link lengths l_1 and l_2, calculate θ_1 and θ_2. In less mathematical language, we can think of the robot trying to find out, "How do I have to bend if I want to get my gripper to a certain desired location and orientation?"

For our simplified two-joint robot, the inverse kinematic problem either has no solution or can always be solved unambiguously. However, for more realistic robots, there can be two or more sets of joint angle values that give rise to the same **E**. More properly stated, the inverse kinematic problem is to find all possible sets of joint angle values that result in the given end effector matrix **E**. For most robots, this is harder than the forward kinematic problem. For our simple two-dimensional, two-link robot, it is manageable if one has some skill at geometry and trigonometry. However, for three-dimensional robots in use today, solving the inverse kinematic problem requires linear algebra and often multivariable calculus.

EXERCISES

**Marks challenging exercises.*

End Effector Position

1. Draw a sketch to illustrate Example 6.19.

2. (a) What are the components of **b**, the position vector of B, on the left side of Figure 6.24? O denotes the origin of the world frame, whose x-vector lies along the dotted segment.

   (b) What are the components of the vector $\mathbf{g} - \mathbf{b}$ on the right side of Figure 6.24? The world frame origin is not shown. Do you need to know anything about it? If so, make a reasonable assumption.

3. Suppose $l_1 = 20$, $l_2 = 10$, $\theta_1 = 90°$, and $\theta_2 = -45°$. Sketch the configuration. Find the coordinates of G.

4. Suppose $l_1 = 10$, $l_2 = 20$, $\theta_1 = 60°$, and $\theta_2 = 30°$. Sketch the configuration. Find the components of $\mathbf{g}$.

DEFINITION

The *workspace* of the robot consists of all points the end effector point can reach as we vary the angles θ_1 and θ_2 through all possibilities from $-180°$ to $+180°$. (Ignore interference of the base with movement of the links.)

5. Suppose $l_1 = 20$ and $l_2 = 10$. Sketch the workspace as a shaded region of the plane.

6. Suppose $l_1 = 10$ and $l_2 = 20$. Sketch the workspace as a shaded region of the plane.

7. Suppose $l_1 = l_2$. Sketch the workspace as a shaded region of the plane.

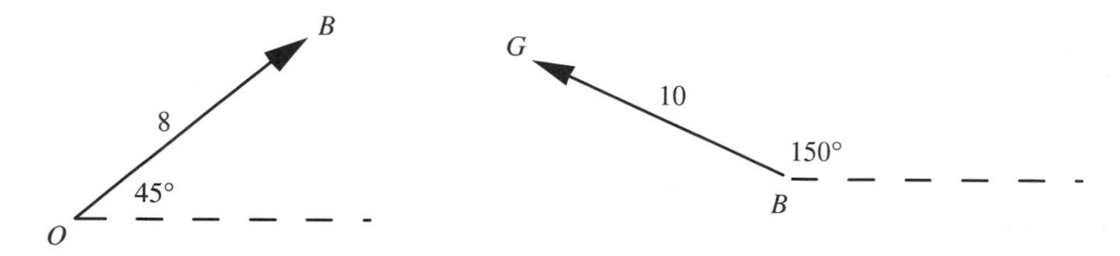

Figure 6.24 Finding components of vectors from lengths and angles.

8. Think of the human arm as a robot-like mechanism for a moment—one with two main links, the upper arm of length l_1 and the forearm with length l_2. Why is it useful for us that l_1 and l_2 are nearly equal?

* 9. Suppose the x and y coordinates of G are specified, and we need to find values for θ_1 and θ_2 that put the end effector point at that specified position. Can you find formulas that give the angles in terms of l_1, l_2, and the x and y components of G?

10. (a) Using the formulas you worked out in the previous exercise to find θ_1 and θ_2, use them to find the two angles under these circumstances: $l_1 = 10$, $l_2 = 7$, $x_g = 15$, $y_g = 2$.

    (b) Plug in the θ_1 and θ_2 values you obtained in part (a) into Eqs. (6.14), and see if you get $x_g = 15$ and $y_g = 2$.

11. Can you find a position for the end effector point G of our two-link robot (Figures 6.19 and 6.21) where there is precisely one pair of angles θ_1 and θ_2 that put G in the required position?

End Effector Frame

12. Find formulas for the x and y coordinates (with respect to the world coordinate system) of the leftmost gripper jaw tip (Figure 6.19a) in terms of l_1, l_2, θ_1, θ_2 and w, the distance from one jaw tip to the other. Assume that the direction from G to this gripper jaw tip is perpendicular to the direction of the second link.

13. Suppose $l_1 = 2$, $l_2 = 3$, $\theta_1 = 45°$, $\theta_2 = 45°$. Find the matrix for the end effector (gripper) frame.

14. Suppose $l_1 = 4$, $l_2 = 2$, $\theta_1 = -30°$, $\theta_2 = 90°$. Find the matrix for the end effector (gripper) frame.

15. With the data in Exercise 13, find the matrix for the frame $\mathcal{M}_1$.

16. With the data in Exercise 14, find the matrix for the frame $\mathcal{M}_1$.

17. With the data in Exercise 13, find the matrix for the frame $\mathcal{M}_2$. First express it as a product of $\mathbf{M}_1$ and another matrix.

18. With the data in Exercise 14, find the matrix for the frame $\mathcal{M}_2$. First express it as a product of $\mathbf{M}_1$ and another matrix.

19. With the data in Exercise 13, find the matrix for the frame $\mathcal{M}_3$. First express it as a product of $\mathbf{M}_1$ and two other matrices. Can Theorem 6.9 help?

20. With the data in Exercise 14, find the matrix for the frame $\mathcal{M}_3$. First express it as a product of $\mathbf{M}_1$ and two other matrices. Can Theorem 6.9 help?

21. Suppose a camera is mounted on the last link of the robot as shown in Figure 6.25. The camera has a frame attached whose matrix can be used to specify where the camera is and how it is pointing. The middle of the camera bottom (the origin of the frame) is located 1/4 of the way from B to G. Find the matrix formula (analogous to Eq. (6.17)) that describes the camera frame. As usual, the link lengths are l_1, l_2, and the joint angles are denoted θ_1, θ_2.

22. (a) Work out a formula for the position of G in the three-link robot shown in Figure 6.26.

    (b) What is the isometry formula (analogous to Eq. (6.18)) of the gripper frame?

*23. Suppose the gripper frame of our two-link robot is installed with clockwise sense. The x-vector still points in the direction of the last link, but the y-vector is pointing opposite to the way it is pointing in Figure 6.23. Find a formula (analogous to Eq. (6.17)) for this frame in terms of the link lengths and the angles.

24. Suppose the end effector of our robot is actually a camera, and it locates a point of interest whose coordinates are (10, 20) when measured in the end effector frame. Assume that $l_1 = 30$, $l_2 = 10$, $\theta_1 = 90°$, and $\theta_2 = 45°$ exactly as in Example 6.20. Find the coordinates of the point of interest in the world frame.

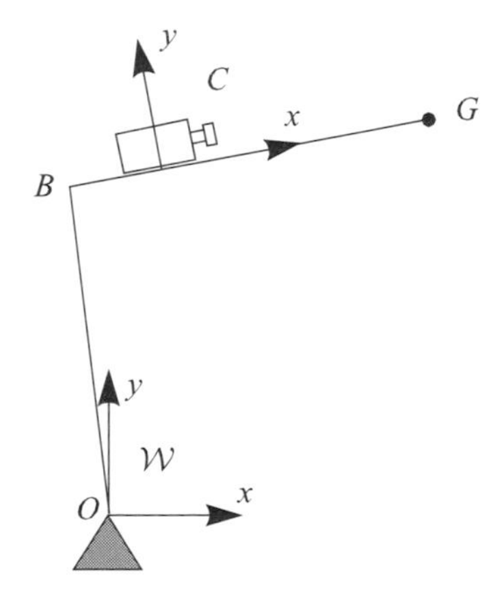

Figure 6.25 Robot with camera.

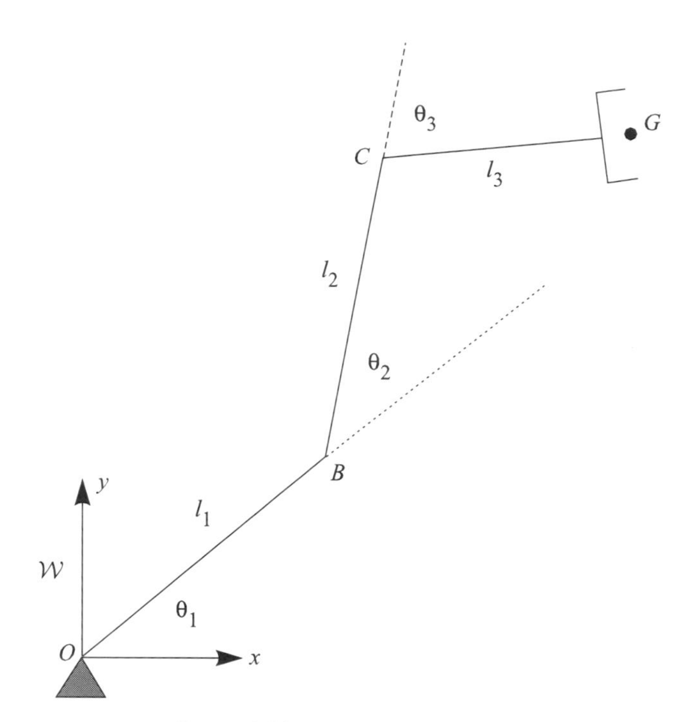

Figure 6.26 A three-joint robot.

*25. There are isometry matrices **E** for which there is no solution to the inverse kinematic problem for our two-joint robot. Explain this. Explain how to construct such an **E** and why there is no solution for that **E**.

Chapter 7

Transformation Geometry III: Similarity, Inversion, and Projection

Prerequisites: Sections 7.1 and 7.2 rely on some facts about similar triangles (material covered in Section 2.4 of Chapter 2). In Section 7.1, composition of transformations (covered in Section 4.2 of Chapter 4) comes up but can be avoided. In any case, only the definition is needed. In Sections 7.3, 7.4, and 7.5 we undertake our most extensive study of three-dimensional geometry, using vector equations of lines and planes (covered in Sections 5.1 and 5.2 of Chapter 5) as well as the axioms pertaining to three dimensions (Section 1.3 in Chapter 1).

Up to this point, we have studied isometries extensively. But isometries are not the only geometric transformations that are theoretically interesting and practically useful. In this chapter we introduce some other types of transformations and their applications in computer graphics, cartography, the theory of mechanical linkages, and computer vision.

7.1 Central Similarity and Other Similarity Transformations

We are surrounded by pictures that are not life size: huge faces on billboards, tiny ones on our drivers licenses, etc. The process by which an image gets to be blown up or squeezed down can be thought of in terms of the geometric transformation called central similarity.

DEFINITION

Let C be any fixed point in the plane and r any positive number. The transformation called[1] *central similarity* $S_{C,r}$ is defined as follows:

(a) The image of C is C itself.

(b) For any other point P, the image P' — also denoted $S_{C,r}(P)$ — is the point with the following two properties:

1. $CP' = rCP$.

2. P and P' are on the same side of C on the line $\overleftrightarrow{CP}$.

The point C is called the *center* of the transformation and r is called the *ratio of similarity*.

It is also possible to define central similarity for a negative value of r: First take the absolute value of r and proceed as above. Finally, apply rotation through 180° to the resulting point. Figure 7.1 shows central similarities with $r = 2$ and -2. Unless we say otherwise, we deal exclusively with central similarities with $r > 0$ in what follows.

The starting point of the theory of central similarities is the following theorem.

✯ THEOREM 7.1

If P and Q are points and P' and Q' are images (see Figure 7.2) under a central similarity $S_{C,r}$ where $r > 0$, then:

(a) $P'Q' = rPQ$.

(b) The lines $\overleftrightarrow{P'Q'}$ and $\overleftrightarrow{PQ}$ are parallel (i.e., direction is preserved).

---

[1] Central similarities are also called *dilations* and *homotheties*.

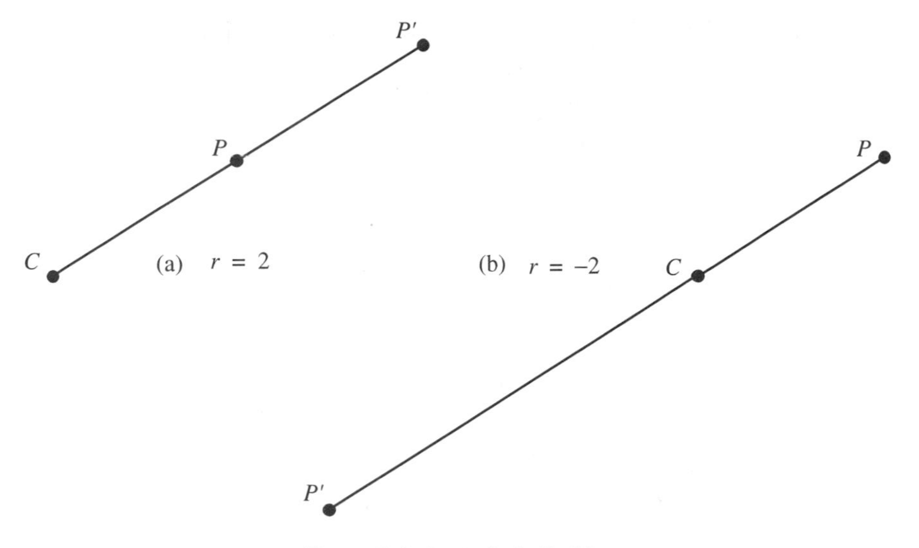

Figure 7.1 Central similarities.

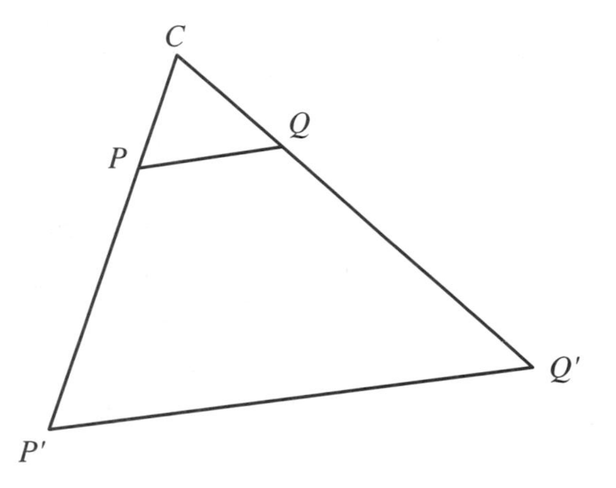

Figure 7.2 Central similarity scales all segments the same and preserves direction.

PROOF

It follows immediately from Theorem 2.22 of Section 2.4 in Chapter 2 (the SAS criterion for similarity) that triangles CPQ and $CP'Q'$ are similar. By the definition of similarity, $P'Q'/PQ = CP'/CP = r$ and part (a) is proven. Part (b) comes from Theorem 2.20 of Section 2.4 in Chapter 2. ■

✯ THEOREM 7.2: THE INVARIANCE OF COLLINEARITY

If P, Q, and R are on a line, then their images under a central similarity, P', Q', and R', also lie on some line.

PROOF

With no loss of generality, suppose R is between P and Q. Then $PR + RQ = PQ$. Multiply this equation through by r and apply the previous theorem to obtain $P'R' + R'Q' = P'Q'$. This means Q' is on a line with P' and Q'; because if this were not true, the triangle inequality (see Exercises 12–14 of Section 2.1 in Chapter 2) would tell us that $P'R' + R'Q' < P'Q'$, contradicting the previous equation. ■

✯ THEOREM 7.3

Circlehood is invariant: The image of a circle under a central similarity is another circle.

PROOF

Let the original circle have center P and radius k. If Q is any point on this circle then $PQ = k$. Consequently, $P'Q' = rk$. Thus, Q' lies on a circle of radius rk around P'. ■

It is also possible to prove that angles are invariant, and we leave this as an exercise.

The effect of central similarity is shown in Figure 7.3 where we use two different positive ratios. For $r > 1$, the image of a figure is what we would loosely describe in everyday English as "zooming in" or "blowing up" or "scaling up" the figure by a factor of 2. For $0 < r < 1$, we get a "squeezing down" or "scaling down" of the figure. Either way, the figure is recognizable because the proportions of the figure are the same and so are the angles. If we have a curved line in the figure, it is also instantly recognizable. (Think of the curve as approximately a whole lot of tiny line segments. The way the curve bends is determined by the angles between these tiny line segments.) We leave it to you to work out what images look like if $r < 0$.

Our next task is to find equations that relate the coordinates of an image point $P'(x', y')$ to the coordinates of the point $P(x, y)$ from which it arose. For simplicity, let's say the center is the origin O (0, 0) (Figure 7.4). Drop a perpendicular from P to the x axis, meeting it at Q. The x coordinates of P and Q are equal, as labeled in the figure. Let Q' be the image of Q under the central similarity. Because Q and Q' are on the x axis, this means that the x coordinate of Q' is r times the x coordinate of Q. By Theorem 7.1, $\overleftrightarrow{P'Q'}$ and $\overleftrightarrow{PQ}$ are parallel so $P'Q'$ is perpendicular to the x axis.

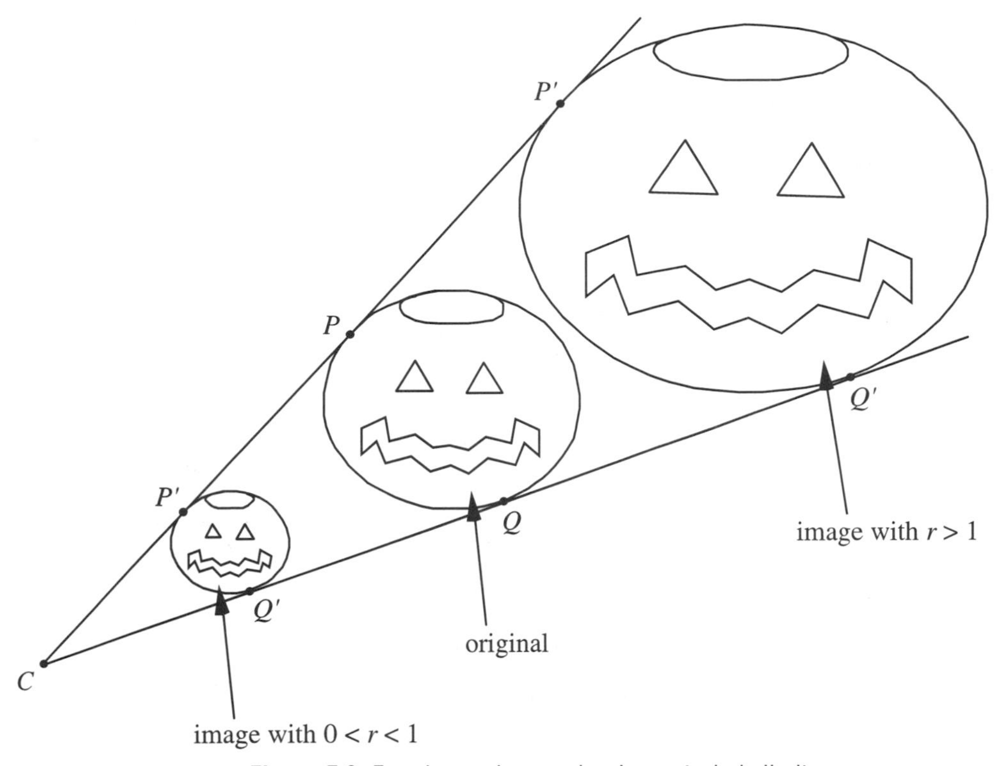

Figure 7.3 Zooming and squeezing by central similarity.

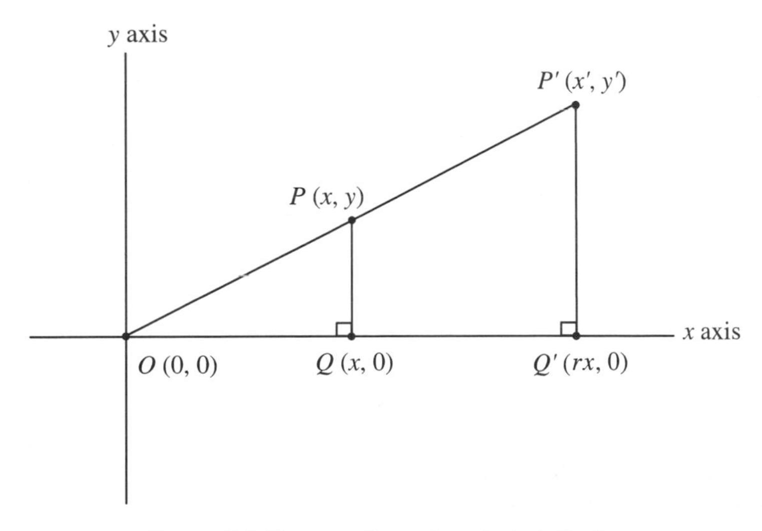

Figure 7.4 The equations of central similarity.

Thus the x coordinates of P' and Q' are equal:

$$x' = OQ' = rOQ = rx.$$

A similar construction and argument involving the y axis instead of the x axis will give the equation for y. Thus, our system of equations is

$$\begin{aligned} x' &= rx, \\ y' &= ry. \end{aligned} \tag{7.1}$$

If the center of the similarity is $C(x_C, y_C)$, then the equations become:

$$\begin{aligned} x' &= r(x - x_C) + x_C, \\ y' &= r(y - y_C) + y_C. \end{aligned} \tag{7.2}$$

EXAMPLE 7.1

Find the images of the points $P(3.2, 3.3)$, $Q(3.2, 5.2)$, $R(4.2, 3.8)$ under $S_{O,2}$ where O is the origin. Then do it under $S_{(4,5),2}$.

In the case where the center is at (0, 0) we obtain:

For P': $x' = 2(3.2) = 6.4$, $y' = 2(3.3) = 6.6$,

For Q': $x' = 2(3.2) = 6.4$, $y' = 2(5.2) = 10.4$,

For R': $x' = 2(4.2) = 8.4$, $y' = 2(3.8) = 7.6$.

In the case where the center is at (4, 5) we obtain:

For P': $x' = 2\ (3.2 - 4) + 4 = 3.4$, $y' = 2\ (3.3 - 5) + 5 = 1.6$,

For Q': $x' = 2\ (3.2 - 4) + 4 = 2.4$, $y' = 2\ (5.2 - 5) + 5 = 5.4$,

For R': $x' = 2\ (4.2 - 4) + 4 = 4.4$, $y' = 2\ (3.8 - 5) + 5 = 2.6$. ■

EXAMPLE 7.2

A computer graphics program needs to use a central similarity with center $O(0, 0)$ and ratio 2 to make a double-size version of the triangle whose vertices are P, Q, and R of the previous example (see Figure 7.5). What makes this problem not completely routine is that a graphics screen cannot show shapes exactly since the smallest parts of the screen that can be lit up are little patches of color called *pixels*, shown as squares in Figure 7.5. For example, to show the triangle PQR, the best the graphics program can do is to turn on all pixels touched by the triangle — the five shaded pixels at the lower left. Finding the pixels touched by some figure is called *scan conversion* of the figure in computer graphics.

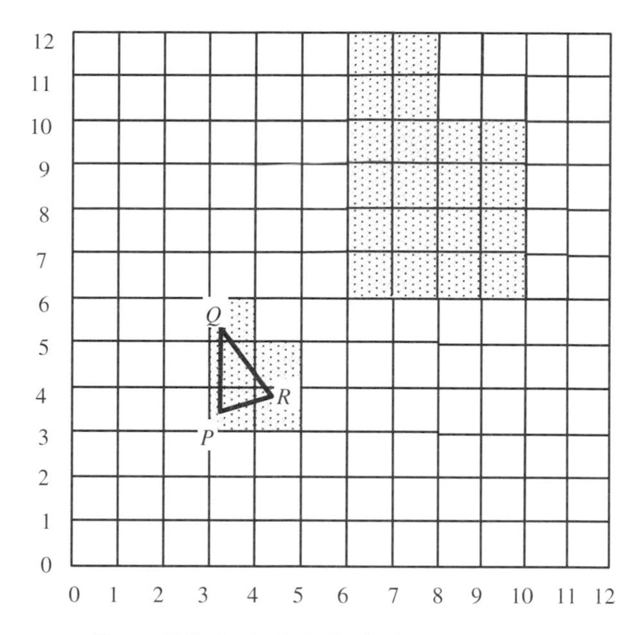

Figure 7.5 Central similarity in computer graphics.

(See Section 1.1 of Chapter 1 for a brief discussion of how scan conversion can be carried out for line segments.)

Scan conversion is relatively time consuming (but necessary in many graphics situations). For this reason, it is tempting to look for methods to avoid it or short-cut it. We present next a method that is sometimes used which avoids scan conversion of the double-size triangle by taking advantage of the fact that scan conversion has already been done on the original small triangle.

Scaling Pixels Method

Apply the central similarity to each pixel used to approximate the original triangle. For example, take the pixel containing the point Q. Its lower left corner is $(3, 5)$ and the upper right is $(4, 6)$. By doubling these coordinates, we find that the image of this pixel is the square stretching from $(6, 10)$ to $(8, 12)$. Notice that this is a 2×2 square (4 pixels altogether), which shouldn't surprise us because $r = 2$. Scaling up a pixel in this way is a simple operation that the computer can do extremely quickly, even if it has to apply the operation to each of the 5 pixels touched by triangle PQR. The result of scaling up the 5 pixels is the collection of 20 shaded pixels at the upper right of Figure 7.5.

Here is the standard method. It takes a little longer. Is it worth the extra time?

SCALING + SCAN CONVERSION METHOD

1. Compute the images P', Q', and R' of points P, Q, and R (as we did in Example 7.1), a very quick computation.
2. Scan conversion: Find the pixels touched by triangle $P'Q'R'$.

Since it is always nice to save computing time, one might suppose that the scaling pixels method would be preferred. But before deciding this we need to compare the quality of the results. Do the methods both give the same results? If not, which gives a more accurate approximation to the true shape of triangle $P'Q'R'$? We leave this to you to explore. Draw the image triangle in Figure 7.5 and find the pixels it touches.

This example is a bit misleading, because a complete graphics screen has hundreds or even thousands of pixels in each direction. Each one is so small that the inaccuracies are usually less annoying than the "boxiness" we see in Figure 7.5. Nonetheless, there are situations where it pays for the programmer to think carefully about which of the two methods to use.

You might suppose that the need to save computing steps is also something we rarely need worry about — after all, computer hardware is getting faster all the time. But as machines get faster, we give them more to do (e.g., more complex pictures needing more scan conversion) so the desire for speed, which is central to our example, is not misleading at all. ■

Not every transformation that scales figures by a factor of r is a central similarity. For example, suppose we precede a central similarity S by a rotation of 90°, $R(90°)$. Here is how we can see that the composite transformation $U = S \circ R(90°)$ is not some other central similarity. $R(90°)$ takes a horizontal segment and makes it vertical. S preserves the vertical direction of this segment [part (b) of Theorem 7.1]. Thus, U transforms horizontal segments to vertical ones. But if U were a central similarity, it would keep horizontal segments horizontal.

DEFINITION

If S is a central similarity and E is any isometry, then $S \circ E$ is a *similarity transformation*, or, more simply, a *similarity*.

☆ THEOREM 7.4

When similarity is applied to a triangle, the result is a triangle similar to the original.

PROOF

When we apply an isometry, the lengths of the sides of the image triangle are the same as in the original. When we follow the isometry by a central similarity we get a third triangle whose sides all have the same ratio to their corresponding sides in the original (by Theorem 7.1). Thus, we can simply apply the SSS criterion for similarity (Exercise 21 of Section 2.4 in Chapter 2). ■

✯ THEOREM 7.5

If triangle PQR is similar to triangle $P^*Q^*R^*$, then there is a similarity that maps PQR to $P^*Q^*R^*$.

PROOF

By a combination of translation and rotation we can move P to $P' = P^*$ and move Q to Q', which lies on ray $\overrightarrow{P^*Q^*}$. Let R' be the image of R under this transformation. Because translation and rotation preserve angles, $m\angle Q'P'R' = m\angle QPR$. By hypothesis and the definition of similarity, $m\angle QPR = m\angle Q^*P^*R^*$. Thus, $m\angle Q'P'R' = m\angle Q^*P^*R^*$. Consequently, R' either lies on ray $\overrightarrow{P^*R^*}$ or will do so if we reflect R' in line $\overleftrightarrow{P^*R^*}$. Finally, apply a central similarity whose center is P^* and whose ratio is the ratio of corresponding sides. ■

EXERCISES

**Marks challenging exercises.*

1. Apply $S_{O,2}$ to the following points: $P(1, -3)$, $Q(3, -4)$, $R(0, 2)$.
2. Apply $S_{(1,2),3}$ to the following points: $P(1, -3)$, $Q(3, -4)$, $R(0, 2)$.
3. Draw a figure illustrating Eqs. (7.2) and use your figure to derive the equations.
4. Draw a sketch like Figure 7.3 to show the effects of central similarity with negative values of r. (You needn't use anything as complicated as a jack-o-lantern. The letter "L" would be good enough.)
5. (a) Apply $S_{(1,0),1/2} \circ S_{O,2}$ to the following points: $P(2, 1)$, $Q(-1, 2)$, $R(3, -1)$. When you compare the image points with the originals, do you see a pattern?

   (b) If you do this composite transformation to (x, y) what point results?

6. Consider the square whose diagonally opposite corners are $(-2, 1)$ and $(0, 3)$. Sketch this square and its image under $S_{O,2}$ where O is the origin. Now do it with $S_{O,0.5}$.

7. Which of the two methods of scaling a triangle mentioned in Example 7.2 is more accurate? Or are they equivalent? Explain with a drawing.

8. Suppose we had a triangle with vertices $P(3.6, 6.9)$, $Q(5.7, 11.4)$, and $R(8.7, 5.7)$.

   (a) Scan convert this triangle (by plotting and drawing on graph paper — not calculation) and show the results by shading pixels in a figure as was done in Figure 7.5.

   (b) If you wanted to apply $S_{O,1/3}$ to triangle PQR and show the results as a set of pixels on a computer screen as done in Figure 7.5, could you apply the scaling pixels method (no further scan conversion)?

   (c) Apply $S_{O,1/3}$ to the triangle PQR and then scan convert the image of triangle PQR (by plotting and drawing on graph paper — not calculation).

9. Suppose you know that A' results from A by central similarity and B' results from B. Is the central similarity (its effects on other points) completely determined by this information? For example, could you find the center by some process of measurement and calculation or some construction? Could you find r?

*10. Suppose P, Q, and R are not collinear. Suppose you know that P' results from P by a similarity transformation (not necessarily a central similarity) and Q' results from Q and R' results from R. Is the similarity (its effects on other points) completely determined by this information? (*Hint*: A similar question for isometries is taken up in Section 4.3 of Chapter 4. See if it gives you a clue.)

11. What is the image of the curve whose equation is $x^2 + y^2 = 16$ under $S_{O,1/2}$? (*Hint*: If (x', y') is the image of a point (x, y) on the curve, what equation must (x, y) satisfy?)

12. What if we follow the central similarity $S_{O,r}$ by a rotation by θ around the origin O?

   (a) Make the most extensive list you can of invariants.

   (b) In addition, list some things that are not invariant.

13. Find equations for the composite transformation described in Exercise 12.

14. Suppose we take each point (x, y) on the curve whose equation is $y = x^2$ and subject it to a central similarity whose center is at the origin and whose ratio is 2. What equation is satisfied by the resulting points (x', y')?

15. Let P, Q, and R be three points and P', Q', and R' their images under a similarity. Explain why $m\angle P'Q'R' = m\angle PQR$.

16. The proof of Theorem 7.3 is incomplete. What we showed is that all points on a given circle wind up on some other circle when central similarity is applied. But we did not show that every point on the second circle is actually the image of some point on the first circle under the central similarity. Fill in this gap in the proof. Show how to construct the desired point and explain why it lies on the first circle.

17. Suppose S scales every segment by ratio r and has a fixed point. Show by example that S need not be a central similarity.

*18. Suppose a transformation U has these two properties:

    (i) There is a ratio r so that for any two points P and Q and images P', Q', $P'Q'/PQ = r$.

    (ii) For any two points P and Q and images P', Q', $\overleftrightarrow{P'Q'}$ is parallel to $\overleftrightarrow{PQ}$.

    Show that U is a central similarity transformation provided $r \neq 1$.

19. (a) What, if anything, can you say about the nature of the composite transformation $S_{C,q} \circ S_{C,p}$ in the case where $pq \neq 1$? (*Hint*: Make use of the result of Exercise 18.)

    (b) What can you say about $S_{C,q} \circ S_{C,p}$ if $pq = 1$?

20. What if we follow a central similarity having center at the origin and $r \neq 1$ by a translation. Could this be a central similarity? If not, give an example to show it isn't true. If it is true, prove it. (*Hint*: Use any of the previous exercises if you can.)

21. Show that $S_{B,q} \circ S_{A,p}$ is a central similarity even when $A \neq B$. How is its ratio of similarity related to p and q? (*Hint*: Examine previous exercises.)

22. Is it true that $S_{C,1/r} \circ S_{C,r} = I$? If not, give an example that shows this. If it is true, prove it. (*Hint*: Examine previous exercises.)

*23. If U is a similarity, show U^{-1} is a similarity.

*24. Suppose a transformation U is such that there is a set of three noncollinear points P, Q, and R fixed by U. In addition, if ABC is any other triangle, its image under U is similar to ABC. Prove that U is the identity.

*25. Suppose a transformation U is such that every triangle is mapped to a similar one. Prove that U is a similarity transformation. (*Hint*: Use the previous exercises.)

7.2 Inversion

The transformations we have considered so far—the various isometries, central similarities, and similarities—all preserve many of the most common properties of geometric figures: collinearity, angles, and circlehood, for example. A triangle transforms into a congruent one in the case of the isometries and transforms into a similar one in the case of the similarities. But more exotic transformations that lack these invariants are also useful and interesting. In this section we consider inversion, a transformation that sometimes turns lines into circles and vice versa. This means that a triangle may not even turn into a triangle, never mind one that is congruent or similar to the original. Despite these surprising features, inversion is useful as well as interesting.

DEFINITION

Choose a circle, called the *circle of inversion*. Its center C will be called the *center of inversion*, and its radius r will be called the *radius of inversion*. For any point P that is not C we find the inverse[2] of P (relative to this particular circle of inversion) by joining C to P and defining P' to be that point on the ray $\overrightarrow{CP}$ so that

$$(CP)(CP') = r^2. \tag{7.3}$$

EXAMPLE 7.3

Suppose the circle of inversion is centered at $O(0, 0)$ and has radius of inversion $r = 2$. How far from O is the inverse of $P(5, 12)$?

SOLUTION

$$OP = \sqrt{5^2 + 12^2} = 13,$$

$$OP' = \frac{r^2}{OP} = \frac{4}{13}.$$

How can one calculate the coordinates of P'? We'll examine this shortly. ■

---

[2] The terminology is a little odd because normally it is transformations that have inverses in geometry.

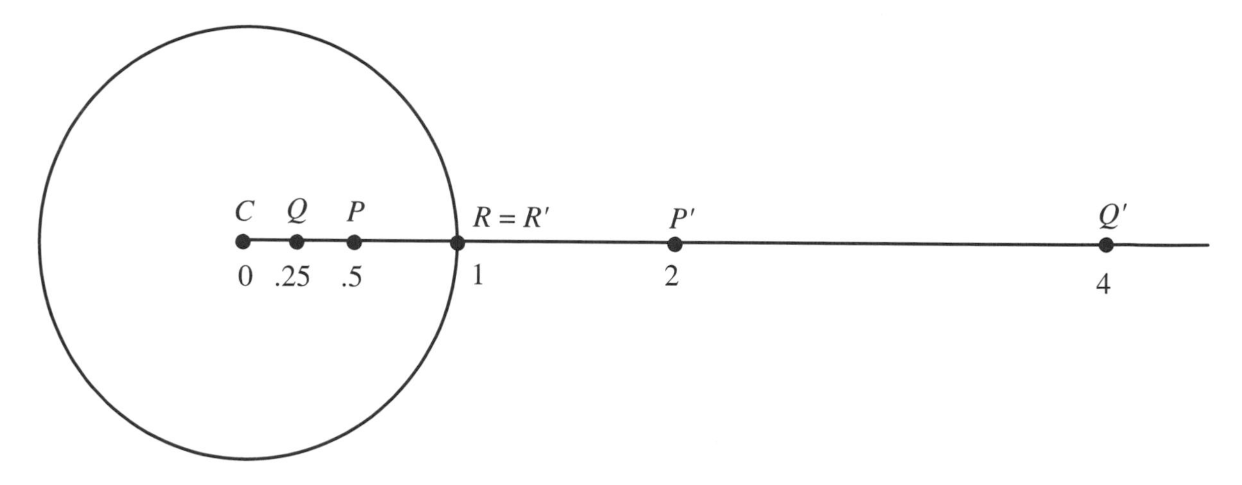

Figure 7.6 Inversion.

EXAMPLE 7.4: INVERTING POINTS ON THE *X* AXIS

Suppose $C = (0, 0)$ and $r = 1$. The image of a point on the positive x axis is also on the positive x axis (see Figure 7.6). This makes it easy to find the coordinates of the image of such a point, given the coordinates of the original.

Suppose $P = (0.5, 0)$. Because P and P' are on the positive x axis, Eq. (7.3) is really just about the x coordinates:

$$\begin{aligned} x_P x_{P'} &= 1 \\ (0.5) x_{P'} &= 1 \\ x_{P'} &= \frac{1}{0.5} = 2\,. \end{aligned}$$

Now let's move closer to the center of inversion, to $Q = (0.25, 0)$. Then $Q' = (4, 0)$. Note that the image is farther from the center of inversion than P'.

If we want P'', the image of P', then

$$\begin{aligned} x_{P'} x_{P''} &= 1 \\ (1/x_P) x_{P''} &= 1 \\ x_{P''} &= x_P = 0.5\,. \end{aligned}$$

Note that $P'' = P$.

Finally, we note that the inverse of $R(1, 0)$ is itself: $R = R'$. Indeed, any point on the circle of inversion will be its own inverse. ■

Here are some lessons we can learn from the previous example:

☆ THEOREM 7.6

1. Inversion does not always preserve distance ($PQ = 0.25$ while $P'Q' = 4$ in the example).
2. The inverse of an inverse is the original point.
3. Points on the circle of inversion remain fixed by inversion.
4. Points inside the circle of inversion have images outside that circle and vice versa. ■

Before continuing to investigate the properties of inversion, we should pay our dues to precision and notice that it is not truly a transformation of the entire plane because Eq. (7.3) does not define an image for C, the center of inversion. Furthermore, no points map into C. In this respect, inversion is a little odd in comparison with isometries and similarities. The oddness can be removed by inventing a new ideal point I_∞ to serve as the image of C. I_∞ is not in the plane—there is no way to visualize its location or make a mark in a diagram to show it. Nonetheless, one often thinks of I_∞ as being infinitely far away from C for the following reason: If we imagine a point P moving ever closer to C, the point P' moves ever farther away. Not only is I_∞ the inverse of C, but we define C to be the inverse of I_∞. Finally, we regard I_∞ as lying on every line.

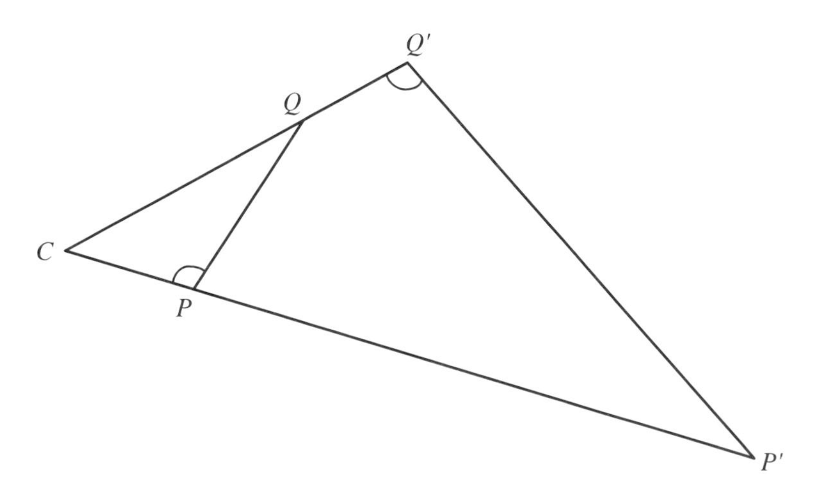

Figure 7.7 Inversion and similar triangles.

All of our results about inversion grow out of the following connection between inversion and similar triangles. In reading this theorem, pay careful attention to the order in which the vertices are listed because this conveys which vertices correspond to which under the similarity.

✯ THEOREM 7.7

If P and Q are different points (see Figure 7.7) and each different from C and not collinear with C, then

(a) triangle CPQ is similar to triangle $CQ'P'$ and

(b) $m\angle CPQ = m\angle CQ'P'$ and $m\angle CQP = m\angle CP'Q'$.

Proof

By Eq. (7.3) we have

$$(CQ)(CQ') = r^2 = (CP)(CP').$$

By dividing we get

$$\frac{CQ}{CP'} = \frac{CP}{CQ'}.$$

Since the triangles share an angle and the sides forming the angle are in proportion, the SAS criterion for similarity (Theorem 2.22 of Section 2.4 of Chapter 2) tells us that the triangles are similar. By the definition of similarity, $m\angle CPQ = m\angle CQ'P'$ and $m\angle CQP = m\angle CP'Q'$. ■

✯ THEOREM 7.8

If L is a line through the center of inversion, then L is invariant under the inversion. In other words, each point of L maps into another point of L.

Proof

This follows immediately from the definition of inversion. ■

✯ THEOREM 7.9

(a) If L is a line that does not pass through the center of inversion, its image is a circle that does pass through the center of inversion (see Figure 7.8).

(b) The image of a circle that passes through the center of inversion is a line that does not pass through the center of inversion.

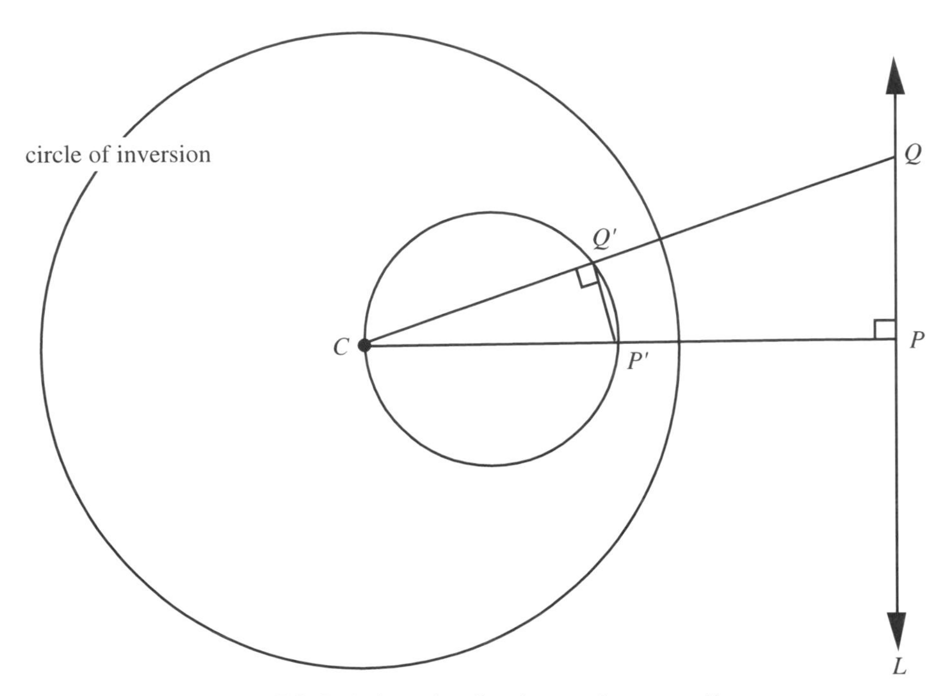

Figure 7.8 A circle and a line inverse to one another.

PROOF

We begin with part (a). Let P be the foot of the perpendicular dropped from C to line L and let P' be its inverse. Let Q be any other point on L and let Q' be its inverse. By Theorem 7.7, $m\angle CQ'P' = m\angle CPQ = 90°$. Furthermore, this is true no matter where on line L the point Q is located. As Q moves along L, Q' traces out the locus of all points where $m\angle CQ'P' = 90°$. By part (b) of Theorem 2.14 of Section 2.3 in Chapter 2, this is a circle whose diameter is CP'.

The proof of part (b) is left as an exercise. ■

We turn now to the question of whether angles are preserved by inversion. If P, Q, R are any points, will we have $m\angle PRQ = m\angle P'R'Q'$? Normally, no. However, there is an important special case where the answer is yes:

☆ THEOREM 7.10

If P and Q are collinear with C, and R is any other point, then $m\angle PRQ = m\angle P'R'Q'$ (see Figure 7.9).

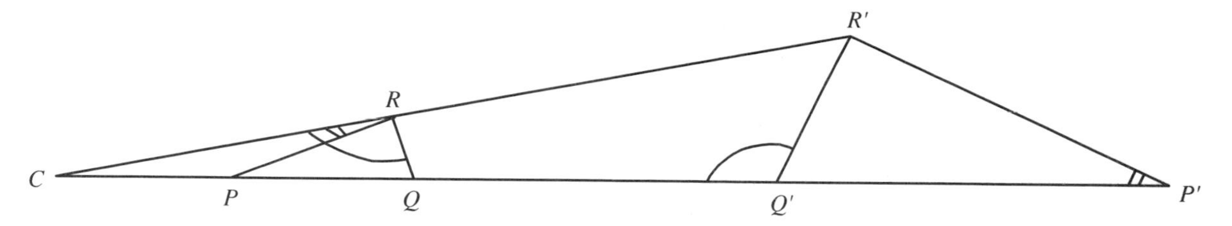

Figure 7.9 A case where angles are invariant.

Proof

By use of part (b) of Theorem 7.7, we deduce that the marked angles are equal to one another:

$$\mathrm{m}\angle CRQ = \mathrm{m}\angle CQ'R',$$
$$\mathrm{m}\angle CRP = \mathrm{m}\angle CP'R'.$$

Therefore,

$$\begin{aligned}\mathrm{m}\angle PRQ &= \mathrm{m}\angle CRQ - \mathrm{m}\angle CRP \\ &= \mathrm{m}\angle CQ'R' - \mathrm{m}\angle CP'R' \\ &= (\mathrm{m}\angle P'R'Q' + \mathrm{m}\angle CP'R') - \mathrm{m}\angle CP'R' \\ &= \mathrm{m}\angle P'R'Q'.\end{aligned}$$

The next to last equality follows because the measure of an exterior angle of a triangle equals the sum of the measures of the interior angles that are remote from it (Exercise 8 in Section 2.3, Chapter 2). ■

☆ THEOREM 7.11

The inverse of a circle that does not pass through the center of inversion is another circle that does not pass through the center of inversion (see Figure 7.10).

Proof

Draw the line connecting C to the center of the circle we wish to invert. Let P and Q be the points where this line intersects the circle. Note that $\overline{PQ}$ is a diameter of the circle. Let R be any other point on the circle. By use of the previous theorem we note that

$$\mathrm{m}\angle PRQ = \mathrm{m}\angle P'R'Q'.$$

But $\angle PRQ$ is inscribed in a semicircle ($\overline{PQ}$ is a diameter) so it is a right angle. Consequently, $\mathrm{m}\angle P'R'Q' = 90°$. As R moves around the circle, R' traces out a locus where

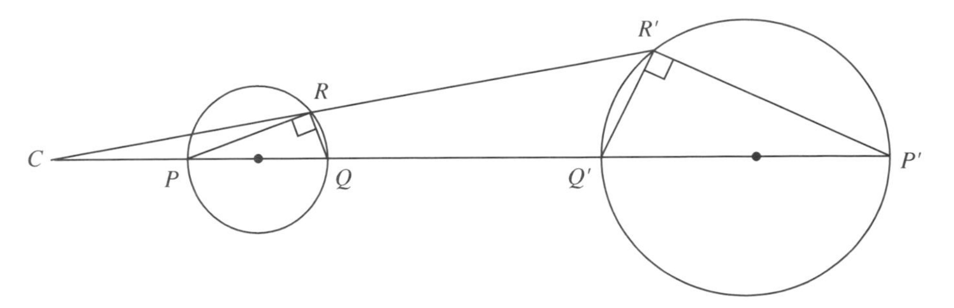

Figure 7.10 A circle not through C inverts to a circle (circle of inversion not shown).

m $\angle P'R'Q'$ is always 90°. This locus is a circle by Theorem 2.14 of Section 2.3, Chapter 2. We leave it to the reader to explain why this circle does not pass through the center of inversion. ■

☆ THEOREM 7.12

The angle between two curves — more precisely, between their tangents — is preserved by inversion (see Figure 7.11).

PROOF

Let γ_1 and γ_2 be any two curves that cross at R and let Q be any other point on γ_2. Let P be the point where $\overrightarrow{CQ}$ crosses γ_1. Let P', Q', and R' be the inverses of P, Q, and R on the inverse curves $\gamma_1{}'$ and $\gamma_2{}'$. By Theorem 7.10, m $\angle PRQ =$ m $\angle P'R'Q'$. As Q moves toward R along γ_2, P will also move toward R on γ_1 and P' and Q' will move toward R' on their curves. Thus the angles, always equal to one another, will get closer and closer to the angles between the tangents (not shown in Figure 7.11) to the curves at R. Thus, the angles between the tangents are equal. ■

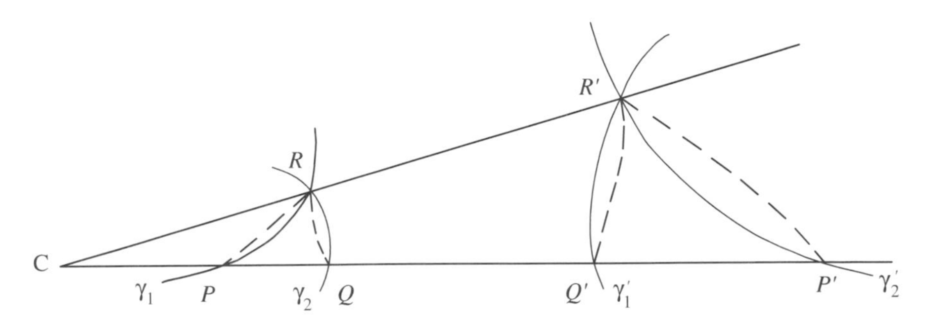

Figure 7.11 The angle between two curves is invariant.

APPLICATION: Inversion in Mechanisms and Map Making

When you're in your car and want the window rolled down, you turn the crank. How is it that the circular motion of the crank makes the window move in a straight line? It turns out that the material world around us is full of instances where back and forth motion and circular motion get transformed one into the other (think of car engines, locomotives, etc.). There are many ways in which mechanical engineers can design devices to do this useful trick. One of them is based on the mathematics of inversion and was invented by A. Peaucellier in 1864.

The boldface parts of Figure 7.12 make up the device. The boldface segments represent rigid rods and there are hinges where they meet at *C, P, Q, R,* and *S*. Points *P, S, Q,* and *R* are free to move but *C* is fixed in place on the circumference of a circle we wish to invert.

Visualize how the linkage can flex (temporarily ignore the circle and the segments that are not part of the linkage). For example, we can pull *S* toward *C*, thereby moving *R* away and flattening the quadrilateral *PSQR*. Instead, we can rotate the whole linkage around *C* without changing its shape. If we do this, *S* pulls away from the circle (visualize a circle centered at *C* with radius *CS*). By combining both motions in just the right proportions we can get *S* to move around on the circle.

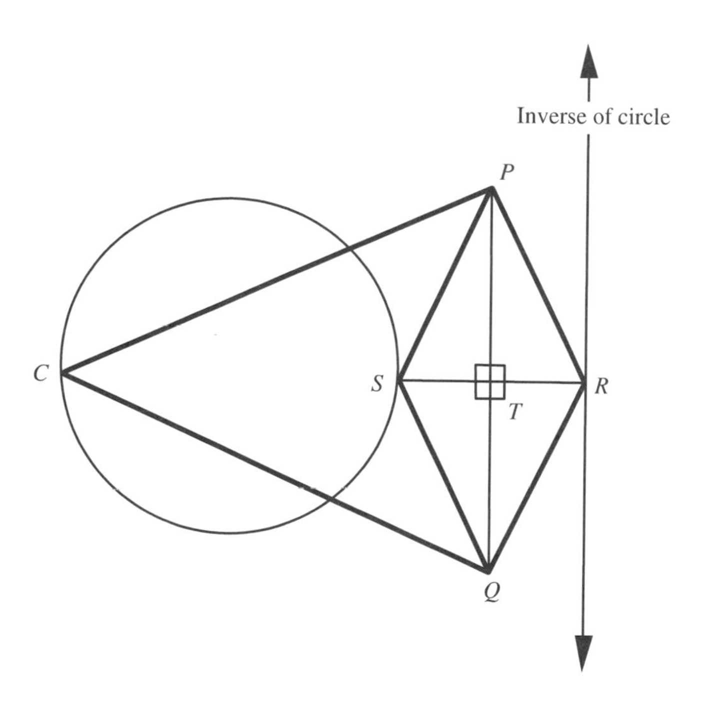

Figure 7.12 Peaucellier's linkage: tracing a straight line by inverting a circle (circle of inversion not shown).

Four of the the rods have equal length: $PS = SQ = PR = RQ$. This makes $PSQR$ a rhombus, and so its diagonals bisect each other and are perpendicular. We will use this and the Pythagorean theorem to show that no matter how the linkage flexes, S and R are inverses of one another with C the center of inversion and a radius of inversion that will emerge from our calculation.

$$\begin{aligned}(CS)(CR) &= (CT - ST)(CT + TR)\\ &= (CT - ST)(CT + ST)\\ &= CT^2 - ST^2\\ &= (CP^2 - PT^2) - (PS^2 - PT^2)\\ &= CP^2 - PS^2.\end{aligned}$$

But CP and PS are both unchanging quantities (the rods are fixed in length) so our equations show $(CS)(CR)$ is constant. If we take a radius of inversion of $\sqrt{CP^2 - PS^2}$ and a center of inversion at C, then our equations show S and R are inverses of one another. By part (b) of Theorem 7.9, if we make S go around a circle through C, R will trace a straight line.

There is a three-dimensional version of inversion that has been used by cartographers to create maps of parts of the earth's surface. The definition of the inverse of a point is exactly as in the plane, except that one applies the idea to all points of the three-dimensional space. Many of the same theorems we have proved in the planar case apply in three dimensions as well. For example, it is possible to prove that

If we take a sphere which passes through the center of inversion, the inverse is a plane.

Now imagine that the earth, which we assume to be a perfect sphere, is placed inside a larger sphere as in Figure 7.13, with the North pole C at the center of the large sphere and the South pole on the surface of the

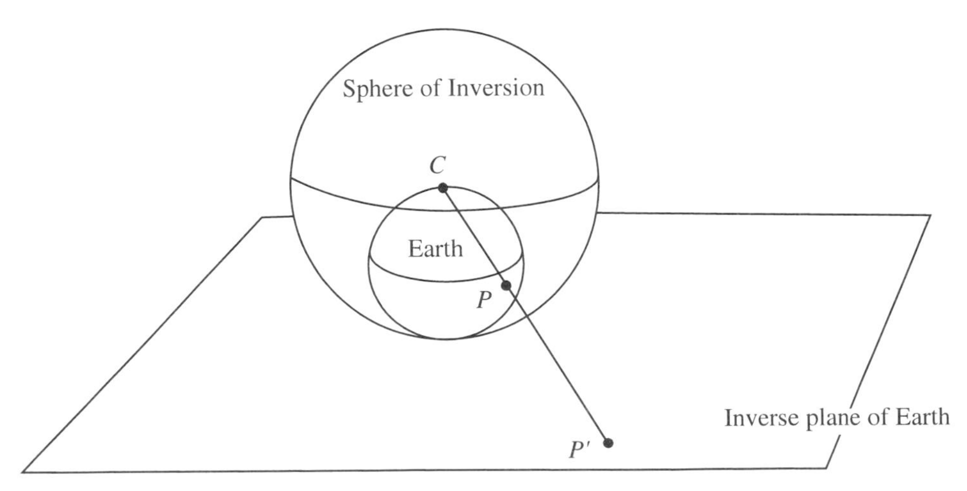

Figure 7.13 Stereographic projection of the earth.

large sphere. The inverse of the earth is a plane—a plane in which our map will be created. The boundary of a country on the earth's surface will invert to a curve in the inverse plane. Experience shows that there is a good deal of distortion of shapes in this process. However, one advantage of this type of map is that the angle at which two curves (say, the boundaries of two countries) meet is preserved. This is far from obvious, but it can be shown that an analogue of Theorem 7.12 holds for curves on the earth and their images. Cartographers call this type of map making *stereographic projection*. We study it again in the next section under the name *perspective projection*.

EXERCISES

**Marks challenging exercises.*

1. In each of the following cases, determine how far from the center of inversion P' is. Assume the center of inversion is the origin O and the radius of inversion is 1.

   (a) $P=(3, 4)$

   (b) $P=(1, 1)$

   (c) $P=(\sqrt{2}/2, \sqrt{2},2)$

2. Suppose we invert the given point in the sphere centered at O having the specified radius. Find how far from the origin the inverse point is.

   (a) Point (3, 1, 2); radius 7

   (b) Point (2, 5, 3); radius 4

   (c) Point (1, 1, 1); radius 3

3. Suppose a square is such that it circumscribes the circle of inversion (each side tangent to the circle). Sketch the inverse of the square boundary.

4. What is the inverse of a line tangent to the circle of inversion?

5. Show by example that angles are not always invariant. To do this, find P, Q, and R so that it is clear that $m\angle PQR \neq m\angle P'Q'R'$.

6. Are there any points P and Q where $PQ=P'Q'$?

7. In general, it is not true that $P'Q'=PQ$. If you wanted to give a very dramatic example of this, where would you locate P and Q?

8. Does a triangle ever invert into a triangle? If so, give an example. If not, explain why not.

9. If we know P and Q are inverses to one another for a certain circle of inversion, could there be another circle of inversion (not necessarily with the same center) for which P and Q are also inverse? If, not prove it. If so, then:

   (a) Show how you would find this other circle of inversion.

   (b) Explain whether you think there could be even more circles of inversion mapping P to Q.

10. Suppose P and Q invert into P' and Q' using a circle of inversion centered at C with radius of inversion r. Prove that $P'Q' = r^2PQ/[(CP)(CQ)]$.

11. Suppose a curve γ intersects its inverse curve γ' in a point P. Prove that unless P lies on the circle of inversion, there is another crossing of γ and γ'.

12. Derive the following equations for inversion in a circle of radius r centered at the origin:

$$x' = \frac{xr^2}{x^2 + y^2} \qquad y' = \frac{yr^2}{x^2 + y^2}$$

   (*Hint*: To derive the x' equation, use Figure 7.14 to relate x to OP and θ. Do something similar for x'. Then show that $xx' = r^2 \cos^2 \theta$.)

13. (a) If (ρ, θ) are the polar coordinates of a point P and (ρ', θ') are the polar coordinates of the image P', express ρ' and θ' in terms of ρ and θ.

   (b) If (ρ, θ, ϕ) are the spherical coordinates of a point P in space and (ρ', θ', ϕ') are the spherical coordinates of the image P', express ρ', θ', and ϕ' in terms of ρ, θ, and ϕ. (*Hint*: See Section 3.2 of Chapter 3 for a description of spherical coordinates.)

14. Suppose V_1 represents inversion where the circle of inversion is O and the radius is r_1 and V_2 represents inversion where the circle of inversion is O and the radius is r_2.

   (a) Apply $V_2 \circ V_1$ to the following points: (2, 0), (0, 1), $(x, 0)$, $(y, 0)$. Do you see a pattern?

   (b) $V_2 \circ V_1$ is the same as another transformation studied elsewhere in this book. What is it?

15. Prove that if P is outside the circle of inversion, the following construction works to find P'. Draw a tangent from P to the circle, meeting it at T. P' is the point where a perpendicular from T meets $\overline{PC}$. (*Hint*: What do you know about $m\angle CTP$? Use similar triangles.)

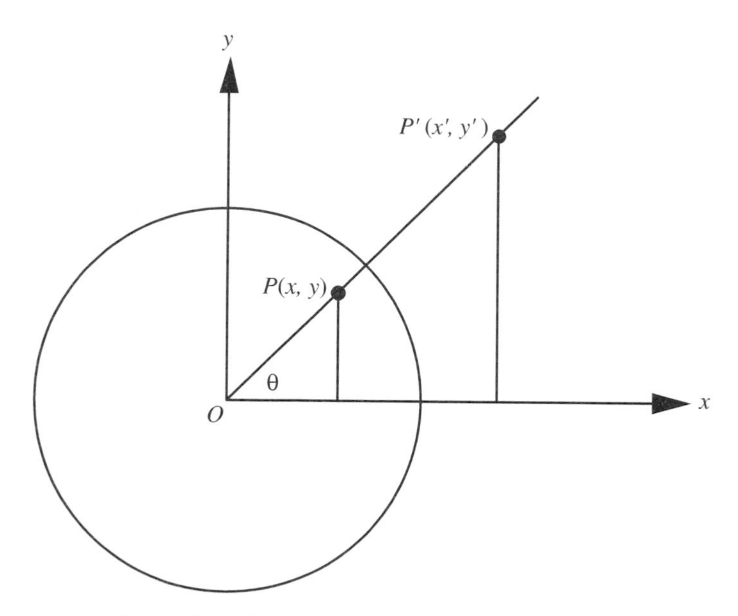

Figure 7.14 Deriving the equations of inversion.

16. Assuming that the construction in Exercise 15 works, does it give you an idea of what to do if P is inside the circle of inversion? What could you do to construct the same configuration (with P and P' interchanged) you end up with in Exercise 15? (The answer is in the next exercise.)

17. Prove that if P is inside the circle of inversion, the following construction works to find P'. Construct a perpendicular to $\overline{PC}$ at P, meeting the circle at T. Construct a tangent at T. P' is the intersection of this tangent with $\overrightarrow{CP}$.

The following three exercises require some facts from high school geometry about tangents and secants to circles that have not been dealt with in this book.

*18. Suppose a circle γ crosses the circle of inversion at right angles. Let P and Q be two points on γ that are collinear with C. Show that P and Q are inverses of one another.

*19. Let P and P' be distinct inverse points with respect to some circle of inversion. Let γ be any circle passing through P and P'. Show that γ is invariant under the inversion.

*20. Suppose two intersecting circles γ_1 and γ_2 have unequal radii and there is a central similarity with center C that transforms one circle into the other. Show that there is a circle γ^* centered at C such that if we invert γ_1 in γ^* we get γ_2.

APPLICATIONS

21. Prove that if *PRQS* is a quadrilateral with all sides equal, its diagonals bisect each other and are perpendicular. (See Peaucellier's linkage, Figure 7.12.)

22. Can Peaucellier's linkage trace the infinite line it is attempting to draw? If not, can you give a precise description of how much it can do?

23. In Peaucellier's linkage, can *S* go around the whole circle, all the way to *C*? If not, can you give a precise description of how much it can trace?

24. In Figure 7.13, how far from the center of the sphere of inversion is the inverse plane of earth's sphere? Answer in terms of earth's diameter.

7.3 Perspective Projection and Image Formation

Overview of Image Formation

We are surrounded by flat pictures of things that are actually three dimensional: pictures on T-shirts, pictures on paper, pictures on a computer screen, pictures small enough to go in our wallets, pictures so huge they take up entire billboards. Furthermore, we see the three-dimensional world with our eyes, which create pictures on the two-dimensional retina at the back of the eyeball. All such two-dimensional pictures, whether they appear on film in a camera, in our eyeball, on paper, or on a computer screen, are called *images*.

In this section we encounter two image formation processes: *Parallel projection* takes a very brief bow (but see Section 7.5 for more details) and *perspective projection* will be studied in detail. One of the things we will discover is that perspective projection (like every image formation process) creates distortions alongside the truth that it tells us about the objects being depicted. We proceed in two styles simultaneously: an algebraic style in which we do calculations to come to conclusions, and a more visual style in which we back up our intuitions with axiomatic arguments. Before proceeding to any type of proof of anything, however, it is important to develop a sound intuition about the projection processes imvolved. Good visualization, including good drawing, is the basis of mathematical understanding in this area.

Figure 7.15 shows an example of perspective projection. Points of the three-dimensional world, such as P_1, P_2, and P_3, that are to be depicted in the image, are called *world points*. There is a surface on which the image is to appear—in the cases we study it will be a plane called the *image plane*, denoted *I*. For each world point *P*,

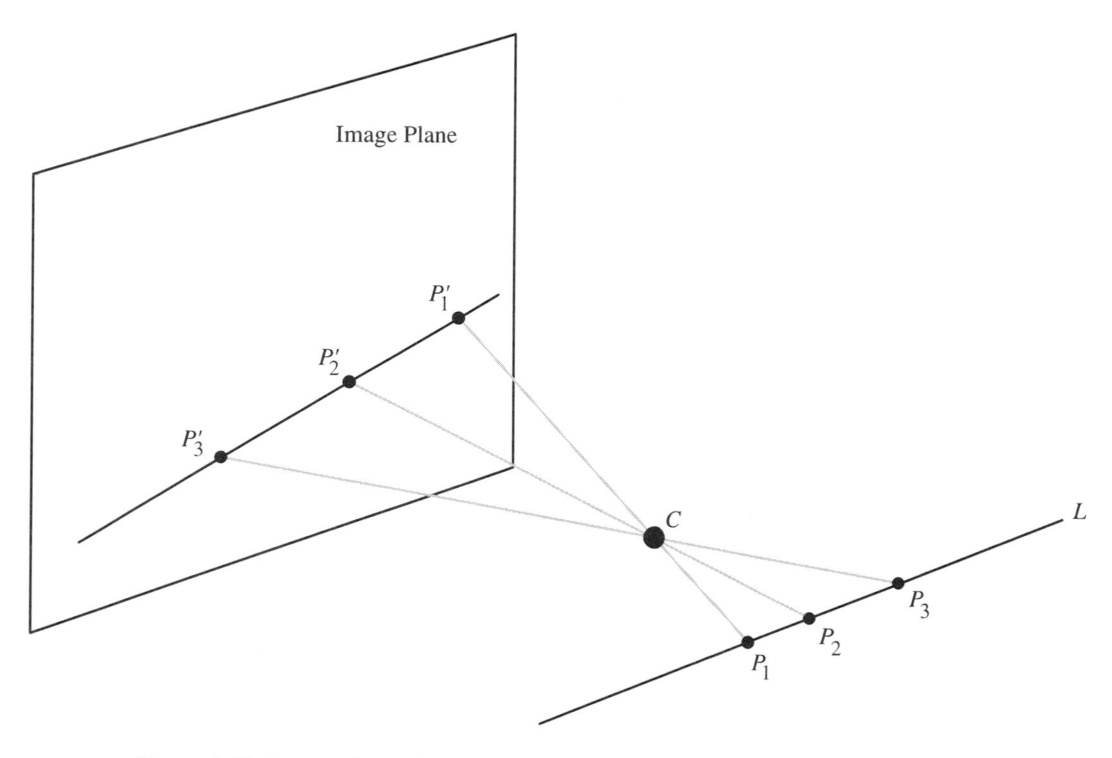

Figure 7.15 Perspective projection of three points on line L onto an image plane.

we create the line connecting it to a certain fixed point C called the *center of projection*. This line is called a *projector line*. The point where the projector line crosses the image plane is the *image* of P, denoted P'.

Perspective projection forms a pretty good model for the image formation process carried out by a camera. Figure 5.7 of Section 5.1 in Chapter 5 shows a very simple kind of camera, called a pinhole camera, in which the image formation process is exactly described by a perspective projection. What we have here is a box that has a pinhole in the center of one side. Light from a world point on a three-dimensional object, traveling in a straight-line path, enters the box and casts a spot of light on the back face of the box. This light ray is a projector for the world point in question and the pinhole is the center of projection. Real cameras have openings larger than pinholes in order that more light can enter. This extra light needs to be controlled by focusing mechanisms and that introduces lots of complications. None of these complications alters the fact that a pinhole camera shows the basic geometry of all real cameras fairly accurately.

There is an analogy between the human eye and a pinhole camera: The pinhole corresponds to the pupil of the eye. However, the eye is spherical rather than boxlike. Despite this, the image is formed on the back side of the eye, a two-dimensional surface called the retina which corresponds to the back face of our pinhole camera.

Since the Renaissance, artists have used perspective projection to create pictures. The woodcut in Figure 7.16 by Albrecht Dürer shows how they viewed the process. The artist paints on a transparent screen. He looks through an unmoving eyepiece at a point on the subject and then puts a dot of paint on the screen right on his line of sight. The dot he paints represents the point he sees and should be the same color. If the painting could be done perfectly, you couldn't tell if you were looking at the seated man or the painting of him as long as you had your eye at the eyepiece. Did artists ever use this apparatus with a screen and eyepiece to paint pictures? Probably rarely if ever. But Dürer's woodcut shows us that they were often aiming to get similar results.

In a different projection process, called *parallel projection*, the projectors do not all pass through a center of projection. Figure 7.25, shown later in Section 7.5, depicts an example of parallel projection. In a parallel projection, a direction is chosen and every point's projector line is parallel to that single direction. The projector lines do not pass through a common center of projection. Parallel projection does not correspond to the way cameras or the human eye work, but it has some advantages that we will study in Section 7.5.

Map makers are extremely interested in the process of making accurate two-dimensional maps of the three-dimensional earth, and this process is also an example of image formation. In Figure 7.17a we see an example of perspective projection, called *stereographic projection* by map makers, of the Northern Hemisphere. A special type of parallel projection, called an *orthogonal projection*, is shown in Figure 7.17b for comparison. The two maps we get from these different projection processes are quite different, especially near the equator. Neither of these projections is exactly accurate, so map makers have worked out yet other ways to make images of the earth, which you can read about in books on cartography.[3] The impossibility of making perfect maps of the earth is discussed in Section 3.2 of Chapter 3.

For the remainder of this section we focus exclusively on perspective projection. One of the most important questions about any projection process is this: What are the invariants, features of an object which are also found in the image of the object? Many common geometric features of objects are not invariant under perspective projection.

[3] The interested reader should consult: Grosvenor, Gilbert, *The Round Earth on Flat Paper*, National Geographic Society, 1947; Mainwaring, James, *An Introduction to the Study of Map Projection*, Macmillan, 1960.

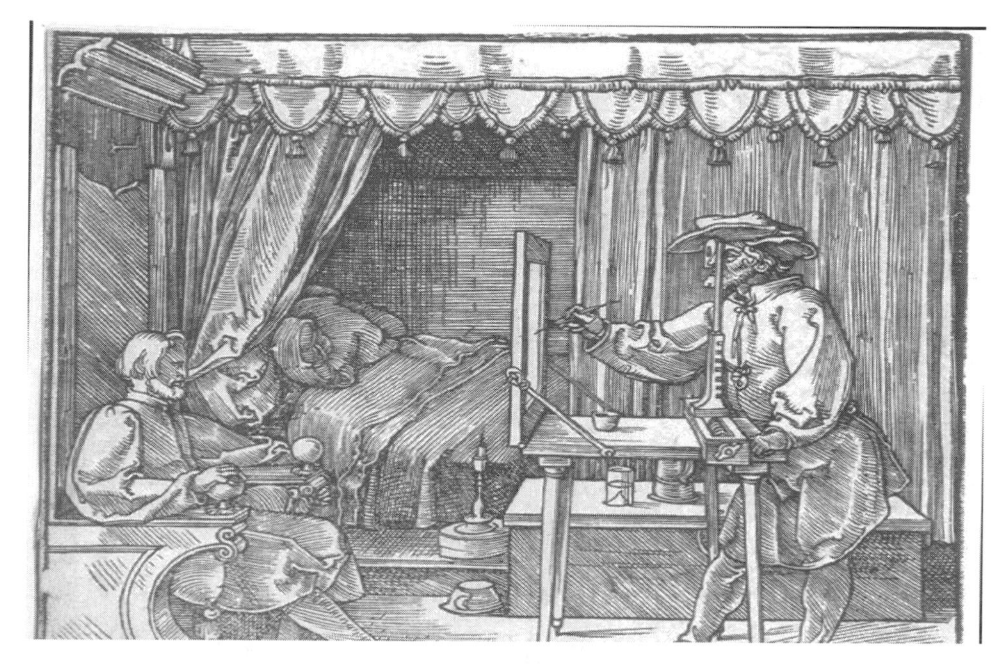

Figure 7.16 Dürer's *Artist Drawing Portrait of a Man* illustrates perspective projection. Courtesy of Metropolitan Museum of Art.

For example, Figure 7.18 shows that a circle does not usually project to a circle. (The shadow of a frisbee is usually an ellipse. Think of the sun as the center of projection and the ground as the image plane.) If we join the points of the circle to the center of projection, the projector lines obtained in this way make a cone. The image of the circle is the intersection of the cone and the image plane. As we tilt the image plane in various ways, we get different cross-sections, most of which are ellipses and not circles. The property of being a circle is not an invariant under perspective projection.

Parallelism is another important feature that is usually not preserved by perspective projection. For example, the human eye sees parallel railroad tracks as coming together in the distance. However, we can show that collinearity is an invariant of perspective projection.

✯ THEOREM 7.13

Suppose points P_1, P_2, P_3 are collinear, lying on line L, and their images under perspective projection are P'_1, P'_2, P'_3. These image points P_i' are also collinear (Figure 7.15).

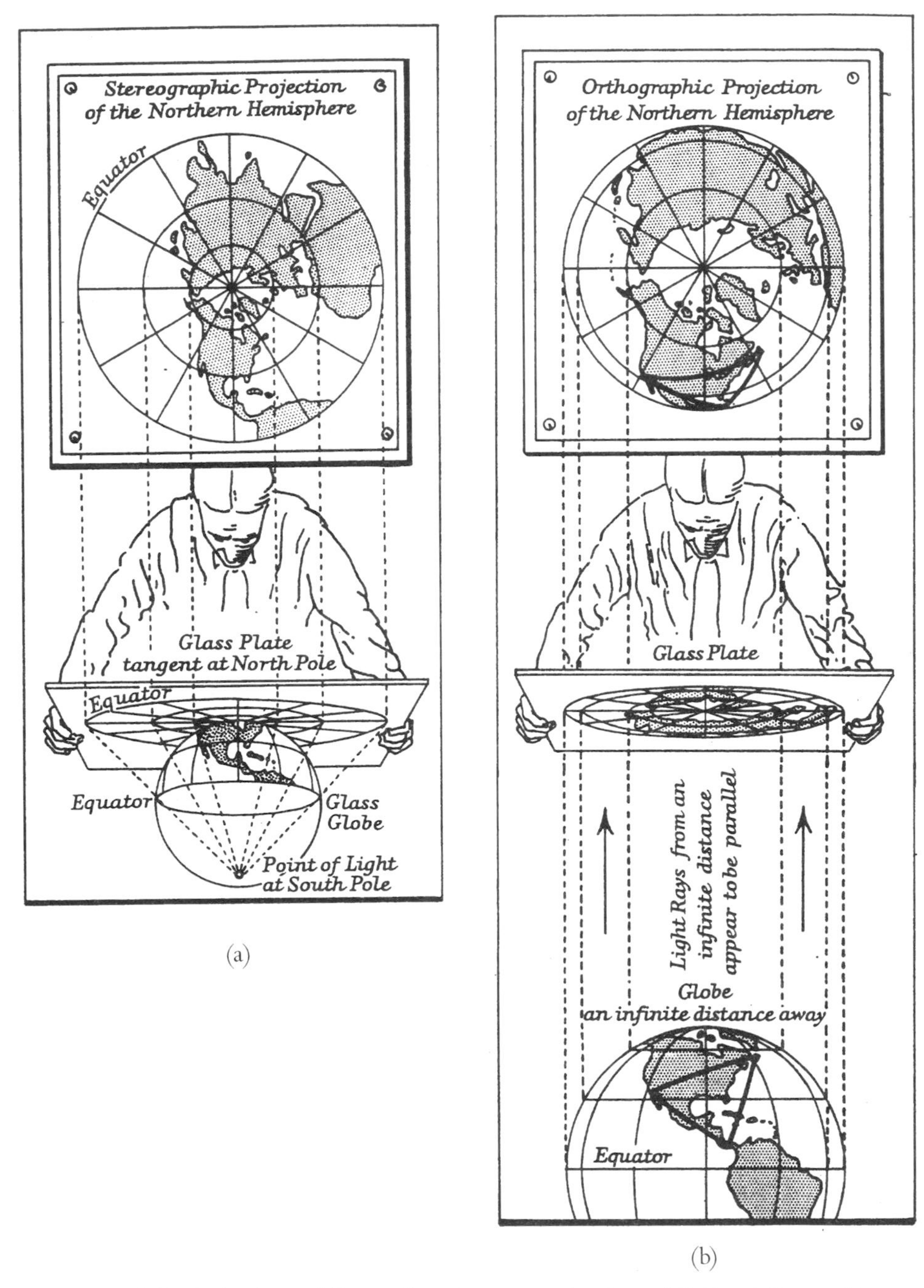

Figure 7.17 Projections of the earth onto a map. Courtesy of National Geographic.

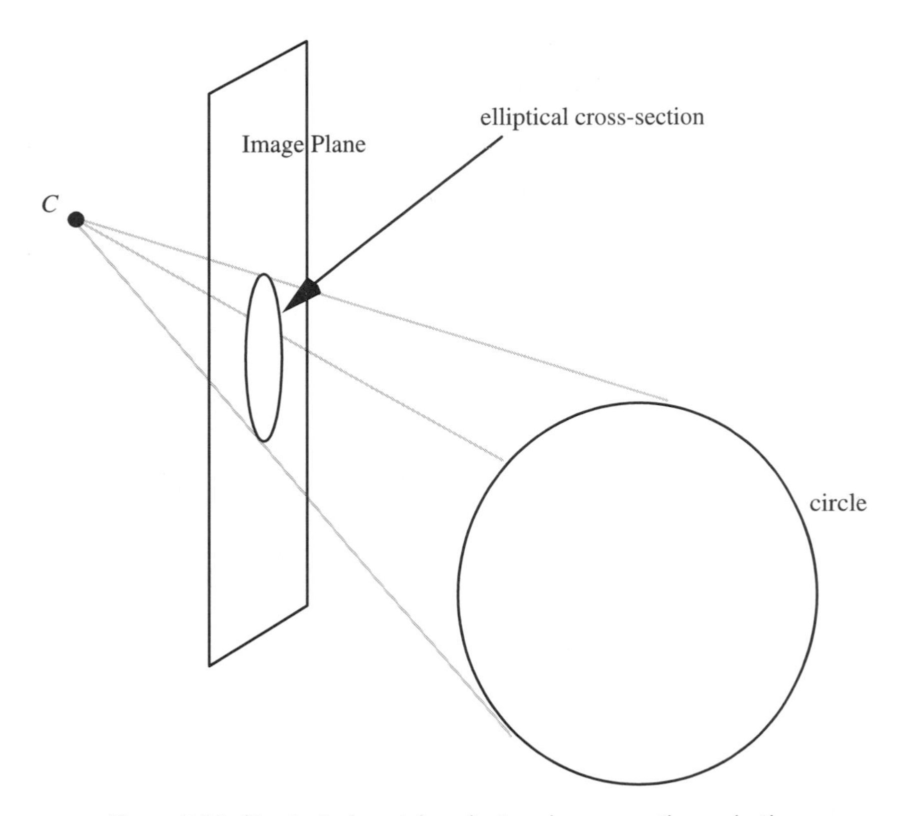

Figure 7.18 Circularity is not invariant under perspective projection.

PROOF

There is a plane containing P_1, P_2, and C (Axiom 10). This plane contains the whole of line L (Axiom 13) and so contains P_3. Again using Axiom 13, we find that this plane contains the whole of the projector lines $\overleftrightarrow{CP_1}$, $\overleftrightarrow{CP_2}$, and $\overleftrightarrow{CP_3}$. As a result, the plane contains the image points P'_1, P'_2, P'_3. But this plane cuts the image plane in a line L' (Axiom 11). Therefore, P'_1, P'_2, P'_3 lie on the line L'. ■

Invariants play a key role in computer graphics and computer vision. Computer graphics artists, like many of their forerunners before the high-tech era, need to know how to portray three-dimensional objects accurately. Theorem 7.13 tells us that perspective projection will not distort the straightness of things. The edge of a building will still be straight in a photograph of a building and will be so in a perspective drawing of a building.

Here is an example of how invariants might be used in computer vision. One goal of computer vision, not yet achieved, is for a computer to analyze a picture and describe the scene in words. Suppose, for example, that we have a dining room table scene. On the table there is a circular dinner plate, which appears as an ellipse in the picture. The computer would have to realize that the actual object need not be an ellipse and could be a circle. In other words, the elliptical appearance of the image of an object does not rule out that the object might be a plate. The computer should also realize that the actual boundary of this object could not be made up of straight-line segments because collinearity is invariant and these boundary segments would appear as straight in the image. Thus the object could not be a rectangular placemat.

Equations of Perspective Projection

We now derive the equations for a particular kind of perspective projection called a *canonical perspective projection*. Being canonical means simply that the image plane is the coordinate plane $z = 0$ and the center of perspective C is on the z axis, at $(0, 0, c)$.

For a world point $P(x, y, z)$, what formula allows us to compute the coordinates of the image point $P'(x', y', z')$? Since the image lies in the x-y plane,

$$z' = 0. \tag{7.4}$$

To derive equations for x' and y', recall (see Section 5.1 of Chapter 5) that the equation of the line through C and P can be written in terms of the constant position vectors $\mathbf{c}$ and $\mathbf{p}$ and the variable position vector $\mathbf{x}$ as $\mathbf{x} = \mathbf{p} + t(\mathbf{c} - \mathbf{p})$. If P' is on this line, there must be a t value for which

$$\mathbf{p}' = \mathbf{p} + t(\mathbf{c} - \mathbf{p}). \tag{7.5}$$

We begin by finding the value of t. To do this we rewrite Eq. (7.5) in terms of the individual x, y, and z components:

$$\begin{aligned} x' &= x + t(0 - x), \\ y' &= y + t(0 - y), \\ z' &= z + t(c - z). \end{aligned} \tag{7.6}$$

The z equation, along with Eq. (7.4), allows us to solve for the t value:

$$\begin{aligned} 0 &= z + t(c - z), \\ t &= \frac{z}{(z - c)}. \end{aligned} \tag{7.7}$$

Substituting this expression for t into the x equation of Eq. (7.6) gives:

$$\begin{aligned}x' &= x + \frac{z}{(z-c)}(-x)\\ &= \frac{x(z-c) - xz}{(z-c)}\\ &= \frac{cx}{(c-z)}.\end{aligned} \tag{7.8}$$

Substituting our expression for t into the y equation of Eq. (7.6) yields:

$$y' = \frac{cy}{(c-z)}. \tag{7.9}$$

Gathering the key equations together:

CANONICAL PERSPECTIVE PROJECTION EQUATIONS

$$\begin{aligned}x' &= \frac{cx}{(c-z)},\\ y' &= \frac{cy}{(c-z)},\\ z' &= 0.\end{aligned} \tag{7.10}$$

EXAMPLE 7.5: PERSPECTIVE PROJECTION OF A POINT

Suppose $c=3$ and the world point is $P(-1, 2, 4)$. Then the image P' has coordinates

$$\begin{aligned}x' &= \frac{3(-1)}{(3-4)} = 3,\\ y' &= \frac{3(2)}{(3-4)} = -6,\\ z' &= 0. \quad \blacksquare\end{aligned}$$

Equations (7.10) only describe canonical perspective projection. But even though canonical perspective projections are special, their specialness lies mostly on the algebraic and numerical side of things. If we are interested in the geometric nature of perspective projection, we can often get all the insight we need by studying a canonical perspective projection.

EXAMPLE 7.6: PERSPECTIVE PROJECTION OF A LINE

Let L be the line $\mathbf{x} = \langle 2, -1, 3\rangle + t\langle 0, 2, 4\rangle$. Find the image of L under canonical perspective projection with $c=2$.

SOLUTION

Our strategy is to pick two points on L, project them and then find the line connecting the images. For our first point pick $t=0$ and get $\mathbf{p}_0 = \langle 2, -1, 3\rangle$. Next pick $t=1$ and get $\mathbf{p}_1 = \langle 2, -1, 3\rangle + \langle 0, 2, 4\rangle = \langle 2, 1, 7\rangle$. The images, obtained from Eqs. (7.10), are

$$\mathbf{p}_0' = \left\langle \frac{(2)(2)}{2-3}, \frac{(2)(-1)}{2-3}, 0\right\rangle = \langle -4, 2, 0\rangle \text{ and } \mathbf{p}_1' = \left\langle \frac{(2)(2)}{2-7}, \frac{(2)(1)}{2-7}, 0\right\rangle = \left\langle \frac{-4}{5}, \frac{-2}{5}, 0\right\rangle.$$

The line connecting these points is $\mathbf{x} = \mathbf{p}_0' + t(\mathbf{p}_1' - \mathbf{p}_0') = \langle -4, 2, 0\rangle + t\langle 16/5, 12/5, 0\rangle$. ■

A perspective projection is definitely not a one-to-one function. For example, if P projects into the image point P' and Q is any other point on the projector line $\overleftrightarrow{PC}$, then Q also projects to P'. Because of this we often see things which "aren't right." For example, look at the edge of the doorway of your room; now hold a pencil up so it appears to cut across the edge of the doorway. The line of the pencil and the line of the doorway don't really intersect in the three-dimensional world, but in the image produced by your eye they do.

EXAMPLE 7.7: PERSPECTIVE PROJECTION IS MANY-TO-ONE

We now verify, using algebra, that all points on a line through C project to the same image point under perspective projection. (We do this not to convince you—it should be obvious—but to see how algebra tells this story.) Suppose the line also passes through a point A. Then its equation may be written in terms of position vectors as $\mathbf{x} = \mathbf{c} + t(\mathbf{c} - \mathbf{a})$. Let's abbreviate $\mathbf{c} - \mathbf{a}$ by $\mathbf{d} = \langle x_d, y_d, z_d\rangle$. Then components of $\mathbf{x}$ may be expressed:

$$\begin{aligned} x &= 0 + tx_d, \\ y &= 0 + ty_d, \\ z &= c + tz_d. \end{aligned}$$

Substituting into the canonical perspective projection equations, Eqs. (7.10), gives the image of $\mathbf{x}$:

$$\begin{aligned} x' &= \frac{ctx_d}{c - (c + tz_d)} = \frac{-cx_d}{z_d}, \\ y' &= \frac{cty_d}{c - (c + tz_d)} = \frac{-cy_d}{z_d}, \\ z' &= 0. \end{aligned}$$

We see that t has dropped out. This means that the image point will be the same regardless of which point on the line we project. ■

However, perspective projections are onto the image plane: Every point U of the image plane arises as the image of some world points. Every point on the line $\overleftrightarrow{UC}$ will project onto U In particular, U will map to U. *Any point on the image plane maps to itself under perspective projection.*

Exceptional Points and the Exceptional Plane

Two of the equations in Eqs. (7.10) have $c - z$ in the denominator. If this denominator is ever 0, the equations are meaningless, because division by 0 is impossible. Thus, Eqs. (7.10) can only be applied if $c \neq z$. The exceptional points P where $z = c$ form a plane parallel to the x-y plane (Figure 7.19) and containing the point C (0, 0, c). The projector line that joins a world point in this plane to C is a line which lies in this plane (Axiom 13) and therefore

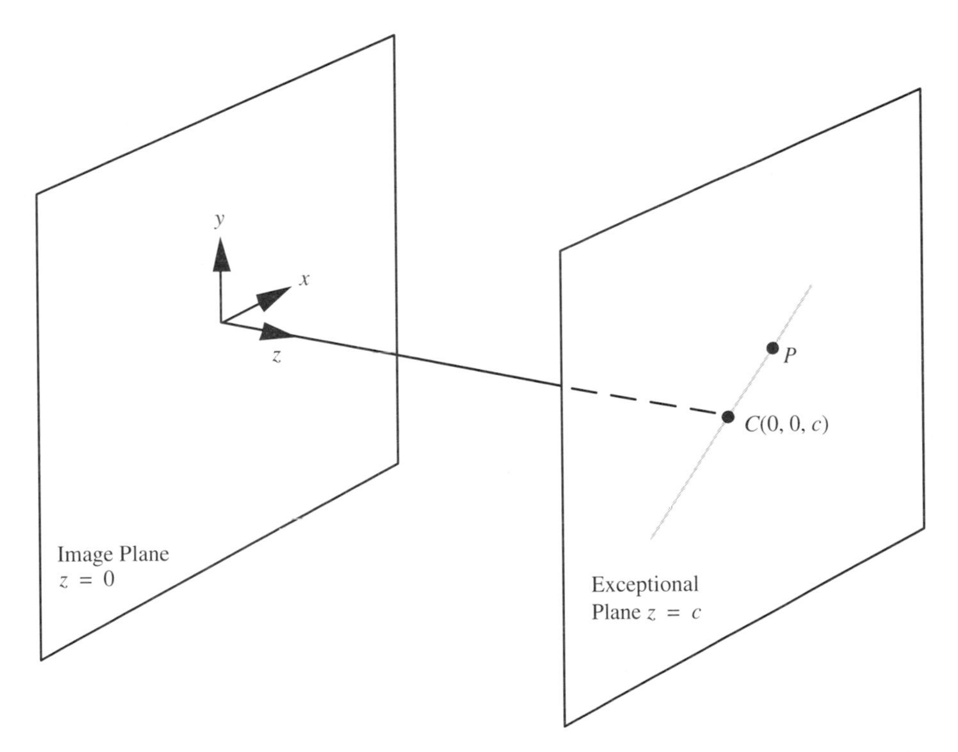

Figure 7.19 Projectors (such as $\overleftrightarrow{CP}$) in the exceptional plane do not cross the image plane.

never intersects the parallel image plane. Consequently, there is no image for a world point whose $z=0$. This plane of points with no image is called the *exceptional plane* and the points on it are called *exceptional points*. These points are not part of the domain of this canonical perspective projection. Noncanonical perspective projections also have exceptional planes but their equations are not so exceptionally simple.

The exceptional plane is not just a mathematical curiosity—it plays a role in devices that produce images. For example, when writing computer graphics programs that include equations such as Eqs. (7.10) it is important to be aware of the existence of the exceptional plane. If the program is asked to find the image of an exceptional point, the attempt to divide by zero will usually produce a run-time error. Normally this will make the program crash (stop) without completing its tasks. If these tasks are important then some way needs to be found to avoid trying to project exceptional points. One way around the problem is for the programmer to write the program so it tests whether $c=z$ before carrying out the division. If $c=z$, an error message can be printed but the program can continue without a crash.

The exceptional plane for the human eye passes through the pupil and is perpendicular to the direction in which we are looking. For the most part, the things we see well are those that are fairly near our line of sight, which means they are not near the exceptional plane. We do have some peripheral vision, but it gets worse and worse as objects get closer and closer to the exceptional plane.

In the case of a camera, the exceptional plane passes through the lens opening and is parallel to the front plane of the camera. Most cameras give good images only of points that are fairly close to the direction in which the camera is pointing, that is, far away from the exceptional plane. A wide-angle lens will come a bit closer to capturing points nearer the exceptional plane, but only at the cost of a fair amount of distortion.

EXAMPLE 7.8: EQUATION OF THE EXCEPTIONAL PLANE

Find the equation of the exceptional plane for the noncanonical perspective projection from an arbitrary center C and with image plane $\mathbf{x}\cdot\mathbf{n}=k$.

SOLUTION

As with canonical perspective projections, the exceptional plane of any perspective projection is parallel to the image plane and passes through C. It seems reasonable that if planes have the same normal they are parallel. (A formal proof: If $\mathbf{x}\cdot\mathbf{n}=k$ and $\mathbf{x}\cdot\mathbf{n}=k'$ had a point in common, $\mathbf{x}_0$, then $k=\mathbf{x}_0\cdot\mathbf{n}=k'$ and the two equations are identical so we don't have two planes, a contradiction.) Therefore, the exceptional plane must have equation $\mathbf{x} \cdot \mathbf{n} = k'$ for some k'. Since C lies on this plane, $\mathbf{c} \cdot \mathbf{n} = k'$ and we can write the plane $\mathbf{x} \cdot \mathbf{n} = \mathbf{c} \cdot \mathbf{n}$, or $\mathbf{x} \cdot \mathbf{n} - \mathbf{c} \cdot \mathbf{n} = 0$. Using the linearity of the scalar product gives another form of the equation of the exceptional plane:

$$(\mathbf{x} - \mathbf{c}) \cdot \mathbf{n} = 0. \tag{7.11}$$

■

EXAMPLE 7.9: IS THIS POINT EXCEPTIONAL?

Given that $C = (2, 5, -3)$ and the image plane is $\mathbf{x} \cdot \langle 5, -6, 0 \rangle = 3$, determine whether $P(1, 2, -1)$ is exceptional.

SOLUTION

We test whether Eq. (7.11) holds:

$$\begin{aligned}(\langle 1, 2, -1 \rangle - \langle 2, 5, -3 \rangle) \cdot \langle 5, -6, 0 \rangle &= \langle -1, -3, 2 \rangle \cdot \langle 5, -6, 0 \rangle \\ &= 13 \\ &\neq 0.\end{aligned}$$

Thus, P is not exceptional. ■

EXERCISES

**Marks challenging exercises.*

Overview of Image Formation

For Exercises 1–17 no proofs, calculations, algebra, or numbers are needed. The verdict of your visual intuition is acceptable.

1. Give a geometric description of which world points can be "seen" by the pinhole camera of Figure 5.7 of Section 5.1 in Chapter 5.
2. What if the pinhole in a pinhole camera is a circle instead of a point (which is, of course, more realistic). What is the image of a point under this circumstance?
3. (a) In a perspective projection, if we take a fixed object, say, a line segment, and move it toward or away from the center of projection, what is the effect on the size of the image?

   (b) Is there any limit to how big or small the image can be?

   (c) What happens when the segment moves to a point where it contains the center of projection?

4. Answer part (a) of Exercise 3 for the case of a parallel projection.

5. Suppose we stretch the pinhole camera (Figure 5.7 in Section 5.1 of Chapter 5) so that the back face (image plane) moves farther from the pinhole, but we leave the face containing the pinhole where it is. We take a picture of an unmoving object before and after the stretch. How will the images compare?

6. In a perspective projection, the image of an upright object may be upside down (as in Figure 5.7 of Section 5.1 in Chapter 5) or may be right side up.

   (a) Draw a picture to show how the right-side-up possibility can occur.

   (b) Can you explain what determines whether the image is right side up or upside down?

7. You are taking a picture of a person with the pinhole camera, a person holding a copy of this book in her left hand. Using both your hands in the normal way, you grasp the resulting snapshot right side up (the person's head at the top of the picture). Will the book in the snapshot be nearer your right hand or your left?

8. Describe an example of three noncollinear points that project to collinear ones under a perspective projection.

9. Answer Exercise 8 for parallel projection.

10. Explain how two lines in space can be mapped into the same line in the image plane under parallel projection.

11. Explain how the image of a four-sided polygon can be a line segment under perspective projection.

12. Answer Exercise 11 for parallel projection.

13. Can a four-sided polygon (not necessarily planar) project into a triangle in the image plane under perspective projection? Explain your answer with an example or proof.

14. Answer Exercise 13 for the case of parallel projection.

15. Under what circumstances does a line segment not project into infinitely many points under perspective projection?

16. Answer Exercise 15 for the case of parallel projection.

17. What consequence is there for our human vision of the fact that perspective projection is not one-to-one?

18. Prove that if three points in three-dimensional space are not collinear and the plane they determine does not contain the center of a perspective projection, then their images are not collinear. (*Hint*: Try a proof by contradiction.)

Equations of Perspective Projection

19. Find the image of $P=(2, -3, 1)$ under the canonical perspective projection equations, Eqs. (7.10). Use $c=4$.

20. Find the image of the line $\mathbf{x}=\langle 1, 2, -5\rangle + t\langle 0, 3, -7\rangle$ under the canonical perspective projection with $c=4$.

21. Find the image of the line $\mathbf{x}=\langle 1, 2, -5\rangle + t\langle 0, 3, -7\rangle$ under the perspective projection with $C(-2, 6, 3)$ and image plane with equation $\mathbf{x}\cdot\langle -1, 3, 0\rangle = 4$.

22. Show by numerical example, or convincing argument with a picture, that distance between points is not an invariant of perspective projection.

23. Show by numerical example, or convincing argument with a picture, that angles are not invariant under perspective projection.

*24. Consider the circle in the plane $z=3$ whose x and y coordinates satisfy $x^2+y^2=16$. Show that the images of points on this circle under canonical perspective projection also fall on a circle in the x-y plane. Find its equation. Use $c=1$.

*25. Carry out the previous exercise for the circle $(x-6)^2+(y-2)^2=25$ in the plane $z=3$. Do you get a circle again or some other curve?

26. What is the effect of a canonical perspective projection with $c=1$ on points in the plane $z=3$ (which is parallel to the image plane)? Show that the image of the point $P(x, y, z)$ in this plane is half as far from the z axis as P is. [*Hint*: The distance of a point (a, b, c) to the z axis is $\sqrt{a^2+b^2}$.]

27. Show that the effect of a canonical perspective projection on points in the plane $z=k$ (which is parallel to the image plane) is the same as a uniform scaling transformation. This means there is a scaling constant r so that, if P and Q are any two points in $z=k$, and P' and Q' are the images, then $P'Q'=rPQ$. How is r related to c and k? (*Hint*: See Exercise 26.)

28. In deriving the perspective projection equations, Eqs. (7.10), we assumed a canonical perspective projection. Derive the equations in the case where C lies on the x axis, that is, is of the form $(c, 0, 0)$, and where the image plane is the plane $x=0$.

*29. In our derivation of the perspective projection equations, Eqs. (7.10), we assumed a canonical perspective projection. Derive a vector equation in the case where $C = (x_C, y_C, z_C)$ and where the image plane has the equation $\mathbf{x} \cdot \mathbf{n} = k$ where $\mathbf{n} = \langle x_n, y_n, z_n \rangle$. Your equation should express $\mathbf{x}'$ in terms of $\mathbf{x}$, $\mathbf{n}$, $\mathbf{c}$, and $\mathbf{k}$.

Exceptional Points and the Exceptional Plane

30. If $c = 6$ in a canonical perspective projection, what is the equation of the exceptional plane?

31. If $c = 6$ in a canonical perpsective projection, is $(0, 5, -6)$ exceptional?

32. If $C = (2, 1, -4)$ and the image plane is $\mathbf{x} \cdot \langle 1, 0, 4 \rangle = 3$, what is the exceptional plane?

33. If $C = (5, 1, 2)$ and the image plane is $\mathbf{x} \cdot \langle -1, 3, 0 \rangle = 10$, is $P = (-1, -1, 4)$ exceptional?

34. Here is a sequence of points that get closer and closer to the exceptional plane: $P_1 = (1, 1, c - 1)$, $P_2 = (1, 1, c - 1/2), \ldots, P_n = (1, 1, c - 1/n), \ldots$. Calculate the images of these points and describe "where they are heading." (What we are trying for here is an understanding of what it means to be "nearly exceptional." The points in our sequence are closer and closer to exceptional as we go along.)

7.4 Parallelism and Vanishing Points of a Perspective Projection

When we look at two railroad tracks, we see an illusion: They seem to come together in the distance. In this section, we will see that the reason is that perspective projection, which is more or less what the eye carries out, maps parallel lines into lines that intersect. The crossing point is called the vanishing point of the parallel lines.

Figure 7.20 shows one way in which this convergence toward a vanishing point can occur. Lines 1 and 2 are parallel but project into lines $\overleftrightarrow{A'B'}$ and $\overleftrightarrow{D'E'}$, which intersect at V. Does this happen only in this one drawing, or is the phenomenon common? How can we calculate a vanishing point?

Theory of Parallelism

To answer these questions, we need to understand direction vectors and parallelism among lines in three dimensions. The story is fairly simple — and you may recall much

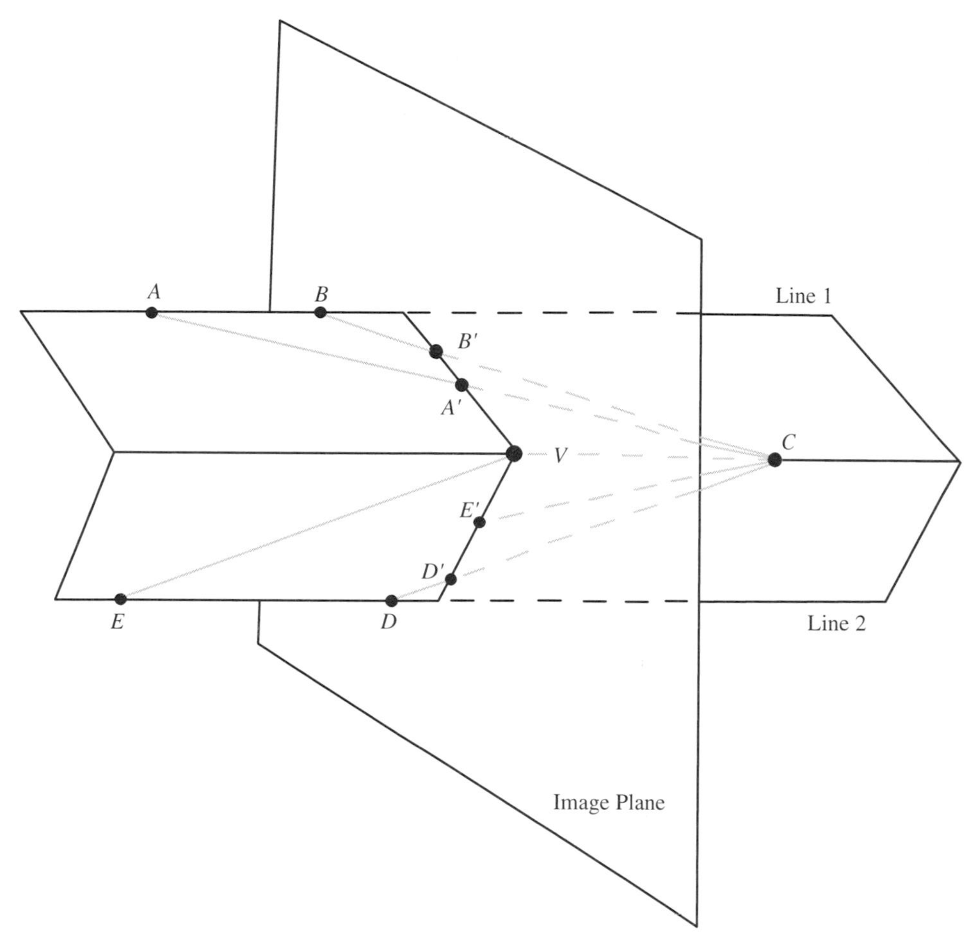

Figure 7.20 Lines 1 and 2 are parallel but their images converge to *V*.

of it from multivariable calculus or some other course — so here is a summary of what we will prove in this subsection. Just as a line in the plane can have its "tilt" described numerically with a number called its slope, we can give a numerical description of the tilt of a line in three dimensions. Instead of using a single number, we have an infinite collection of direction vectors, any of which can serve the purpose. If we multiply the components of a direction vector through by a nonzero constant we get another vector, which also signifies the tilt for the line. If a line has direction vector $\mathbf{d}$ (or any multiple of $\mathbf{d}$), all lines parallel to the line also have $\mathbf{d}$ as a direction vector.

Conversely, if two lines share a direction vector, they are either the same or are parallel. As in the plane, parallelism is a transitive relationship, by which we mean that if L_1 is parallel to L_2 and L_2 is parallel to L_3, L_1 is parallel to L_3.

A line may have three relationships to a plane: It may lie in it, it may cross it in a single point, or it may be parallel (never cross it). Each of these cases can be easily detected by a simple computation.

In the remainder of this subsection we provide the details to flesh out what we have just summarized. Let X be a variable point on the line passing through A and B. We can write the line's parametric equation in terms of position vectors $\mathbf{x} = \mathbf{a} + t(\mathbf{b} - \mathbf{a})$. We call $\mathbf{d} = \mathbf{b} - \mathbf{a}$ a *direction vector* for the line. Instead of A and B, we could have picked two other points P and Q on the line, leading to a different direction vector, $\mathbf{q} - \mathbf{p}$. The following definition describes the relationship between any two direction vectors of a line such as $\mathbf{b} - \mathbf{a}$ and $\mathbf{q} - \mathbf{p}$.

DEFINITION

Two vectors $\mathbf{d}_1$ and $\mathbf{d}_2$ are *proportional* if there is a nonzero scalar r so that $\mathbf{d}_1 = r\mathbf{d}_2$ (equivalently $\mathbf{d}_2 = s\mathbf{d}_1$ where $s = 1/r$); r is called the *proportionality factor.*

EXAMPLE 7.10: PROPORTIONAL VECTORS

(a) $\langle 2, 1, -5\rangle$ is proportional to $\langle 4, 2, -10\rangle$ since $\langle 4, 2, -10\rangle = 2\langle 2, 1, -5\rangle$.

(b) $\langle 0, 0, 0\rangle$ is not proportional to any other vector since a proportionality factor needs to be nonzero.

(c) To see if $\langle -3, 2, 4\rangle$ is proportional to $\langle -10.2, 6.4, 12.8\rangle$ we need to find r where

$$-10.2 = -3r,$$

$$6.4 = 2r,$$

$$12.8 = 4r.$$

The r which satisfies the first of these equations ($r = 3.4$) does not satisfy either of the other two so the vectors are not proportional. ■

☆ THEOREM 7.14

(a) If $\mathbf{d}$ is a direction vector for line L, then so is $r\mathbf{d}$ for any nonzero r (Figure 7.21).

(b) Any two direction vectors for a line are proportional.

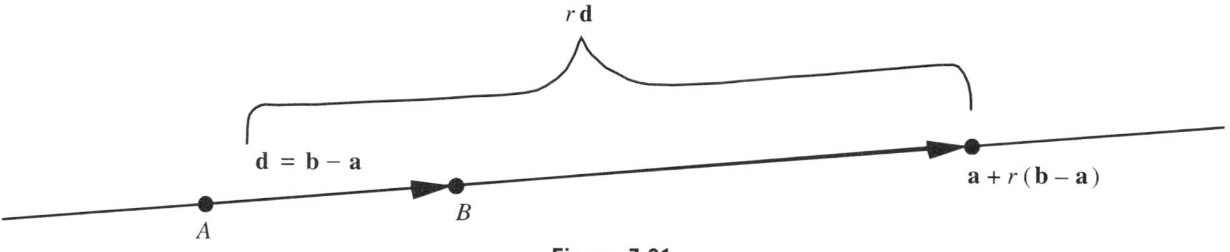

Figure 7.21

Proof

To prove part (a), we need to show that since $\mathbf{d}$ connects two points on line L, so does $r\mathbf{d}$. Since $\mathbf{d}$ is a direction vector for L, there are points A and B on the lines where $\mathbf{b} - \mathbf{a} = \mathbf{d}$. Thus, one equation for the line is $\mathbf{x} = \mathbf{a} + t(\mathbf{b} - \mathbf{a})$ and, substituting r for t, we get that $\mathbf{a} + r(\mathbf{b} - \mathbf{a})$ is on L. Calculating a direction vector from $\mathbf{a}$ and $\mathbf{a} + r(\mathbf{b} - \mathbf{a})$ gives $\mathbf{a} + r(\mathbf{b} - \mathbf{a}) - \mathbf{a} = r(\mathbf{b} - \mathbf{a}) = r\mathbf{d}$. Thus, $r\mathbf{d}$ is a direction vector for L.

We leave the proof of part (b) as an exercise. ■

Notice that Theorem 7.14 does not require r to be positive. For example, $\langle 1, 0, 0\rangle$ is a direction vector for the x axis, so $(-1)\langle 1, 0, 0\rangle = \langle -1, 0, 0\rangle$ is also, even though it points in the opposite direction.

To proceed, we need some facts about parallel lines, including a definition. (When working in three dimensions, we can't simply say that parallel lines are lines that don't intersect. Skew lines don't intersect but are not parallel; see Figure 5.8.)

DEFINITION

Lines L_1 and L_2 are *parallel* if:

(a) they are different lines, and

(b) there is a plane which contains both, and

(c) they don't intersect.

☆ THEOREM 7.15

(a) Suppose two different lines L_1 and L_2 share a direction vector. Then L_1 and L_2 are parallel.

(b) Suppose L_1 and L_2 are parallel. Then there is a vector $\mathbf{d}$ which is a direction vector for both lines.

Proof

See the exercises. ■

COROLLARY 7.16

If two lines have the same direction vector and the lines share a point, then they are the same line.

Proof

Suppose the lines were different. Then part (a) of the previous theorem implies that they are parallel. But parallel lines, by definition, have no point in common. This contradiction proves the lines can't be different. ■

EXAMPLE 7.11: COMPARING LINE EQUATIONS

Below we give four parametric equations for lines.

(a) Do equations (i) and (ii) describe the same line or different lines? If different, are they parallel?

(b) Answer the same questions for equations (i) and (iii).

(c) Answer the same questions for equations (i) and (iv).

(i) $\mathbf{x} = \langle 2, 1, 2\rangle + t\langle -3, 2, 2\rangle$,

(ii) $\mathbf{x} = \langle 8, -3, -2\rangle + t\langle -6, 4, 4\rangle$,

(iii) $\mathbf{x} = \langle 5, -1, 2\rangle + t\langle -1.5, 1, 1\rangle$,

(iv) $\mathbf{x} = \langle 5, -1, 2\rangle + t\langle -10.2, 6.4, 6.4\rangle$.

SOLUTION

(a) Comparing direction vectors for (i) and (ii), we see that $\langle -6, 4, 4\rangle = 2\langle -3, 2, 2\rangle$ so the lines are either parallel or the same. $(8, -3, -2)$ is a point on the second line. If it also lies on the first line, the lines are the same. Otherwise they are different. It lies on the first line if and only if there is a t satisfying

$$\langle 8, -3, -2\rangle = \langle 2, 1, 2\rangle + t\langle -3, 2, 2\rangle,$$
$$\langle 6, -4, -4\rangle = t\langle -3, 2, 2\rangle.$$

Also, $t = -2$ satisfies this equation so the lines are the same.

(b) The direction vectors are proportional: $\langle -1.5, 1, 1\rangle = \frac{1}{2}\langle -3, 2, 2\rangle$. To see if $(5, -1, 2)$ lies on line (i) we seek to solve

$$\langle 5, -1, 2\rangle = \langle 2, 1, 2\rangle + t\langle -3, 2, 2\rangle,$$
$$\langle 3, -2, 0\rangle = t\langle -3, 2, 2\rangle.$$

There is no solution for t, so the lines are parallel.

(c) The direction vectors are not proportional: $-10.2/-3 = 3.4$ but $6.4/2 = 3.2$. Thus the lines are different and not parallel. ■

✯ THEOREM 7.17

If two lines are each parallel to a third, then they are parallel to each other.

Proof

Let $\mathbf{d}_1$ be a direction vector L_1 and L_3 have in common and let $\mathbf{d}_2$ be a direction vector L_2 and L_3 have in common. Since $\mathbf{d}_1$ and $\mathbf{d}_2$ are both direction vectors for L_3, part (b) of Theorem 7.14 tells us they are proportional vectors. By part (a) of Theorem 7.14, $\mathbf{d}_2$ is a direction vector for L_1 as well as for L_2. By part (a) of Theorem 7.15, L_1 and L_2 are parallel. ■

DEFINITION

Let L be any line. The *parallel class* of L is the set of all lines parallel to L, together with L itself. Alternatively, let $\mathbf{d}$ be any nonzero vector. The parallel class of $\mathbf{d}$ is the set of all lines that have $\mathbf{d}$ as a direction vector.

By Theorem 7.17, all the lines in a parallel class are parallel to one another.

✯ THEOREM 7.18

If P is any point off line L, then there is exactly one parallel to L through P.

Proof

Pick two points A and B on L. By Axiom 10 there is exactly one plane containing A, B, and P. By Axiom 13 this plane contains all of L. By Euclid's parallel axiom (Axiom 9), there is exactly one line lying in this plane, passing through P, parallel to L. ■

A consequence of this theorem is that if P is any point and we choose any parallel class, there will be a member of this parallel class passing through P. To see this let L be any line in the class. If it does not pass through P, the previous theorem implies that there will be a line parallel to L that does.

We turn now to the question of how a line and plane may be related.

✯ THEOREM 7.19

The line $\mathbf{x} = \mathbf{a} + t\mathbf{d}$ and the plane $\mathbf{x} \cdot \mathbf{n} = k$ are related as follows:

1. If $\mathbf{d} \cdot \mathbf{n} \neq 0$, then the line has one point in common with the plane.
2. If $\mathbf{d} \cdot \mathbf{n} = 0$ and A does not lie in the plane ($\mathbf{a} \cdot \mathbf{n} \neq k$), then the line is parallel to the plane (i.e., does not intersect the plane).
3. If $\mathbf{d} \cdot \mathbf{n} = 0$ and A lies in the plane ($\mathbf{a} \cdot \mathbf{n} = k$), then the line lies completely in the plane.

PROOF

To attempt to find a point which the line has in common with the plane, we substitute the parametric expression for $\mathbf{x}$ into the plane equation. We get

$$\begin{aligned}(\mathbf{a} + t\mathbf{d}) \cdot \mathbf{n} &= k \\ \mathbf{a} \cdot \mathbf{n} + t(\mathbf{d} \cdot \mathbf{n}) &= k \\ t(\mathbf{d} \cdot \mathbf{n}) &= k - \mathbf{a} \cdot \mathbf{n}. \end{aligned} \tag{7.12}$$

The number of points the line has in common with the plane will be shown by how many solutions Eq. (7.12) has for t.

CASE 1. If $\mathbf{d} \cdot \mathbf{n} \neq 0$, then there is exactly one solution for t:

$$t = \frac{k - \mathbf{a} \cdot \mathbf{n}}{\mathbf{d} \cdot \mathbf{n}}.$$

CASE 2. If $\mathbf{d} \cdot \mathbf{n} = 0$ but $k - \mathbf{a} \cdot \mathbf{n} \neq 0$ (A does not lie in the plane), then there are no solutions of Eq. (7.12) for t. The line is parallel to the plane.

CASE 3. If $\mathbf{d} \cdot \mathbf{n} = 0$ and $k - \mathbf{a} \cdot \mathbf{n} = 0$ (A does lie in the plane), then every t satisfies Eq. (7.12), which means that every point of the line lies in the plane. ■

We say the *direction* $\mathbf{d}$ is parallel to the plane $\mathbf{x} \cdot \mathbf{n} = k$ provided $\mathbf{d} \cdot \mathbf{n} = 0$. According to the theorem just proved, this means that any line with direction parallel to a given plane is either parallel to or lies in the plane.

EXAMPLE 7.12: HOW LINES RELATE TO PLANES

Explain how the following lines are related to the plane $\mathbf{x} \cdot \langle 1, -2, 5 \rangle = 9$.

(a) $\mathbf{x} = \langle 1, 2, 3 \rangle + t\langle 2, -4, -2 \rangle$,

(b) $\mathbf{x} = \langle 2, -1, 1\rangle + t\langle -5, 5, 3\rangle$,

(c) $\mathbf{x} = \langle 1, -3, 4\rangle + t\langle 4, 1, -1\rangle$.

SOLUTION

(a) $\langle 2, -4, -2\rangle \cdot \langle 1, -2, 5\rangle = 2 + 8 - 10 = 0$ so the line either lies in the plane or is parallel. We check whether $(1, 2, 3)$ lies in the plane. $\langle 1, 2, 3\rangle \cdot \langle 1, -2, 5\rangle = 12 \neq 9$ so the line is parallel to the plane.

(b) $\langle -5, 5, 3\rangle \cdot \langle 1, -2, 5\rangle = -5 - 10 + 15 = 0$ so the line either lies in the plane or is parallel. We check whether $(2, -1, 1)$ lies in the plane. $\langle 2, -1, 1\rangle \cdot \langle 1, -2, 5\rangle = 9$ so the line lies in the plane.

(c) $\langle 4, 1, -1\rangle \cdot \langle 1, -2, 5\rangle = -3 \neq 0$ so the line intersects the plane in a single point. ■

Vanishing Points

DEFINITION

Let L be any line that intersects the image plane but does not lie in it. The *vanishing point, V*, for the parallel class of L is the intersection of the image plane with the line L_C in the parallel class of L which passes through C.

To make this a proper definition we would need to be sure that there is only one line in the parallel class of L which passes through C. This follows by Theorem 7.18 (see Figure 7.22). In addition, we need to be sure that this line L_C intersects the image plane in exactly one point (i.e., that it does not lie parallel or within the image plane). We leave this as an exercise.

EXAMPLE 7.13: FINDING THE VANISHING POINT

(a) For a canonical perspective projection with $c = -5$, find the vanishing point V of the parallel class for which $\mathbf{d} = \langle -1, 4, -2\rangle$ is a direction vector.

(b) Find formulas for the coordinates of the vanishing point V in terms of c and $\mathbf{d}$, assuming a canonical perspective projection.

SOLUTION

(a) The line L_C is $\mathbf{x} = \langle 0, 0, -5\rangle + t\langle -1, 4, -2\rangle$. The z component of $\mathbf{x}$ is $-5 - 2t$. At the crossing with the image plane (the vanishing point), $-5 - 2t = 0$ so $t = -5/2$. Substituting t into the equation for L_C: $\mathbf{v} = \langle 0, 0, -5\rangle - 5/2\ \langle -1, 4, -2\rangle = \langle 5/2, -10, 0\rangle$.

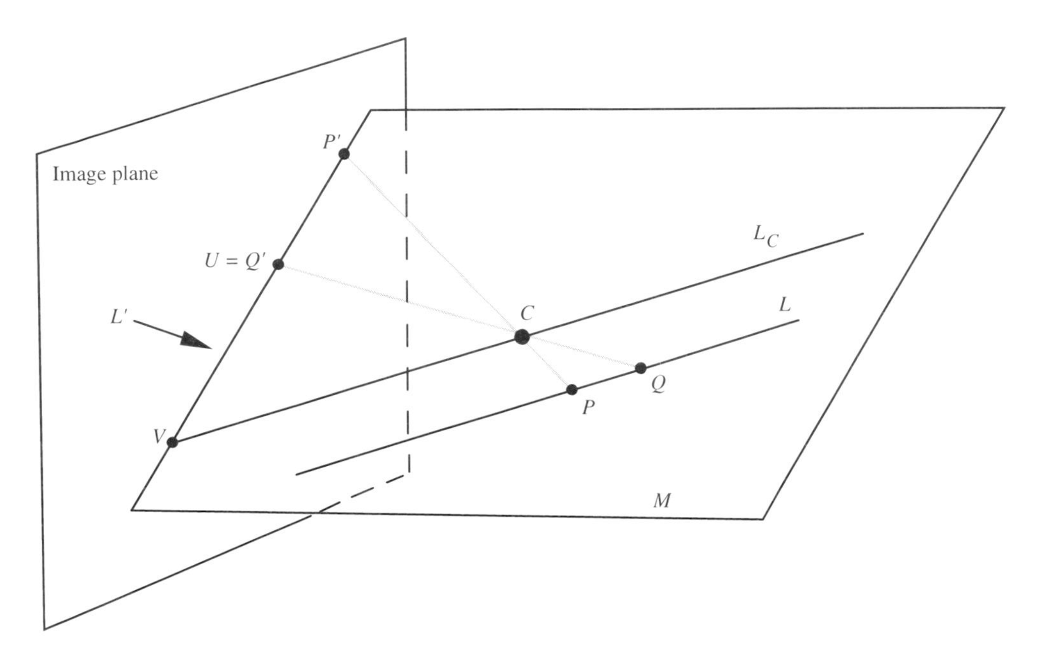

Figure 7.22 The image of line L is a line minus the vanishing point.

(b) The line through C in direction $\mathbf{d}$ has equation

$$\mathbf{x} = \mathbf{c} + t\mathbf{d}.$$

Formulas for the point where this line intersects the image plane were worked out in Example 7.7 of the previous section:

$$V = \left(\frac{-cx_d}{z_d}, \frac{-cy_d}{z_d}, 0 \right).$$

This is the vanishing point for $\mathbf{d}$. Notice that if we replace $\mathbf{d}$ by any multiple, say, $k\mathbf{d}$, where $k \neq 0$, then the vanishing point is now $(-ckx_d/kz_d, -cky_d/kz_d, 0)$. Cancelling k's, we see that this is the same point as before. This should not be surprising because the vector $k\mathbf{d}$ corresponds to the same parallel class as $\mathbf{d}$. ■

In the previous section, we saw that when we apply perspective projection to a line L, the image is a subset of some line L' in the image plane. But we did not consider what the nature of that subset was. For example, would it be the whole line L'? The next theorem gives the answer: "Almost."

☆ THEOREM 7.20: IMAGE OF A LINE

Suppose L is a line in three-dimensional space which

1. intersects the image plane but does not lie in it (Figure 7.22) and
2. does not pass through C.

Then the images of the points of L under perspective projection are collinear and the collection of these images makes up an entire line L' in the image plane except for the vanishing point of the parallel class of L.

PROOF (FIRST ATTEMPT)

Let M be the plane determined by C and L and let L' be the line which is the intersection of M and the image plane. If P is a point on L, P' lies on L'. Let U be any point other than V on L'. The line joining U to C lies in M and is not parallel to L since U is not the vanishing point of L. Let Q be the point where this line intersects L. Notice that $Q' = U$. The vanishing point V is not the image of any point on L. To see this, note that the line $\overleftrightarrow{VC}$ contains the only world points that might project into V. But this line does not cross L. Thus, no point on L can project to V. ■

Our proof of Theorem 7.20 never used the hypothesis that L does not pass through C. In the exercises, we ask you to explain why the theorem is not true without that hypothesis and to correct the proof by making use of the hypothesis at the crucial point.

We will call the image of a line L under perspective projection a *deleted line* in recognition of the "one point missing" aspect of Theorem 7.20. We'll use the notation L' for the whole line consisting of the deleted line plus the vanishing point V. Thus the image of L is $L' - \{V\}$.

To get a better understanding of the vanishing point, imagine a series of world points on L, say, $P, Q, R, \ldots$, heading off to infinity on L. The images $P', Q', R', \ldots$, lie on the deleted line $L' - \{V\}$. But instead of heading off to infinity, they converge toward the vanishing point. Our next example provides an algebraic confirmation of this.

EXAMPLE 7.14: MARCHING OFF TO THE VANISHING POINT

Let X be a variable point on the line through point A with direction $\mathbf{d}$. The line's parametric equation is $\mathbf{x} = \mathbf{a} + t\mathbf{d}$. What happens to the image of X as $t \rightarrow \infty$ if we project X with the canonical perspective projection?

SOLUTION

Using Eqs. (7.10), the image of X is

$$(x', y', 0) = \left[\frac{c(x_a + tx_d)}{(c - z_a) - tz_d}, \frac{c(y_a + ty_d)}{(c - z_a) - tz_d}, 0\right].$$

Divide numerator and denominator by t in each coordinate of this point:

$$(x', y', 0) = \left[\frac{c(x_a/t + x_d)}{(c - z_a)/t - z_d}, \frac{c(y_a/t + y_d)}{(c - z_a)/t - z_d}, 0\right].$$

As $t \to \infty$, the terms with t in their denominators approach 0 and

$$(x', y', 0) \to \left(\frac{-cx_d}{z_d}, \frac{-cy_d}{z_d}, 0\right).$$

According to part (b) of Example 7.13, this is the vanishing point for the direction **d**. ■

Our next theorem shows that parallelism is as badly noninvariant as it possibly could be under perspective projection. Not only are the images of the lines in a parallel class not parallel, but they display every conceivable direction in the image plane.

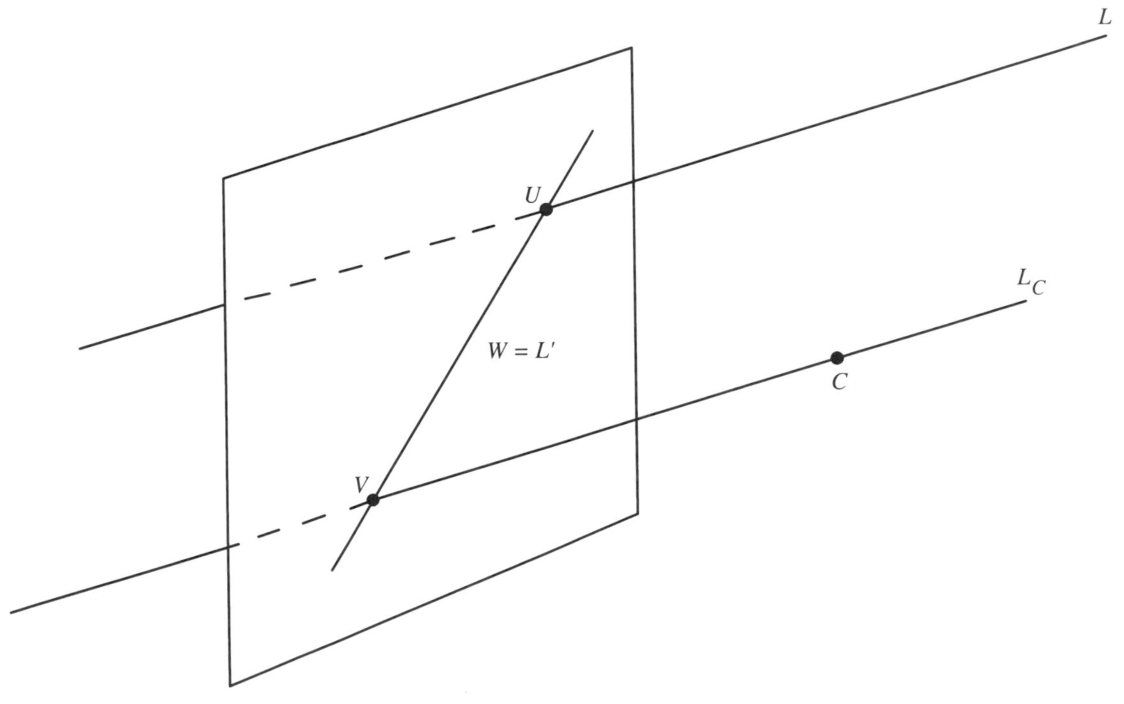

Figure 7.23 Lines of all directions arise as images of members of a parallel class.

☆ THEOREM 7.21: NON-INVARIANCE OF PARALLELISM

Let L_C be a line through C which intersects the image plane of a perspective projection at a single point V (which is the vanishing point of the parallel class of L_C; see Figure 7.23). Let W be any line in the image plane that passes through V. Deleted line $W - \{V\}$ is the image of some line in the parallel class of L_C.

PROOF

Let U be any point on W other than V. Let L be the line of the parallel class which passes through U (see the comment following Theorem 7.18) and let L' be the line containing the image of L. Point U is its own image under the perspective projection so $U = U'$ is on L'. But V is on L' by Theorem 7.20. Thus L' is the line W since it contains two points of W, namely, U and V. ■

EXERCISES

**Marks challenging exercises.*

Theory of Parallelism

1. Which of the drawings in Figure 7.24 could be a perspective projection of a right-angled box? For the ones that could not be, can you say why they can't?

2. In each case, state whether the two vectors are proportional.

   (a) $\langle 4, -2, 3 \rangle$ and $\langle 28, -14, 21 \rangle$

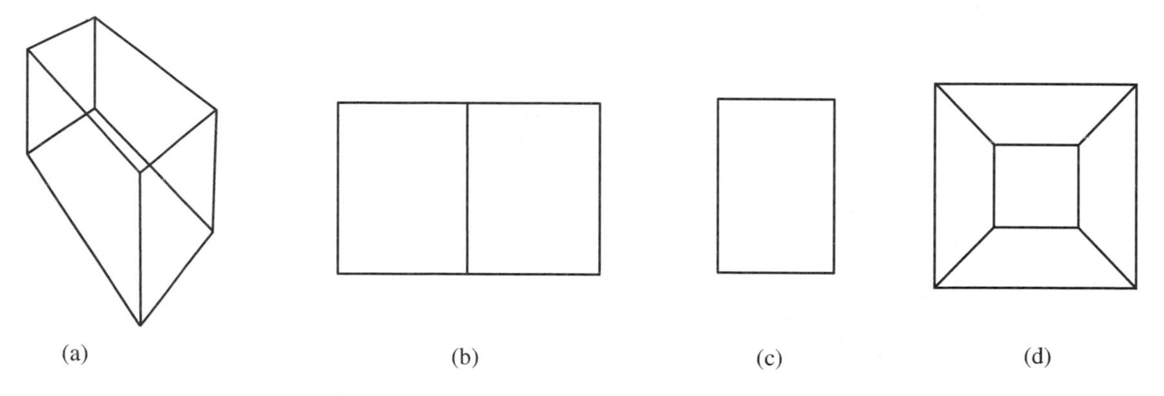

Figure 7.24 Which of these are views of a box?

(b) $\langle 1.5, -4.5, 6.0\rangle$ and $\langle -1, -3, -4\rangle$

(c) $\langle 0, 0, 1\rangle$ and $\langle 1, 0, 0\rangle$

(d) $\langle 5, -3, 7\rangle$ and $\langle -65, 39, -91\rangle$

3. In each case state whether the two equations describe the same line or not. If the lines are different, say whether they are parallel.

   (a) $\mathbf{x}=\langle 1, 0, 4\rangle+t\langle -2, 1, 3\rangle$ and $\mathbf{x}=\langle -1, 1, 7\rangle+t\langle 18, -9, -27\rangle$

   (b) $\mathbf{x}=\langle 1, 0, 4\rangle+t\langle -2, 1, 3\rangle$ and $\mathbf{x}=\langle -2, 2, 10\rangle+t\langle -34, 17, 51\rangle$

   (c) $\mathbf{x}=\langle 1, 0, 4\rangle+t\langle -2, 1, 3\rangle$ and $\mathbf{x}=\langle 1, -2, 3\rangle+t\langle 8, -4, 12\rangle$

4. For the following lines, say whether the line is in the plane whose equation is $\mathbf{x}\cdot\langle 3, -4, 1\rangle=10$, or is parallel to the plane, or intersects the plane once.

   (a) $\mathbf{x}=\langle 1, -1, 3\rangle+t\langle 5, 4, 1\rangle$

   (b) $\mathbf{x}=\langle 4, -5, 2\rangle+t\langle 3, 2, -1\rangle$

   (c) $\mathbf{x}=\langle 2, 2, -3\rangle+t\langle -10, -12, -3\rangle$

5. For each of the following lines, find a line parallel to that line and passing through the given point.

   (a) $\mathbf{x}=\langle 4, 1, 0\rangle+t\langle 7, 3, -8\rangle$; $(1, 0, 0)$

   (b) $\mathbf{x}=\langle 0, 1, 1\rangle+t\langle 1, 1, -1\rangle$; $(2, -3, 4)$

   (c) $\mathbf{x}=\langle 4, 4, -8\rangle+t\langle 3, 4, 5\rangle$; $(3, 6, -3)$

6. Can a line be in two different parallel classes? If so, give an example. If not, explain why not.

* 7. Give a proof of Theorem 7.17 that does not use equations or coordinates. (It is quite possible to give a reasonably short proof just from the axioms of Section 1.3 in Chapter 1 without invoking any theorems except Theorem 7.18.)

8. Show that if two planes have normal vectors $\mathbf{n}$ which are proportional, then these planes are either identical or have no points in common.

9. Prove part (b) of Theorem 7.14.

The next two exercises, together, are meant to prove part (a) of Theorem 7.15.

10. Suppose $\mathbf{d}$ is a direction vector for line L_1, which passes through A_1, and also for line L_2, which passes through A_2. Show that if we define $\mathbf{n}$ to be any vector perpendicular to both $\mathbf{d}$ and to $\mathbf{a}_2 - \mathbf{a}_1$, and if we set $k = \mathbf{a}_1 \cdot \mathbf{n}$, then the plane $\mathbf{x} \cdot \mathbf{n} = k$ contains both lines.

11. Suppose the lines L_1 and L_2 with equations $\mathbf{x} = \mathbf{a}_1 + t\mathbf{d}$ and $\mathbf{x} = \mathbf{a}_2 + t\mathbf{d}$ have a point in common. Let its representation on L_1 be $\mathbf{a}_1 + t_1\mathbf{d}$ and its representation on L_2 be $\mathbf{a}_2 + t_2\mathbf{d}$. Show that A_2 lies on L_1 and, further, each point of L_2 lies on L_1.

12. Prove part (b) of Theorem 7.15: If L_1 and L_2 are in a plane and don't meet, there is a vector that is a direction vector for both lines.

Vanishing Points

13. Suppose we have a noncanonical perspective projection where $C = (1, -2, 1)$ and the image plane is $\mathbf{x} \cdot \langle 1, 2, -4 \rangle = 3$. Determine whether or not the following lines intersect the plane in a single point. In cases where the intersection is one point, find the vanishing point.

    (a) $\mathbf{x} = \langle 2, -1, 0 \rangle + t\langle 2, -1, 0 \rangle$

    (b) $\mathbf{x} = \langle 4, 3, -2 \rangle + t\langle -4, 2, 3 \rangle$

    (c) $\mathbf{x} = \langle 1, 1, 1 \rangle + t\langle 5, -2, 3 \rangle$

14. In each of the following cases, as point X moves to infinity along the line, what point do the images X', under canonical perspective projection with $c = 5$, converge to?

    (a) $\mathbf{x} = \langle 2, 8, -7 \rangle + t\langle 1, 1, -2 \rangle$

    (b) $\mathbf{x} = \langle 100, 1000, 100{,}000 \rangle + t\langle 1, 1, -2 \rangle$

    (c) $\mathbf{x} = \langle 1, 1, 1 \rangle + t\langle 4, -3, 2 \rangle$

15. Rework the previous problem, but with a noncanonical perspective projection where $C = (1, 2, -5)$ and the image plane has equation $\mathbf{x} \cdot \langle 2, 2, -4 \rangle = 3$.

16. Find the vanishing point for the line $\mathbf{x} = \langle 2, -4, 3 \rangle + t\langle 6, 0, 2 \rangle$ using the canonical perspective projection having $c = 1$. Next do it for $c = 2$. Can you formulate a conjecture about what happens as c increases?

17. For perspective projection onto the image plane $\mathbf{x} \cdot \langle -2, 4, 3 \rangle = 7$, give an example of a direction for which there is no vanishing point.

18. Let L be a line with vanishing point V. V divides L' into two rays, R_1 and R_2, where $R_1 \cap R_2 = \{V\}$. Consequently the points of L are partitioned into those which project to $R_1 - \{V\}$ and those which project to $R_2 - \{V\}$. Describe these two subsets of L as specifically as you can.

*19. Prove that if line L crosses a plane I at point P but does not lie in I, and L_C is parallel to L, then L_C intersects I in a single point.

20. Explain why Theorem 7.20 would not be true if we left off the hypothesis that L does not pass through C. Determine what the image would be in that case.

21. Show where, in the proof of Theorem 7.20, we should have used the hypothesis that L does not pass through C.

22. In Example 7.14, explain what happens if the line happens to pass through C.

23. In the proof of Theorem 7.20 we did not justify the assertion that M crosses the image plane. Why does it?

24. In discussions of paintings, one sometimes sees the phrase "the vanishing point of the painting." Mathematically speaking, is it reasonable to suppose that there is just one vanishing point in a painting? Explain what meaning this phrase may have.

7.5 Parallel Projection

Parallel projection is an alternative to perspective projection as a means of making a two-dimensional image of a three-dimensional object. We will see that it has more invariants than perspective projection, which can be considered an advantage. On the other hand, perspective projection has its own advantages in that it creates images rather like those the eye creates, which is not true of parallel projection. As an example, we will see that a pair of parallel railroad tracks would be shown as parallel lines in a parallel projection. However, the human eye sees them as coming together in the distance, and they would be depicted that way by a perspective projection, as we saw in the previous section.

Creating a parallel projection is different from creating a perspective projection. In a parallel projection we first select a direction of projection and all projector lines are parallel to that direction instead of passing through some center of projection. In a parallel projection, there is no center of projection.

Our objective in this section is to study parallel projection from two points of view:

1. An algebraic point of view in which we come to conclusions based on calculations about vectors and the equations of lines and planes.
2. A visual point of view in which we come to intuitive conclusions "by eye" and verify them by nonalgebraic proofs based on the axioms of Section 1.3 in Chapter 1.

Since the algebraic techniques rest on the axioms, neither approach has greater validity than the other. But from a practical point of view, it is helpful to be able to have both methods at one's disposal. In this double-barreled way we concentrate particularly on the invariants of parallel projection.

Basic Concepts and Algebra of Parallel Projection

DEFINITION

1. Let the image plane be denoted I and have equation $\mathbf{x} \cdot \mathbf{n} = k$. Let $\mathbf{d}$ be any nonzero vector not parallel to I (i.e., $\mathbf{d} \cdot \mathbf{n} \neq 0$). For any given point P, let L_P denote the line passing through P with direction vector $\mathbf{d}$ and let P' be the intersection of I with L_P (Figure 7.25). *Parallel projection* is the transformation which associates P' with P. P' is called the *image* of P.

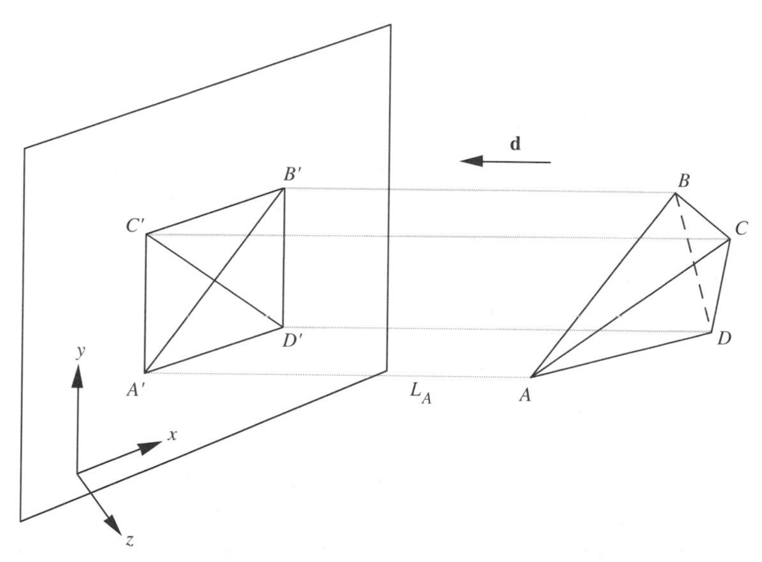

Figure 7.25 Canonical parallel projection of the edges of a four-sided solid *ABCD*.

2. A *canonical parallel projection* is one in which the image plane is $z = 0$.

The name of this projection method, of course, derives from the fact that since the projector lines all have the same direction **d**, they are parallel according to Theorem 7.15.

From the parametric equation for L_P, we see that there must be a t value for which $\mathbf{p}' = \mathbf{p} + t\mathbf{d}$. This is actually three equations, one each for the x, y, and z components. Suppose $\mathbf{p}$ has components $\langle x, y, z\rangle$ and suppose $\mathbf{p}'$ has components $\langle x', y', z'\rangle$. Then

$$\begin{aligned} x' &= x + t x_d, \\ y' &= y + t y_d, \\ z' &= z + t z_d. \end{aligned}$$

But, since P' lies in the plane $z = 0$, we know that $z' = 0$. Set $z' = 0$ in the last equation, and solve for t:

$$t = \frac{-z}{z_d}.$$

Substituting into the equations for x' and y':

CANONICAL PARALLEL PROJECTION EQUATIONS

$$\begin{aligned} x' &= x - \frac{x_d}{z_d} z, \\ y' &= y - \frac{y_d}{z_d} z, \\ z' &= 0. \end{aligned} \tag{7.13}$$

EXAMPLE 7.15: ORTHOGRAPHIC PROJECTION

Let us take $\mathbf{d} = \langle x_d, y_d, z_d\rangle = \langle 0, 0, 1\rangle$ and let I be the x-y plane. Our direction **d** is parallel to the z axis and therefore orthogonal to the image plane. A parallel projection of this sort, in which the projection direction is orthogonal to the image plane, is called an *orthographic projection*. Substituting into Eqs. (7.13) gives

$$\begin{aligned} x' &= x, \\ y' &= y, \\ z' &= 0. \end{aligned}$$

Figure 7.26a shows an orthographic projection of a cube whose front face is parallel to the image plane. ■

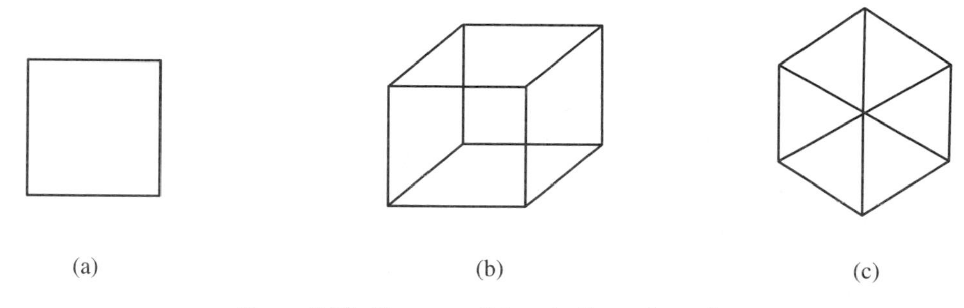

Figure 7.26 Three parallel projections of a cube.

EXAMPLE 7.16: MATRIX REPRESENTATION

We can express Eqs. (7.13) as a single vector–matrix equation

$$\begin{pmatrix} x' \\ y' \\ z' \end{pmatrix} = \begin{pmatrix} 1 & 0 & -\frac{x_d}{z_d} \\ 0 & 1 & -\frac{y_d}{z_d} \\ 0 & 0 & 0 \end{pmatrix} \begin{pmatrix} x \\ y \\ z \end{pmatrix}. \tag{7.14}$$

In abbreviated form $\mathbf{p}' = \mathbf{M}\mathbf{p}$. ■

In mathematics, matrix representations are quite important and are not normally left to an example as we have just done. We downplay this matrix representation because it lacks universality: it can be shown that it only works for parallel projections when the image plane passes through the origin.[4]

In one of the exercises we provide some hints that will allow you to prove the following theorem, which applies to noncanonical and canonical parallel projections.

✯ THEOREM 7.22

For any parallel projection using direction $\mathbf{d}$ onto plane $\mathbf{x} \cdot \mathbf{n} = k$, the image of an arbitrary point P is

$$\mathbf{p}' = \mathbf{p} + \frac{(k - \mathbf{p} \cdot \mathbf{n})}{\mathbf{d} \cdot \mathbf{n}} \mathbf{d}. \tag{7.15}$$

[4] By introducing "points at infinity" and homogeneous coordinates, matrices can be made more widely useful. See the first two chapters of Maxwell (1951) for the theory and Chapter 5 of Foley, van Dam, Feiner, and Hughes (1990), for the usefulness of this in computer graphics.

Proof

Omitted. ■

EXAMPLE 7.17: A NONCANONICAL PARALLEL PROJECTION

Suppose the image plane has equation $x-y+2z=4$; in vector form this is $\mathbf{x}\cdot\langle 1,-1,2\rangle=4$. Suppose the projector direction $\mathbf{d}$ is parallel to the x axis, for example, $\mathbf{d}=\langle 1,0,0\rangle$. In the notation of Theorem 7.22, $\mathbf{n}=\langle 1,\ -1,\ 2\rangle$ and $k=4$. Substituting into Eq. (7.15) gives

$$\mathbf{p}' = \mathbf{p} + [4-(x-y+2z)]\langle 1,0,0\rangle,$$
$$\langle x',y',z'\rangle = \langle x,y,z\rangle + \langle 4-x+y-2z,0,0\rangle.$$

Resolving this into x, y, and z components:

$$x' = x+(4-x+y-2z) = 4+y-2z,$$
$$y' = y,$$
$$z' = z.$$

The last of these two equations tells us something we could have foretold merely by examining $\mathbf{d}$. Because the y and z components of $\mathbf{d}$ are 0, as a point moves along a projector line from a point to its image, there is no change in y or z.

What if we repeat the previous calculation with $\mathbf{d}=\langle -3,0,0\rangle$? This direction vector describes the same parallel class of lines as $\mathbf{d}=\langle 1,0,0\rangle$ so the same equations should result. It works out exactly as we expect. We leave this as an exercise. ■

Just like perspective projection, parallel projection is not a one-to-one mapping because all points along a projector line map to the same image point. In our next example we show how this fact is revealed by algebra in the case of a canonical parallel projection.

EXAMPLE 7.18: PARALLEL PROJECTION IS MANY-TO-ONE

Suppose a canonical parallel projection has direction $\mathbf{d}$. Let A be some point. We now show that any X on the line through A with direction $\mathbf{d}$ projects to the same image. Our line has equation $\mathbf{x}=\mathbf{a}+t\mathbf{d}$. To find the image $\mathbf{x}'$, substitute this into Eqs. (7.13).

$$x' = (x_a+tx_d) - \frac{x_d}{z_d}(z_a+tz_d) = x_a - \frac{x_d z_a}{z_d},$$
$$y' = (y_a+ty_d) - \frac{y_d}{z_d}(z_a+tz_d) = y_a - \frac{y_d z_a}{z_d},$$
$$z' = 0.$$

The parameter t is missing from the simplified expressions for x', y', and z'. This shows that all points on the line $\mathbf{x} = \mathbf{a} + t\mathbf{d}$ project to the same point. ■

In contrast with perspective projection, where we had exceptional points, the domain of parallel projection consists of all of three-dimensional space. Every point has an image. This is because the projection direction is not parallel to the image plane, so all projector lines cross the image plane in a single point. There is nothing comparable to the exceptional plane.

Continuing our comparison with perspective projection, there is one feature shared by both projection methods: Collinearity is preserved. Here are some preliminary results that will lead to this conclusion.

✯ THEOREM 7.23

Let M be a plane containing a line L_1 and also containing a point Q that is not on L_1. Let L_2 be a line through Q and parallel to L_1. Then L_2 is contained in M.

PROOF

There is a plane N containing L_1 and L_2 (definition of parallel lines). N and M are the same since N and M both contain the line L_1 and point Q and since a line and a point off of the line are contained in exactly one plane (a simple consequence of Axioms 10 and 13 of Section 1.3 in Chapter 1). Because L_2 lies in N and $N = M$, L_2 lies in M. ■

COROLLARY 7.24

For any given line L and any given parallel projection, there is a plane that contains all projector lines through the various points of L.

PROOF

Let P be a point on L and let M be the plane determined by L and L_P. Let Q be any other point on L. Because L_P and L_Q have the same direction $\mathbf{d}$ they are parallel. The previous theorem implies L_Q lies in M. ■

The plane mentioned in the previous corollary is called the *projector plane* of the line L.

EXAMPLE 7.19: ALGEBRA YIELDS THE PROJECTOR PLANE

We can also carry out the proof of the previous corollary using an algebraic approach. Let P be a point on L and let M be the plane containing L and L_P. Suppose the equation of M is $\mathbf{x} \cdot \mathbf{n} = k$. Since P lies on M:

$$\mathbf{p} \cdot \mathbf{n} = k. \tag{7.16}$$

Since any point on L_P lies on M, then for any t at all

$$(\mathbf{p} + t\mathbf{d}) \cdot \mathbf{n} = k.$$

From this and Eq. (7.16) it follows that $t(\mathbf{d} \cdot \mathbf{n}) = 0$ for every t. This can only be true if

$$\mathbf{d} \cdot \mathbf{n} = 0. \tag{7.17}$$

Now let Q be any point on L different from P. Since Q lies on M:

$$\mathbf{q} \cdot \mathbf{n} = k. \tag{7.18}$$

We need to show that all points of L_Q lie in M. L_Q has equation $\mathbf{x} = \mathbf{q} + t\mathbf{d}$. We need to show that $(\mathbf{q} + t\mathbf{d}) \cdot \mathbf{n} = k$ for all t. But

$$\begin{aligned}(\mathbf{q} + t\mathbf{d}) \cdot \mathbf{n} &= \mathbf{q} \cdot \mathbf{n} + t(\mathbf{d} \cdot \mathbf{n}) \\ &= k\end{aligned}$$

by Eqs. (7.17) and (7.18). ■

Invariants of Parallel Projection

✯ THEOREM 7.25: INVARIANCE OF COLLINEARITY

Suppose P, Q, R are collinear. Their images under parallel projection, P', Q' and R', are collinear as well.

PROOF

Let L be a line containing P, Q, R and let M be the projector plane of L (Corollary 7.24). P', Q', and R' lie in M. By Axiom 11 the intersection of M and the image plane I is a line. P', Q', and R' lie in this line since they lie in both M and I. ■

It can happen that P', Q', and R' of the previous theorem are all the same point. We leave the explanation of this as an exercise.

EXAMPLE 7.20: PARALLEL PROJECTION OF A LINE

We now calculate the parametric equations of the line that results from projecting $\mathbf{x} = \langle 1, 4, -2\rangle + t\langle 0, 3, 1\rangle$ using the projection equations of Example 7.17. The individual equations for the line are

$$\begin{aligned}x &= 1, \\ y &= 4 + 3t, \\ z &= -2 + t.\end{aligned}$$

Substituting these into the projection equations gives:

$$x' = 4 + (4 + 3t) - 2(-2 + t) = 12 + t,$$
$$y' = 4 + 3t,$$
$$z' = -2 + t.$$

These equations can be written as $\mathbf{x}' = \langle 12, 4, -2\rangle + t\langle 1, 3, 1\rangle$, which we recognize to be the parametric equation of some line. ■

✯ THEOREM 7.26: INVARIANCE OF PARALLELISM

The images of parallel lines are parallel or coincident (Figure 7.27).

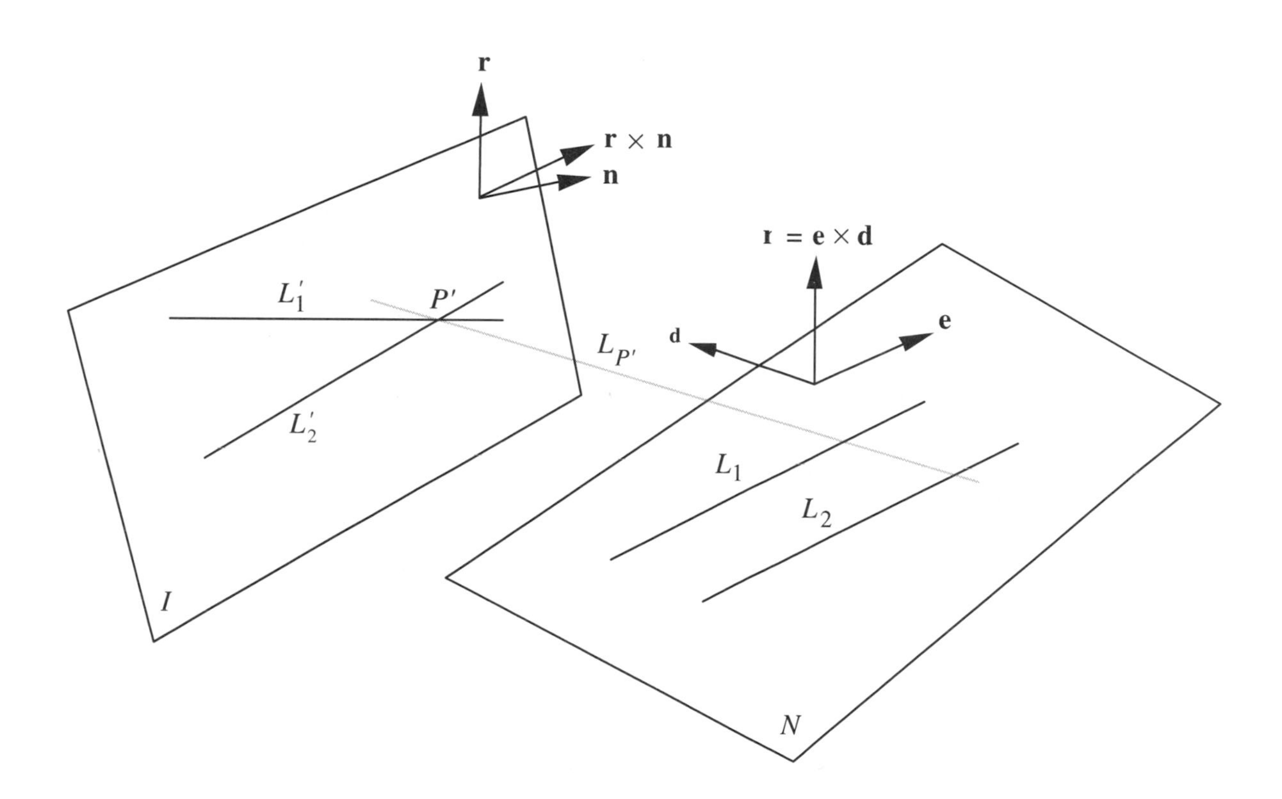

Figure 7.27 An impossibility: parallel lines whose images cross.

AXIOMATIC PROOF OF THEOREM 7.26

Let the parallel lines be called L_1 and L_2 and let the images be L_1' and L_2'. We will show that if L_1' and L_2' are not parallel and have a point P' in common then $L_1' = L_2'$. Since L_1 and L_2 both contain points that project to P', $L_{P'}$ crosses both L_1 and L_2. Let N be the plane determined by the parallel lines L_1 and L_2. Since $L_{P'}$ has two points in common with N (one on L_1 and one on L_2), N contains the line $L_{P'}$ (Axiom 13). Since all points on L_1 or L_2 have projector lines parallel to $L_{P'}$, by Corollary 7.24 these projector lines are contained in N. Therefore the images of all points on L_1 and L_2 lie in the intersection of N and I, which is a single line (Axiom 11). L_1' and L_2' must therefore be the same line. ■

Next we give an algebraic proof of the previous theorem. This has the advantage that it not only tells us that the image lines have a common direction, but it allows us to compute the direction. To carry this out, we need to recall that the cross-product of two vectors $\mathbf{r} \times \mathbf{n}$ is perpendicular to both $\mathbf{r}$ and $\mathbf{n}$. In addition, if planes with normals $\mathbf{r}$ and $\mathbf{n}$ intersect in a line, this line will be perpendicular to both $\mathbf{r}$ and $\mathbf{n}$ because a line lying in a plane is perpendicular to the plane's normal. Thus we can compute a direction vector for the line of intersection by computing $\mathbf{r} \times \mathbf{n}$.

ALGEBRAIC PROOF OF THEOREM 7.26

As usual, let $\mathbf{n}$ denote a normal to the image plane, and let L_1 and L_2 have direction $\mathbf{e}$. Consider the projector plane formed by L_1 and the projector lines through points of L_1. Its normal vector is perpendicular to $\mathbf{e}$ and $\mathbf{d}$ and so we can take it to be the cross-product $\mathbf{r} = \mathbf{e} \times \mathbf{d}$. L_1' is the intersection of this projector plane and the image plane so it is orthogonal to $\mathbf{r}$ and $\mathbf{n}$. We can take the direction vector of L_1' to be $\mathbf{r} \times \mathbf{n}$. Notice that our calculation only needed the direction of L_1 and would come out the same for L_2 since it also has direction $\mathbf{e}$. That is, we get $\mathbf{r} \times \mathbf{n}$ as a direction vector for the image of L_2 as well. Because L_1' and L_2' share a direction vector, they are parallel. ■

EXAMPLE 7.21: DIRECTION OF THE IMAGE LINE

We illustrate the ideas in the algebraic proof of Theorem 7.26. If the direction vector of a parallel projection onto the plane $\mathbf{x} \cdot \langle 3, -5, 0 \rangle = 3$ is $\mathbf{d} = \langle 1, 2, -3 \rangle$, find the direction of the image of the line $\mathbf{x} = \langle 1, -7, 9 \rangle + t\langle -1, 3, 2 \rangle$. (We have solved more or less the same kind of problem in Example 7.20, but now we show a different method.)

SOLUTION

A normal to the projector plane can be computed by $\mathbf{r}=\mathbf{e}\times\mathbf{d}=\langle -1, 3, 2\rangle \times \langle 1, 2, -3\rangle$.

$$\begin{aligned}
\mathbf{r} &= \begin{vmatrix} \mathbf{i} & \mathbf{j} & \mathbf{k} \\ -1 & 3 & 2 \\ 1 & 2 & -3 \end{vmatrix} \\
&= -13\mathbf{i} - \mathbf{j} - 5\mathbf{k} \\
&= \langle -13, -1, -5\rangle; \\
\mathbf{r}\times\mathbf{n} &= \begin{vmatrix} \mathbf{i} & \mathbf{j} & \mathbf{k} \\ -13 & -1 & -5 \\ 3 & -5 & 0 \end{vmatrix} \\
&= -25\mathbf{i} - 15\mathbf{j} + 68\mathbf{k} \\
&= \langle -25, -15, 68\rangle.
\end{aligned}$$

■

If we are looking at lines in a drawing of some object or if a computer vision system is analyzing this drawing, what can be deduced if the two lines are parallel? Theorem 7.26 does not tell us that the original lines in the object must have been parallel. Could we perhaps prove this in a separate theorem? Figure 7.25 shows that the answer is no: The skew lines $\overleftrightarrow{AC}$ and $\overleftrightarrow{BD}$ have parallel images. However, the following theorem tells us that if the original lines are not skew then we can conclude that they were parallel.

☆ THEOREM 7.27

If L_1 and L_2 project to parallel lines L_1' and L_2' by parallel projection and L_1 and L_2 are not skew lines (there is a plane containing them) then L_1 and L_2 are parallel.

PROOF

If L_1 and L_2 are not parallel we will deduce a contradiction. Since L_1 and L_2 lie in a plane and are not parallel, they have a point in common, say, P. P' would lie in both L_1' and L_2'. This contradicts the fact that L_1' and L_2' are parallel and proves the theorem. ■

One reason parallel projections are often used is that it is possible to show one face of an object with complete accuracy. Here is what this means in mathematical terms.

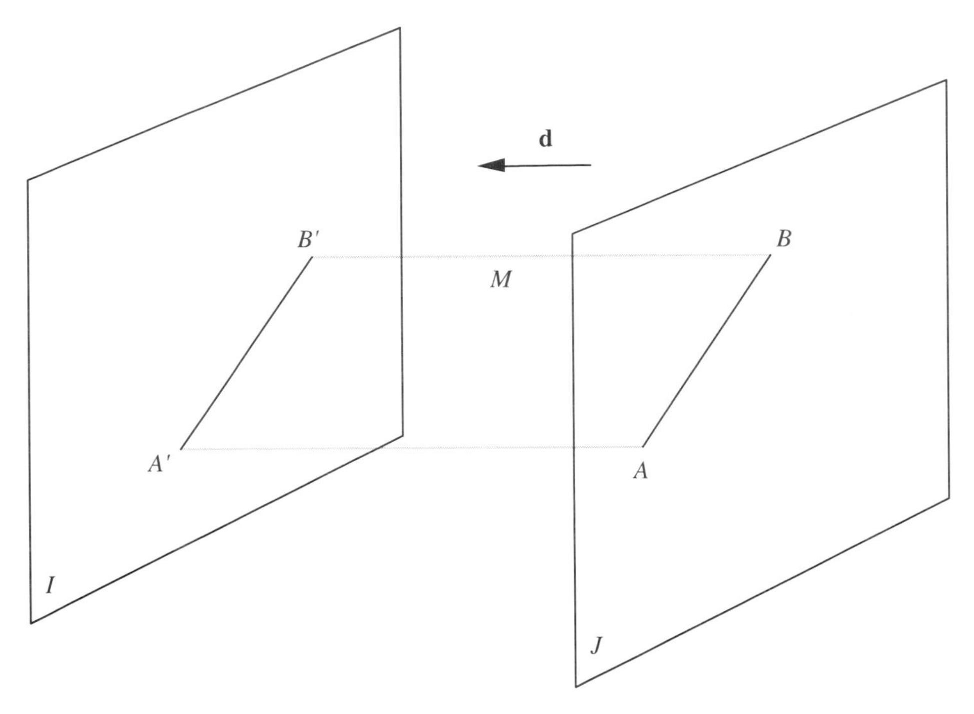

Figure 7.28 Parallel projection preserves distances in a plane parallel to the image plane.

☆ THEOREM 7.28: PARTIAL DISTANCE PRESERVATION

Suppose A and B are any two points in a plane J that is parallel to the image plane I (Figure 7.28). Then the images A' and B' under parallel projection are such that $AB = A'B'$.

PROOF

We will show that ABB′A′ is a parallelogram. Since opposite sides of a parallelogram are equal, this will prove the theorem.

Projector lines L_A and L_B are parallel so there is a plane M containing them. $\overleftrightarrow{A'B'}$ lies in this plane (Axiom 13) and is parallel to $\overleftrightarrow{AB}$ since I and J are parallel. Thus $ABB'A'$ is a parallelogram since opposite sides are parallel. ■

Theorem 7.28 can be rephrased to assert that a parallel projection, when restricted to J, is an isometry. It can be shown (Chapter 4) that isometries preserve angles as well as distance. As a result, any feature of the object that lies in J will be depicted exactly in I.

EXAMPLE 7.22: CERTAIN DISTANCES ARE PRESERVED BY PARALLEL PROJECTION

We check Theorem 7.28 for the particular parallel projection of Example 7.17. For J we take the plane

$$x - y + 2z = -1, \tag{7.19}$$

which has the same normal vector as the image plane and is therefore parallel to it.

Let $A(x_A, y_A, z_A)$ and $B(x_B, y_B, z_B)$ be two points on the plane J. Their distance is

$$AB = \sqrt{(x_A - x_B)^2 + (y_A - y_B)^2 + (z_A - z_B)^2}.$$

Their images under the parallel projection of Example 7.17 are $A'(4 + y_A - 2z_A, y_A, z_A)$ and $B'(4 + y_B - 2z_B, y_B, z_B)$.

The coordinates of A and B satisfy the plane Eq. (7.19) for J. From those equations we can deduce

$$4 + y_A - 2z_A = x_A + 5,$$
$$4 + y_B - 2z_B = x_B + 5.$$

Using these equations we can rewrite the coordinates of A' and B': $A'(x_A + 5, y_A, z_A)$ and $B'(x_B + 5, y_B, z_B)$. Their distance is

$$A'B' = \sqrt{[(x_A + 5) - (x_B + 5)]^2 + [y_A - y_B]^2 + [z_A - z_B]^2},$$

which is clearly equal to our earlier formula for AB. ■

Often when we are creating a picture of an object we find that the main features of interest are contained in a single face — for example, the keyboard is the interesting part of a hand calculator. We naturally want to show the interesting face with as much accuracy as possible. Inspired by the previous theorem, we might use a parallel projection onto an image plane parallel to the interesting face. By Theorem 7.28 and the properties of isometries, this should produce an exact-sized copy of the interesting face and its features in the image plane. But if the object is very large or very small, an exact copy may be too small or too big for the screen or page. For this reason, parallel projection is usually followed by some central similarity to scale the image up or down to fit the space provided.

We conclude with a table comparing perspective projection and parallel projection with respect to their invariants (Table 7.1). As we see, the invariants of parallel projection are more numerous. However, two things prevent us from declaring it to be a better method in all cases. The first is that duplicating what the eye sees is very valuable even though distances, angles, and areas are distorted and parallelism is not preserved. The human visual system is clever enough to figure out what it is looking at despite these apparent shortcomings. Second, it can be proved that in many of the most common viewing situations what perspective projection shows is very close to what a parallel projection (followed by scaling up or down) would show. (Exercise 27 at the end of this section elaborates on this.)

TABLE 7.1 COMPARING PROJECTION METHODS

	Perspective projection	Parallel projection
Preserves collinearity	Yes	Yes
Preserves parallelism	No	Yes
Preserves distance	No	In plane parallel to image plane
Preserves angles	No	In plane parallel to image plane
Preserves areas	No	In plane parallel to image plane
Duplicates what the eye sees	Yes	No

EXERCISES

**Marks challenging exercises.*

Basic Concepts and Algebra of Parallel Projection

1. Which of the drawings of Figure 7.24 (previous section) could be a parallel projection followed by scaling up or down of a right-angled box? For the ones which could not be, can you say why not?

2. Find the image of the point $(4, 1, -3)$ under a canonical parallel projection with $\mathbf{d} = \langle 2, 0, 3 \rangle$.

3. Do Exercise 2, but with a different projection, that of Example 7.17.

4. Do the calculation suggested at the end of Example 7.17.

5. Show that if we replace $\mathbf{d}$ by a scalar multiple, $k\mathbf{d}$, then the same canonical parallel projection equations, Eqs. (7.13), result. Is this intuitively reasonable?

6. Work out the projection equation if the image plane is the y-z plane and the projection direction is perpendicular to it. Can you find the matrix form of the projection equations?

7. Suppose the image plane is $x + 2y + z = k$ where k is an unspecified constant. Find the parallel projection equations in terms of k and the direction $\mathbf{d} = \langle 1, -2, -1 \rangle$.

8. Under what circumstances can the equations you obtained in Exercise 7 be writtten in terms of a matrix–vector multiplication as in Example 7.16?

9. Explain why two skew lines L_1 and L_2 cannot project into the same line in the image plane.

*10. Prove that if two lines lie in a plane that contains the projection direction (this means that if we take an instance of **d** with tail in the plane, the head of the vector lies in the plane as well), then both lines project into the same line.

11. In Corollary 7.24, what if the line L were parallel to the projection direction **d**? Is the corollary still true? Is a stronger statement possible?

12. Derive the formula in Theorem 7.22. (*Hint*: First you need to find the t value of the intersection of the projector line and the image plane. This is similar to what was done in case 1 of Theorem 7.19 of Section 7.4.)

*13. Let $\mathbf{d}'$ represent a direction vector written as a column vector $\mathbf{d}' = \begin{pmatrix} x_d \\ y_d \\ z_d \end{pmatrix}$. Let $\mathbf{n} = \langle x_n, y_n, z_n \rangle$ be the normal to the image plane, written as a row vector. Let $\mathbf{M}$ be the 3×3 matrix $\mathbf{d}'\mathbf{n}$.

(a) Write $\mathbf{M}$ showing its nine entries.

(b) If P is any point and $\mathbf{p}$ its position vector, show that $(\mathbf{p} \cdot \mathbf{n})\mathbf{d}' = \mathbf{d}'\mathbf{n}\mathbf{p}$. $\left[\text{For the right-hand side, } \mathbf{p} \text{ is to be written as a column vector } \begin{pmatrix} x \\ y \\ z \end{pmatrix}.\right]$

(c) Show that if $k = 0$, then Eq. (7.15) can be written

$$\mathbf{p}' = \left(\mathbf{I} - \frac{1}{\mathbf{d} \cdot \mathbf{n}}\mathbf{M}\right)\mathbf{p}$$

where $\mathbf{I}$ is the 3×3 identity matrix and where $\mathbf{p}$ and $\mathbf{p}'$ are written as column vectors.

*14. In Examples 7.15 and 7.17, we wound up with equations which stated that some coordinates did not change when we go from a point $\mathbf{p}$ to its image (e.g., $y' = y$). What can you say about the circumstances when this happens?

Invariants of Parallel Projection

15. Walter is writing a geometry book and creates the following exercise. What's wrong with it?

Suppose we have a cube whose faces lie in the planes $x = -1$, $x = 1$, $y = -1$, $y = 1$, $z = 5$, $z = 7$. We intend to subject it to a canonical parallel projection with projection direction $\mathbf{d} = \langle x_d, y_d, z_d \rangle$. Explain how to choose $\mathbf{d}$ so that the face in the plane $z = 5$ projects to something very small. Then explain how to make it vanish completely (project to a line segment). You can answer numerically or in terms of the relation of $\mathbf{d}$ to the cube's features.

16. There is a circumstance under which A', B', and C' of Theorem 7.25 are all the same point. Can you think of the circumstance?

17. Find the parametric equations of the image of the line $\mathbf{x} = \langle 1, 3, -4 \rangle + t\langle 2, -1, 0 \rangle$ under the projection equations of Example 7.17.

18. Do Exercise 17 but with a different projection, that of Example 7.15.

19. Do Exercises 17 and 18 by this method: Pick two convenient points on the given line; project them using the projection equations; find the parametric equation of the line connecting the image points.

20. Let L be a line through A with direction vector $\mathbf{e}$, that is, one with equation $\mathbf{x} = \mathbf{a} + t\mathbf{e}$. Show that the image of this line, under the parallel projection in direction $\mathbf{d}$ onto the plane $\mathbf{x} \cdot \mathbf{n} = k$, is

$$\mathbf{x}' = \mathbf{a} + \left(\frac{k - \mathbf{a} \cdot \mathbf{n}}{\mathbf{d} \cdot \mathbf{n}}\right)\mathbf{d} + t\left(\mathbf{e} - \frac{\mathbf{e} \cdot \mathbf{n}}{\mathbf{d} \cdot \mathbf{n}}\mathbf{d}\right).$$

21. Using the projection of Example 7.15, find the direction of the image of a line whose direction vector is $\langle -2, 3, 4 \rangle$. Use the result of Exercise 20. Then use the method of Example 7.21 and compare your answers.

*22. Consider an orthogonal projection onto the plane $x + 2y + 3z = 0$ and how it portrays the unit cube whose face planes are $x = 0$, $x = 1$, $y = 0$, $y = 1$, $z = 0$, $z = 1$. Find the directions, in the image plane, of the images of the edges meeting at vertex $(1, 1, 1)$. (*Hint*: Examine Exercises 20 and 21.)

23. Is Corollary 7.24 still true if we replace the words "parallel projection" by "perspective projection"? Explain your answer.

*24. Provide an alternate proof of Theorem 7.26 for canonical parallel projection, using the following outline. Let A_1 and B_1 be points on line L_1 and let A_2 and B_2 be points on line L_2. Since they are parallel, L_1 and L_2 have the same direction so that there is a scalar r where $\mathbf{b}_2 - \mathbf{a}_2 = r(\mathbf{b}_1 - \mathbf{a}_1)$. Using the matrix representation of Example 7.16, express the images, A_1', B_1', A_2', and B_2', of the four points under a canonical parallel projection with direction **d**. Next express the direction vectors for L_1' and L_2'. Show that one direction vector is r times the other.

*25. Prove that if three points in three-dimensional space are not collinear and the plane they determine does not contain the direction of a parallel projection, then their images are not collinear.

*26. Carry out an algebraic verification of the partial distance preservation of parallel projection (as in Example 7.22) in the following circumstance. The image plane is $\mathbf{x} \cdot \langle 1, -1, -2 \rangle = 4$. The projection direction is $\mathbf{d} = \langle 1, 4, -2 \rangle$. Show that if A and B lie in plane J whose equation is $\mathbf{x} \cdot \langle 1, -1, -2 \rangle = 0$, then $A'B' = AB$.

*27. This exercise illustrates the last sentence of this section. The idea is to show that if an object is skinny in the z direction and not too close to the center of a perspective projection, it makes little difference for any of the object's points P whether one uses the perspective projection or instead uses parallel projection in the z-axis direction followed by a suitable central similarity to scale up or down. We examine just the x coordinates of an image point, since the calculation for y is similar. So here is the problem: Let $P(x, y, z)$ be such that $10 < z < 12$. Suppose P is subjected to a canonical perspective projection with $c = 1$, resulting in $(x', y', 0)$. Next suppose P is subjected to a parallel projection that is both canonical and orthogonal, followed by the central similarity in which each coordinate is divided by -10, the result being $(x'', y'', 0)$.

(a) Find formulas for x' and x'' in terms of x, y, and z.

(b) The relative error in approximating x' by x'' is defined to be $(x' - x'')/x'$. Show that $-0.1 <$ relative error < 0.1.

(c) Redo the problem with $w < z < w + 2$ (in place of $10 < z < 12$, and the central similarity being division by $-w$. What happens to the relative error as w gets large?

Chapter 8

Graphs, Maps, and Polyhedra

Prerequisites: High school mathematics

Graphs (in the sense of networks) are all about the idea of connections among a discrete set of things. Connection is a very old idea — as soon as there were roads, there must have been maps of sorts and a road map is a kind of graph. But the study of graphs is surprisingly modern, having been developed mostly in the twentieth century. In our presentation, we concentrate on the parts of graph theory where the graphs have an intimate connection to some geometric structure such as the plane or some polyhedron.

In this chapter, we show graph theory applied to subjects in pure mathematics, such as map coloring and polyhedra. In addition, we display a variety of applications, including chemistry, scheduling, computer vision, and computer chip design.

Some of the first section of this chapter will be familiar to students who have taken a discrete mathematics course.

8.1 Introduction to Graph Theory

Vertices, Edges, and Valence

In Euclid's geometry, the shapes of objects are quite important. If you bend the edge of a triangle into a curve, then you can't use the methods you have learned to compute its area, check its congruence, etc. But we are surrounded by geometrical objects where the exact shape is not crucial. For example, Figure 8.1b is an inexact free-hand drawing but you can tell from it that you can't drive from Wyoming to New Mexico without going through some other state. In Figure 8.1a, we have drawn a crude map of some roads. Since the drawing is not to scale, the lengths of the various roads can't be judged from the drawing. Numbers on the roads give that information.

What the previous examples have in common is that they can be modeled by a mathematical structure called a graph.[1] A graph consists of vertices and edges. In Figure 8.1a, the vertices are the points $x_1, \ldots, x_6$ and the edges are the connecting roads. In Figure 8.1b, the edges are the boundaries between states and the vertices are the heavy dots.

DEFINITIONS

A *graph* consists of a finite number of elements, called *vertices*, together with another finite set of elements called *edges*. Each edge is associated with a pair of vertices, called the *endpoints* of the edge. The edge is said to connect the endpoints. In some cases, the endpoints of an edge may be the same vertex.

We may wish to label each edge of a graph with a number that measures an important aspect of the edge (as in Figure 8.1a). In some applications this number might be a length, while in others it might be the strength of a chemical bond or something else entirely. Because of the varied meanings the numbers can have in different applications, we give such numbers the overall designation of *weights*, and we call the graph a *weighted graph*.

A graph need not have a geometric character. For example, a graph might have as its vertices a set of four people on a committee: Jill, Bill, Lil, and Gil. If two of these people are already friends, we can indicate that by forming an edge associated with those two.

[1] Some authors call this a *network* to distinguish it from the graph of a function, which is quite different from the meaning of "graph" intended here.

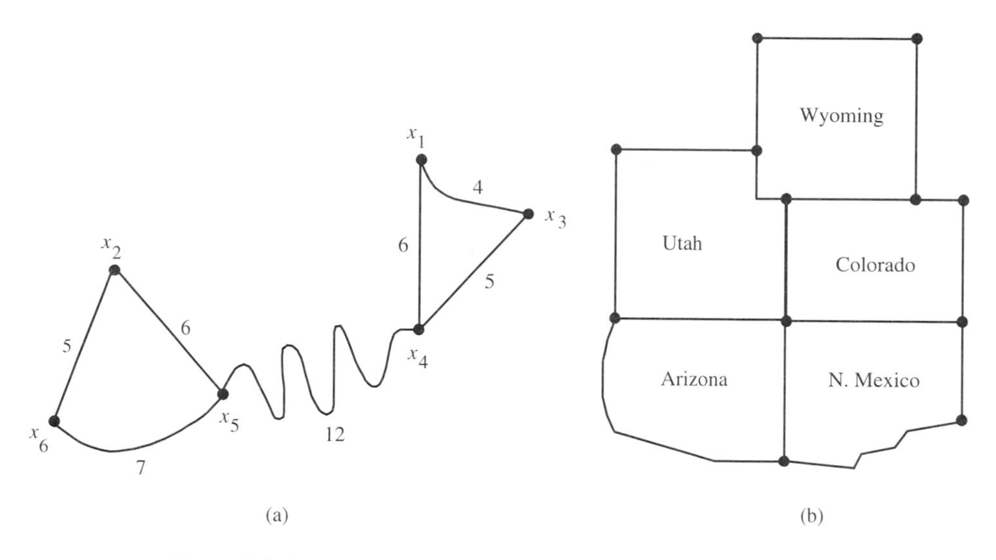

Figure 8.1 (a) A road network. (b) A map of some western states.

These edges can be conveyed by a list. For example the edge list {(Bill, Lil), (Gil, Jill)} shows that there are only two friendships, one between Bill and Lil and the other betweeen Gil and Jill.

Despite the fact that many graphs are not inherently geometric, it is extremely common to make geometric pictures of them by representing each vertex by a point and each edge by an arc. Our interest in this chapter is mostly in graphs where the vertices and edges *do* have geometric meaning, but at times we will deal with nongeometric graphs as well (graph theory has a way of spilling across mathematical categories).

Figure 8.2 shows some more examples of graphs. An edge of a graph may start and end at the same vertex (Figure 8.2c). Such an edge is called a *loop*. Figure 8.2a shows that there may be more than one edge connecting the same two vertices—such a set of edges is called a *set of multiple edges*. In Figure 8.2b we see that it is possible for a graph to fail to "hang together." Such a graph is called *disconnected* into separate *components*.

DEFINITION

The *valence* of a vertex of a graph is the number of edges that touch that vertex. However, if there is a loop at a vertex, this edge is counted twice in determining the valence.

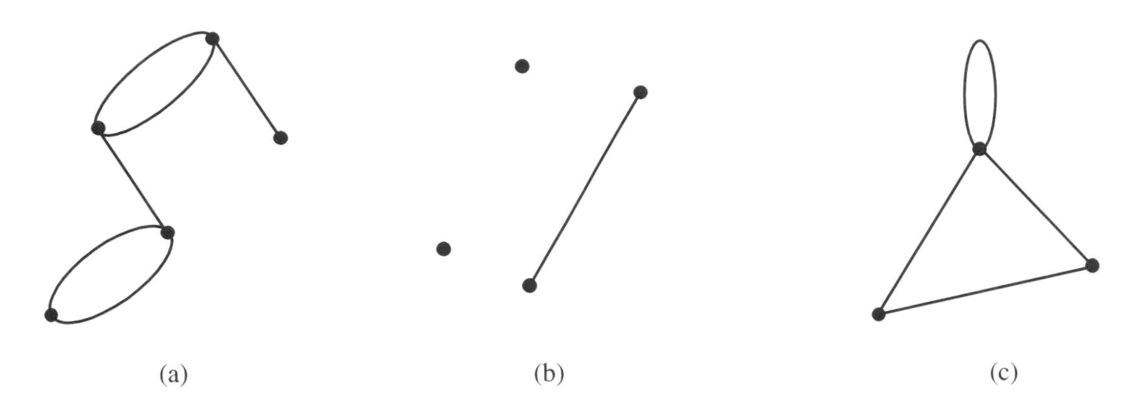

Figure 8.2 (a) A graph with two sets of multiple edges. (b) A graph with three components. (c) A graph with a loop.

In Figure 8.1a, x_2 has valence 2 while x_5 has valence 3. The isolated vertices of Figure 8.2b have valence 0. The graph shown in Figure 8.2c has one vertex of valence 4 (the vertex with the loop) and two of valence 2.

Let v_i stand for the number of vertices whose valence is i and let m be the largest valence in the graph. For example, in Figure 8.3a we have $v_1 = 4$ and $v_4 = 2$, all other

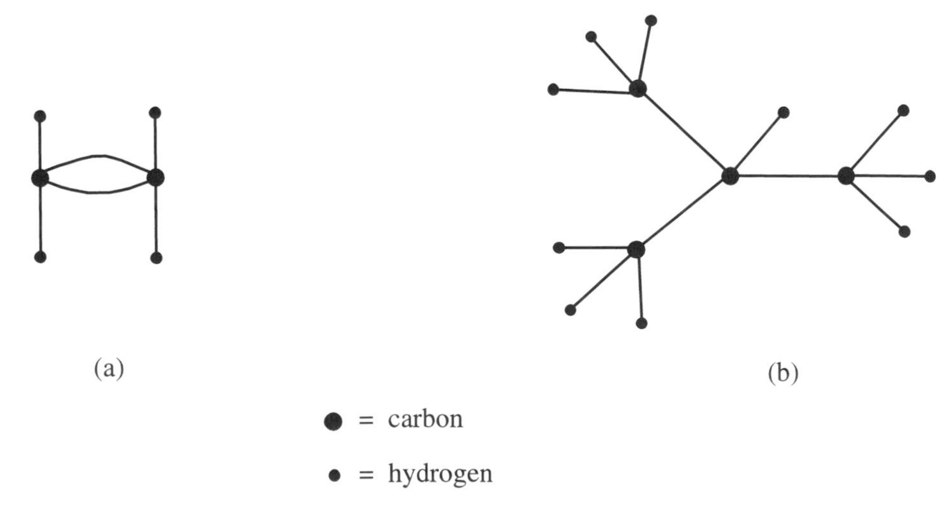

Figure 8.3 Two graphs representing hydrocarbons.

$v_i = 0$, while in Figure 8.3b we have $v_1 = 10$, $v_4 = 4$, all other $v_i = 0$. Furthermore, $m = 4$ in both graphs. An immediate consequence of this notation is

$$v_0 + v_1 + v_2 + v_3 + \cdots + v_m = \text{number of vertices of the graph.} \qquad (8.1)$$

✯ THEOREM 8.1

In any graph,

$$v_1 + 2v_2 + 3v_3 + \cdots + mv_m = 2(\text{number of edges}). \qquad (8.2)$$

PROOF

Imagine visiting every vertex of the graph. At each vertex, write down the valence, but also put a check mark on each edge touching the vertex, right near that vertex. (If the vertex had a loop, it would get two check marks, one at each end of the loop.) The number 1 will be written v_1 times, 2 will be written v_2 times, and so on. If you add the numbers you have written down, you'll get the left side of Eq. (8.2). This will also be the number of check marks. But each edge will get two check marks, one when you visit one end of the edge and a second when you visit the vertex at the other end. Thus, the sum of the numbers written down is twice the number of edges. ■

One of the earliest uses of graphs was to model chemical molecules, the vertices representing atoms such as carbon, hydrogen, and oxygen, and the edges representing chemical bonds between them. Chemists had been using the term valence to describe how many chemical bonds an atom could have and the term carried over into graph theory. The previous theorem sets some limits on the kinds of chemical molecules that can exist, as we see in the next example.

EXAMPLE 8.1

A hydrocarbon is a molecule with just hydrogen and carbon atoms. Each carbon atom has valence 4. Double or triple bonds—multiple edge sets—are possible. Each hydrogen atom has valence 1. Can a hydrocarbon exist that has an odd number of hydrogen atoms?

SOLUTION

Using the notation of Theorem 8.1, v_1 is the number of hydrogen atoms and v_4 is the number of carbon atoms. Equation (8.2) reduces to

$$v_1 + 4v_4 = 2(\text{number of edges}).$$

The right side is even and so is $4v_4$. Therefore, v_1 must be even. ■

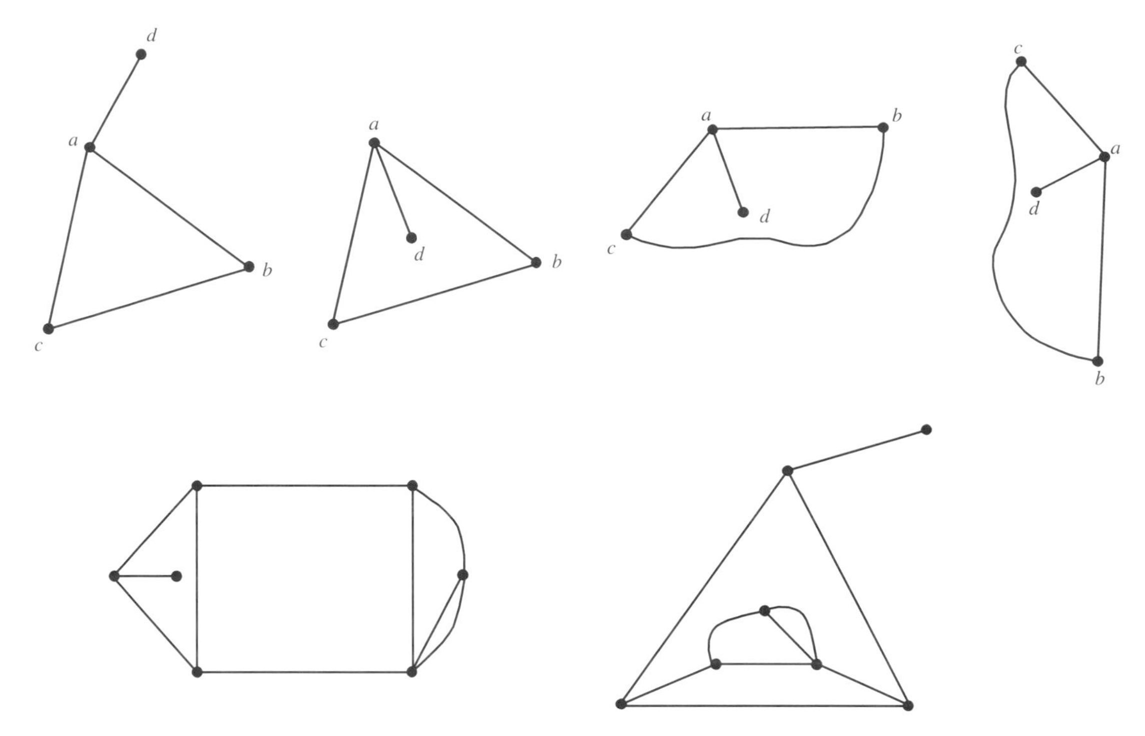

Figure 8.4 Isomorphism.

Isomorphism

In the theory of graphs, we make the assumption that the positions of the vertices and shapes of the edges—whether they are straight or crooked—doesn't matter. (Of course, for some applications, shape is important. A road engineer cares how sharply roads curve. Graph theory needs to be supplemented with other ideas for such applications.) For example, the first and last graphs in the top row of Figure 8.4 are considered to be essentially the same. One proof of this is that we can slowly deform one into the other, as shown by the middle two drawings. In doing such a deformation, we are allowed to shift or rotate the drawing, and we are allowed to change positions of vertices and shapes of edges. But we are not allowed to add or delete vertices. Likewise, we are not allowed to add or delete edges or cut an edge in pieces. The technical term we use for the deformability of one graph into another is that the graphs are *isomorphic*. It is not always easy to tell when two graphs are isomorphic. Are the two graphs at the bottom of Figure 8.4 isomorphic? Can you deform one of them, moving vertices around and reshaping edges, so that it looks just like the other?

When you have deformed one graph G into another, G', showing that they are isomorphic, each vertex x of G winds up moved to a vertex x' of G'. Thus the isomorphism involves a matching up of vertices. Furthermore, if x and y are two vertices of G, then the corresponding vertices x' and y' in G' are connected by a certain number of edges if and only if x and y are connected by that same number of edges. This gives rise to an alternate, less geometric, definition of isomorphism:

DEFINITION

An isomorphism between graphs G and G' is a function f from the vertices of G to those of G' such that

1. f is one to one and onto, and
2. the number of edges connecting x and y in G equals the number connecting $f(x)$ and $f(y)$ in G'.

When looking for an isomorphism, you may proceed geometrically, by drawing a series of deformations of one graph turning it into the other, or you may look for a function of the sort just described.

Drawings of Graphs

The applications we have in mind for graphs mostly arise in geometric contexts. For this reason, graphs usually come to us pictorially and we deal with them that way. But there are questions about exactly how to draw the picture.

Figure 8.5 shows three drawings of a wiring diagram, indicating that each of the four electronic components a, b, c, and d must be connected to each of the other three. In Figure 8.5a, one edge goes under another. Such a drawing invites us to try to visualize the graph in three dimensions, which is not usually one of the purposes of graph theory.

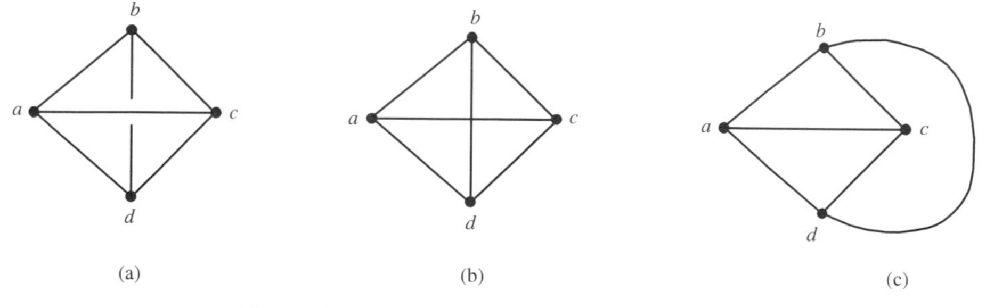

Figure 8.5 Three drawings of a wiring diagram.

In the present case we are just trying to display which component is connected to which, not how the edges are situated in space. Consequently, we rarely draw a graph with one edge shown going under another. Instead, we might draw the graph as in Figure 8.5b. Here two edges cross at a place that is not a vertex — notice that there is no heavy dot there. This is called a *crossing* and cannot be used as if it were a vertex. For example, you can't go from d to c by going up to the crossing and turning right. We could avoid a crossing by redrawing one edge so that it loops around, as in Figure 8.5c. Often the choice between Figures 8.5b and 8.5c depends on nothing more than personal preference — whether we are more annoyed by crossings or edges which are not straight. However, sometimes a problem calls for avoiding or minimizing crossings as in the next example.

DEFINITION

If a graph has an isomorphic version drawn in the plane with no crossings, we call it a *planar* graph.

EXAMPLE 8.2: PLANARITY (TWO PUZZLES)

Three houses each need to be hooked up to the Water Company, the Gas Company, and the Electric Company. The connections must lie in the plane, so no crossings are allowed. Figure 8.6 shows such a hookup except that the connection from the Electric

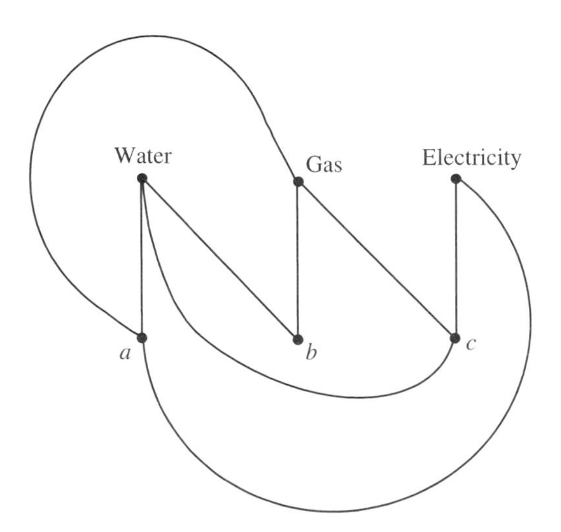

Figure 8.6 Can you connect b to the Electric Company without a crossing?

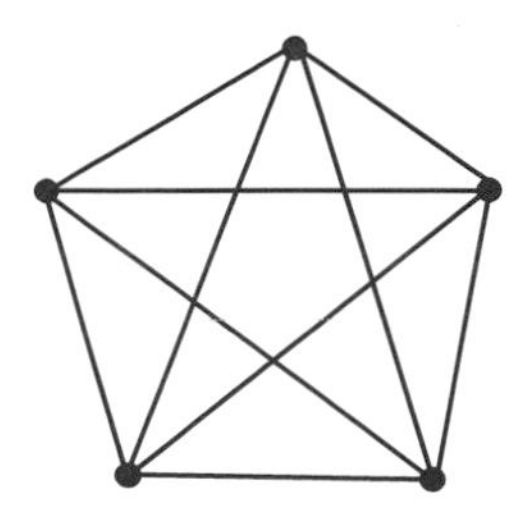

Figure 8.7 K_5, the complete graph on five vertices.

Company to house *b* is missing. If we keep the other edges as drawn, it is not possible to put in this last edge without a crossing. What if we rearranged the locations of the vertices and the shapes of the other edges? In other words, can we draw an isomorphic version of the graph in Figure 8.6 that has no crossings and for which the missing edge can be added without a crossing?

Figure 8.7 shows a problem with a similar flavor. In this graph, we have five vertices, every vertex being connected to all of the other four. We have shown all the connections straight and this causes five crossings. You should be able to find an isomorphic version with fewer crossings. Can you create one where there are no crossings? ■

The water–gas–electricity problem is really more of a puzzle than an applied problem. But a variant of this puzzle is practical in the field of computer chip design. Imagine that the six vertices of the problem are electrical components on the surface of a computer chip. Each of the three components on the top row needs to be hooked up to each of the three on the bottom. The connections will be tiny electrical "roadways" etched on the surface of the chip (think about scratching edges in a table top). These connections scratched into the surface need not be straight. However, it is important that they don't cross because electricity that flowed along a scratch that crossed another scratch would have a chance to take an undesired turn at a crossing. The requirement that connections lie in a plane and not cross is quite restrictive because the graph that needs to be etched into the chip is often not planar. For example, the pattern of connections in the water–gas–electricity puzzle could not be put on a "one-layer" chip of the type we have just described.

A common solution to this dilemma is to have an additional surface in which connections can be etched. Think of a slice of bread in which the components are sitting on top. Think of a connection as a skinny trail of mustard. It might lie completely on the top surface, or it might tunnel vertically through the slice to the bottom surface and run along the bottom surface. A connection might also switch from one surface to the other over and over again, but we'll rule this out to make things simpler. If we had two surfaces, we could solve the water–gas–electricity problem by putting the edges shown in Figure 8.6 on the top surface, and using the bottom surface for the one remaining edge from the Electric Company to house *b*.

The graph in Figure 8.7 is called a *complete graph* on five vertices. A complete graph is one in which every vertex is connected to every other vertex with an edge. If the number of vertices in a complete graph is n, then we denote this graph by K_n. Figure 8.7 shows K_5.

The graph we would like to draw in the water–gas–electricity problem is an example of a *complete bipartite* graph. Being bipartite means that its vertices fall into two sets in such a way that there are no edges between vertices in the same set (in our example, none among the top vertices and none among the bottom vertices.) A bipartite graph is called complete if every vertex in one set is connected to every one in the other. We denote a complete bipartite graph by $K_{m,n}$ where m is the number of vertices in the first set and n is the number in the second. The water–gas–electricity graph is $K_{3,3}$ and can be obtained from Figure 8.6 by adding an edge from the Electric Company to vertex *b*.

Chip manufacturers are usually dealing with many more components than the six shown in Figure 8.6, so it is natural to expand our water–gas–electricity problem. Suppose we need to put $K_{4,4}$ on a chip. Could we still manage with two levels? What if we had $K_{5,5}$? In experimenting with these problems, use dotted lines for one level and solid for the other. Just make sure that no dotted edges cross each other and no solid ones cross each other. Solid lines can cross dotted lines as often as needed.

Suppose you wanted each edge of a graph to consist of straight pieces whose directions were restricted to being vertical or horizontal. An edge could consist of many such vertical and horizontal pieces meeting at right angles (see Figure 8.8). Such a graph is called *orthogonal*. Can you redraw the graph in Figure 8.6 (without the missing edge from *b* to the Electric Company) in this fashion without crossings? If you want to include

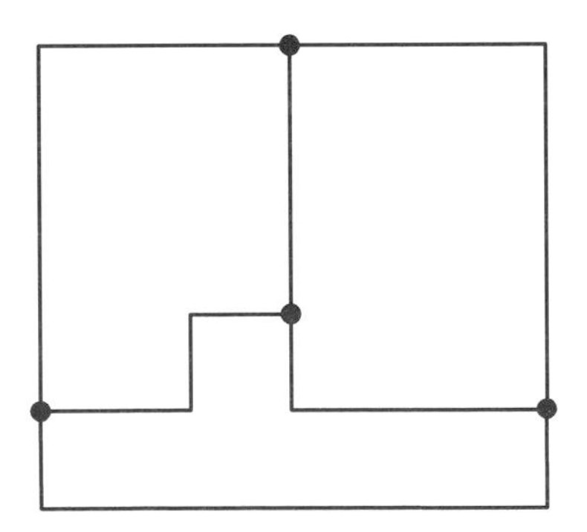

Figure 8.8 K_4 in orthogonal form.

the missing edge, can you do the connections in this rectangular fashion using two levels?

Matrices for a Graph

It is possible to represent the essential information about a graph nonpictorially, using one or another type of matrix. This becomes vital if you wish to analyze or manipulate the graph with computer software. In the following, assume that the vertices of the graph are called $x_1, x_2, \ldots, x_n$. The number of vertices is, therefore, n.

DEFINITION

1. The *adjacency matrix* of a graph without multiple edge sets is created in this way: Prepare a matrix with as many rows as the graph has vertices (n), and the same number of columns. In row i and column j place a 1 if you can get from x_i to x_j by traversing one edge. If there is no edge from x_i to x_j, enter a zero for the ij entry of the matrix.
2. The *weight matrix* (also called the *weighted adjacency matrix*) of a graph without multiple edge sets and without loops is created in this way: Prepare a matrix with as many rows as the graph has vertices (n), and the same number of columns. If there is an edge connecting x_i to x_j, enter the weight of this edge as the entry in the ith row and jth column of the matrix. If there is no edge connecting x_i to x_j, enter ∞ as the ij entry. Finally, put 0's on the main diagonal.

These definitions of matrices representing a graph require that the vertices of the graph be labeled $x_1, x_2, \ldots, x_n$. Often graphs come to us without vertex labels (see Figure 8.1b), or the labels do not have numerical subscripts (e.g., a, b, m, x). In such a case, choose any ordering for the vertices. This means that different matrices could be devised for a graph, depending on the ordering you choose.

EXAMPLE 8.3: MATRICES FOR GRAPHS

The adjacency matrix of the road network graph in Figure 8.1a is

$$\mathbf{A} = \begin{pmatrix} 0 & 0 & 1 & 1 & 0 & 0 \\ 0 & 0 & 0 & 0 & 1 & 1 \\ 1 & 0 & 0 & 1 & 0 & 0 \\ 1 & 0 & 1 & 0 & 1 & 0 \\ 0 & 1 & 0 & 1 & 0 & 1 \\ 0 & 1 & 0 & 0 & 1 & 0 \end{pmatrix}. \tag{8.3}$$

The weight matrix of this same road network graph is

$$\mathbf{D} = \begin{pmatrix} 0 & \infty & 4 & 6 & \infty & \infty \\ \infty & 0 & \infty & \infty & 6 & 5 \\ 4 & \infty & 0 & 5 & \infty & \infty \\ 6 & \infty & 5 & 0 & 12 & \infty \\ \infty & 6 & \infty & 12 & 0 & 7 \\ \infty & 5 & \infty & \infty & 7 & 0 \end{pmatrix}. \tag{8.4}$$

■

Notice that both matrices in Example 8.3 are *symmetric*. This means that the *ij* entry is the same as the *ji* entry for each *i* and *j*. For example, the 1,3 entry and the 3,1 entry of **D** are both 4. Both of these entries refer to the edge of length 4 connecting x_1 with x_3.

A matrix is a convenient form to use to describe a graph to a computer. Consequently, it is practical to know how graph theory questions can be answered without looking at the graph, just by manipulations carried out by the computer on a matrix representing the graph. For example, if you want to determine the valence of x_1, see how many entries there are in the first row of the adjacency matrix which are not 0. Can you figure out how to find the number of edges of the graph?

Paths, Circuits, and Connectivity

One of the things we are most concerned about in graphs, in both theory and applications, is the question of whether some pair of vertices can be linked by a series of edges called a path.

If you took a walk on the edges of a graph, starting with vertex *s* and ending at *g*, you would encounter the start vertex (*s*), an edge, a vertex, another edge, etc., till you got to *g*. This motivates the following definition.

DEFINITION

A *path* from a start vertex *s* of a graph (or weighted graph) to a goal vertex *g* is an alternating series of vertices and edges where each edge contains the vertices before and after it in the sequence.

EXAMPLE 8.4

In Figure 8.9, *iFhEi* is a path. Note that just listing the vertices, *ihi*, would be ambiguous because it doesn't tell us which edge was used first. But in many cases (if there are no multiple edge sets) there is only one way to interpret a sequence of vertices, so we leave out the edges when listing a path. For example, *acdf* is a path in Figure 8.9. A path may repeat vertices, as in *abcae*. It may repeat edges as in *abcab*. ■

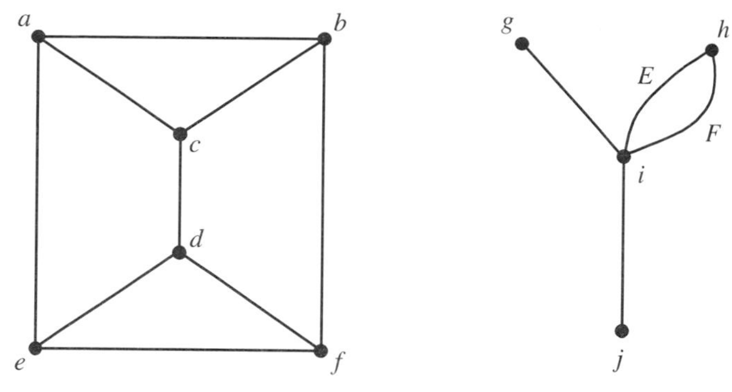

Figure 8.9 A disconnected graph with 10 vertices.

DEFINITION

A graph is *connected* if, for any pair of vertices you choose, it is possible to find a path that starts at one vertex and ends at the other.

There is no path from a to g in Figure 8.9. Therefore, the graph is not connected. The graph has two *components*. To understand what a component is, imagine starting at some vertex and determining all vertices and edges that can be reached with paths starting at your chosen vertex. Those vertices and edges (including the start vertex) make up a component. In Figure 8.9, the component we get if we start with vertex a is simply the left-hand part of the picture. We would get the same component if we started at b, c, d, e, or f. This component does not contain g, h, i, or j. The right-hand component can be gotten by starting with g, h, i, or j. Two different components never have any vertices in common.

EXAMPLE 8.5: CONNECTIVITY AND SHORTEST PATHS

Nowadays, some cars have built-in navigation systems that can help you get around in an unfamiliar place. These systems know exactly where you are at any moment, will show you this location on an electronic map on the dashboard, and will calculate the best route to any destination you specify, showing you the route on the dashboard map and giving you voice instructions when you have reached an intersection where a turn is required.

In the road network shown in Figure 8.1a, can we get from vertex x_5 to vertex x_1? We can easily see by eye that the answer is yes. But how would the car's computerized navigation system get the answer if all it had were the adjacency matrix [Eq. (8.3)] or the weight matrix [Eq. (8.4)]?

Here is one example of how we might reason about the adjacency matrix of Figure 8.1a to find a series of edges connecting x_5 to x_1. First we put two copies of the matrices side by side as if we were going to multiply them (see following matrices). Run the index finger of your left hand along row 5 of the left-hand matrix and, moving in step with the left finger, run the index finger of your right hand down the first column. (See the boldfacing in the following matrices.) For example, when your left finger is pointing to the third entry of its row, the right finger should be pointing to the third entry of its column. You are looking for an entry where both fingers point to a 1. This happens at the fourth entry. The "1" your left-hand finger points to is telling you there is an edge from x_5 to x_4. The "1" your right-hand finger points to is telling you there is an edge from x_4 to x_1. Putting these two pieces of information together tells us that there is a two-step path from x_5 to x_1, using x_4 as an intermediate "stepping stone."

$$\begin{pmatrix} 0 & 0 & 1 & 1 & 0 & 0 \\ 0 & 0 & 0 & 0 & 1 & 1 \\ 1 & 0 & 0 & 1 & 0 & 0 \\ 1 & 0 & 1 & 0 & 1 & 0 \\ \mathbf{0} & \mathbf{1} & \mathbf{0} & \mathbf{1} & \mathbf{0} & \mathbf{1} \\ 0 & 1 & 0 & 0 & 1 & 0 \end{pmatrix} \begin{pmatrix} \mathbf{0} & 0 & 1 & 1 & 0 & 0 \\ \mathbf{0} & 0 & 0 & 0 & 1 & 1 \\ \mathbf{1} & 0 & 0 & 1 & 0 & 0 \\ \mathbf{1} & 0 & 1 & 0 & 1 & 0 \\ \mathbf{0} & 1 & 0 & 1 & 0 & 1 \\ \mathbf{0} & 1 & 0 & 0 & 1 & 0 \end{pmatrix}$$

What we have done here seems like a specialized trick insofar as it only tells us about two-step paths. One of our exercises extends the idea a bit. ■

DEFINITION

A *circuit* in a graph G is a path that comes back to its starting vertex. We normally describe it by listing the vertices in the order in which they occur on the circuit (e.g., $x_1, x_2, x_3, \ldots, x_n, x_1$). The circuit is called *simple* if the vertices $x_1, x_2, \ldots, x_n$ are all different and if the edges traversed by the circuit are all different.

DEFINITION

If x_i and x_j lie on a simple circuit and are joined by an edge that is not part of the circuit, then that edge is called a *diagonal* of the circuit.

EXAMPLE 8.6

In Figure 8.9, *abfea* is a simple circuit. It has no diagonals. Edge *cb* is a diagonal of the simple circuit *abfdca*; *dcbacd* is a circuit but it is not simple. ■

Coloring Maps and Graphs

The publisher of the map in Figure 8.1b wants to color the states so that two different states that share a border get different colors (so you can tell them apart). In mathematical language, we call the states *faces* and we call the task *face coloring* the map, or sometimes simply *coloring* the map. One way to do this would be to give each state its own special color, used only for that one state. This would require five colors. If we wanted to reduce the number of colors used, we could color Wyoming the same as New Mexico because they do not share an edge. Alternatively, we could color Utah and New Mexico the same because they have only a vertex in common. Using both strategies would lead to a face coloring with three colors. Can you see why you can't get away with two colors? What if we had a different set of states, or the entire United States, or some completely new arrangement of imaginary states or countries? Is there some number of colors that will always suffice, no matter what map we might be asked to face color?

In face coloring maps, it is conventional to also color the region surrounding the states — what we might call the border around the map. Because we like to think about mathematical maps as lying in the infinite plane, instead of a rectangular page of a book, this border extends to infinity in all directions and we call it the *infinite face*. If we want to color the infinite face of Figure 8.1b as well as the states, what is the minimum number of colors needed? Henceforth, unless we say otherwise, we will include the infinite face as one to be colored.

Geographical maps give rise to the idea of mathematical maps, which we define in terms of graphs as follows.

DEFINITION

If G is a planar graph that has been drawn in the plane with no crossings, the structure of vertices, edges, and faces (sometimes called *regions*) makes up a *planar map*. The infinite face is considered to be a face of the map. Just as with graphs, we do not consider a map to be changed in any essential way if we reposition the vertices and change the shapes of the edges. For example, Figure 8.2a shows a planar map with three faces, six edges, and five vertices. Figure 8.2b shows a planar map with one face (the infinite face), one edge, and four vertices.

Just as there are planar maps, there are also maps that reside on a donut-shaped surface (toroidal maps) or on other mathematical surfaces. Because we are only going to deal with planar maps we sometimes refer to them simply as maps. The study of face coloring maps has been an exciting challenge in modern mathematics and one of the most intriguing elementary results relates face coloring to a special type of circuit.

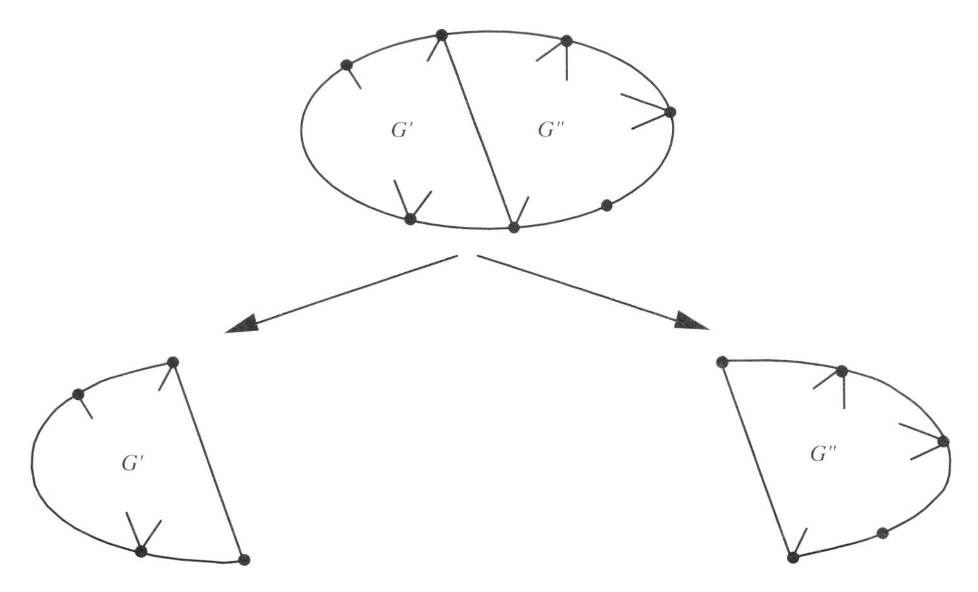

Figure 8.10 Cutting apart a Hamilton circuit without exterior diagonals.

DEFINITION

A circuit is called a *Hamilton circuit* if it includes each vertex of the graph and does not revisit any vertex or edge until it returns to its start (it is simple).

A graph that has a Hamilton circuit is clearly connected — to get from vertex a to any other vertex b, just follow the circuit around, starting at a, and you'll eventually get to b. Consequently, Figure 8.9 has no Hamilton circuit. However, the left-hand component has *abcdfea* as a Hamilton circuit. The right-hand component is connected, but still has no Hamilton circuit. Can you see why?

If a graph has a Hamilton circuit, any edge that is not part of the circuit must be a diagonal of the circuit (since both of its vertices are on the circuit). If there are no diagonals in the exterior of the Hamilton circuit, we call the map a *Hamilton circuit without exterior diagonals* (see Figure 8.10). If there are no diagonals in the interior of the Hamilton circuit, we call the map a *Hamilton circuit without interior diagonals*. The following series of theorems shows that these concepts are closely related to map coloring.

✯ THEOREM 8.2

If G is a Hamilton circuit without exterior diagonals (Figure 8.10), then the faces inside the Hamilton circuit can be face colored with no more than two colors.

Proof

We will give a proof by contradiction. If there are some Hamilton circuits without exterior diagonals where the inside cannot be face colored with one or two colors, then let $G_{\min}$ be such a counterexample which has the minimum number of diagonals inside the circuit. This means that any Hamilton circuit without exterior diagonals that has fewer interior diagonals than $G_{\min}$ can be face colored with one or two colors.

If $G_{\min}$ had no interior diagonals, then it would have just one face inside, and we could color it with one color, contrary to the fact that $G_{\min}$ needs more than two colors for its interior. So $G_{\min}$ has at least one diagonal. Cut $G_{\min}$ along any of its diagonals, into two maps, G' and G'', as follows (see Figure 8.10). G' consists of everything inside or on the Hamilton circuit on one side of the diagonal and G'' is everything inside or on the circuit on the other side. The outer boundaries of G' and G'' are Hamilton circuits for those maps, and neither map has exterior diagonals. Furthermore, G' and G'' both have fewer interior diagonals than $G_{\min}$, so both can be face colored with no more than two colors. Next we intend to paste G' and G'' together again and try to get a coloring of $G_{\min}$ out of the two individual colorings. If the faces on either side of the cutting diagonal have different colors, this works fine. If these faces touching the diagonal have the same color, we can revise the coloring of G'' by interchanging colors 1 and 2 on all the faces of G'', and then put G' and G'' together as before. Either way, we contradict the assumption that $G_{\min}$ cannot have its inside faces colored with one or two colors. ■

The proof we have just given can be easily adapted to deal with Hamilton circuits without interior diagonals. Just replace the words "inside" and "interior" with "outside" and "exterior" in the proof! Thus:

✯ THEOREM 8.3

If G is a Hamilton circuit without interior diagonals, then the faces outside the Hamilton circuit can be face colored with no more than two colors. ■

✯ THEOREM 8.4

If G is a planar map with a Hamilton circuit, then the whole planar map can be face colored with four colors.

Proof

By Theorem 8.2 we can color the inside with colors 1 and 2, and by Theorem 8.3 we can color the outside with colors 3 and 4. ■

When mathematicians first began studying face colorings of planar maps, it was conjectured that four colors would suffice to color any planar map, not just one with a Hamilton circuit. Trying to prove this turned out to be a long saga, with many odd twists and turns. This "four-color conjecture" was originally suggested to the British mathematician Augustus De Morgan by a student, Frederick Guthrie, who got it from his brother Francis Guthrie. Professor De Morgan was stumped by the students' challenge and circulated the problem. The first written reference to it occurs in a letter from De Morgan to the Irish mathematician and physicist William Rowan Hamilton. In 1879, A. B. Kempe published a proof that four colors were enough to color a planar map. But in 1890, P. J. Heawood showed that Kempe's proof was wrong! It was, however, possible to salvage a "five-color theorem" from the wreckage of Kempe's "proof." (In Section 8.5 we present a new proof of this five-color theorem.) For more than half a century, until he reached the age of 88, Heawood attempted to provide a correct proof of the four-color conjecture, but in vain. Scores of other mathematicians joined the battle but with no success. Finally, in 1976 Haken and Appel were able to show that, with the possible exception of a list of many thousands of maps, the four-color conjecture is true. It did not seem practical to check the thousands of possible exceptional maps, many of them dauntingly large, to see whether or not they could be four-colored. To cope with this obstacle, Haken and Appel wrote a computer program to generate these maps and look for four-colorings of them. After a considerable amount of computer time, four-colorings were found for all the possible exceptions. The four-color conjecture became the four-color theorem.

Some mathematicians objected that we couldn't really trust that the computer didn't make an error since no human being was able to devote the time to checking what the computer did. Others said that even if the computer made no mistakes, it was still only a minor advance in mathematics since the point of doing mathematics is to enhance *human* understanding. In this point of view, just knowing that something is true is not as important as understanding why.

This idea of face coloring, which was proposed in the nineteenth century as a puzzle, gave rise in the twentieth century to a related concept of *vertex coloring*, which has useful applications.

DEFINITION

A vertex coloring of a graph is an assignment of colors (or labels) to the vertices of a graph in such a way that neighbors (different vertices connected by an edge) always get different colors.

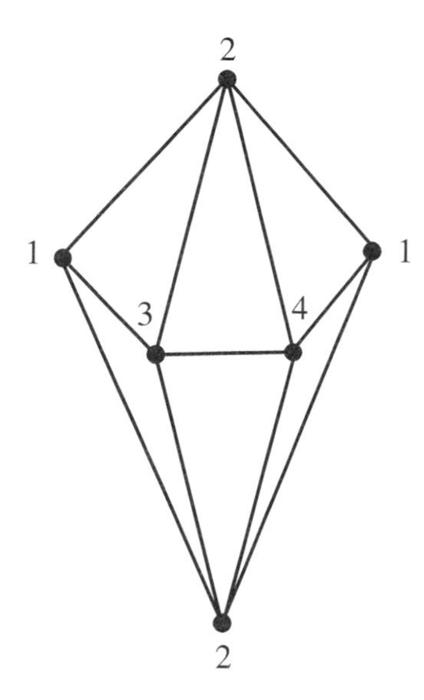

Figure 8.11 A vertex coloring with four colors.

EXAMPLE 8.7: VERTEX COLORING

Figure 8.11 shows a vertex coloring of a graph using four colors. In this case, there is also a vertex coloring with three colors. Can you find it? ■

EXAMPLE 8.8: APPLICATION OF VERTEX COLORING

A college is sponsoring a symposium featuring six speakers, *a, b, c, d, e,* each of whom is going to give a one-hour lecture. If the speakers are scheduled in six different time slots, the day's events will take too long. On the other hand, having them all speak at the same time in different rooms would force members of the audience to miss all but one of the speakers. An "in between" solution is sought, based on the fact that some pairs of speakers may conflict, because few in the audience would want to hear both, while some pairs of speakers should not conflict because many listeners would want to hear both. Table 8.1 is drawn up, showing which pairs of speakers could speak simultaneously and which should not. The X's indicate conflicts that are not allowed to occur. For example, the X in the column labeled *a* and the row labeled *e* indicates that speakers *a* and *e* should not speak in the same time slot.

Would three time slots be enough to schedule the speakers? What is the minimum number of time slots?

TABLE 8.1 SPEAKERS WHO SHOULD NOT BE IN THE SAME TIME SLOT

	a	b	c	d	e	f
a					X	
b						
c				X	X	X
d			X		X	X
e	X		X	X		X
f			X	X	X	

It is interesting and useful to notice that this problem can be made to fall into the category of vertex coloring of a graph. To do this, create a graph in which there is a vertex for each of the speakers (see Figure 8.12). We connect two vertices if the speakers they represent cannot conflict. In place of colors, we'll place numerical labels on the vertices. These labels will represent the time slots. The vertex coloring requirement that vertices connected by an edge must get different labels is an exact reflection of the requirement that speakers which cannot conflict must get different time slots. Figure 8.12 shows a vertex coloring with four colors. ■

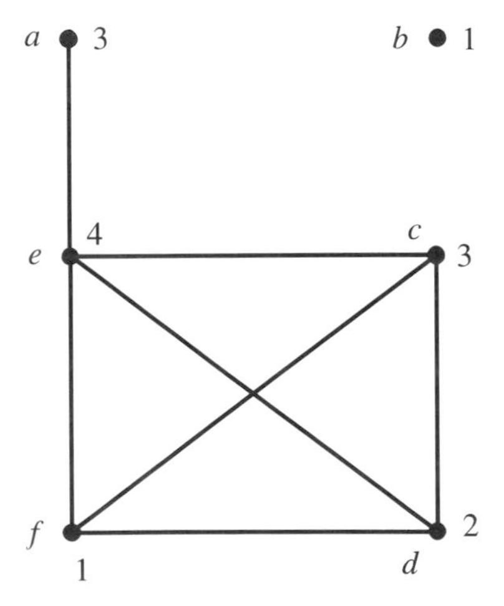

Figure 8.12 Can you color this symposium graph with fewer colors?

EXERCISES

**Marks challenging exercises.*

Valence and Isomorphism

1. (a) Draw a graph with $v_1 = 1, v_2 = 1, v_3 = 3$. (Keep in mind that loops and multiple edge sets are possible.)

   (b) Now draw another graph that is not isomorphic to the first.

   (c) Can you find an example with no loops or multiple edge sets?

2. (a) Draw a graph with $v_1 = 1$, $v_2 = 1$, $v_3 = 3$, $v_4 = 4$.

   (b) Now draw another graph that is not isomorphic to the first.

3. Draw a graph representing a hydrocarbon with three carbon atoms and six hydrogen atoms.

4. Draw a graph representing a hydrocarbon with two carbon atoms and four hydrogen atoms.

5. Find two different (nonisomorphic) graphs representing hydrocarbons with four carbon atoms and eight hydrogen atoms.

6. Show that the two graphs of Figure 8.13 are isomorphic, by deforming one into the other.

7. Show that the two graphs of Figure 8.14 are isomorphic, by deforming one into the other.

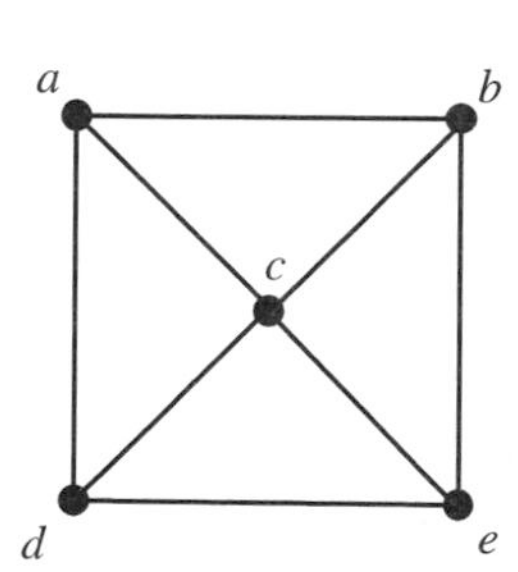

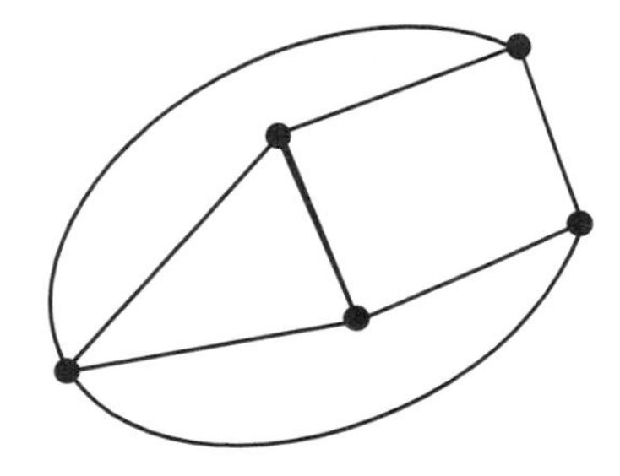

Figure 8.13

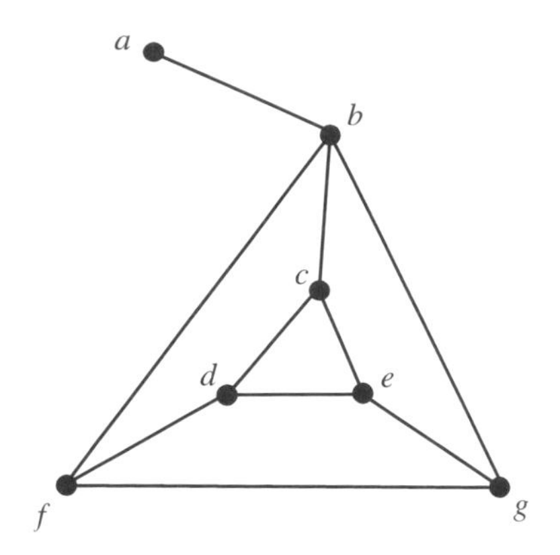

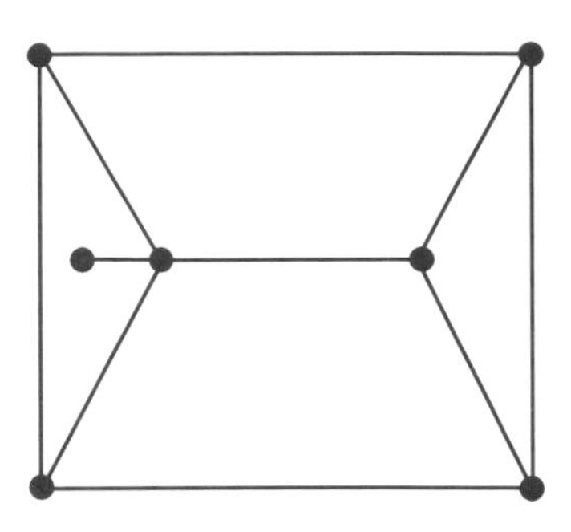

Figure 8.14

8. Are the graphs of Figure 8.15 isomorphic or not? If not, why not? (It's not enough to say you tried but weren't able to deform one into the other. Maybe you didn't try long enough. Some simpler and more persuasive evidence is wanted here.)

Drawings of Graphs

9. Can you draw a graph isomorphic to the one in Figure 8.5 in which there are no crossings, but all the edges are straight? Remember, you are allowed to move the vertices around.

10. Do Exercise 9 for Figure 8.6.

11. (a) Draw a version of K_5 with as few crossings as possible. Keep in mind that edges don't need to be straight.

    (b) What if edges need to be straight?

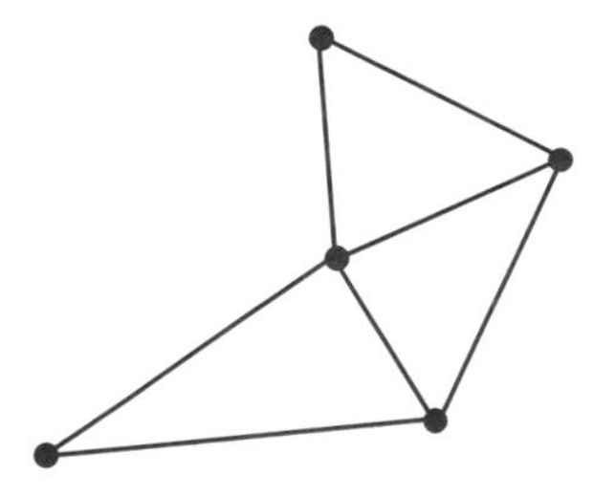

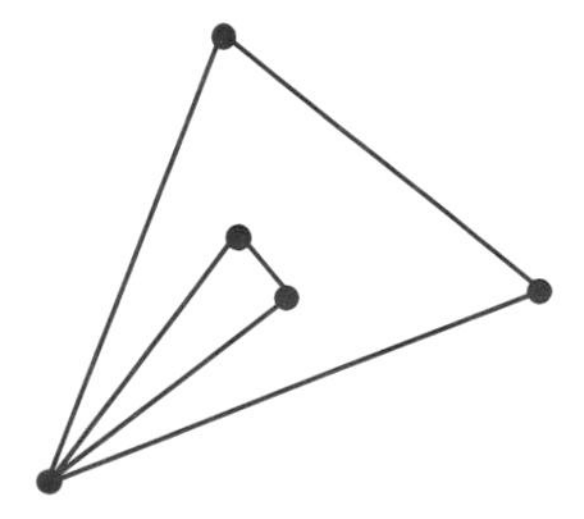

Figure 8.15

12. Suppose you are allowed to remove some edges from K_5 in the hopes of making it planar. What is the fewest number you could remove and have it be planar? Demonstrate your answer with a drawing.

13. Answer Exercise 12 if you want the graph not only to be planar, but orthogonal too.

14. Can you draw $K_{4,4}$ on two levels?

15. $K_{2,4}$ is planar. Show this with a drawing.

16. Show that $K_{4,8}$ fits on two levels. (*Hint*: Exercise 15 might help.)

Matrices for a Graph

17. Write an adjacency matrix for Figure 8.2b.

18. Write an adjacency matrix for Figure 8.5.

19. Write an adjacency matrix for $K_{3,3}$.

20. Find the adjacency matrix of the graph which has the following weight matrix. (You should be able to do this without drawing the graph.)

$$\begin{pmatrix} 0 & 5 & 3 & 5 & \infty \\ 5 & 0 & 3 & \infty & 5 \\ 3 & 3 & 0 & 3 & 3 \\ 5 & \infty & 3 & 0 & 5 \\ \infty & 5 & 3 & 5 & 0 \end{pmatrix}$$

21. Draw a graph with no multiple edge sets whose weight matrix is that given in Exercise 20. If multiple edges sets were allowed, could other graphs be drawn?

22. Find the adjacency matrix of the graph which has the following weight matrix. (You should be able to do this without drawing the graph.)

$$\begin{pmatrix} 0 & 4 & 2 & 4 \\ 4 & 0 & 2 & \infty \\ 2 & 2 & 0 & 2 \\ 4 & \infty & 2 & 0 \end{pmatrix}$$

23. Draw a graph whose weight matrix is that given in Exercise 22.

24. Suppose you want to determine whether or not there was a two-edge connection from x_5 to x_1 using the weight matrix **D** from Exercise 20. How would you do it? (*Hint*: You might get an idea from the discussion in Example 8.5.)

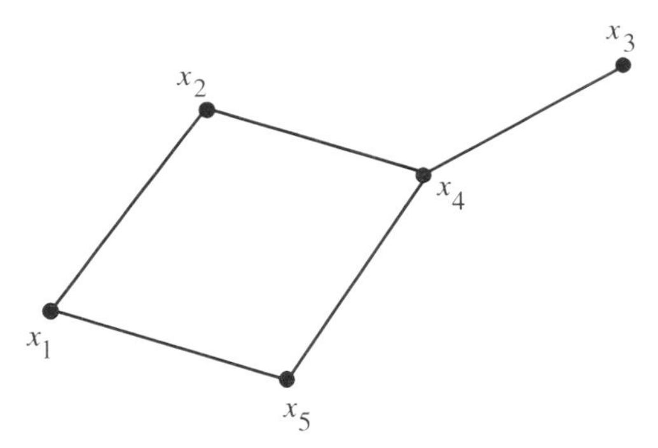

Figure 8.16

25. How would you find the number of edges in a graph, using the adjacency matrix of the graph (but without drawing the graph)?

*26. (a) Find the adjacency matrix **A** of the graph shown in Figure 8.16.

 (b) Compute $\mathbf{A}^2$.

 (c) Each entry of $\mathbf{A}^2$ tells you something about paths in the graph. Can you figure out what it tells you? (*Hint*: Study Example 8.5.)

 (d) Determine the significance of the entries of $\mathbf{A}^3$.

Paths, Circuits, and Connectivity

27. Find Hamilton circuits in the following graphs.

 (a) Figure 8.6

 (b) $K_{3,3}$

 (c) Figure 8.7

 (d) Figure 8.11

28. Which of the following has a Hamilton circuit?

 (a) Figure 8.13

 (b) Left part of Figure 8.14

 (c) Left part of Figure 8.15

 (d) Right part of Figure 8.15

29. (a) Explain why K_n always has a Hamilton circuit, regardless of the value of n.

    (b) Does $K_{m,n}$ ever have a Hamilton circuit? For what values of m and n does it have a Hamilton circuit?

30. (a) Find a graph that has more than one Hamilton circuit. (We don't consider two Hamilton circuits different if they have the same edges, but one circuit starts at a different vertex, or the circuits go around in opposite directions.)

    (b) Find a graph that has exactly one Hamilton circuit. Can you find one that has one or more diagonals?

31. (a) Explain why a graph that has even a single edge will have a circuit (but not necessarily a simple circuit).

    (b) Draw some connected graphs with no simple circuits and having at least one edge. Is there any generalization you can make about v_1 and how it is related to $v_2, v_3, \ldots, v_m$? A good guess with evidence is acceptable here.

32. Suppose a graph G is connected and is such that every edge lies on at least one simple circuit. Prove that if any individual edge is removed from G, you will still have a connected graph.

Coloring Maps and Graphs

33. Can you find a face coloring of the map shown in Figure 8.17 with three colors? What is the minimum number of colors needed for a vertex coloring?

34. (a) Suppose you want to face color a map, including the infinite face, and you notice that each vertex is three-valent. In addition, no face meets itself

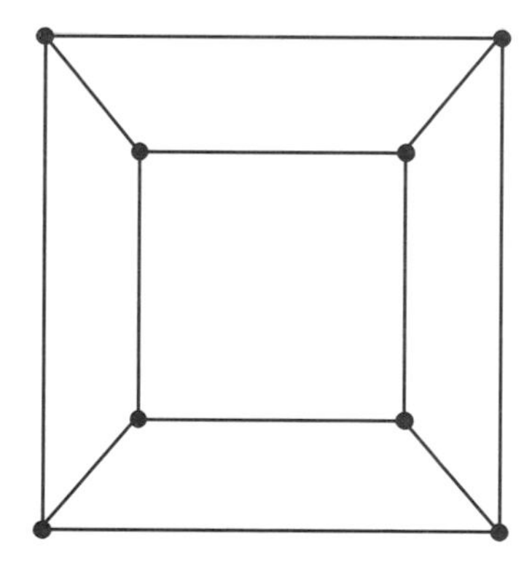

Figure 8.17

along an edge. You have begun an attempt at face coloring with three colors by placing colors 1 and 2 on two faces that meet at a particular vertex. Now add colors to other faces, each time coloring a face that touches two already colored faces. Do you ever have any choice about what color to put down? Show by example that if there is even one face with an odd number of neighboring faces, then you can't complete a face coloring with three colors by proceeding in this way.

(b) Suppose every face, including the infinite face, has an even number of neighbors. Do you think this guarantees that the face coloring with three can be completed? (In the next section we will establish some theorems that will allow us to settle this question definitely; see the exercises at the end of the next section.)

35. (a) Find a vertex coloring using just three colors for the graph in Figure 8.9.

(b) Find a vertex coloring using just three colors for the graph in Figure 8.11.

(c) Find a vertex coloring using just two colors for $K_{2,3}$.

36. For the following graphs, find vertex colorings. In each case, use the minimum number of possible colors.

(a) Figure 8.1a

(b) Figure 8.5

(c) Figure 8.6

(d) Figure 8.7

37. A mathematics department has the courses shown below to schedule in time slots. Some courses have others as prerequisites. If A is a prerequisite for B then they may go in the same time slot, since no one would ever want to take both. But if there is no prerequisite relation between courses, they have to go in different time slots. Draw a graph that shows each course as a vertex and where courses that cannot conflict are joined by an edge. Find a vertex coloring of this graph with the minimum number of colors (time slot labels).

Course	Prerequisites
Calculus	None
Linear Algebra	Calculus
Geometry	Linear Algebra, Calculus
Modern Algebra	Calculus
Differential Equations	Calculus, Linear Algebra

Miscellaneous Exercises

38. Can you determine how many edges $K_{4,4}$ has without drawing the graph?

39. Can you determine how many edges $K_{m,n}$ has as a function of m and n?

40. Can you determine how many edges K_6 has without drawing the graph?

41. Can you determine how many edges K_n has as a function of n?

8.2 Euler's Formula and the Euler Number

Euler's Formula

In this section we discuss Euler's formula, one of the most remarkable formulas in mathematics. It is easy to understand, simple to prove, and very important in elementary and advanced branches of mathematics. Adding a bit of intrigue is the fact that it could easily have been discovered by the early Greek geometers, but no proof of this formula appeared until 1758.[2]

Euler's formula concerns planar maps, which we defined earlier as being created by planar graphs, graphs which can be drawn in the plane without crossings. In such a map, the vertices and edges divide the plane into regions called *faces*. To understand the nature of a face intuitively, imagine cutting along the edges of the graph with scissors. The "chunks" of paper that remain are faces.

For example, in Figure 8.18a, when we remove the edges and vertices we are left with two faces: a square portion of the plane and an infinite portion outside the square. Notice that a face is a connected piece of the plane, which means that you can get from any point inside a face to any other in that face by drawing some type of curve that does not cross any edge or vertex. In Figure 8.18b there is only one face, the infinite face, and it has no edges touching it.

As you can see from Figure 8.18, our notion of a planar map includes many things (like parts b and c) you would never see in a "real map" in an atlas. However, Figure 8.18e is an example that might be found in an atlas but which does not qualify as a mathematical map under our definition: an island nation such as Ireland or Sri Lanka, surrounded by water. The drawing of such an island consists of a closed curve with no

[2] The formula may have been known to Descartes a century or so earlier, as it is a direct corollary of another theorem he proved.

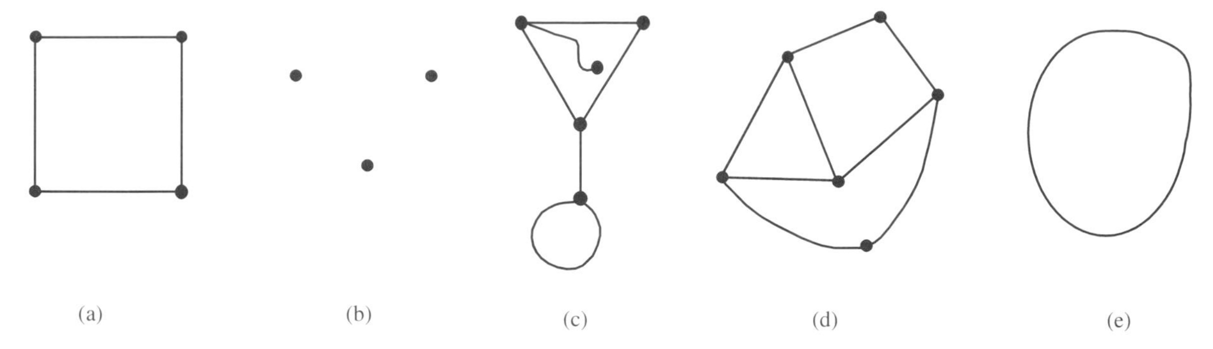

Figure 8.18 Some planar maps (a, b, c, d) and part (e), which isn't a map.

vertices. This closed curve does not qualify as a graph. (Each edge in a graph has to have at least one vertex.) Thus, a mathematical map is not quite the same thing as a geographical map.

In Figure 8.18a, both faces share a polygonal boundary, but polygonal boundaries are not essential to the concept of a face: One or more edges could be curved. Figure 8.18b has just one face, the infinite face, and it is the whole plane minus three points. It has no edges touching it at all. In Figure 8.18c, we have a map with three faces. One of them has a curved edge poking into it — that face touches four edges altogether.

We will refer to a planar map as *connected* if the graph that gives rise to it is connected. Figure 8.18b is not a connected map (it has three components), whereas the maps in Figures 8.18a, c, and d are connected. For both the proof and some applications of Euler's formula, we need to understand a special category of edge that we might find in a planar map, a disconnecting edge. If a map has more components after a certain edge has been removed, then we call that edge a *disconnecting edge*. Figure 8.18c shows a map with a single component which can be disconnected by the removal of the vertical edge. There is even a second way to disconnect this map by removing one edge: If we remove the edge poking into the triangle, we leave an isolated vertex inside the triangle disconnected from the rest of the graph. You should investigate whether K_5 or $K_{3,3}$ can be disconnected by removing one edge, because this will come up later.

As you can see from Figure 8.18c, a disconnecting edge in a planar graph has the same face on both sides. We say this face *meets itself along an edge*. This example illustrates the following theorem, whose proof we omit.

✫ THEOREM 8.5

(a) If a planar map has a disconnecting edge, then there is a face that meets itself along that edge.

(b) Conversely, if there is a face that meets itself along an edge, that edge disconnects the component of the graph containing that edge.

(c) In a planar map that has no disconnecting edge, each edge lies on two faces. ■

✯ THEOREM 8.6: EULER'S FORMULA FOR PLANAR MAPS

If G is a map in the plane with v vertices, e edges, f faces (including the infinite face) and c components, then

$$v - e + f - c = 1. \tag{8.5}$$

Proof

First note that the theorem is clearly true for maps that have no edges. For in such a map every component consists of just one vertex, so $v = c$. In addition, $f = 1$, $e = 0$, so, by simple substitution, Eq. (8.5) holds.

If the map does have edges, we perform a reduction on the map that results in a new map, with fewer edges but where $v - e + f - c$ is the same as it was in the original map. We carry out the reduction process over and over again until there are no edges left and we therefore know that Eq. (8.5) holds. Because the value of the left side of Eq. (8.5) has not changed during any of the reductions, it equals 1 for the original map as well.

The reduction process involves removing any edge of the map, but not the vertices at the end of the edge (Figure 8.19). The effect on the map depends on what kind of edge has been removed:

1. Suppose you have chosen an edge ab that is not a disconnecting edge (top of Figure 8.19). Before its removal the two sides of ab were different faces. But after the removal, these faces are joined to make one face, so f has gone down by 1. In addition, e has gone down by 1. Neither c nor v has changed, so $v - e + f - c$ hasn't changed.

2. Suppose you have removed an edge that disconnects some component of the map, as in the bottom of Figure 8.19. Then we increase c by 1 but decrease e by 1, while v and f stay the same. Therefore, $v - e + f - c$ hasn't changed. ■

One of the applications of Euler's formula that we wish to discuss is to prove that K_5 and $K_{3,3}$ are not planar. To do this, we introduce the idea of counting how many faces of a planar map fall into various "size" categories according to how many edges touch them. For example, in Figure 8.18d, we have one face touching three edges, two faces touching

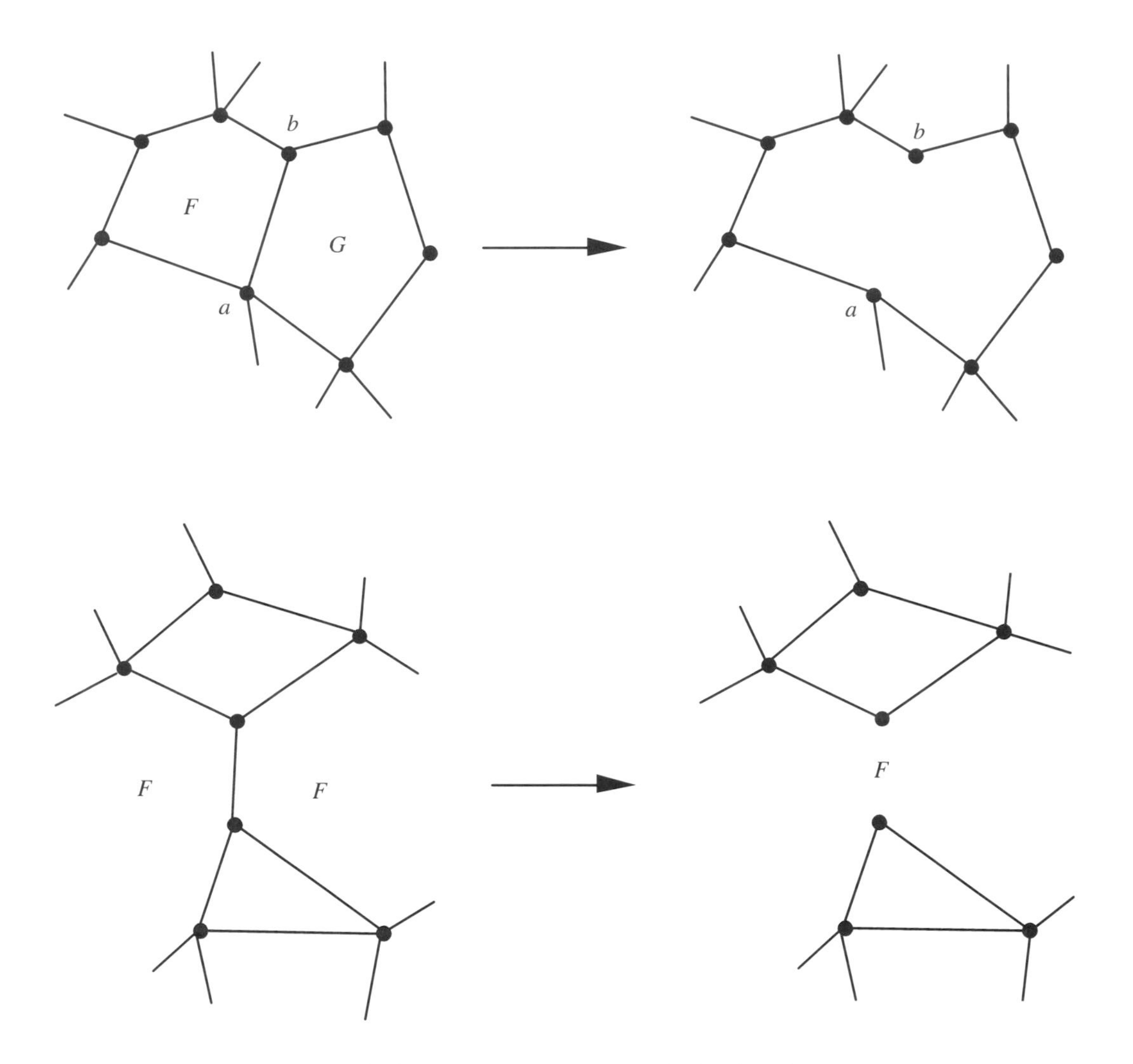

Figure 8.19 Reductions used in proof of Euler's formula.

four edges, and the infinite face touching five. Let f_i denote the number of faces that touch i edges and let n be the largest number of edges touching any face. In Figure 8.18d, $f_3 = 1$, $f_4 = 2$, and $f_5 = 1$, all other $f_i = 0$, $n = 5$. An immediate consequence of this notation is

$$f_1 + f_2 + f_3 + f_4 + \cdots + f_n = f. \tag{8.6}$$

☆ THEOREM 8.7

If a planar map has no disconnecting edge (no face meets itself along an edge), then

$$f_1 + 2f_2 + 3f_3 + 4f_4 + \cdots + nf_n = 2e. \tag{8.7}$$

Proof

Imagine visiting each face of the map and counting the edges which touch that face, keeping a running total. First we visit the f_1 faces having a single edge (the loops) — this starts the count off with f_1. Then we visit faces with two edges, which adds $2f_2$, and so on. [For example, in Figure 8.18d, if we visit first the three-sided faces then the four-sided faces, etc., then our final total is $3(1) + 4(2) + 5(1)$.] As we tally an edge, we place a check mark on it.

$$\text{number of marks} = f_1 + 2f_2 + 3f_3 + 4f_4 + \cdots + nf_n.$$

By Theorem 8.5c, each edge gets two check marks, therefore

$$\text{number of marks} = 2e.$$

The result follows from comparing these two equations. ■

The previous theorem is not true without the hypothesis that there is no disconnecting edge. However, the following related result applies to any planar graph whatever. We leave the proof to the reader (but see the exercises for a hint).

☆ THEOREM 8.8

For any planar map,

$$f_1 + 2f_2 + 3f_3 + 4f_4 + \cdots + nf_n \leq 2e. \tag{8.8}$$

■

☆ THEOREM 8.9

$K_{3,3}$ is not planar. (See Figure 8.6 of the previous section, but with the edge from b to the Electric Company added.)

Proof

We proceed indirectly, assuming that we had a planar drawing of $K_{3,3}$. There are three main parts to the proof.

1. From Euler's formula and the fact that $K_{3,3}$ has $v=6$, $e=9$, and $c=1$ ($K_{3,3}$ is connected), we can compute that the number of faces in this assumed planar drawing would be

$$\begin{aligned} f &= 1+e+c-v \\ &= 1+9+1-6 \\ &= 5. \end{aligned}$$

2. Our next goal is to show that some of the terms of Inequality (8.8) are 0. Each face has an even number of edges. This follows because if we take a trip around a face, we alternate between "house" vertices (bottom of Figure 8.6) and "utility" vertices (top) since there are no edges from a house vertex to another house vertex, or from a utility vertex to another utility vertex. If we start at a house vertex (say) then after 1, 3, or any odd number of steps (edges) we will be at a utility vertex. When we finally get back to our start vertex, we must, therefore, have gone an even number of steps. Thus $f_i=0$ if i is odd. Consequently, Inequality (8.8) applies with odd terms dropped out. In addition, there are no multiple edge sets in $K_{3,3}$ so $f_2=0$. Therefore, Inequality (8.8) reduces to

$$4f_4 + 6f_6 + 8f_8 + \cdots + \leq 2e = 18. \tag{8.9}$$

3. Alongside Inequality (8.9) we bring the following consequence of step 1:

$$f_4 + f_6 + f_8 + \cdots = f = 5.$$

Multiply this equation by 4 and subtract from the Inequality (8.9) to obtain

$$2f_6 + 4f_8 + 6f_{10} + \cdots \leq -2.$$

But the left-hand side cannot be negative because none of the terms are negative. This is a contradiction that proves the theorem. ∎

A similar style of proof will show that K_5 is not planar. We leave this as an exercise.

The Euler Number and Computer Vision

One of the key problems of computer vision is to identify important characteristics of a two-dimensional image. For example, to determine that the letter "i" in Figure 8.20a has two pieces, or to determine that the letter "o" of Figure 8.20b has just one piece, but encloses a hole. We shall see that Euler's formula can be useful in finding the so-called *Euler number* which can help distinguish some letters from others.

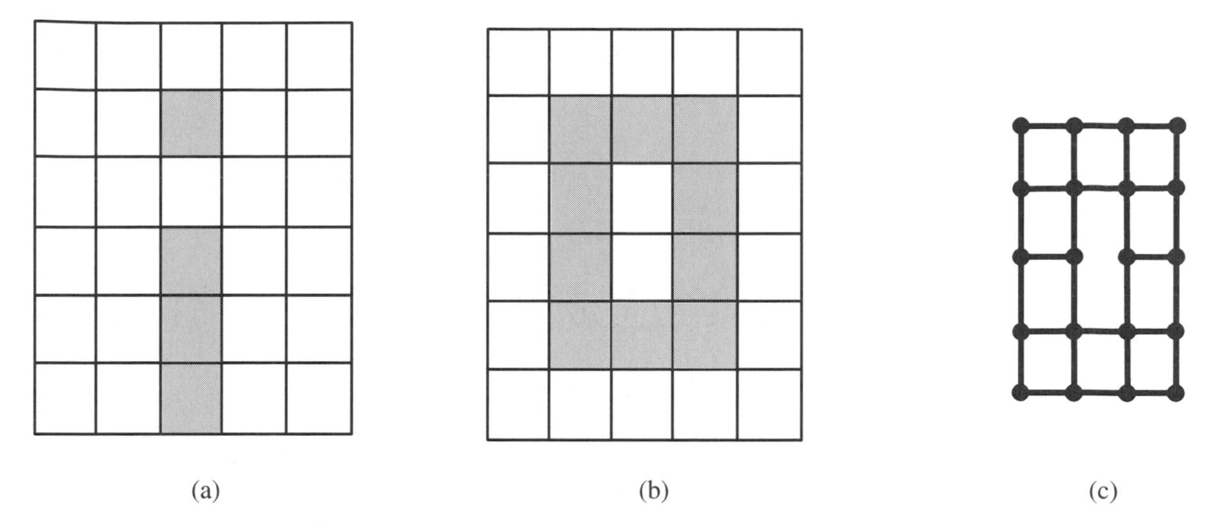

Figure 8.20 (a) Digital image of letter "i." (b) Digital image of letter "o." (c) Map derived from digital image in part (b).

DEFINITION

For a planar region:

$$\text{Euler number} = \text{number of pieces} - \text{number of holes}.$$

For example, the letter "o" has a Euler number of 0, while "i" has a Euler number of 2.

Our human vision system can count holes and pieces so quickly and easily, we might guess that there should be no problem in having a computer do it. But it isn't that easy, and to see why we must understand how computer vision systems typically work. We also need to clarify a fine point about pieces: Two diagonally opposite cells with just a corner in common are counted as making one piece.

A vision system starts with a special type of camera, but instead of focusing light on film it focuses light on a photo-sensitive surface that is divided into a lot of little cells called *pixels*, which we depict as squares in Figure 8.20. Pixels on which lots of light fall develop a high electrical charge, which we symbolize as white in our pictures. Pixels that don't get much light get a small charge, which we symbolize as dark pixels in our drawings.[3] This set of charges obtained by a computer vision system is called a

[3] There are typically hundreds of thousands of pixels in a camera, so smoothly curved regions can usually be approximated quite nicely.

digital image. From the electrical charges, the computer makes a list of the pixels along with the charges. Table 8.2 shows how the list would start for Figure 8.20b. The computer is capable of checking this pixel list:

For any row and column combination (r, c), the computer can find out whether the pixel in row r and column c is black or white.

TABLE 8.2 THE PIXEL VISIT LIST FOR FIGURE 8.20B

Pixel address	Charge (color)
1st column	
(1, 1)	High (white)
.	.
.	.
.	.
(6, 1)	High (white)
2nd column	
(1, 2)	High (white)
(2, 2)	Low (black)
etc.	

For the "serial computers" mostly in use today, the computer does not "see" the whole of the digital image at once in the way we seem to. The way computers work is reminiscent of a folk tale about some blind people examining an elephant. One is examining the trunk and believes he is dealing with a snake. Another is feeling the leg and she thinks she is touching a tree. A third feels the side of the elephant and thinks he is touching the side of a house. If they pooled their narrow bits of information, they could discover the truth. In our computer vision example, it is also necessary to cleverly pool the information obtained from the individual pixels in order to compute the Euler number of the digital region.

The theory behind this pooling starts by associating a planar map with our digital region (Figure 8.20c). The vertices are the corners of the black pixels. The edges are the sides of the black pixels. Each black pixel then becomes a face. In addition, each hole is a face. Finally we have the infinite face. Adding these three kinds of faces:

$$f = \text{number of black pixels} + \text{number of holes} + 1. \tag{8.10}$$

For example, in Figure 8.20c we have $v = 20$, $e = 30$, $f = 12$.

✯ THEOREM 8.10

The Euler number of a digital region $= v - e +$ black pixels (Figures 8.20b and c).

PROOF

First, notice that the number of pieces of the digital region is the same as the number of components of the graph of the region.[4] Euler's formula:

$$v - e + f = c + 1.$$

Substituting from Eq. (8.10),

$$\begin{aligned} v - e + \text{black pixels} + \text{holes} + 1 &= c + 1, \\ v - e + \text{black pixels} &= c - \text{holes} \\ &= \text{Euler number.} \quad \blacksquare \end{aligned}$$

The usefulness of this theorem in computing the Euler number depends on whether we have an algorithm (software) to compute $v +$ black pixels $- e$ by means of "visiting" pixels one at a time (checking entries in the pixel list) to determine whether they are black or white, and while we are at it keeping a running total of v and e. To devise such an algorithm, we might start by setting up a "pixel visit list" in this order: Start at the upper left (see Figure 8.21) and proceed down the first column. Continue with the second column, from top to bottom, and then the third and so on. This is illustrated in Table 8.2.

It is easy enough to count the number of black pixels as we go through the list. Counting vertices and edges requires care to make sure we do not count any more than once.

Let's examine the edge counting problem (Figure 8.21). If you are visiting the (r, c) pixel — the one in row r and column c — and it is black, since it has four edges you might suppose that our tally of edges should be increased by 4 (e := e + 4 in many programming languages). But what if the $(r-1, c)$ pixel above it is also black? The horiozontal edge these pixels share has already been counted on our earlier visit to pixel $(r - 1, c)$. Therefore, our increase by 4 needs to be corrected by deducting 1. Likewise, if the $(r, c - 1)$ pixel to the left is black, the vertical edge it has in common with pixel (r, c) must not be counted again. Therefore, deduct another 1 from the edge tally. We can express these activities during our "visit" in a style similar to many programming languages like this:

```
If pixel (r, c) is black then
    { e := e + 4;
      if pixel (r − 1, c) is black then e := e − 1;
      if pixel (r, c − 1) is black then e := e − 1 }
```

[4] In fact, the word *pieces* is often replaced by *components* in many discussions of this subject.

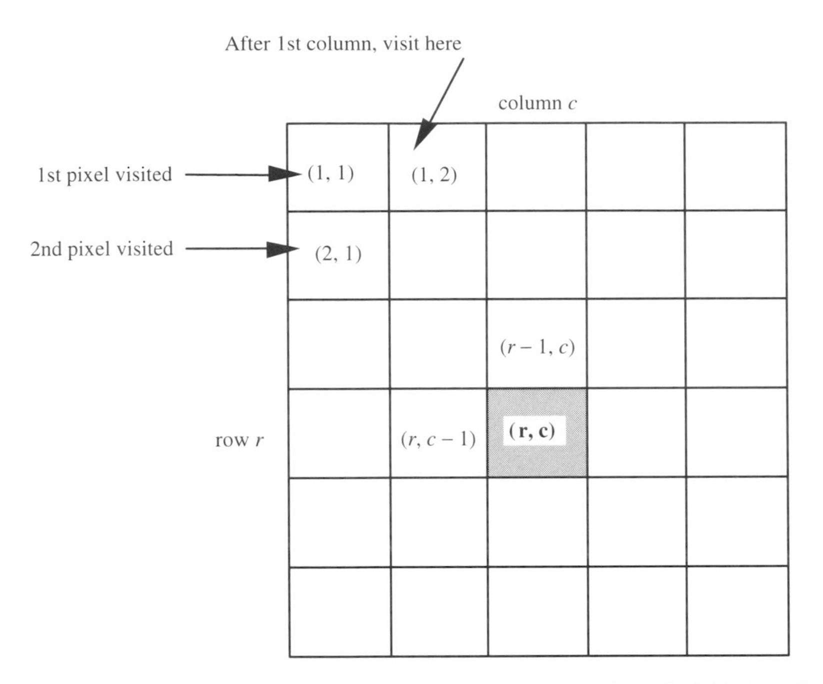

Figure 8.21 Two neighbors of pixel (r,c) that share an edge with it and which have been visited earlier.

We leave the problem of counting the vertices on the black pixels to the reader (but see the exercises for hints).

All of this assumes that it is useful to know the Euler number. It does help us distinguish between some letters of the alphabet in optical character recognition (an important application), but most letters of the alphabet have the same Euler number, so it only gives a little help here. In combination with a program to find the number of components, it gives us the number of holes immediately, and this can be helpful in some computer vision applications.

EXERCISES

**Marks challenging exercises.*

Euler's Formula

1. Suppose G is a planar map that has no disconnecting edge and where each face has three edges. Show that $3f=2e$.

2. If G is a planar map with $f_1 = 0$ and $f_2 = 0$, but which may or may not have disconnecting edges, show that $3f \leq 2e$.

3. (a) Draw a planar map that has no loops but $f_1 \neq 0$. (*Hint:* There is just one example and it is simple.)

   (b) Draw a planar map that has no multiple edge sets but $f_2 \neq 0$. Now draw another that is not isomorphic to the first. Is there a third example?

4. If G is a planar map that has no disconnecting edge and which has no loops or multiple edge sets, show that $3f \leq 2e$. (*Hint:* Use Exercises 2 and 3.)

5. Suppose G is a planar map where each vertex has valence 3. Show that $3v = 2e$. (*Hint:* Examine Section 8.1 of this chapter for results about valences.)

6. Let G be a planar map where every valence is at least 3.

   (a) Show $3v \leq 2e$.

   (b) Use part (a) and Euler's formula to show that if G is connected and each valence is at least 3, then $3f \geq 6 + e$.

7. Suppose G is a connected planar map with no loops or multiple edge sets and with all valences at least 3 and where no edge disconnects a component of the graph.

   (a) Use Exercises 2 and 6 to show $6f \geq 12 + 2e \geq 12 + 3f$ and, from this, that there are at least four faces in the map.

   (b) Show by example that the hypothesis that each valence is at least 3 is really necessary.

8. For a connected planar graph with no disconnecting edge and no loops or multiple edge sets, show that $3v \geq 6 + e$. (*Hint:* Euler's formula and one of the previous exercises will help.)

9. A certain connected planar map is such that no edge disconnects a component of the graph and each valence is exactly 3. Furthermore, each face is either four or six sided.

   (a) Show $3f - e = 6$.

   (b) Show that $4f_4 + 6f_6 = 2e$.

   (c) Use parts (a) and (b) and some simple algebra to show that there are exactly six 4-sided faces.

(d) Draw an example of such a map. Do you think there is more than one such map?

10. (a) Draw an example of a connected map whose graph can be disconnected by removing one edge and where Eq. (8.7) does not hold.

    (b) Let $e' =$ the number of edges that lie on just one face, and e'' the number that lie on exactly two. Show that the left side of Eq. (8.7) equals a formula involving e' and e''. What is that formula? Prove it. (*Hint:* Reread the proof of Theorem 8.7.)

    (c) Prove Theorem 8.8.

11. Suppose a planar map has the property that every edge is part of some simple circuit. Show that Eq. (8.7) holds for this map.

*12. Suppose a planar map has the property that no face meets itself at an edge. Prove that every edge is part of some simple circuit.

*13. Show that K_5 is not planar. (*Hint:* Follow the general strategy of Theorem 8.9.)

*14. If G is a connected planar map with no disconnecting edge and where each valence is at least 3, then:

    (a) Derive $12 \leq 5f_1 + 4f_2 + 3f_3 + 2f_2 + f_5 - (f_7 + 2f_8 + 3f_9 + \cdots)$. (*Hint:* Among other things, use Exercise 6.)

    (b) Prove that there is at least one face with five or fewer edges.

*15. Can you draw $K_{3,3}$ with no crossings on a doughnut-shaped surface (called a *torus* in mathematics)? How about K_5? What is the largest n so that K_n can be drawn with no crossings on the torus?

*16. Suppose a rectangle is tiled into other rectangles by horizontal and vertical lines as shown in Figure 8.22. Now examine an edge of the graph, such as ab in the figure. If there is another edge next to it with the same direction, put the two together to make a longer line segment. Continue doing this till you can't extend your segment any more in either direction. You have just created what we call a *border*. For example, the segment stretching from c to d is a border, as is fg. Let B denote the number of borders. Let R be the number of smaller rectangles inside the big one, and let C be the number of places where two borders cross (4-valent vertices such as e). In our figure, $B = 10$, $R = 8$, $C = 1$. Show that $B + C - R = 3$ for all such tilings.

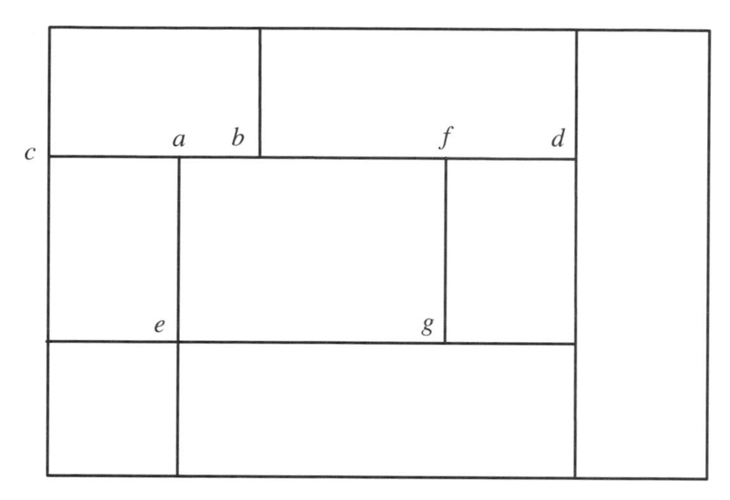

Figure 8.22

The Euler Number and Computer Vision

17. Suppose we have a digital region where pixels (1,2), (2, 1), (2, 3), and (3, 2) are black.

    (a) Draw a pixture of this.

    (b) Would you say there were four pieces or just one? How many holes? What is the Euler number?

18. Devise formulas that take as input an integer k and compute the row index (r) and column index (c) of the kth pixel on the pixel visit list (Table 8.2).

In the following two exercises pertaining to the Euler number, assume that the image surface has m *rows and* n *columns of pixels. "Devise an algorithm" means write a set of instructions that could be turned into a computer program. If you wish to actually write these algorithms as programs in a computer language of your choice, pick* m *and* n *to be small so you can easily create some test data by hand.*

19. Devise an algorithm to update e as a result of a visit to a pixel. (The incrementation was discussed in the text, but you need to add the loop control.)

20. (a) Suppose you want to devise an algorithm to update v as a result of a visit to a pixel. The order of visiting pixels should be as described earlier: down the columns, doing the columns left to right. When you come to a new pixel that is black, you increase e by 4. But how many previously visited pixels

do you need to check in order to "second guess" (modify) this change in e? Feel free to turn this analysis into an algorithm, but if it seems too tiresome, try part (b).

(b) Here is an alternative to part (a) as a means of counting v. The idea is to visit vertices and not pixels. Let V be a counter for the vertical lines on which the vertices lie: $V = 0, 1, 2, \ldots$, with 0 designating the leftmost border. H counts the horizontal lines, starting at the top with $H = 0$. Set up a double loop on H and V. For any combination (H, V) encountered in the loops, you need to check the four pixels that the (H, V) vertex touches. Can you turn this into an algorithm?

*21. Suppose you had two digital images I_1 and I_2 and none of the pixels of the first touches any of the pixels of the second (for example, two separate letters). Suppose we designate the Euler number of a digital image I as $E(I)$.

(a) Can you predict $E(I_1 \cup I_2)$ from knowledge of $E(I_1)$ and $E(I_2)$?

(b) What if $I_1 \cap I_2 \neq 0$? If it is not possible, do you think you could do it if you had some knowledge of $I_1 \cap I_2$ or of $E(I_1 \cap I_2)$? Try some easy cases, cases where the intersection is a single point, or a single edge of one cell, or a small set of cell edges.

22. Some researchers think that two pixels should not be considered to be connected to one another if they touch just at one corner. For example, a digital image consisting of the (1,1) pixel and the (2,2) pixel would be thought of as having two components. If we adopt this new point of view, Theorem 8.10 couldn't be true. In what step would the proof break down?

8.3 Polyhedra, Combinatorial Structure, and Planar Maps

Polyhedra and Convexity

In the study of three-dimensional objects, geometers have always paid special attention to flat-sided solids called polyhedra. Our objective in this section is to show how some questions about convex polyhedra can be studied by studying an associated planar map, which is easier to visualize. In the next section we apply this way of thinking to the so-called regular polyhedra, which have been objects of fascination to geometers since antiquity, and then to some molecules discovered in the 1980s that have been exciting chemists.

DEFINITION

A *polyhedron* is a three-dimensional solid bounded by a finite number of polygons called *faces*. Points where three or more faces meet are called *vertices*. Line segments where exactly two faces meet at an angle are called *edges*. The vertices and edges of the polyhedron make a graph called the *graph of the polyhedron*.

A box (technically a rectangular parallelopiped), bounded by six rectangles meeting at right angles, is an example of a polyhedron. Figure 8.23 shows other examples, the regular polyhedra. Cylinders and spheres are not polyhedra.

Although we do see polyhedra in the world around us, most of the shapes we see have curved or irregular boundaries. Nonetheless, polyhedra have lately become closely tied to applications. When a computer graphics program needs to show a curved solid, it does so by approximating the solid by a polyhedron with so many small faces that the eye is fooled into perceiving a smoothly curved surface (Figure 8.24 and Figure 5.11a). The field of chemistry gives another example of polyhedra in applications. When chemists contemplate a complicated molecule, they sometimes find it useful to think of the polyhedron whose corners are the atoms of the molecule. Modern examples such as these show that the ancient geometers showed good judgment in beginning the study of polyhedra.

Our definition of polyhedron may strike you as a little unsatisfactory (and it is!) because it is not clear whether some objects, such as Figures 8.25b and c and e, conform to it or not. These questions are taken up in the study of cell complexes in a branch of mathematics called topology. Our interest is in the subcategory of *convex* polyhedra where doubtful cases do not arise. We won't pursue the task of defining the wider class of polyhedra in an unambiguous way.

DEFINITION

A set is convex if, given any two points A and B contained in the set, the entire line segment AB is also contained within the set (Figure 8.26).

The definition applies to sets in any dimension and to sets that are not polyhedra as well as to polyhedra. A circle plus its interior is convex. Likewise a sphere plus its interior is convex. All the solids in Figure 8.23 are convex, but in Figure 8.25, only the cylinder (provided we include the interior) is convex. Henceforth, we will deal only with convex polyhedra. Convex polyhedra are often called *polytopes*.

Convex figures can "go to infinity." For example, an entire line is convex. If two rays meet at a point, making an angle, the infinite "inside of the angle" is convex. However, a polyhedron, whether or not it is convex, cannot be infinite in extent, since it must be bounded by a finite number of polygons according to our definition.

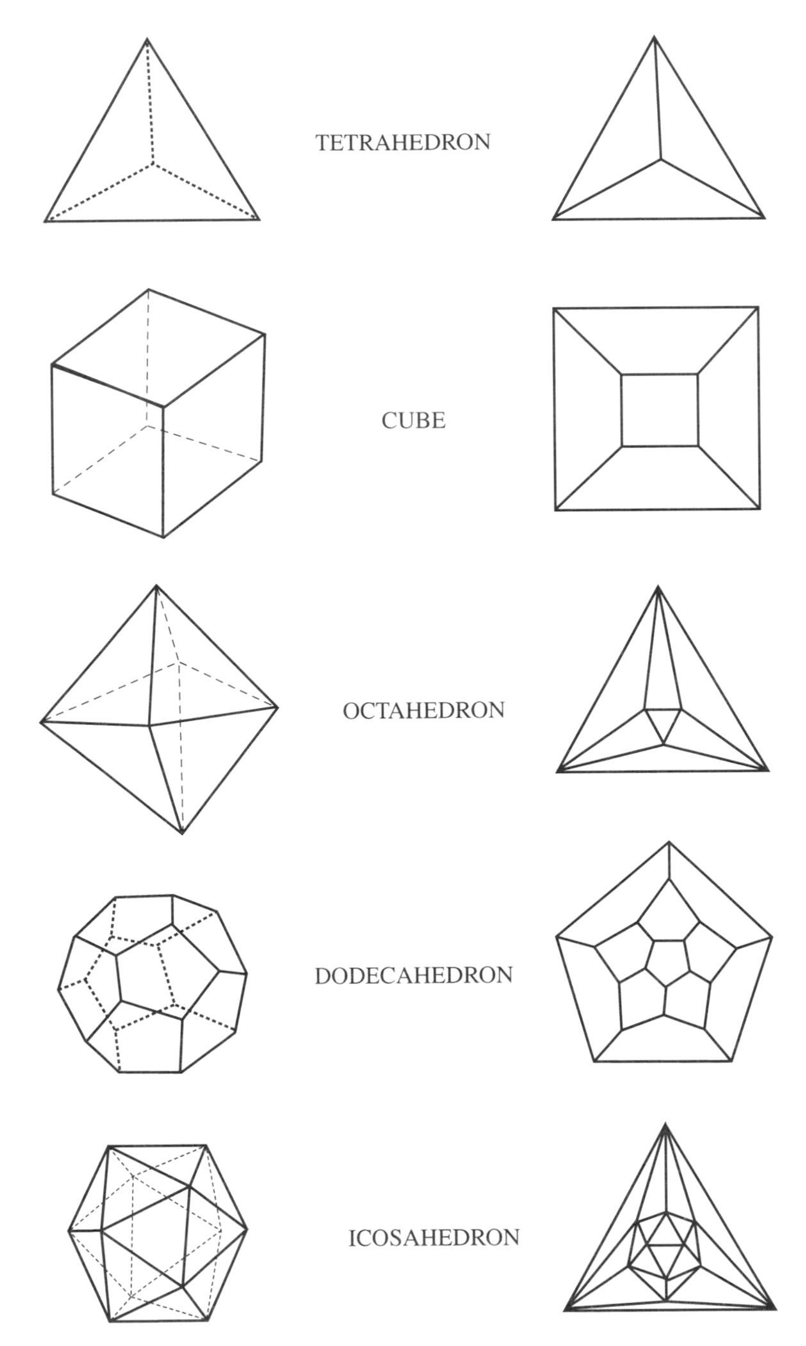

Figure 8.23 The five regular polyhedra and their planar maps.

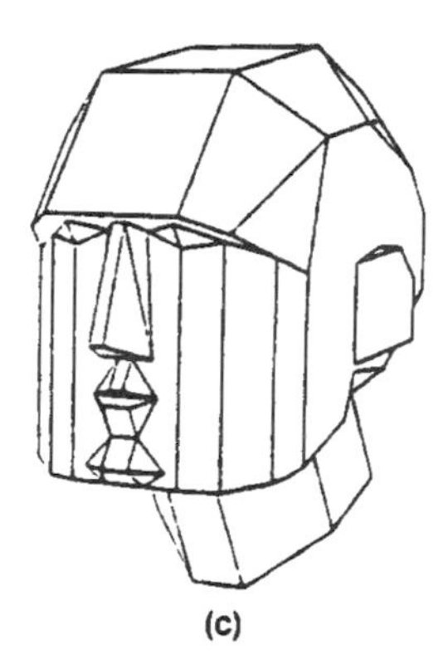

(c)

Figure 8.24 Curved surface approximated by a polyhedron. Courtesy of IBM.

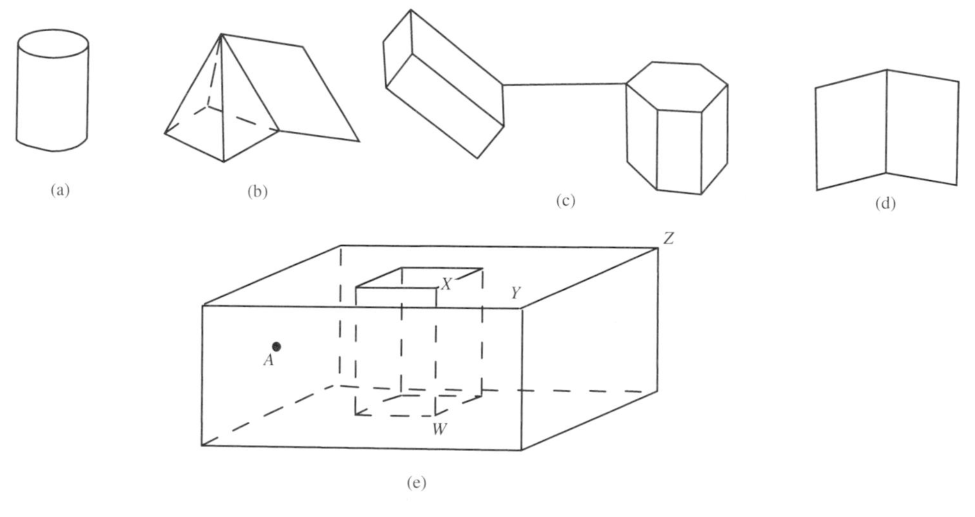

Figure 8.25 Are these polyhedra?

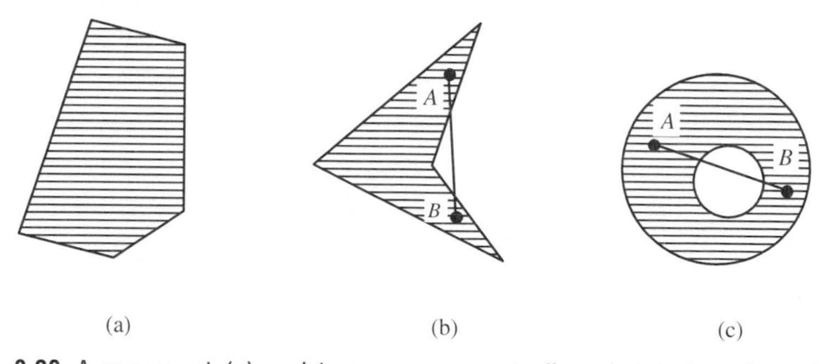

Figure 8.26 A convex set (a) and two nonconvex sets (b and c) in two-dimensional space.

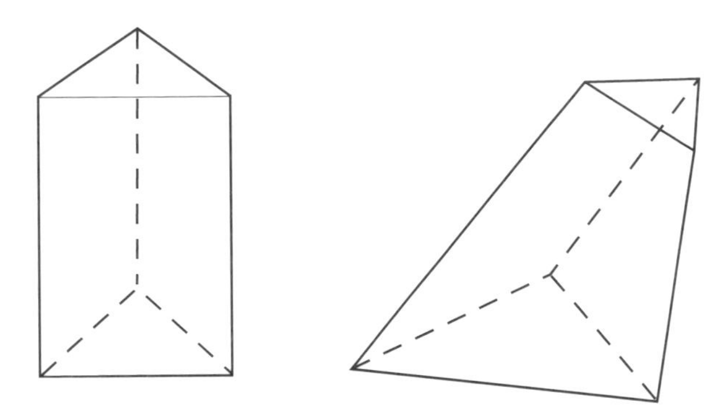

Figure 8.27 Polyhedra with the same combinatorial structure.

Planar Maps of Convex Polyhedra

What do the two convex polyhedra in Figure 8.27 have in common? They differ in regard to lengths of edges, areas of faces, and measures of angles. On the other hand, the surface of each consists of two triangles, separated by a belt of three 4-sided polygons. In addition, all vertices have valence 3 in each polyhedron. We would say they have the same *combinatorial structure*. In thinking about the combinatorial structure, we ignore things that need to be measured, like lengths, areas, and angles, in favor of things that can be counted, such as how many three-sided faces there are, and how many vertices there are with a certain valence. Most importantly, we concern ourselves with how the polyhedron is "put together." For example, the two polyhedra in Figure 8.27 have two 3-sided faces, which have no edge or even a vertex in common. If we had some third polyhedron with two 3-sided faces that touched one another, this polyhedron would not have the same combinatorial structure as the ones in Figure 8.27. Here is a more precise definition.

DEFINITION

Two convex polyhedra have the same combinatorial structure if

1. there is a one-to-one correspondence between the faces of the two polyhedra,
2. another one-to-one correspondence between the edges, and
3. a third one-to-one correspondence between the vertices so that:
   whenever two elements (vertex, edge or face) of one polyhedron touch (a vertex is on an edge or face, or two edges share a vertex, or an edge is on a face, or two faces share an edge or vertex), the corresponding elements of the other polyhedron touch in the same way.

Because polyhedra are sometimes hard to visualize and tinker with, it would be nice if we could see the combinatorial structure of a convex polyhedron in some simpler way than the three-dimensional drawings shown in Figure 8.27. The combinatorial structure of a polyhedron reminds us of a planar map since it also consists of vertices, edges, and faces. This raises the natural question of whether we can make a planar map whose pattern of faces, edges, and vertices shows us the combinatorial structure of a polyhedron.

Here is a "mind movie" you can visualize to convince yourself that the answer is yes. Imagine that the polyhedron boundary is made of rubber that can be stretched or compressed. The vertices and edges are painted on this rubber. Cut one face out of the polyhedron with scissors and discard it. Then grab what remains and stretch it out flat so it lies in the plane with the painted edges on the visible side of the rubber. Make sure nothing rips during the stretching. The flattened sheet you have, together with the infinite face surrounding its outer boundary, is the planar map of the convex polyhedron. Think of the infinite face as taking the place of the face you cut out and discarded. Such a planar map of a polyhedron is sometimes called a *Schlegel diagram.*

There is a lot of leeway in creating this planar map: You can stretch by various amounts, you can keep edges straight or let them curve, etc. But, for the combinatorial structure, we are uninterested in the sizes and shapes of the edges and faces, so stretch any way you like.

When we create the planar map of the polyhedron as just described, we can clearly match up the elements (faces, edges, and vertices) of the map with those of the polyhedron with one-to-one correspondences as described in our definition of when two polyhedra have the same combinatorial structure. Just keep track of an element of the polyhedron as you twist and stretch it. Therefore, it seems reasonable to say that the planar map and the polyhedron have the same combinatorial structure. For a precise definition of what this means, just take our earlier definition, which was phrased just for polyhedra, and replace one of the polyhedra with a planar map.

This rubber sheet manipulation always enables us to find a planar map with the same combinatorial structure as a given convex polyhedron, but it may not work for a nonconvex polyhedron. For example, the doughnut-shaped polyhedron in Figure 8.25 cannot be turned into a planar map in this way without some parts of the sheet getting torn or overlapping once the sheet is flattened. (A proof of this requires concepts from a subject called topology.)

Given a drawing of a polyhedron like those shown in Figure 8.27, or the even more complicated examples in the left hand column of Figure 8.23, it can be tricky to draw the planar graph by visualizing this rubber sheet manipulation. You may wish to think of the rubber sheet manipulation as just a guarantee that a planar map with the same combinatorial structure can be drawn rather than a method for actually doing the drawing. The next example shows a drawing method that may be easier.

EXAMPLE 8.9: DRAWING THE PLANAR MAP OF A CONVEX POLYHEDRON

Select one face of the polyhedron to remove (this face will turn into the infinite face of the planar map). For example, say we select *abc* in Figure 8.28. We would draw a three-sided region $a'b'c'$ in the plane that will wind up bounding the infinite face (bottom left of Figure 8.28). Now select a neighboring face of the polyhedron, for example, *adeb*. In the planar map there must be a neighboring four-sided face $a'd'e'b'$, so draw it (face number 2 in our figure). In the same way, draw $b'e'f'c'$, touching both

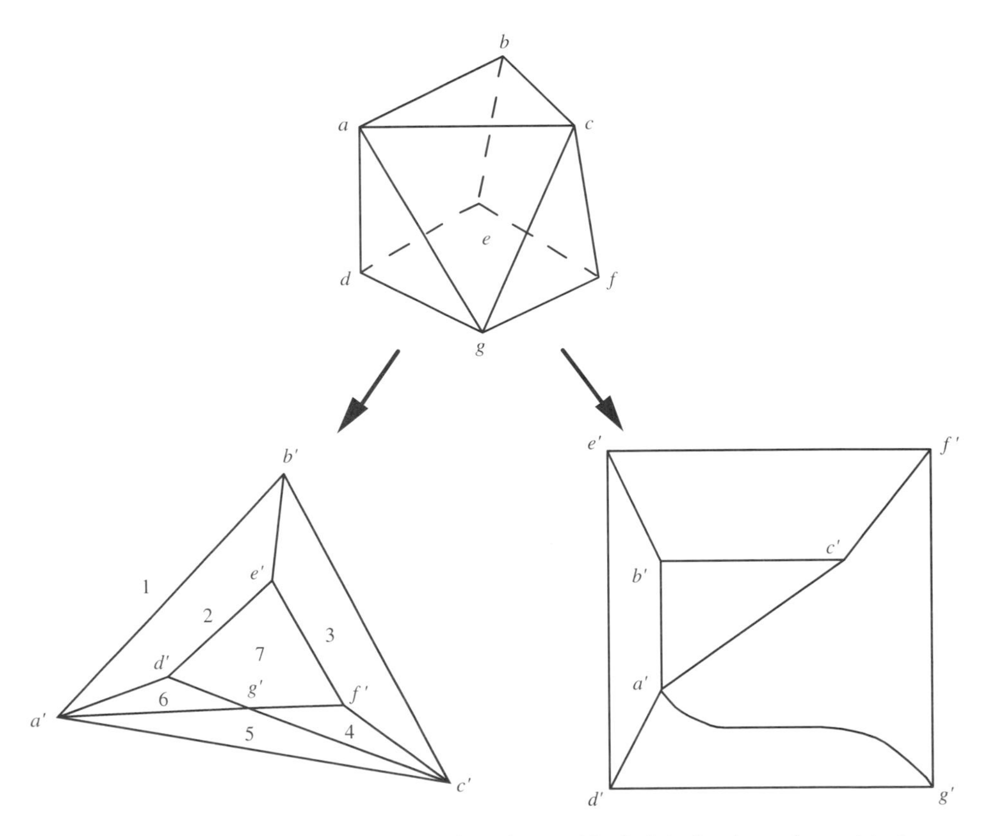

Figure 8.28 Two planar maps representing the combinatorial structure of a polyhedron.

previous faces. Continue until all faces are drawn with the same "touching" relationships to one another as in the polyhedron. The numbers in our figure show the order in which we added the faces, but other sequences are possible.

If we had started this process by removing the four-sided face *efgd*, we would obtain a planar map such as the one on the right side of Figure 8.28. Both of these planar maps convey the combinatorial structure of the polyhedron. In the second map, we have made edge $a'g'$ curved, just as a reminder that this has no effect on the combinatorial structure being shown. ■

The problem of making a planar drawing of the combinatorial structure of a convex polyhedron is similar to the problem cartographers face in making maps of the earth on a flat piece of paper. If the methods we have described so far for drawing a planar map for a convex polyhedron do not seem clear or convincing, we can borrow one of the cartographer's solutions to give us yet another method. That solution is called *stereographic projection* and works like this (Figure 8.29). Pick a projection point S outside the polyhedron, near the center of a face. Parallel to that face, on the other side of the polyhedron, place a plane M in which our planar map will appear. For each point P on an edge of the polyhedron define the projection of P to be $\overleftrightarrow{SP} \cap M$. The set of all such projected points forms the edges of a planar map which (if we include the infinite face) can be proved to have the combinatorial structure of the original polyhedron.

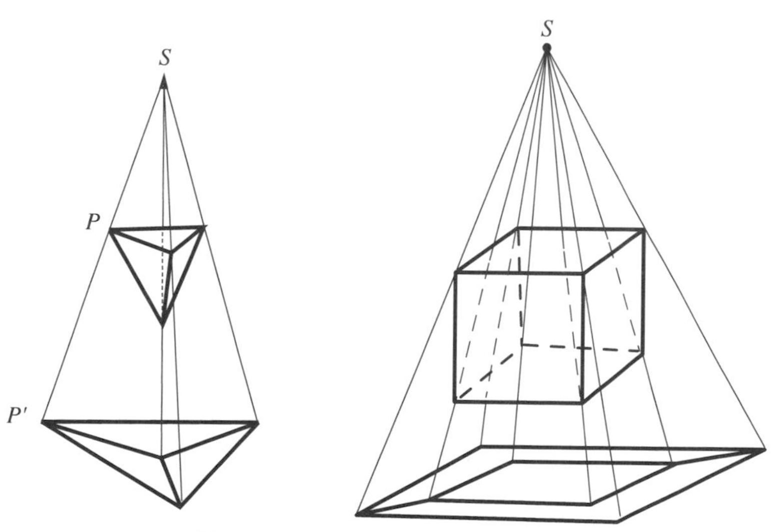

Figure 8.29 Stereographic projection.

This set of projected edges can be thought of as a shadow picture of the polyhedron's edges. Imagine the polyhedron is made of glass with the edges painted black. Install a light bulb at point S and project the shadow of the edges onto a screen. Alternatively, instead of a light bulb place your eye at S. If you ignore your depth perception, you would see the blackened edges as forming a planar map.

Figure 8.30 shows what might go wrong with stereographic projection if the point S is not taken close enough to the center of a face. The images of two faces overlap one another (and two edge images cross as well) and we do not have a planar map. If you imagine moving S closer and closer to the center of the top triangle, you will see that eventually the image of $ABCD$ "flips over" and no longer overlaps the image of ABE. We do not give a formal proof that this works for any convex polyhedron, but if we were to do so, we would need to use the convexity of the polyhedron. If the polyhedron is badly nonconvex, no matter where you put S, you can't get a planar map with the same combinatorial structure. The doughnut-shaped polyhedron in Figure 8.25 is a good example of this. If you place the projection point S near A and the projection plane vertical and to the right of the polyhedron, edges WX and YZ will project to edges that cross in the plane. But in a planar map, edges are not allowed to cross. Can you find a better place to put S?

Although every convex polyhedron has a planar map with the same combinatorial structure, the converse is not true: There are planar maps for which there is no convex polyhedron with the same combinatorial structure. If a planar map has a face that meets itself along an edge, like the edge ab in Figure 8.31a, there is no convex polyhedron with that combinatorial structure. This is because an edge of a polyhedron is, by definition, where *two different* faces meet.

Likewise, there is no convex polyhedron with the combinatorial structure of Figure 8.31b. Can you prove this? The necessary and sufficient condition for a map to have a convex polyhedron with the same combinatorial structure is given by the following theorem, whose proof we omit.[5]

★ THEOREM 8.11: STEINITZ

Given a planar map, there will be a convex polyhedron with the same combinatorial structure if and only if the graph of the map is connected and cannot be disconnected by the removal of any one or two vertices. ■

[5] For a proof see Lyusternik (1963), Ziegler (1994), or Grünbaum (1967).

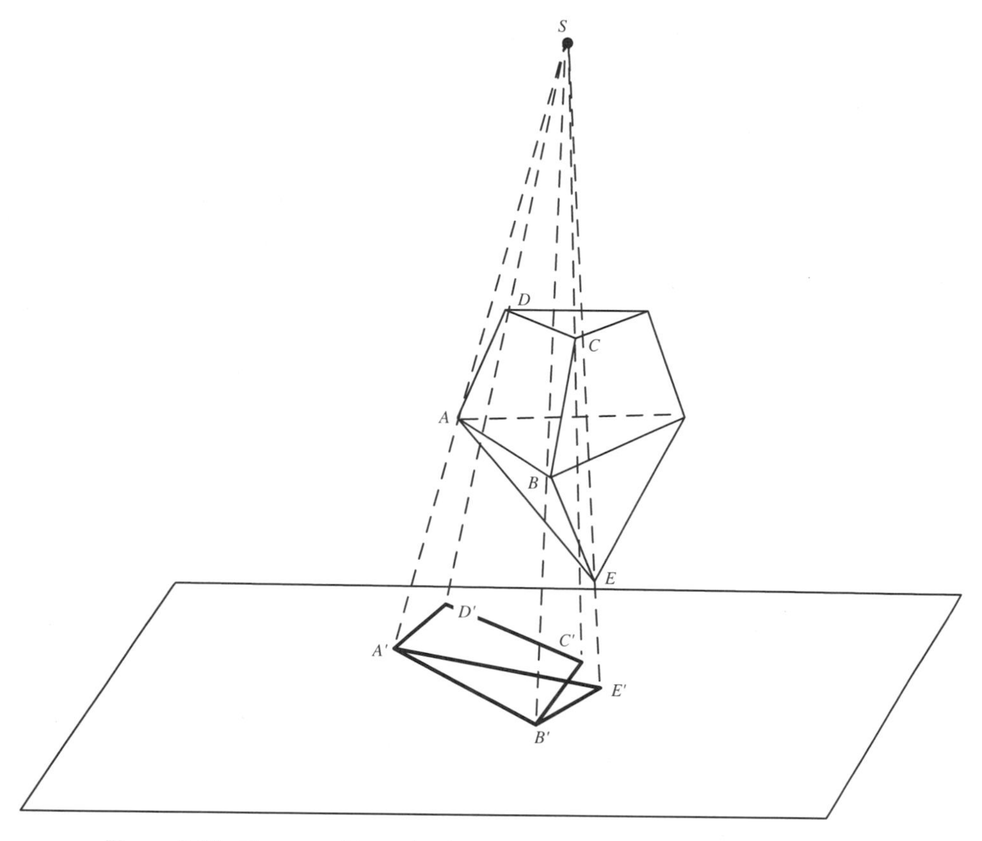

Figure 8.30 Stereographic projection needs a good center of projection (S).

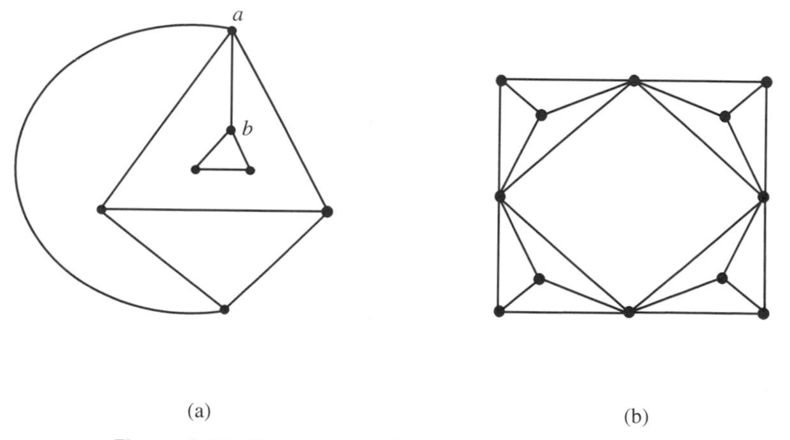

Figure 8.31 Planar maps that don't arise from polyhedra.

✯ THEOREM 8.12: EULER'S POLYHEDRAL FORMULA

If a convex polyhedron has f faces, v vertices, and e edges, then

$$f + v - e = 2. \tag{8.11}$$

Proof

Consider a planar map with the same combinatorial structure as the convex polyhedron. The number of faces in the planar map (including the infinite face) is the same as the number in the original polyhedron. The same is true for the edges and vertices. Applying Euler's formula to the planar map gives

$$f + v - e - c = 1.$$

But the graph of a convex polyhedron is connected. (This is part of what Steinitz's theorem tells us, but it is also fairly obvious on intuitive grounds.) Therefore, the planar map is connected also. Thus, $c = 1$. Making this substitution gives Eq. (8.11). ■

If two planar maps have the same combinatorial structure, then if one has just one triangle so does the other. If one has exactly two triangles then so does the other. Using the notation described in the previous section, the numbers $f_1, f_2, f_3, \ldots,$ for one map are the same as for the other map. It is natural to wonder whether these "counts" determine the combinatorial structure. Our next example shows that this is not always the case.

EXAMPLE 8.10: FACE STATISTICS DO NOT ALWAYS DETERMINE COMBINATORIAL STRUCTURE

Can you build two combinatorially different planar maps with $f_3 = 2, f_4 = 2, f_6 = 1$, and all other $f_i = 0$?

SOLUTION

Building planar maps is a trial-and-error process, a bit like doing a jigsaw puzzle: You sometimes have a choice of where to build on next. Let's begin by putting down a triangle (Figure 8.32). Where can the second triangle go? We show two possibilities. (Is there a third?) Now there is no guarantee that either of these can be completed to form a planar map with the desired face statistics. However, as the figure shows, we are in luck because in each case we can add the two required four-sided faces, thereby creating the six-sided infinite face as needed. The maps have the same face statistics but are not combinatorially equivalent because in one case the two 3-sided faces have an edge in common while in the other they do not. ■

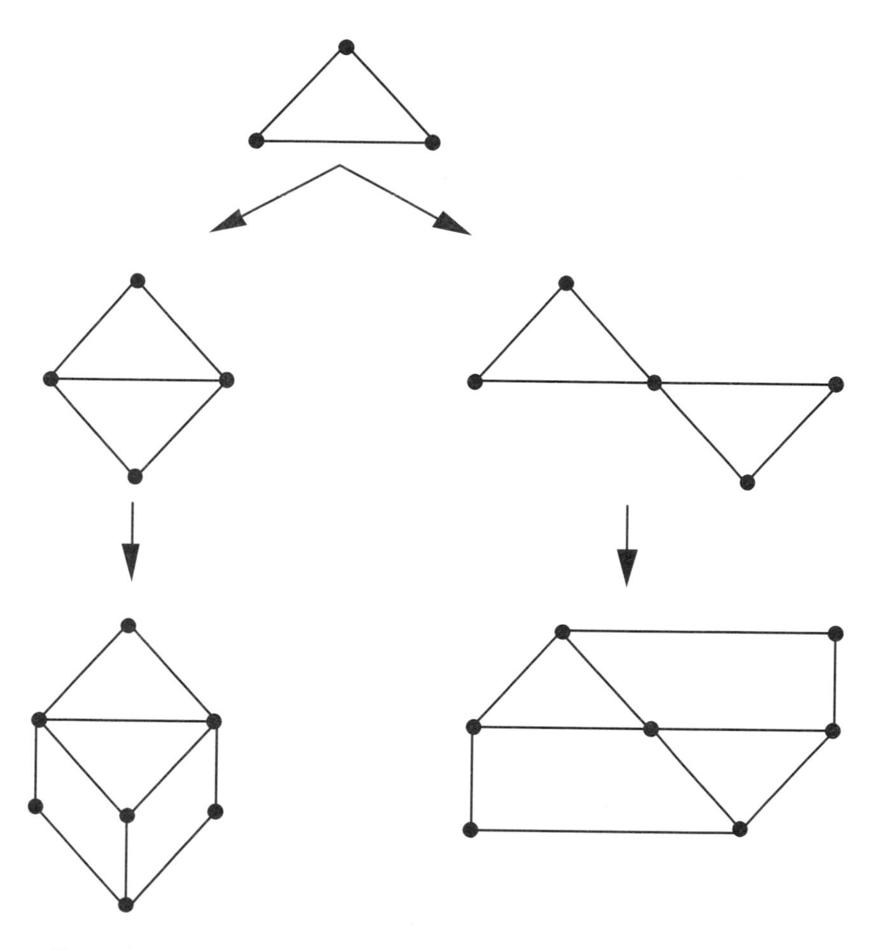

Figure 8.32 Building two combinatorially different maps with the same f_i's.

Despite the example we have just given, there are some circumstances where the face statistics do allow only one map. For example, if you try to build a planar map with $f_3 = 2$, $f_4 = 1$, and all other $f_i = 0$, you will find that all such maps are combinatorially equivalent. The kind of choice we had in the previous example does not exist. For example, if you start with a triangle, and try to put the other triangle down so it only has a vertex in common with the first, you will find that you cannot complete the map with the right statistics.

You might wonder whether the face statistics, together with the vertex statistics (the v_i) would determine a map completely.

EXERCISES

**Marks challenging exercises.*

Polyhedra and Convexity

1. Read the definition of a polyhedron carefully and explain why Figure 8.25d is not a polyhedron.

2. What extra conditions could we add to our definition of a polyhedron to make it clear that the objects in Figures 8.25b and c were not polyhedra?

3. Is Figure 8.25e a polyhedron according to our definition? Would it be according to the definition you proposed in Exercise 2?

4. Which of the following seems to be convex according to your visual intuition?

   (a) The boundary of a tetrahedron

   (b) A line segment

   (c) A plane

   (d) The union of two line segments making the shape of a letter V

*5. Prove that a circle (including its interior) is convex. (*Hint*: Try to apply Exercise 13 of Section 2.1 of Chapter 2.)

6. Let A and B be convex sets. Prove that $A \cap B$ is convex as well.

7. Is the union of two convex sets necessarily convex? If so, give a proof. If not, give a counterexample.

8. Explain why a finite set of points can never be convex.

9. Explain by reference to specific points why the objects in Figures 8.25b–e are not convex.

The following three exercises, along with Exercise 6, show how to prove that the graph of a convex polyhedron is connected—one of the easier parts of Steinitz's theorem.

*10. Prove that if a plane intersects a polyhedron, the intersection is either a point, an edge, a face, or a convex polygon that is not a face.

*11. Prove that if U and V are vertices of a convex polyhedron, then there is a connected series of line segments stretching from U to V where each line segment lies in a face of the polyhedron. (*Hint:* You might need Exercise 6. Feel free to assume the following: A convex set in the plane bounded by a finite number of line segments is a convex polygon.)

*12. Prove that if U and V are vertices of a convex polyhedron, then there is a connected series of edges stretching from U to V.

Planar Maps of Convex Polyhedra

13. Label the vertices of the polyhedra in Figure 8.27 and then indicate the three one-to-one correspondences in the definition of when two polyhedra have the same combinatorial structure.

14. Draw planar maps that have the same combinatorial structure as the polyhedron shown in Figure 8.33

    (a) with the top face turning into the infinite face;

    (b) with the bottom face turning into the infinite face;

    (c) with the triangular face on the right turning into the infinite face.

15. Draw a planar map that has the same combinatorial structure as the polyhedron shown in Figure 8.34

    (a) with *ABCD* turning into the infinite face;

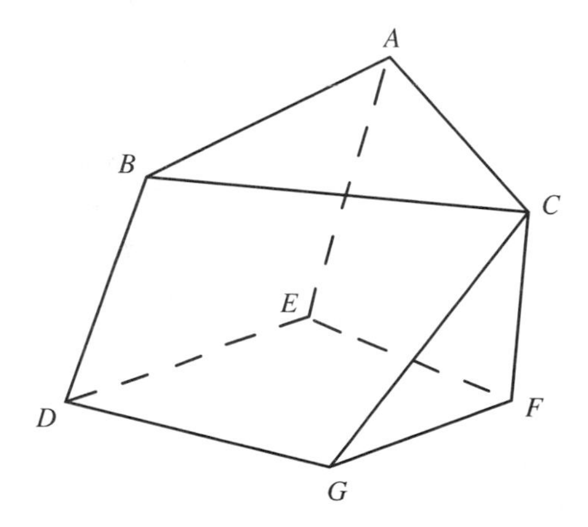

Figure 8.33

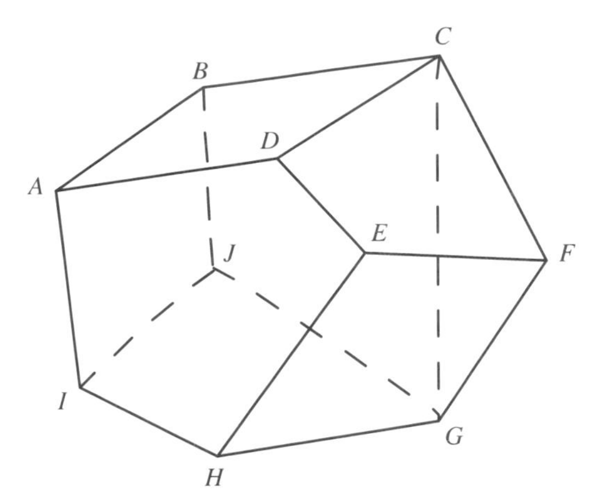

Figure 8.34

(b) with CFG turning into the infinite face;

(c) with $BCGJ$ turning into the infinite face.

16. If a convex polyhedron has four faces and four vertices, how many edges must it have? Is there such a convex polyhedron?

17. Explain why there is no convex polyhedron with the combinatorial structure of Figure 8.31b. (*Hint*: Read the axioms about planes in Section 1.3.)

18. In Figure 8.30, as S moves toward the center of the top triangular face, there will be a position where $A'B'C'D'$ no longer overlaps $A'E'B'$, but it has not yet "flipped." Describe (in terms of the features of the polyhedron) where such an "in-between" position for S might be.

19. Figure 8.25e can be obtained by starting with a convex polyhedron (a big box) and "drilling" a convex polyhedral hole through it that "breaks" the boundary in just two faces. (The hole does not cut through an edge, thereby altering two faces at the same end of the hole.)

    (a) What relation exists between the number of vertices of the original polyhedron, the number of vertices of the hole polyhedron, and the number of vertices in the final result? Answer the same question for edges and faces.

    (b) For such "drilled" polyhedra, find a relation like Euler's formula between the vertices, edges, and faces.

20. What is the least number of vertices a convex polyhedron can have? Explain your answer.

21. What is the least number of faces a convex polyhedron can have? Explain your answer.

22. If a polyhedron has exactly four vertices, it is possible to give a fairly exact description of how many faces and edges it has, and the number of edges on the various faces. Can you do this?

23 Build two combinatorially different planar maps with $f_3 = 2, f_4 = 3$, and all other $f_i = 0$.

24. Build two combinatorially different planar maps with $f_3 = 8, f_4 = 2$, and all other $f_i = 0$.

25. Can you build two combinatorially different planar maps with $f_3 = 4, f_4 = 1$, and all other $f_i = 0$?

26. Can you build two combinatorially different planar maps with $f_3 = 3$, $f_4 = 3$, $f_5 = 1$, and all other $f_i = 0$?

8.4 Special Kinds of Polyhedra: Regular Polyhedra and Fullerenes

Regular Polyhedra

Looking at things from an applications point of view, it is not easy to see why geometers of ancient times found polyhedra so interesting. There were only a few polyhedra to be found in nature or the man-made world at that time in history: rooms of buildings, which are basically rectangular parallelopipeds; the pyramids of Egypt; and perhaps a few other examples. Today, polyhedra are important in many areas of pure and applied mathematics, but it is hard to believe that the ancient geometers foresaw this. Most likely, they were motivated by the aesthetic charm of polyhedra. They were particularly interested in a subcategory of convex polyhedra called regular polyhedra.

The notion of a regular polyhedron is inspired by the idea of a *regular polygon*. A polygon in the plane is regular if all of its edges are the same length and all of its angles are congruent. A square is regular, as is an equilateral triangle. We can make a regular polygon of any number of sides using a protractor and a circle. Divide the 360 degrees at the center of the circle into as many equal angles as we want sides in the polygon, rather like cutting up a pie. Mark the points where the radii making these angles intersect the circle. These points are the vertices of the polygon.

DEFINITION

A convex polyhedron is *regular* if

1. all of its faces are regular polygons,
2. all faces have the same number of sides, $p \geq 3$, and
3. the same number of faces meet at each vertex $q \geq 3$.

Our definition does not explicitly say that the faces of a regular polyhedron are congruent to one another, but this is a consequence you can try to deduce as an exercise.

Figure 8.23 shows five examples of regular polyhedra. Are there any others? Since there are infinitely many regular polygons, we might suppose that there are infinitely many regular polyhedra, but it turns out that every regular polyhedron is a scaled-up or scaled-down version of one of the five in Figure 8.23. This seems remarkable, and is surely one reason geometers have been fascinated by the regular polyhedra since the time of Euclid.

If we create the planar map of a regular polyhedron, what properties would it have? As stated in Steinitz's theorem (Theorem 8.11), the graph of any convex polyhedron is connected. In addition, since the graph cannot be disconnected by removing one or two vertices, no face meets itself along an edge. Finally, the planar map inherits conditions 2 and 3 of our definition of regular convex polyhedron. We call a planar map with these conditions a *regular planar map*. Summarizing:

DEFINITION

A planar map is called *regular* if

1. its graph is connected,
2. no face meets itself along an edge,
3. each face has the same number of edges $p \geq 3$, and
4. each vertex has the same valence $q \geq 3$.

The essential numerical features of a regular planar map are the numbers p and q, so we often describe a regular planar map by this pair (p, q). For example, the (p, q) combination for a cube is (4, 3), meaning that faces are 4-sided and vertices have valence 3.

✯ THEOREM 8.13: THE FIVE REGULAR MAPS

Every regular planar map has a (p, q) combination that is one of the ones given in Table 8.3. (Also see Figure 8.23 for maps exhibiting these possibilities.)

TABLE 8.3 THE FIVE REGULAR POLYHEDRA

Face size, p	Valence, q	f	v	e	Name
3	3	4	4	6	Tetrahedron
3	4	8	6	12	Octahedron
3	5	20	12	30	Icosahedron
4	3	6	8	12	Cube
5	3	12	20	30	Dodecahedron

Proof

We begin with Eq. (8.7) of Section 8.2. In the case of a regular map, $f_p = f$ and all other $f_i = 0$, so this equation becomes

$$pf = pf_p = 2e.$$

Likewise, Eq. (8.2) reduces to

$$qv = qv_q = 2e.$$

Substituting these into Euler's formula, and taking into account that the map is connected so $c = 1$, gives

$$\begin{aligned} 2e/p + 2e/q - e &= 2, \\ 1/p + 1/q &= 1/2 + 1/e. \end{aligned} \tag{8.12}$$

We would like to conclude that this equation has only five solutions in which p, q, and e are all integers and where p and q are at least 3. The basic idea is that if p and/or q are too large, then the left side is too small to be greater than 1/2, as required by the right side. In particular, if p is as large as 6, $1/p \leq 1/6$, so $1/q > 1/3$ in order that the left side exceed 1/2. But $1/q > 1/3$ means $q < 3$, which is ruled out since at least three faces meet at every vertex. Thus $p = 3$, 4, or 5. In the same way we prove that $q = 3, 4, 5$. Now there are nine possible combinations for p and q, where each is either 3, 4, or 5. Five of them correspond to regular planar maps (see Figure 8.23 and Table 8.3). The other four combinations — (4, 4), (4, 5), (5, 4), and (5, 5) — are ruled out because if we make these substitutions into Eq. (8.12) we don't get a positive integer for e. ■

Table 8.3 shows the (p, q) values that satisfy our equations. But, unless we are already familiar with the polyhedra mentioned in the last column of the table, there is no guarantee that there are regular planar maps with these values (all we have proved is that these are the only possibilities). To see if there are any, you need to do some pencil-and-paper experimentation, trying to draw planar maps with the statistics specified in the table. Of course, this has already been done, long ago, and Figure 8.23 shows the examples cited in the last column of the table. After studying these examples, you should try to draw your own regular planar maps with the five different (p, q) combinations in the table. As you work on a particular (p, q) combination, you should ask yourself whether two essentially different maps could be drawn—that is, as you add edges, vertices, and faces to build the map, do you ever have any real choice about what to do next? In each case, you should conclude that the answer is no. This is illustrated in the next example.

EXAMPLE 8.11: DRAWING A REGULAR MAP

Here is how such a drawing experiment might work out for $p=4$, $q=3$. We have to start somewhere so draw a four-sided face (left side of Figure 8.35). We can make it any size and the edges can be straight or curved, but these are inessential features of a map—we really have no choice about how to draw a four-sided face. At each vertex the valence must end up 3, so we can draw edges sticking out (middle of Figure 8.35). Could two of these edges lead to the same vertex; for example, could $a=b$? That would create a three-sided face, which is not possible since $p=4$. So we come to the middle part of Figure 8.35. Because all faces are four-sided, we are forced to draw in the edges ab, bc, cd, da, leading to the right-hand part of

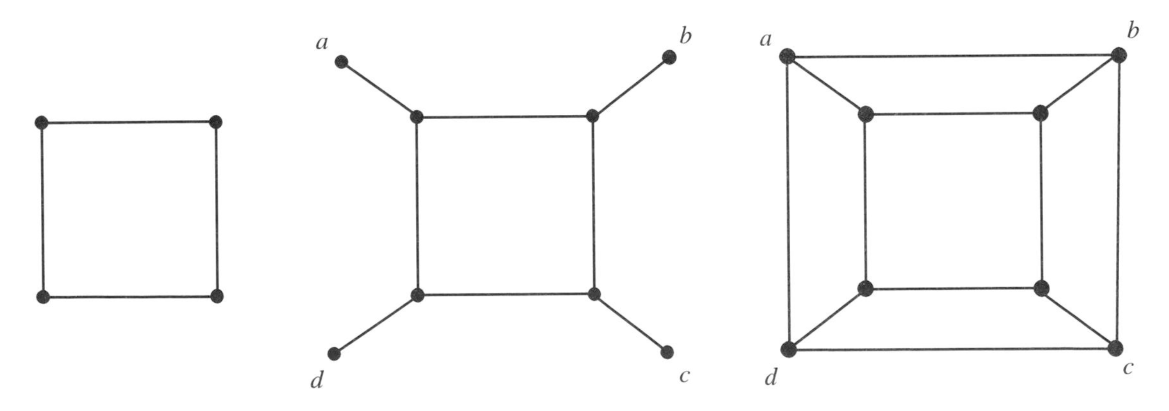

Figure 8.35 Building a regular map with $p=4$, $q=3$.

Figure 8.35. Because all valences are 3, there is no opportunity to draw any additional edges leading from the vertices in the map so far. Since the map must be connected, there is no opportunity to place any edges that are not connected to what we have drawn so far. In other words, what we have drawn so far is it! This same kind of argument can be made for each of the other four solutions in the table, showing that there is exactly one regular map with the required p and q values. More precisely, you can draw more than one map, but they will all be isomorphic. ■

One of the most remarkable examples of the lure of the regular polyhedra is the attempt by Johannes Kepler (1571–1630) to explain the distances between the six planets known in his time (Uranus, Neptune, and Pluto had not yet been discovered). In Kepler's day, it was thought that the planets had circular orbits with the sun at the center. The radii of these orbits had been estimated from measurements made from the earth. Kepler believed that the five gaps between the six known planets could be connected with the fact that there were precisely five types of regular polyhedra. He imagined a large outer sphere (see Figure 8.36), with the same center as the sun, on which Saturn, the farthest planet he knew of, moves. Inscribed in this sphere he placed a cube, and within this cube he inscribed a sphere to contain Jupiter's orbit. He continued in this way, with a tetrahedron containing an inscribed sphere for Mars's orbit. Next came a dodecahedron with a sphere inside for Earth, an icosahedron with a sphere inside for Venus, and an octahedron with a sphere inside for Mercury, the planet closest to the sun. The particular order Kepler picked for the polyhedra in this arrangement was the one that created radii for the concentric spheres fairly close to the distances of the planets from the sun.

In the modern view, the fact that this gives distances that are approximately correct would be regarded as a meaningless coincidence, especially since the scheme doesn't deal with all of the nine planets known today. Furthermore, there are $5! = 120$ different ways to arrange the order of the regular polyhedra and it is not surprising that one of these orderings would give approximately the right radii. Most importantly, we have learned to understand the solar system through concepts such as mass, gravitation, and velocity, so this scheme involving polyhedra seems to "come out of left field" and be out of place. But Kepler was so pleased by his model that he proposed that the Duke of Württemberg have a craftsman build a gold decanter in the shape of this model, and that each of the various spheres be a separate container holding a separate beverage. What seems like good science in one era can look silly later.

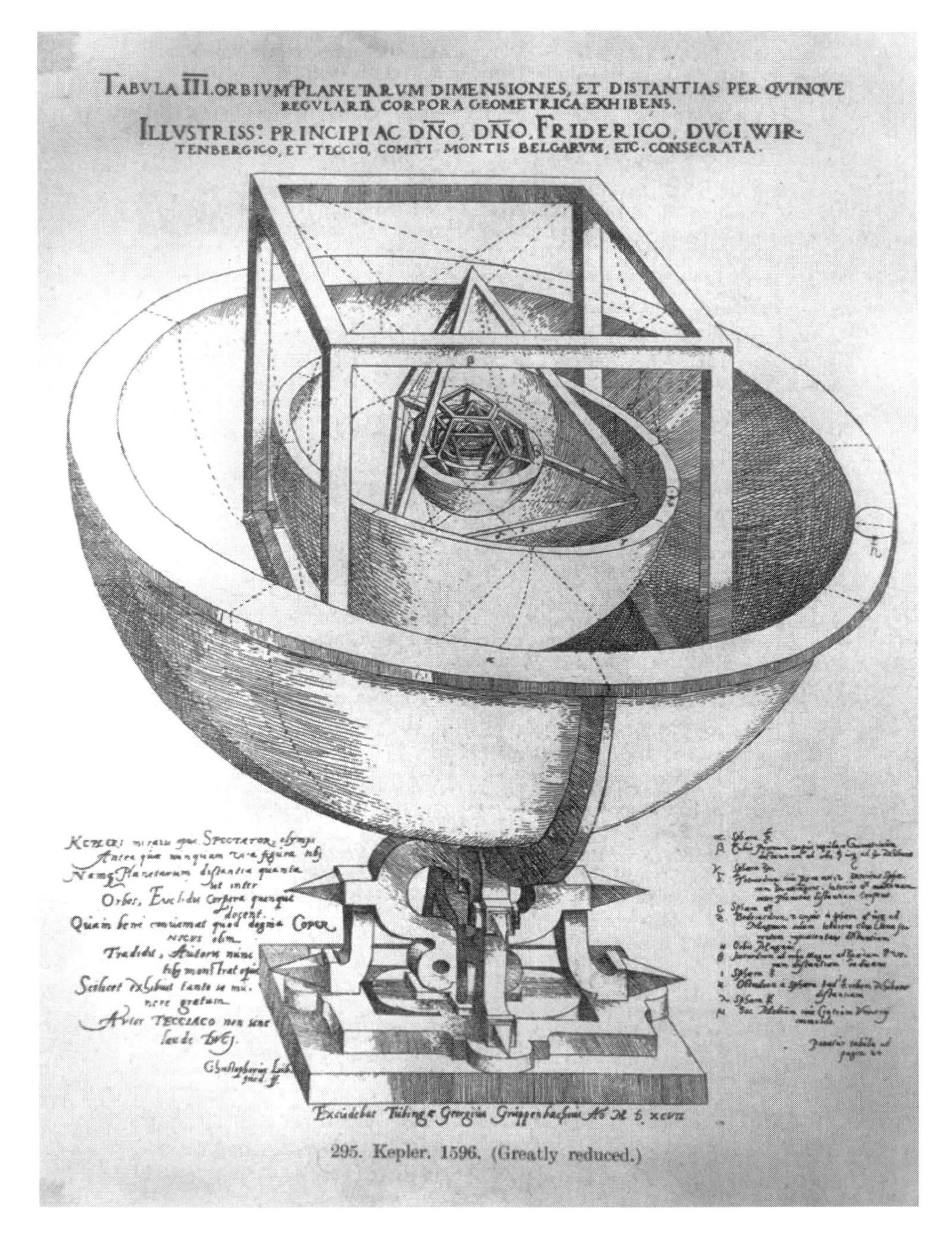

Figure 8.36 Kepler's model of the solar system. Courtesy of Corbis-Bettman.

Fullerene Molecules and Maps

In 1985 chemists discovered the buckyball.[6] This is a molecule consisting entirely of carbon atoms arranged as the vertices of a polyhedron shown in Figure I.5 of the Introduction to this book. The map of this polyhedron is exactly that of the traditional soccer ball, with 12 pentagonal faces and 20 hexagonal faces. This was a big surprise because chemists hadn't thought that carbon atoms could arrange themselves in this way. This discovery stimulated the search for other molecules of a similar type. Chemists are always interested in finding new kinds of molecules in the hopes of using them to create new drugs or products.

What chemists were looking for was molecules that consisted entirely of carbon atoms and that followed these rules:

F_1. The carbon atoms are arranged as the vertices of a convex polyhedron.

F_2. Each carbon is linked to three others by chemical bonds (edges of the polyhedron).

F_3. Each face of the polyhedron consists of either a pentagon (five carbons in a ring) or a hexagon (six carbons in a ring.)

Molecules of this type are called *fullerene molecules* or just *fullerenes.*

The restrictions F_1, F_2, and F_3 were not chosen arbitrarily or in consultation with a mathematician. In each case, there is a sound chemical reason for the restriction. For example, rings with more than six carbon atoms are unstable and break apart and this is part of the motivation for F_3.

Chemists were aware that the convex polyhedron of a fullerene molecule could have its combinatorial structure represented by a planar map, as described in the previous section. This planar map would have the following properties:

M_1. The map is connected. (Because of F_1 and part of Steinitz's theorem.)

M_2. No face meets itself along an edge. (F_1 and part of Steinitz's theorem.)

M_3. Each vertex has valence 3. (F_2.)

M_4. Each face has either 5 or 6 edges. (F_3.)

Figure 8.37 shows a planar map with these characteristics, the map of the buckyball.

[6] Named for the architect Buckminster Fuller, who promoted the idea of "geodesic domes," unusual buildings in the shape of polyhedra.

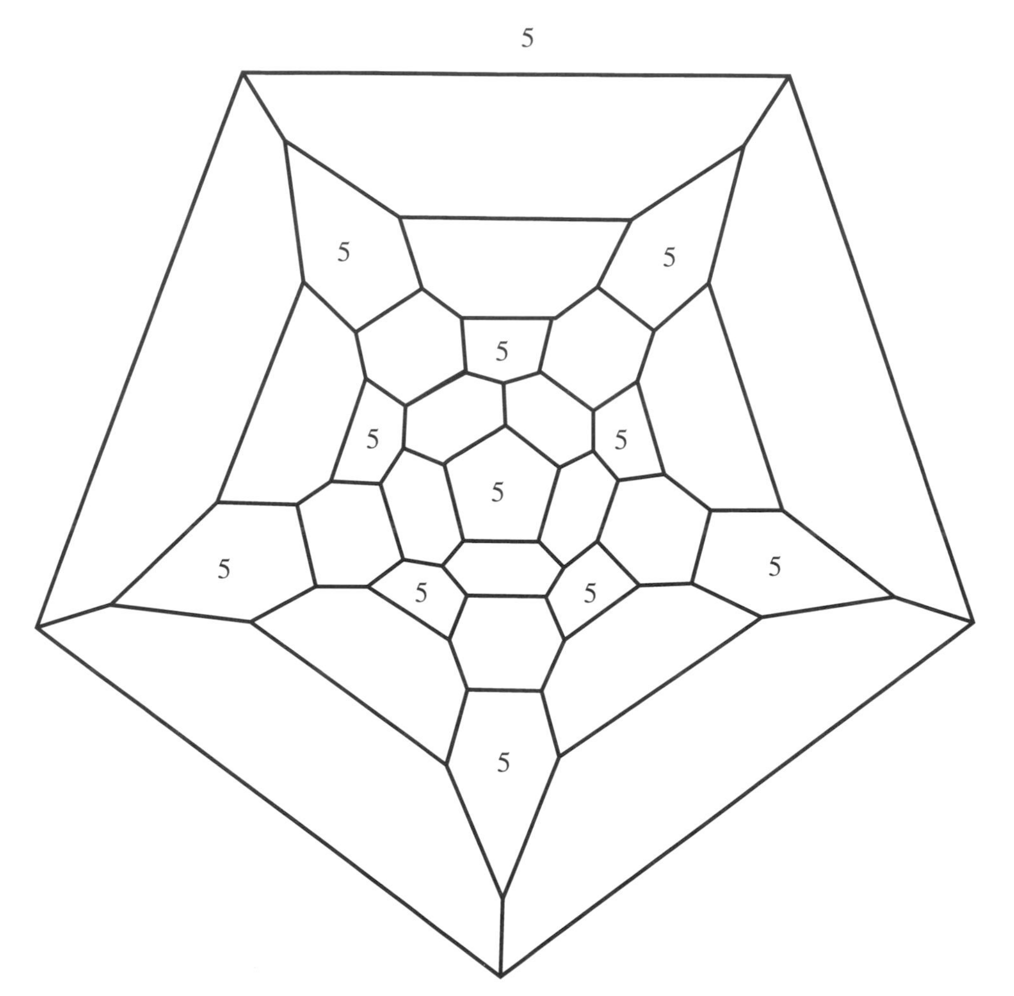

Figure 8.37 Planar map of the buckyball.

We call a map of a polyhedron with these characteristics a *fullerene map*. Since planar maps are easier than polyhedra to tinker with using pencil and paper, some chemists asked themselves what kinds of fullerene maps could exist. Surprisingly, there is one hard and fast rule about fullerene maps.

✯ THEOREM 8.14

A fullerene map must have exactly 12 pentagons.

Proof

We start with simple consequences of Euler's formula (Theorem 8.12) and Eq. (8.2) in combination with condition M_3.

$$6v + 6f - 6e = 12,$$
$$3v = 2e.$$

Using the second equation to substitute for $6v$ in the first gives:

$$6f - 2e = 12. \tag{8.13}$$

Because condition M_2 holds, Eq. (8.7) of Section 8.2 applies. Because a fullerene map has only pentagons and hexagons, Eqs. (8.6) and (8.7) of Section 8.2 give

$$f = f_5 + f_6,$$
$$2e = 5f_5 + 6f_6.$$

Making these substitutions into Eq. (8.13) and simplifying gives $f_5 = 12$. ■

Notice that Theorem 8.14 says nothing about how many hexagons there are. The map of the dodecahedron is a fullerene map and this shows that there could be no hexagons. Could we devise a fullerene map with one hexagon? The best way to study this is by experiment (Figure 8.38). Let's begin by drawing one hexagon. If there is only one hexagon, all the faces surrounding our hexagon are pentagons. This brings us to the top part of Figure 8.38 (keep in mind that all valences are 3). At each of the six 2-valent vertices, we add edges in order to make the valences 3 (bottom part of the figure). Now we have no choice but to connect the ends of these edges, making pentagons around the outer rim. In the drawing we have now reached (not shown), we have no place to "build more" onto the graph. All vertices are 3-valent, so there is no place to add an edge. The map is done! But notice that the infinite face is a hexagon. We have been unable to avoid this second hexagon. Thus, it is impossible to have just one hexagon.

What our experiment has shown us, however, is that there is a fullerene with two hexagons. In fact, it is now known that for any number of hexagons you please, provided it is greater than 1, you can build a fullerene map with that many hexagons and 12 pentagons. Thus, there are a lot of potential fullerenes chemists could look for in nature or try to synthesize in the laboratory. Figure 8.39 shows a model of a fullerene that may be the basis of a new computer revolution based on "micro-microchips." This tubelike molecule has some of the same properties as the silicon devices now used for computer chips, but is much smaller (50,000 of these "nanotubes" would be needed to make a bundle the thickness of a human hair). If chips can be made of these nanotube fullerenes, they would be much smaller and faster than the silicon chips that created the microchip revolution of the 1980s.

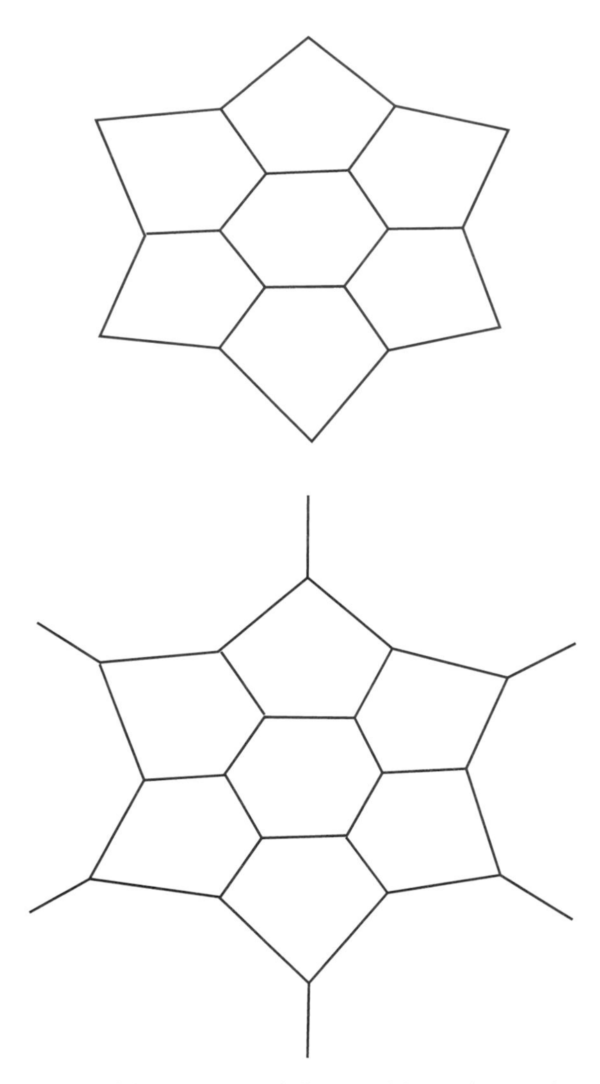

Figure 8.38 Is there a fullerene with one hexagon?

Note that just because you can build a fullerene *map* with a certain number of hexagons doesn't mean you can get carbon atoms to form themselves into a polyhedral *molecule*, a fullerene, with that structure. But if you *cannot* build a fullerene map with a certain number of hexagons, then you cannot build a fullerene molecule with that number of hexagons. In this way, by trading molecules for maps, chemists use mathematics to avoid searching for, or trying to synthesize, molecules that couldn't possibly exist.

Figure 8.39 Alex Zettl holds a model of a fullerene with useful electronic properties. Courtesy of Lawrence Berkeley Labs.

EXERCISES

**Marks challenging exercises.*

Regular Polyhedra

1. Draw a regular map with $p = 3$, $q = 4$ (without looking at Figure 8.23). Can you draw two essentially different examples?

2. Do Exercise 1 for $p = 5$, $q = 3$.

3. Do Exercise 1 for $p = 3$, $q = 5$.

4. Show why the (p, q) combinations (4, 4), (4, 5), (5, 4), and (5, 5) are ruled out for regular maps.

5. Find formulas for f, v, and e in terms of p and q that hold for the regular maps.

6. (a) Show that if two pentagons $ABCDE$ and $A'B'C'D'E'$ are both regular and have the same side length, then they are congruent. (To show them congruent, divide them into triangles by diagonals drawn from one vertex. Then show that each triangle in one pentagon is congruent to a similarly situated one in the other pentagon.)

   (b) Can you prove the analogous result for polygons with an arbitrary number of sides n?

7. Explain why a regular polyhedron cannot have two faces that are not congruent.

8. In our definition of a regular map, for the case where $p = q = 3$, are conditions (a) and (b) necessary or are they implied by conditions (c) and (d)? Try to draw an example of a map satisfying conditions (c) and (d) but failing to satisfy either condition (a) or condition (b). If you can't, can you prove that a map satisfying conditions (c) and (d) must also satisfy conditions (a) and (b)? What about other values of p and q?

9. Imagine an infinite checkerboard. It is very like a regular map, except that it is not finite. It is an example of a *regular tiling* of the plane. Here is the definition of a regular tiling of a plane: an infinite set of polygons covering the plane, not overlapping except at edges, where each polygon is congruent to every other, and where each vertex has the same number of polygons meeting there.

Although we have not proved this, regular tilings satisfy Eq. (8.12) but with e set to infinity, that is,

$$1/p + 1/q = 1/2.$$

What integer solutions can you find to this equation where p and $q \geq 3$? Use them to find drawings of regular tilings besides the infinite checkerboard.

10. This exercise shows, in formulas, exactly how the size of a cube inscribed in a sphere is determined by the size of the sphere. (See Figure 8.36 and the discussion surrounding it.) Suppose the cube has side length s and the diameter of the sphere is d.

    (a) Find a formula, in terms of s, for the length of the diagonal of a face of the cube and then a formula for the length of a diagonal passing through the center of the cube.

    (b) Find a formula relating d to s. Solve for s in terms of d.

11. Find the diameter of a sphere inscribed in a cube of side length s. Make a drawing and use your intuition. What facts about spheres, planes, etc., would have to be proven to make your intuition into a proof?

Fullerene Molecules and Maps

12. Show that there are no fullerenes with two hexagons where these touch each other along an edge. (Try to draw its map and show that you cannot complete it with the right numbers.)

13. Show that there are no fullerenes with two hexagons where these do not touch each other along an edge, but where there is an edge that has one end on each hexagon. (Try to draw its map and show that you cannot complete it with the right numbers.)

*14. Find a fullerene map with three hexagons. (*Hint*: This would be "close to" a dodecahedron map, so start with that. Next, identify three faces all touching one vertex and the ring of six pentagons which encircle that group of three faces. Try to cut some or all of those ring faces in two by extra edges in such a way that you wind up with three hexagons but the same number of pentagons.)

15. Find formulas for the numbers of faces and edges and vertices of a fullerene in terms of f_6.

8.5 List Coloring and the Five-Color Theorem

List Coloring

A common activity in colleges and many other organizations is scheduling. For example, in a college, times of day need to be worked out for classes; in a corporation, meetings need to be scheduled; in a hospital, we have to schedule surgical procedures, diagnostic procedures, meetings, and so on. In Section 8.1 we saw an example of a scheduling problem and how it could be solved by coloring the vertices of a graph. In this section, we present a variation of the idea of vertex coloring, a recently developed concept called *list coloring*, and some of its theory and applications. We also demonstrate a connection between face coloring planar maps and vertex coloring. Finally, we prove the five-color theorem: Every planar map can be face colored with five colors. Let's begin our work with a course scheduling problem, scaled down to deal with just four courses in order to keep things simple.

EXAMPLE 8.12: SCHEDULING

A certain college needs to work out the scheduling of four courses, a, b, c, d. Each course has just one section. There are four time slots $\{1, 2, 3, 4\}$ available, but there are some other complications:

1. For any course, only some of these time slots are really acceptable (see Table 8.4). In a typical college, these restrictions come from a variety of sources. Some education courses are given only in the morning so students can be free for practice teaching in the afternoon. Other courses must be given in the evening for students with daytime jobs. These restrictions are displayed in the second column of the table.

2. Some pairs of courses cannot be given in the same time slot . For example, Calculus I and Physics I are often taken by the same students, so they shouldn't conflict. In addition, if two courses are being taught by the same professor, they can't be in the same time slot. We use X's in Table 8.4 to show conflicts that are not allowed. For example, the X in row a and column b means courses a and b can't conflict.

The many and varied reasons for the restrictions of types 1 and 2 are not important to us. However, it is important to be aware that these two types of restriction have a very different character and will be dealt with differently in our method of solution. Whereas restrictions of type 1 rule out some time slots for a course, restrictions of type 2 are more subtle. For example, there is no restriction on which time of day courses *a* and *b* are given — as long as it is not the same time.

TABLE 8.4 COURSE SCHEDULING RESTRICTIONS

	Type 1 restrictions	Type 2 restrictions			
	Set of acceptable time slots	*a*	*b*	*c*	*d*
a	{3, 4}		X	X	
b	{3, 4}	X		X	
c	{2, 3}	X	X		X
d	{1, 2}			X	

SOLUTION

We draw a graph (Figure 8.40) in which each course is represented by a vertex. Restrictions of type 1 are dealt with by placing the set of possible time slots next to the vertex (ignore the check marks for the moment). Restrictions of type 2 are dealt with by connecting two vertices with an edge if the courses they represent cannot conflict.

Now we need to do a vertex coloring, just as we did in Section 8.1, except that the color for a vertex must come from the appropriate list for that vertex.

For small-scale problems such as this one, trial and error is a reasonable solution method. In Figure 8.40 we might begin by arbitrarily choosing color 2 for *d*, indicated by the check mark in the left-hand version of the graph. Now go to neighbor *c*, which cannot use 2 (since *d* and *c* are connected) and so must use 3. Now *b* is forced to use 4 (since *b* and *c* are connected). This forces *a* to use 3, but this is impossible because *a* and *c* cannot have the same color. This doesn't prove that no list coloring is

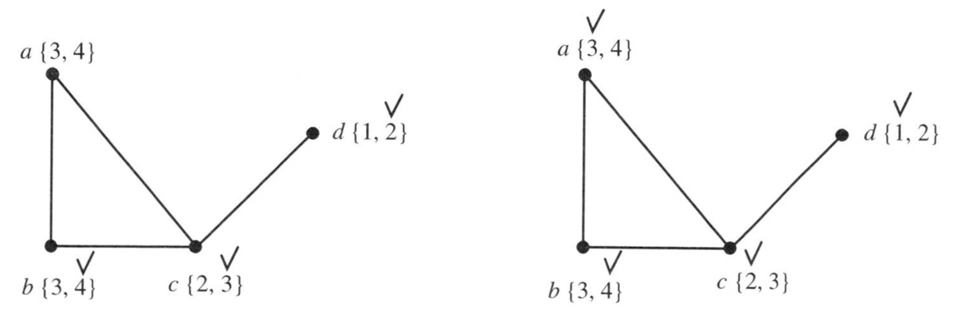

Figure 8.40 The course scheduling problem: A failed attempt at list coloring (left) and a successful list coloring (right).

possible — it only shows we can't start with 2 for d. If we start with 1 for d and pick 2 for c, we will be able to complete a list coloring, as shown in the right-hand version of the graph. ■

DEFINITION

Let G be a graph and suppose that each vertex x has a list L_x of colors associated with it (the lists for the different vertices may be different). To *list color* G means to assign to each vertex a color so that

1. the color for a vertex is contained in the list for that vertex, and
2. if two different vertices are connected by an edge, they get different colors.

If a graph has a multiple edge set — many edges connecting vertices u and v, say — for the purpose of coloring it we can remove all the edges but one connecting u and v. For this reason, in this section we assume that our graphs have no multiple-edge sets.

We have now introduced three kinds of coloring problems: In Section 8.1 we studied face coloring and vertex coloring and now we have introduced list coloring. It is important to keep these coloring concepts straight but also to understand that there are some connections between them. Every vertex coloring problem in which the number of colors is specified can be thought of as a list coloring problem. For example, if we want to vertex color a graph with three colors, we could assign the color set $\{1, 2, 3\}$ to each vertex and look for a list coloring.

But there are differences between list coloring problems and ordinary vertex coloring problems as illustrated in Figure 8.41. This graph can be vertex colored with two colors if we don't have to observe the lists. However, we cannot list color it with the lists given, even though the total number of colors in all the lists is two.

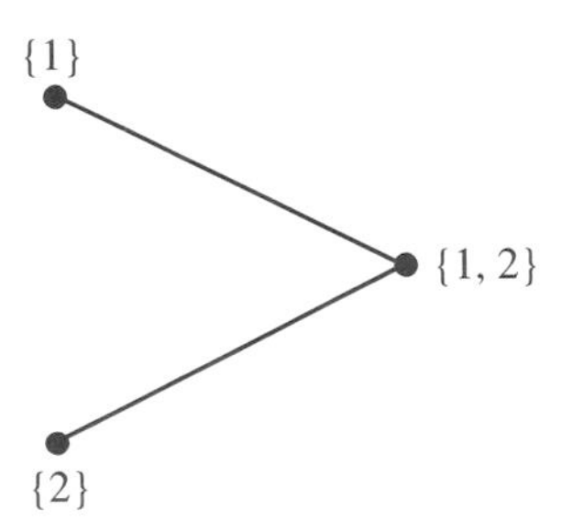

Figure 8.41 A graph that cannot be list colored.

★ THEOREM 8.15

Let G be a graph whose maximum valence is d. If each vertex list has at least $d + 1$ colors, the graph can be list colored using those lists.

PROOF

Arrange the vertices in any order you like and color them using the following procedure. For the first vertex, pick any color from its list. For each following vertex v, examine those of its neighbors which have been previously colored. We need to find a color for v different from all those colors. But v has at most d neighbors, so there are at most d colors to avoid. Since the list for v has $d+1$ colors, there will be at least one possibility. ■

EXAMPLE 8.13

Theorem 8.15 tells us that the graph of Figure 8.42 can be list colored since the maximum valence is 2 and each color list has at least three colors. Thus, it is not a waste of time to search for a list coloring. Can you find one?

Although the applications of list coloring have little to do with geometry, it is interesting that one of the earliest theorems proved about list coloring concerns a category of graphs that share a geometric feature, namely, that they are planar. This theorem, which we state and prove next, also provides a simple route to proving that planar maps can be face colored with five colors. Here is the category of graphs we are concerned with:

DEFINITION

A graph T is a *triangulated simple circuit* (TSC) if

1. it is planar,

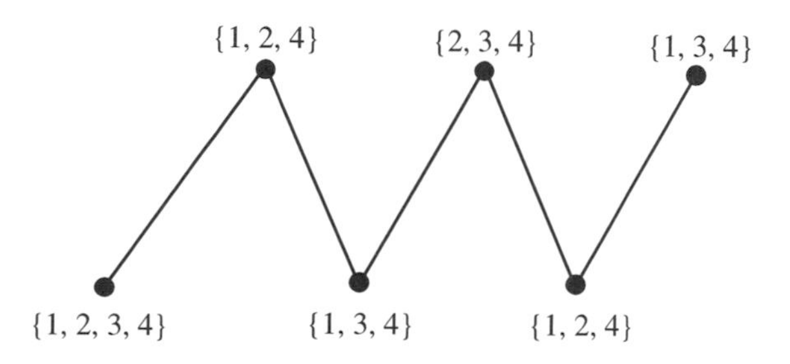

Figure 8.42 Theorem 8.15 says this can be list colored.

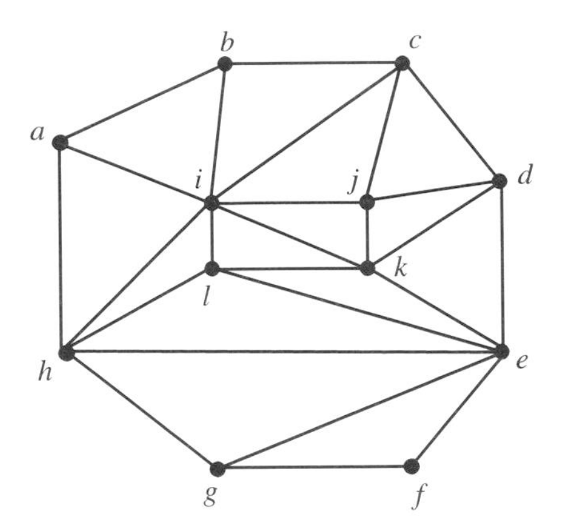

Figure 8.43 A triangulated simple circuit.

2. the boundary of the infinite face is a simple circuit, henceforth called the *boundary circuit*, and
3. all faces, except perhaps the infinite face, are triangles (i.e., simple circuits consisting of three edges), and
4. there are no multiple edge sets (Figure 8.43).

☆ THEOREM 8.16: (THOMASSEN)

Suppose T is a TSC whose boundary circuit vertices, in clockwise order, are $x_1, x_2, \ldots, x_m, x_1$ (Figure 8.44). Suppose further that each vertex has a color list such that

(a) the list for x_1 contains just one color—let's call it 1;

(b) the list for x_2 contains just one color, different from 1, so let's call it 2;

(c) lists for other vertices on the boundary circuit contain at least three colors; and

(d) lists for vertices interior to the circuit contain at least five colors.

Then it is possible to list color G.

PROOF

The proof is by induction on the number of vertices. The smallest triangulated simple circuit is a triangle $x_1\, x_2\, x_3$ with three vertices and no vertices inside the triangle. After coloring x_1 with 1 and x_2 with 2, there is bound to be a color for x_3 since its list has three colors in it.

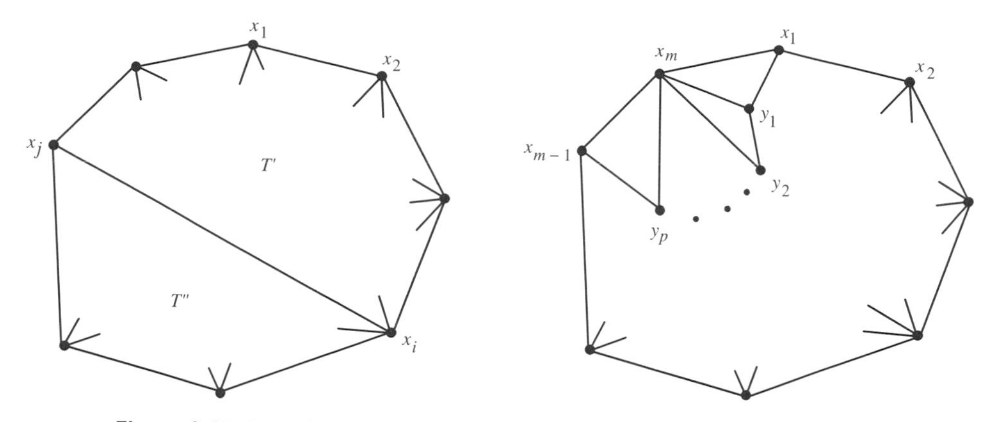

Figure 8.44 Case 1 (left) and case 2 (right) of the proof of Theorem 8.16.

Now suppose the theorem is true for all TSCs of up to n vertices and let us now consider a graph T that is a TSC of $n + 1$ vertices. There are two cases, in each of which we reduce the problem to problems involving smaller TSCs where we apply the inductive hypothesis.

Case 1: There is a diagonal to the boundary circuit which connects a vertex on the circuit to another vertex on the circuit (see the left part of Figure 8.44). Let x_i and x_j be the vertices joined by the diagonal. The diagonal divides the graph into two parts which meet along the diagonal, and each part is a TSC. Let T' be the part containing x_1 and x_2 and let T'' be the other part. Since T' has n or fewer vertices, the inductive hypothesis implies we can list color T' so that x_1 gets color 1 and x_2 gets color 2. Let a and b be the colors used on x_i and x_j, respectively. Then $a \neq b$ since x_i and x_j are connected. We now formulate a new list coloring problem for T''. All the lists are the same as the original lists for T, except that the list for x_j has every color except a removed and the list for x_j has every color except b removed. Because T'' has n or fewer vertices, the inductive hypothesis (with x_i and x_j in the role of x_1 and x_2, and using colors a and b in place of 1 and 2) implies T'' can be list colored with this list. The list colorings of T' and T'' are the same on x_i and x_j, the only vertices where these TSCs overlap. Thus we can reassemble a list colored version of T by putting the list colored graphs T' and T'' together.

Case 2: There is no diagonal for the boundary circuit of T. In this case, x_m must have at least one neighbor other than x_1 and x_{m-1}. If it didn't, since all faces inside the boundary circuit are triangles, there would be an edge from x_{m-1} to x_1 and this would be a diagonal since x_{m-1} and x_1 are different (since there are no multiple-edge sets, $m \geq 3$). Let the neighbors of x_m, starting at x_1 and going around clockwise be $x_1, y_1, y_2, y_3, \ldots, y_p,$

x_{m-1}. Since the faces inside the boundary circuit are triangles, each vertex on this sequence is connected by an edge to the one before and the one after it in the sequence. Thus, the sequence $x_1, y_1, \ldots, y_p x_{m-1}$ is a path. Consequently, $x_1, x_2, \ldots, x_{m-1}, y_p, y_{p-1}, \ldots, y_1, x_1$ is a circuit which we will call C. Because the boundary circuit of T has no diagonals, none of the y_i lie on the boundary circuit. For this reason, and the fact that the boundary circuit of T is simple, it follows that C is a simple circuit. Let $T^\star$ be the graph consisting of the circuit C and all the parts of T that lie inside C. Because all the faces of T (except possibly the infinite face) are triangles, the same is true for $T^\star$. Thus, $T^\star$ is a TSC. We now form lists for $T^\star$ as follows:

(a) All vertices have the lists they had in the original problem involving T, except for $y_1, \ldots, y_p$.

(b) Let a, b, c be any three different colors contained in the list for x_m. At least two of these colors are different from 1, say, a and b. If any of the lists for vertices $y_1, \ldots, y_p$, contain color a or b, remove them from those lists. The original lists had five colors, and at most two have been removed, so now the list for each y_i has at least three colors.

Because $T^\star$ has n vertices the inductive hypothesis applies to $T^\star$ with this list and we can list color $T^\star$. This list coloring is almost a list coloring of T, except that x_m has no color. For x_m use color a unless x_{m-1} got color a, in which case use b. In either case, there is no conflict with x_1, since x_1 gets 1 and neither a nor b is 1. Likewise, there is no conflict with any y_i since no y_i gets a or b, because colors a and b have been removed from the lists of the y_i. Thus, we have a list coloring of T. ■

COROLLARY 8.17

If M is a planar map with all faces triangles and if each vertex has a color list with at least five colors, then M can be list colored.

Duality and the Five-Color Theorem

We next show that the preceding corollary has implications for face coloring a map (as described in Section 8.1). Recall that in face coloring there are no lists of colors. Each face can be given any color as long as faces that have an edge in common get different colors. We wish to show that any planar map can be face colored with five colors. We'll transform this problem to a certain list coloring problem for a related planar map $M^\star$, called the *dual* of M.

We build $M^\star$ as follows: Within each face of M (including the infinite face) place a vertex, called a *dual vertex* (shown bold in Figure 8.45); whenever two faces have an edge in common, connect the dual vertices just placed in those faces with a *dual edge* (shown bold in the figure) crossing the common edge. If two faces have more than one edge in common, as in the middle map of Figure 8.45, then the dual vertices will have more than one dual edge connecting them. If the graph has a face that meets itself along an edge E (right-hand map of Figure 8.45), then the dual vertex for that face has a dual edge which is a loop crossing E.

Two features of $M^\star$ are important for us. First, if we take a little care in drawing the edges of $M^\star$ we can avoid any crossings. Thus *the dual of a planar map is another planar map.* Our second feature applies in the special case where the planar map M is connected and has no face which meets itself along an edge and where each valence is 3 (e.g., graphs like the first one in Figure 8.45, but not like the last two). For such graphs, all faces of $M^\star$ arise from three dual edges "going around" a vertex of valence 3 and are therefore triangular faces. (This may seem obvious, but in the proof, which we leave to the reader, it is necessary to make use of the fact that M is connected and has no face that meets itself along an edge.)

Now suppose we had a vertex coloring of $M^\star$. We can color the faces of M by giving each face the color of the vertex within it. Two such faces which have an edge in common will not get the same color because the vertices within them were connected by an edge and couldn't have gotten the same color in the vertex coloring of $M^\star$. These are some of the key ideas we'll use in our next theorem.

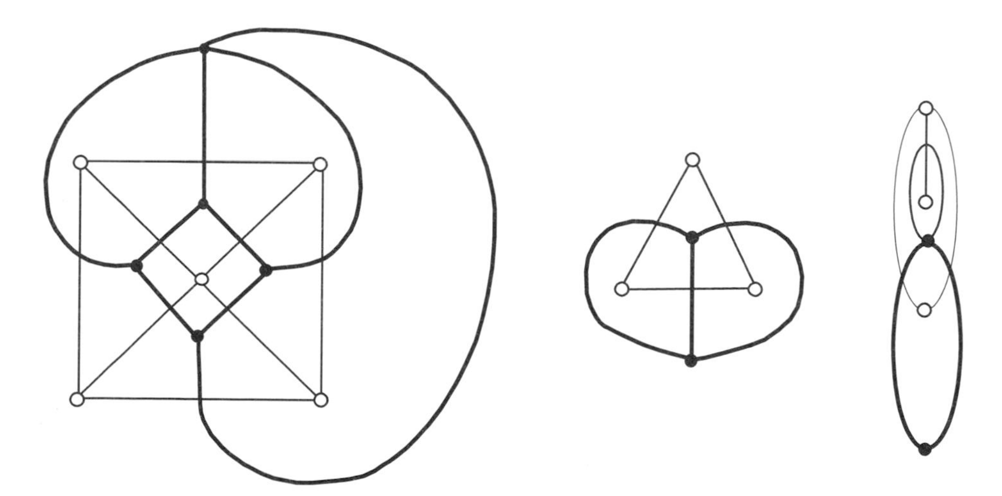

Figure 8.45 Three maps (hollow vertices) and their duals (bold).

The first part of our next theorem, which contains restrictive hypotheses, is merely a stepping stone to prove the less restrictive second part. But, you may wonder if it is really possible to color a map that has an edge where a face meets itself. In a coloring, we normally expect two different colors to appear on the two sides of an edge. This, of course, is impossible if the two sides of an edge are the same face. But if you examine our definition of coloring carefully, you will note that it only says that if the two sides of an edge are *two distinct faces*, then the two sides have different colors.

☆ THEOREM 8.18

1. If M is a connected 3-valent planar map, and has no face which meets itself along an edge, it can be face colored with five colors.
2. If M is any 3-valent planar map, it can be face colored with five colors.

PROOF

1. First create the dual planar map, $M^\star$. As we have seen, because M has no face which meets itself along an edge, and M is 3-valent, each face of $M^\star$ is a triangle. Assign to each vertex of $M^\star$ the same list $\{1, 2, 3, 4, 5\}$. By Corollary 8.17, $M^\star$ can be list colored. Use this coloring to create a face coloring of M with five colors, as in the discussion preceding this theorem.
2. We proceed by induction on n = number of components of M + number of faces meeting themselves on an edge. If $n=1$, we have case 1 and the base of the induction is established. If $n > 1$, reduce n by one of two means. If there is a face F meeting itself along an edge, add two vertices and an edge uv as shown in Figure 8.46a. By the inductive hypothesis, the new map can be face-colored with five colors. When the added vertices and edge are removed to recreate M, leave the colors on the faces to get a face coloring for M with five colors. If no such face can be found but M has more than one component, join two components to make one as in Figure 8.46b. By the inductive hypothesis, the new map can be face-colored with five colors. When the added vertices and edges are removed to recreate M, leave the colors on the faces to get a face coloring for M with five colors. ■

To see how to create N, suppose first that M has just one edge xy (Figure 8.46) where some face meets itself. Add two new vertices, u and v, and the new edge uv, so that we have a 3-valent map in which no face meets itself along an edge. This is the N we want because if we face color it and then remove the added items we get a face coloring of M. In the event that M has more than one edge where a face meets itself, this construction needs to be done once for each such edge. ■

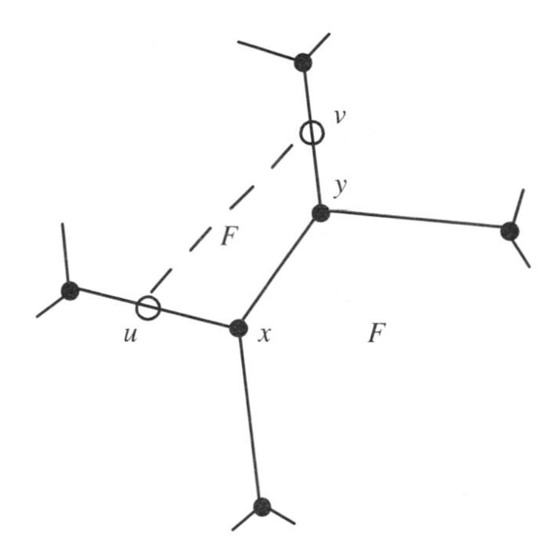

Figure 8.46 Eliminating a face meeting itself (a); reducing components (b).

✯ THEOREM 8.19: THE FIVE-COLOR THEOREM

Every planar map can be face colored with five colors.

PROOF

Let G be any planar map. We might as well assume that there are no isolated vertices (vertices of valence 0) since such vertices can be ignored in face coloring. Now truncate each vertex (Figure 8.47): This means draw a little circle around such a vertex, erase what's in the circle, and create new vertices where the circle crossed edges. The circular arcs between the new vertices are new edges. Now we have a map that is 3-valent and, according to the previous theorem, can be face colored with five colors. Now shrink these truncating circles back to single vertices, allowing the original graph to reappear. The colors in the circles disappear. If two faces had an edge in common in the original map, they had one in common in the truncated map. Because the truncated map

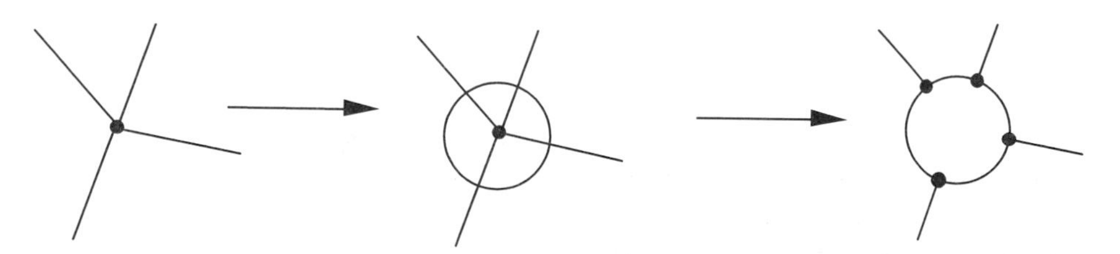

Figure 8.47 Truncating a vertex to produce 3-valence.

is properly colored, these two faces have different colors. Thus, we have a face coloring of the original map. ■

EXERCISES

**Marks challenging exercises.*

List Coloring

1. Find a list coloring of the graph shown in Figure 8.48.

2. In Exercise 1, remove color *e* from the list containing it. Prove that it is now not possible to list color the graph.

3. (a) Suppose G is a graph with five vertices and four edges, consisting of a single path from x_1 to x_5. Suppose each list has two colors (but not necessarily the same two in each case). Show by examples that we can pick the color for x_1 at will and still complete a list coloring. Can you explain why it is always possible?

   (b) Suppose you want to pick the color of the first and last vertex at will. Can you always complete the list coloring?

   (c) Answer parts (a) and (b) if the path has more or fewer than five vertices.

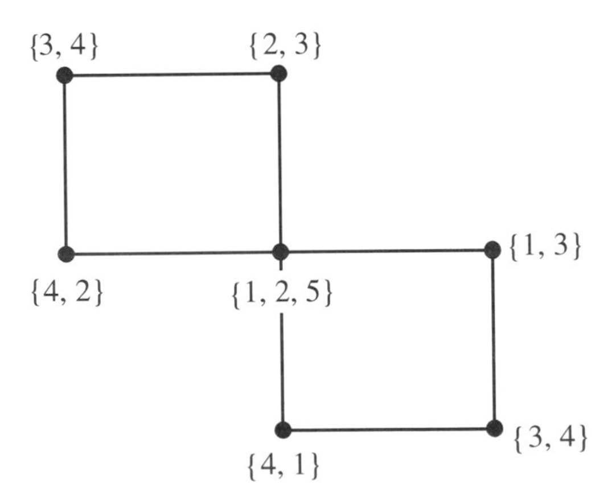

Figure 8.48

4. Draw an example of a graph that contains no simple circuit. (Such a graph is called a *tree*.) Now pick any lists for the vertices where each list has two colors (not necessarily the same two for each list). Show that you can color the graph according to that list. Can you prove that this will always be possible provided the graph has no simple circuit and each list has two colors?

5. Suppose a graph consists entirely of a single simple circuit of four edges (like the corners and sides of a square). Suppose each list is $\{1, 2\}$. Show a list coloring.

6. How many list colorings are possible in the previous exercise?

7. If G consists of a single simple circuit of even length and each list is the same and has size 2, show that G can be list colored.

8. Suppose a graph has simple circuits, but all of them are of even length (the number of edges in each circuit is even). Prove that if each list is $\{1, 2\}$ a list coloring is possible.

9. In Exercise 8, show by example that if the number of vertices is odd and all lists are the same, G cannot be list colored.

10. If G consists of a single simple circuit and each list has size 2, but not all lists are the same, is it true that G can be list colored? Prove this, or show by example that it is not true.

11. Find an example of a list coloring problem on some graph where the union of the lists has three colors and the list coloring cannot be done, but the graph can be vertex colored with three colors. (*Hint:* Examine Figure 8.41, which shows how to do this for two colors.)

12. Can Exercise 11 be carried out for all numbers higher than 3?

13. Suppose x and y are two vertices in a graph where lists have been specified for each vertex. We remove x and y and all edges touching them. We now add a new vertex z, and connect it to each remaining vertex x used to be connected to and to each remaining vertex y used to be connected to. We set $L_z = L_x \cup L_y$. Is it true that if the original graph could be list colored, then the new graph can be list colored? If so, prove it. If not, give an example to show it.

Duality and the Five-Color Theorem

14. Draw the dual map of the map of the octahedron (Figure 8.23). Draw the dual map of the map of the cube. Do you see a relationship?

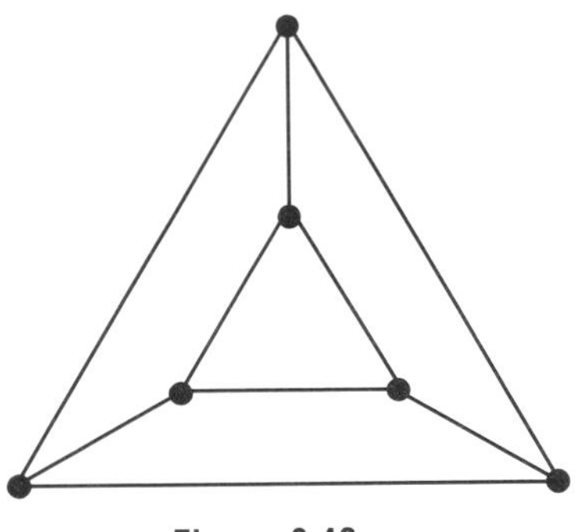

Figure 8.49

15. Find an example of a map that is self-dual, that is, $M^{\star} = M$.

16. Is it true that the dual of any regular map is a regular map? How would you convince someone of your answer?

17. Suppose $M^{\star}$ is the map of the cube. What is M?

18. Suppose G consists of two vertices connected by an edge. Is G the dual of any planar map? If so, what? If not, why not?

19. Let M denote the map shown in Figure 8.49. Find $M^{\star}$. Then find $M^{\star\star}$. Compare $M^{\star\star}$ to M. What does your comparison suggest?

20. Let G be a 3-valent planar map whose faces all have no more than five edges. Each face has a list of at least six colors. Is it possible to find a face coloring of G where the color for a face comes from the list for that face? If so, prove it. If not, show a counterexample.

21. In Theorem 8.15, if we assume instead that each list has d colors, could we still prove that a list coloring is possible? If so, do the proof. If not, find a counterexample.

22. Suppose we relax the hypothesis of Theorem 8.16 so that the faces inside the boundary circuit are not necessarily three sided. We retain hypotheses (a)–(d) about the lists. Prove that such a map can be list colored.

23. Prove that if every 3-valent planar map can be face colored with four colors, then all planar maps can be face colored with four colors.

Bibliography

Chapters 1 and 2

Cederburg, Judith N., *A Course in Modern Geometries*, Springer Verlag, 1989.

Dickerson, Matthew, *Voronoi Diagrams and Proximity Problems*, Geomap Module, available from COMAP, 57 Bedford St., Lexington, MA 02173.

Heath, Thomas L., *The Thirteen Books of Euclid's Elements*, Dover Publications, Inc., 1956.

Heath, Thomas L., *A History of Greek Mathematics*, Dover Publications, Inc., 1981.

Katz, Victor, *A History of Mathematics*, Harper Collins, 1993.

Loomis, Elisha Scott, *The Pythagorean Proposition*, National Council of Teachers of Mathematics, 1968, Reston, VA.

Moise, Edwin E., *Elementary Geometry from an Advanced Standpoint*, 2nd ed., Addison Wesley, 1974.

O'Rourke, Joseph, *Computational Geometry in C*, Cambridge University Press, 1994.

Prenowitz, Walter, and Jordan, Meyer, *Basic Concepts in Geometry*, Ardsley House Publishers, 1989.

Servatius, Brigitte, *Rigidity and Braced Grids*, Geomap Module, available from COMAP, 57 Bedford St., Lexington, MA 02173.

Trudeau, Richard J., *The Non-Euclidean Revolution*, Birkhauser, 1987.

Chapter 3

Gans, David, *An Introduction to Non-Euclidean Geometry*, Academic Press, 1973.

Greenburg, Marvin Jay, *Euclidean and Non-Euclidean Geometry*, 2nd ed., W. H. Freeman and Co., 1980.

Prenowitz, Walter, and Jordan, Meyer, *Basic Concepts of Geometry*, Xerox College Publishing, 1965.

Trudeau, Richard, J., *The Non-Euclidean Revolution*, Birkhäuser, 1987.

Chapter 4

Bentley, W.A., and Humphries, W.J., *Snow Crystals*, Dover, 1962.

Crowe, Donald W., *Symmetry, Rigid Motions and Patterns*, HIMAP Module, available from COMAP, 57 Bedford St., Lexington, MA 02173, 1986.

Martin, George E., *Transformation Geometry*, Springer-Verlag, 1982.

Seneschal, Marjorie, *Crystalline Symmetries*, Adam Hilger, 1990.

Tana, Pradumna, and Tana, Rosalba, *Traditional Designs from India for Artists and Craftsmen*, Dover, 1988.

Washburn, Dorothy K., and Crowe, Donald W., *Symmetries of Culture: Theory and Practice of Plane Pattern Analysis*, University of Washington Press, 1991.

Yale, Paul B., *Geometry and Symmetry*, Dover, 1988.

Chapter 5

Hoffmann, Banesh, *About Vectors*, Prentice-Hall, 1966.

Rogers, David F., and Adams, J. Alan, *Mathematical Elements for Computer Graphics*, 2nd ed., McGraw-Hill, 1990.

Strang, Gilbert, "The Mathematics of GPS," *SIAM News*, vol. 30, no. 5, June 1997, p. 1.

Strang, Gilbert, and Borre, Kai, *Linear Algebra, Geodesy, and GPS*, Wellesley-Cambridge Press, 1997.

Chapter 6

Craig, John J., *Introduction to Robotics*, Addison Wesley, 1986.

Gardan, Yvon, *Numerical Methods for CAD*, MIT Press, 1986.

Hoggar, S. G., *Mathematics for Computer Graphics*, Cambridge University Press, 1992.

Rogers, David F., *Mathematical Elements for Computer Graphics*, McGraw-Hill, 1990.

Chapter 7

Foley, James, Van Dam, Andries, Feiner, Steven, and Hughes, John, *Computer Graphics*, Addison Wesley, 1990.

Grosvenor, Gilbert, *The Round Earth on Flat Paper*, National Geographic Society, 1947.

Mainwaring, James, *An Introduction to the Study of Map Projection*, Macmillan, 1960.

Maxwell, E. A., *General Homogeneous Coordinates in Space of Three Dimensions*, Cambridge University Press, 1951.

Smart, James, *Modern Geometries*, Brooks/Cole, 1994.

Wallace, Edward C., and West, Stephen F., *Roads to Geometry*, Prentice-Hall, 1992.

Wilford, John Noble, *The Mapmakers*, Vintage Books, 1982.

Wylie, C. R., *Introduction to Projective Geometry*, McGraw-Hill, 1970.

Chapter 8

Agarwal, Pankaj K., and Pach, János, *Combinatorial Geometry*, Wiley-Interscience, 1995.

Beck, Anatole, Bleicher, Michael, and Crowe, Donald W., *Excursions into Mathematics*, Worth, 1967.

Behzad, Mehdi, and Chartrand, Gary, *Introduction to the Theory of Graphs*, Allyn and Bacon, 1971.

Cromwell, Peter R., *Polyhedra*, Cambridge University Press, 1997.

Grünbaum, Branko, *Convex Polytopes*, John Wiley and Sons, 1967.

Lyusternik, L. A., *Convex Figures and Polyhedra*, Dover, 1963.

Roberts, Fred S., *Applied Combinatorics*, Prentice Hall, 1984.

Thomassen, Carsten, "Every Planar Graph is 5-Choosable," *Journal of Combinatorial Theory*, Series B 62, 1994, pp. 180–181.

Ziegler, Gunter M., *Lectures on Polytopes*, Springer-Verlag, 1994.

Answers to Odd-Numbered Exercises

Chapter 1

Chapter 1, Section 1.1

1.

Triangle legs	True hypotenuse	Hypotenuse by 1/3 rule	% Error
1,2	$\sqrt{5} = 2.24$	2.33	−4.0
1,3	$\sqrt{10} = 3.16$	3.33	−5.4
1,4	$\sqrt{17} = 4.12$	4.33	−4.1
1,5	$\sqrt{26} = 5.10$	5.33	−4.5

3. The derivative of the percentage error function is not 0 for any value of b in $[1, 3]$—it is always negative. Thus the maximum is at the endpoint $b = 1$ (recall that $b \geq 1$). The maximum absolute value of the % error is about 5.74%. Figure A.1 is a graph showing the percent error for b in $[1, 5]$.

5. Without measurement, you could buy a box of seed, use it, then make an "eyeball" estimate of what fraction of the field you covered. Use that to see how much more seed to buy. This requires two trips to the store. If you can measure the length and width of the field and compute the area, and if the seed box tells how many square feet it covers, then you can calculate how many boxes will be needed and make just one trip.

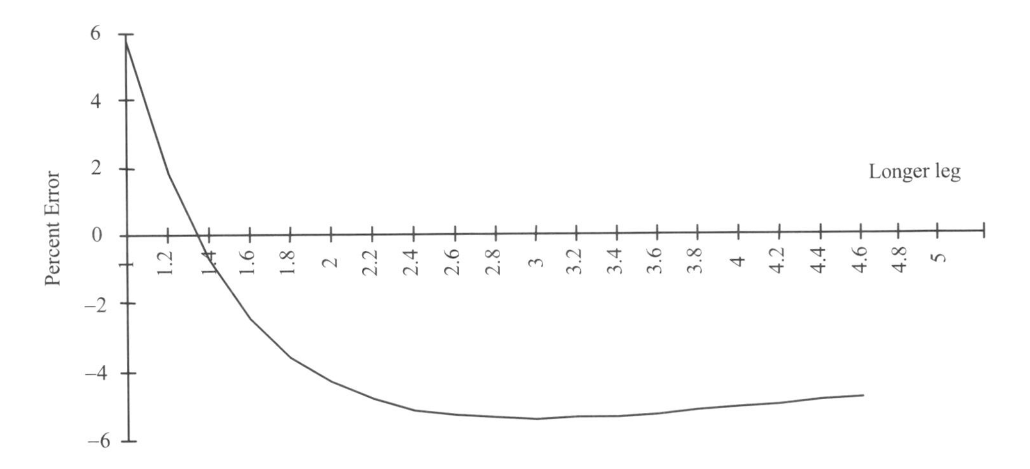

Figure A.1 Percent error in hypotenuse estimation [100(true – est)/true].

7. $(\text{altitude})^2 + (\text{walk})^2 = (\text{string})^2$. Solve for altitude.

9. The author found the Pythagorean theorem surprising, but the theorem that base angles of isosceles triangles are congruent is not surprising at all. Unfortunately, you can't prove the Pythagorean theorem without first proving the base angles theorem or some other "obvious" theorems first.

11. The equation of the line connecting (5, 12) to (10, 14) is $y = (2/5)x + 10$. A_1 is the intersection of this line with $x = 6$, namely, (6, 12.4). A_2 is (7, 12.8), A_3 is (8, 13.2), and A_4 is (9, 13.6).

13. Equation of the line:

$$y - s_2 = (x - s_1)\,\frac{e_2 - s_2}{e_1 - s_1}.$$

Substitute $x = s_1 + i$ to get

$$y = s_2 + i\,\frac{e_2 - s_2}{e_1 - s_1}.$$

15. Suppose the line is very steep, say, with slope 10. The algorithm we gave would choose points that are far apart in their y coordinates (although just one unit in x). For example, if the start is (0, 0) and the end is (1, 10), the algorithm will pick no points between start and end. A lot of close-together points are what is needed.

17. (a) $(s_1 + 1,\ s_2)$, $(s_1 - 1,\ s_2)$, $(s_1,\ s_2 + 1)$, $(s_1,\ s_2 - 1)$.

    (b) There is more arithmetic to do in this distance measurement algorithm than in the vertical grid line algorithm.

    (c) You will somethimes get different pixels. For example, in Figure 1.3, the vertical grid line algorithm gave diagonally opposite pixels for the second and third pixels. This never happens with the distance algorithm because a "neighbor" of a pixel (according to our definition) is on a grid line with it.

Chapter 1, Section 1.2

1. (a) $(OQ)(OP) = r^2$ yields $d(d + 2a) = r^2$, which yields $d^2 + 2ad = r^2$. Add a^2 to both sides and factor to get $(d + a)^2 = r^2 + a^2$.

   (b) $(d + a)^2 = r^2 + a^2 \geq r^2$. By taking the square roots of both sides of the inequality we get $d + a \geq r$.

3. It is necessary to show that the circles in (i) intersect and that the circles in (ii) intersect.

Chapter 1, Section 1.3

1. Condition 3 ensures that no matter how large a number you take, there will always be a point with that coordinate.

3. Still true: parts 1 and 3 of Theorem 1.2, Theorem 1.3, and Theorem 1.4. No longer true: part 2 of Theorem 1.2.

5. "Segments" and "rays" will be the same as before. In our definition of a segment we would not have to specially specify that A and B are in $\overline{AB}$ because, with the new meaning of "between," we have A-A-B and A-B-B.

7. If $f(C) < f(B) < f(A)$ then $AB + BC = |f(B) - f(A)| + |f(C) - f(B)| = f(A) - f(B) + f(B) - f(C) = f(A) - f(C) = |f(C) - f(A)| = AC$.

9. For M to be the midpoint we need $|f(M) - f(A)| = |f(B) - f(M)|$. If we take M to be the point where $f(M) = [f(A) + f(B)]/2$, we can check by substitution that this holds.

11. By part 3 of Theorem 1.2 we have that A-C-B, which means C is in $\overline{AB}$. Thus we have shown $\overrightarrow{AB} \cap \overrightarrow{BA} \subset \overline{AB}$. For the other containment, let C be in $\overline{AB}$. By

Theorem 1.5, C is in $\overrightarrow{AB}$. Since $\overline{AB} = \overline{BA}$, C is in $\overline{BA}$ and Theorem 1.5 tells us C is in $\overrightarrow{BA}$. Thus C is in $\overrightarrow{AB} \cap \overrightarrow{BA}$. Thus we have shown $\overline{AB} \subset \overrightarrow{AB} \cap \overrightarrow{BA}$.

13. Let the coordinates of A, B, and C be a, b, c. A-B-C means $a < b < c$ or $c < b < a$. We'll assume $c < b < a$ (the proof is similar for the other case). Now suppose X is in $\overline{AB} \cup \overline{BC}$ and let x be the coordinate of X. Either X is in $\overline{AB}$ or X is in $\overline{BC}$. Assume X is in $\overline{AB}$ (the proof for the other case is similar). Then $b \leq x \leq a$. But since $c < b$ this means $c < x \leq a$ so X is in $\overline{AC}$. Thus we have shown $\overline{AB} \cup \overline{BC} \subset \overline{AC}$.

    Now suppose Y is in $\overline{AC}$. Then $a \leq y \leq c$. Now there are two cases according to whether $y \leq b$ or $b < y$. If $y \leq b$ then $a \leq y \leq b$ so Y is in $\overline{AB}$. If $b < y$ then $b < y \leq c$ so Y is in $\overline{BC}$. Since Y is in $\overline{AB}$ or Y is in $\overline{BC}$, Y is in $\overline{AB} \cup \overline{BC}$. We've shown $\overline{AC} \subset \overline{AB} \cup \overline{BC}$.

15. Let ABC and $A'B'C'$ be two triangles that are congruent under the original ruler functions with A corresponding to A', B to B', and C to C'. Suppose further that $\overleftrightarrow{A'B'}$ is the line L but none of the other triangle vertices lie on L. (We omit the proof that such congruent triangles exist.) Now if we replace f_L by g_L, $\overline{A'B'}$ will double its original length, but no other lengths or angles of the triangles will change. Since $A'B' = 2AB$, the triangles are not congruent, but by SAS they must be. This is a contradiction.

17. (a) Self-intersecting means that at least two of the segments intersect at a point that is not a vertex. Not-self-intersecting and convex means that no two segments intersect except at vertices and each diagonal determines a line that has the other two vertices on opposite sides. A quadrilateral is not self-intersecting and not convex if no two segments intersect except at vertices and at least one diagonal determines a line which has the other two vertices on the same side.

    (c) Use the interior of a triangle as a building block concept. How would you describe the triangles to be used?

19. We need to show that D and C are on the same side of $\overleftrightarrow{AB}$ (and that D and B are on the same side of $\overleftrightarrow{AC}$). This is equivalent to showing that $\overline{DC}$ does not intersect $\overleftrightarrow{AB}$. If there is an intersection X, X would have to lie on $\overleftrightarrow{DC}$ and would have to be B since $\overleftrightarrow{AB}$ and $\overleftrightarrow{DC}$ can only intersect once. But if B is in $\overline{DC}$, since B is not D or C, then D-B-C. This is incompatible with B-D-C by part 3 of Theorem 1.2.

21. Suppose $\overleftrightarrow{AB}$ intersected $\overline{BC}$ in a point X that was not on $\overline{AB}$. Then $X \neq B$. X lies on both $\overleftrightarrow{AB}$ and $\overleftrightarrow{BC}$. But so does B. These two lines can have at most one point in common, so $X = B$, a contradiction. A similar proof shows $\overleftrightarrow{AB}$ cannot intersect $\overline{AC}$.

Chapter 2

Chapter 2, Section 2.1

1. There are six correspondences:

(a) $A \to P$	(b) $A \to P$	(c) $A \to Q$	(d) $A \to Q$	(e) $A \to R$	(f) $A \to R$
$C \to Q$	$C \to R$	$C \to P$	$C \to R$	$C \to P$	$C \to Q$
$B \to R$	$B \to Q$	$B \to R$	$B \to P$	$B \to Q$	$B \to P$

(a) and (b) are congruences.

3. From the first congruence $m\angle B = m\angle P$ and from the second $m\angle B = m\angle Q$. Thus, $m\angle P = m\angle Q$ and by Exercise 18, $RP = RQ$.

5. From Exercise 4, we know ABC has at least two angles equal, say, $m\angle A = m\angle C$. We will show that $m\angle B \neq m\angle A$ implies a contradiction. Let P be the vertex corresponding to A in the first correspondence, and let Q correspond to B and R to C. Q is the only vertex that could correspond to B in any congruence, because the angles at P and R, being congruent to those at A and C, are not congruent to $\angle B$. Now there are only two possibilities for what vertex A matches up with: P and R. And once A has been matched, the match for C is determined. Thus we only have two possible congruences, contrary to the hypothesis.

7. Triangles ABC and DCB are congruent by SAS. The diagonals are corresponding parts.

9. Unless the figure is a rectangle, $m\angle BAC \neq m\angle EDC$. This becomes visually obvious if you draw a parallelogram $ABED$ in which $m\angle A$ is quite small and $m\angle D$ is close to $180°$.

11. Extend $\overline{BC}$ past C. In place of the angle marked θ in Figure 2.6, we deal with its vertical angle at C. Instead of choosing M to be the midpoint of $\overline{BC}$ pick it to be the midpoint of $\overline{AC}$. Now do the same construction: Extend $\overline{BM}$ to E, connect E to C, splitting the angle of interest, etc.

13. If $AB = AC$ then the angles would be congruent, which they are not. So the sides have different lengths. If $AB < AC$ then Exercise 12 shows $m\angle B > m\angle C$, which contradicts our hypothesis. The only remaining possibility is $AB > AC$.

15. If $AB < A'B'$ we can find a point D' on $\overline{A'B'}$ so that $AB = B'D'$. Triangles $D'B'C'$ and ABC are congruent by SAS. Therefore $m\angle C'D'B' = m\angle CAB = m\angle C'A'B'$. This contradicts the exterior angle inequality.

17. Suppose the triangles are ABC and $A'B'C'$ with $m\angle A = m\angle A'$, $AB = A'B'$ and $m\angle B = m\angle B'$. If we also had $AC = A'C'$, we could apply SAS to get congruence. So suppose $AB < A'B'$. Then find a point D' on $A'C'$ where $AB = A'D'$. Triangles ABC and $A'D'C'$ are congruent by SAS. Thus $m\angle B = m\angle A'B'D' < m\angle B'$, a contradiction. If $AB > A'B'$ a similar proof works, in which we build a copy of $A'B'C'$ in ABC with D on $\overline{AC}$.

19. Triangle ABD is isosceles so $m\angle ABD = m\angle ADB$. Likewise $m\angle DBC = m\angle BDC$. By adding these equations we get $m\angle B = m\angle D$. This allows us to apply SAS.

21. Having a machine check congruence is different from us doing it because we can easily "eyeball" the triangles and see which vertex in one should be matched up with which vertex of the other. One way for the machine to overcome this problem is to test each of the six possible correspondences (see Exercise 1). In testing a correspondence, use the distance formula to find the lengths of corresponding sides. If all three corresponding sides are congruent, SSS tell us we have a congruence. If not, go on to the next correspondence. A small savings in computation can be made by not working out the square roots in the distance formula (just compare squares of distances).

23. We proceed by contradiction. Suppose a circle, centered at C, crossed the line at three or more points. If we pick out three then one is between the other two and we can call the end ones X and Z with the in-between one being called Y. Since $XC = YC = ZC$, we have some isosceles triangles to analyze. Since base angles of isosceles triangles are equal, $m\angle CXY = m\angle XYC$. Likewise $m\angle CXY = m\angle YZC$. From these two equations we get $m\angle XYC = m\angle YZC$. But $\angle XYC$ is an exterior angle of triangle CYZ and $\angle YZC$ is one of its remote interior angles. By the exterior angle inequality, they can't have equal measure. This contradiction proves the result.

25. Let the quadrilaterals be $ABCD$ and $A'B'C'D'$ with $AB = A'B'$, $m\angle B = m\angle B'$, etc. Connect A to C with a diagonal and connect A' to C'. Apply SAS to triangles ABC and $A'B'C'$ to get the diagonals congruent. Then SSS gives the other triangles congruent. The two congruences allow us to prove that all corresponding parts of the quadrilaterals are congruent.

27. A diagonal divides the quadrilateral into congruent triangles (SSS), each of which is isosceles and so has congruent base angles. Thus, we get that opposite angles of the quadrilateral are congruent and each diagonal bisects an angle at both of its ends. The four triangles created by the diagonals are congruent by ASA. Thus, the angles where the diagonals meet are all equal and so are all right angles.

29. Carry out the same procedure for each ship, using the same spike. Say the ships are at E_1 and E_2. Say that sighting the first ship, after walking according to the method and lining up with the spike, brings you to D_1 and sighting the second brings you to D_2. You can prove that $E_1E_2 = D_1D_2$.

Chapter 2, Section 2.2

1. Triangles BMA and BMC are congruent by SAS. $BA = BC$ since they are corresponding parts.

3. The first arc has to have radius big enough to cut L. The second and third radii have to be equal.

5. No.

7. $m\angle i > m\angle r$. As an example, take the extreme case where the speed after the bounce is infinite (i.e., it takes no time at all to get from any point on the mirror to Q). Then the quickest path to Q involves going to the mirror with the shortest path—the pependicular dropped from P to the mirror. In this case $m\angle i = 90° > m\angle r$.

9. The Voronoi diagram is formed by two lines crossing at the center of the rectangle. Each of these lines passes through the midpoints of two opposite sides.

11. The key is that all the perpendicular bisectors of the sides of the pentagon meet in the center of the circle, so the five Voronoi regions are like pie slices, meeting at the center, going out to infinity. Each Voronoi region is an angle plus its interior.

13. $Q(65, 40)$ lies in $V(B)$. $QA = \sqrt{(65 - 40)^2 + (40 - 40)^2} = 25$. $QB = \sqrt{125} = 11.18$, $QC = \sqrt{325} = 18.02$, $QD = \sqrt{425} = 20.62$. Q is closest to B.

15. After studying Figure 2.17 we see that a point with a very small width (y coordinate) could have a length greater than C's but still lie in $V(D)$ [e.g., a point like (10, 85)].

17. The points in all the H_{1i} are as close to S_1 as to any of $S_2, S_3, \ldots, S_n$. Thus, the intersection is $V(S_1)$.

19 and 20. Let A and B be in $U \cap V$. Then A and B are in U and, since U is convex, $\overline{AB}$ lies wholly in U. Likewise, $\overline{AB}$ lies wholly in V. Since $\overline{AB}$ lies wholly in U and $\overline{AB}$ lies wholly in V, it follows that $\overline{AB}$ lies wholly in $U \cap V$.

21. Each Voronoi region is the intersection of half-planes (Exercise 17) and so is convex by Exercise 20.

23. We'll show that M, the midpoint of $\overline{S_1S_2}$, is in $V(S_1)$ and $V(S_2)$. If $i > 2$, $2MS_1 = S_1S_2 < S_1S_i$ by hypothesis. The triangle inequality implies $S_1S_i < S_iM + MS_1$. Thus, $2MS_1 < S_iM + MS_1$ from which we obtain $MS_1 < S_iM$. The same argument shows $MS_2 < S_iM$. Thus, the closest sites to M are S_1 and S_2 (both equally close) and all others are farther, so M is in $V(S_1) \cap V(S_2)$.

25. See Figure A.2.

27. (a) Take any equilateral triangle with a horizontal base and center (point where the perpendicular bisectors of the sides meet) at the origin. There are many such triangles and all have the same Voronoi diagram.

    (b) If we start with a nonequilateral triangle, we can construct a similar triangle with the same Voronoi diagram as follows. Connect I to S_1 and extend an equal amount beyond S_1 to S_1'. Obtain S_2' and S_3' similarly. Now prove that L_1, L_2, and L_3 are perpendicular bisectors of the sides of triangle $S_1'S_2'S_3'$.

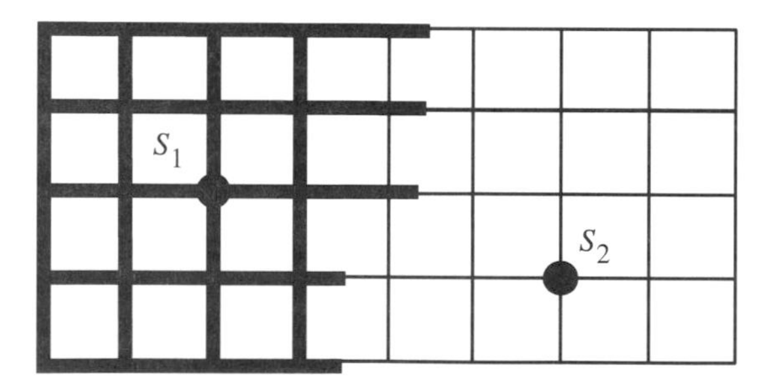

Figure A.2

Chapter 2, Section 2.3

1. If the lines were to meet, we would have a triangle in which one of the alternate interior angles is an exterior angle and the other is a remote interior angle. This contradicts Theorem 2.2, the exterior angle inequality.

3. Let the quadrilateral be $ABCD$ with M being the intersection of the diagonals AC and BD. The hypothesis, along with the equality of vertical angles, makes triangles BMA and DMC congruent with $\angle MDC$ corresponding to—and therefore having the same measure as— $\angle ABM$. These are alternate interior angles for opposite sides of the quadrilateral so these sides are parallel. The other pair of opposite sides is shown to be parallel with the same type of proof.

5. (a) If $\angle 1$ and $\angle 2$ are corresponding angles, then the vertical angle to $\angle 1$ will be alternate interior to $\angle 2$. These alternate interior angles are congruent. Therefore $\angle 1$ and $\angle 2$ are congruent also by Theorem 2.10, part (a).

   (b) If $\angle 1$ and $\angle 2$ are corresponding angles which are congruent, then the vertical angle to $\angle 1$ is congruent to $\angle 2$. But these are alternate interior so we have that alternate interior angles are congruent. By part (b) of Theorem 2.10, the lines are parallel.

7. Let the quadrilateral be $ABCD$ and let M be where the diagonals meet. $AM = BM = CM = DM$. $\angle AMB$ and $\angle CMD$ are vertical angles and therefore congruent. $\angle ABM$ and $\angle DCM$ are corresponding parts and therefore congruent. Since triangle BMC is isosceles, $\mathrm{m}\angle MBC = \mathrm{m}\angle MCB$. Adding this to the previous equality gives $\mathrm{m}\angle ABC = \mathrm{m}\angle DCB$. Repeating this type of argument gives all the angles congruent. Now use Exercise 3 to see that all measure $90°$.

9. Using the vertex labels as in Figure 2.25, $\mathrm{m}\angle DAB + \mathrm{m}\angle ABC = (1/2)(\mathrm{m}\angle DAB + \mathrm{m}\angle ABC + \mathrm{m}\angle BCD + \mathrm{m}\angle CDA) = (1/2)(360°) = 180°$. Now consider the opposite sides passing through A and B. Our calculation shows that the angle alternate interior to $\angle DAB$ has the same measure. Thus these opposite sides are parallel. In the same way, the other opposite sides are parallel.

11. Instead of connecting B to D, draw the other diagonal. Now show the triangles are congruent using the same ideas as we used to show $\mathrm{m}\angle A = \mathrm{m}\angle C$.

13. Opposite sides may not be congruent. Make any such hexagon and think of one of its sides as movable (but staying parallel to its original direction). You can move it in and out, changing its length.

15. In part (b), let R' be the point other than P where $\overleftrightarrow{PR}$ crosses the circle. By Exercise 6 or the exterior angle inequality (Theorem 2.2 of Section 2.1) we can show that $m\angle PQR \neq m\angle PQR = 90°$.

17. (a) Let the angles of the right triangle which are not right angles measure α and β. $\alpha + \beta + 90° = 180°$ so $\alpha + \beta = 90°$. When we fit triangles 1 and 2 together, we get one angle of measure α and one of measure β at the corner where they meet.

    (b) Each corner of the inner figure is supplementary to a right angle.

    (c) $b - a$

    (d) $c^2 = 4(\frac{1}{2}ab) + (b - a)^2$ which gives $c^2 = 2ab + b^2 - 2ab + a^2 = a^2 + b^2$.

19. Tie a string to the top of a vertical pole (e.g., a flagpole). Pull the string tight and attach it to a spike driven into the ground at the point where the shadow of the pole ends. Climb the pole and measure the angle ϕ between pole and the string with your protractor. If you don't want to climb, measure the angle γ between the ground and the string. $\phi = 90° - \gamma$.

21. $\phi = \theta + \Psi$ so $\theta = \phi - \Psi = 7.2° - 1° = 6.2°$. Now use Eq. (2.6).

23. Skew: $\overleftrightarrow{DC}$. Parallel: none.

25. The persistent parallels principle allows us to mark all original vertical segments in the top row. Next, the braced square principle allows us to mark all horizontal segments in the top row. In each column, the persistent parallels principle allows us to go down and mark horizontal segments. The braced square principle allows us to mark verticals in the remaining two cells of the first column. Finally, use the persistent parallels principle to mark remaining verticals in the second and third rows.

27. (a) Mark verticals in the first row using the persistent parallels principle (PP).

    (b) Mark horizontals in upper left and upper right cells using the braced square (BS) principle.

    (c) Mark horizontals in the first and last columns (PP).

    (d) Mark verticals in bottom left and bottom right cells (BS).

    That's all that can be marked. The framework is not rigid.

29. See Figure A.3.

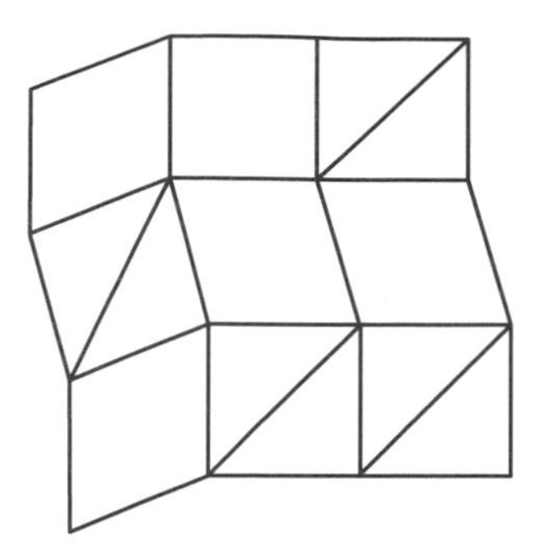

Figure A.3

31. After adding a dotted diagonal in the center cell, you can add one to the cell to the right and to the cell below and, finally, in the lower right cell.

33. Brace all cells in the first row and all in the first column.

35. See Figure A.4.

Chapter 2, Section 2.4

1. A different approach is needed. One way is to put together a bunch of these tall skinny parallelograms in a horizontal sequence, making a parallelogram where the construction of Theorem 2.15 works. Or just rotate the parallelogram and use the existing proof.

3. Make a new copy of the trapezoid and rotate it by 180° (so the top side $\overline{BC}$ becomes the bottom). Slide (without turning) the two copies together, meeting along $\overline{CD}$. A parallelogram results and its area is $(b_1 + b_2)h$. Each trapezoid is half that.

5. Yes. Split the quadrilateral into triangles with a diagonal through the vertex from which the altitudes are dropped.

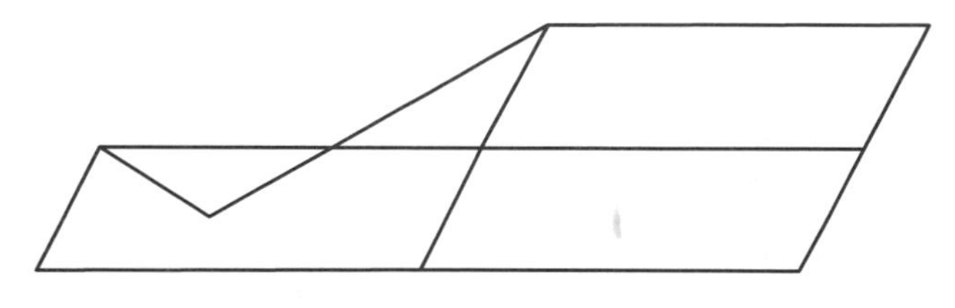

Figure A.4

7. Drop an altitude from a vertex to the opposite side to create two right triangles. Depending on whether the altitude is inside or outside the triangle, you either add the areas or subtract one from the other.

9. Divide the triangle into three smaller triangles by connecting the center of the incircle to the vertices. Add these areas.

11. Area $= |-10| = 10$.

13. Since the figure is a rectangle, we expect area $= bh$.

$$\text{area } ABC = |\tfrac{1}{2}[(0)(0) - (b)(0) + bh - (b)(0) + (b)(0) - (0)(h)]| = \tfrac{1}{2}bh,$$

$$\text{area } ADC = |\tfrac{1}{2}[(0)(h) - (0)(0) + (0)(h) - bh + (b)(0) - (0)(h)]| = \tfrac{1}{2}bh,$$

so area $ABCD =$ area $ABC +$ area $ADC = bh$.

15. $(1)(5) - (2)(3) + (2)(y) - (x)(5) + (x)(3) - (1)(y) = 0$, which simplifies to $y = 2x + 1$.

17. The two striped triangles ADE and CDE. Let A' and B' be the feet of the perpendiculars from A and B, respectively, to line M. $\overleftrightarrow{AA'}$ is parallel to $\overleftrightarrow{BB'}$. Thus $ABB'A'$ is a parallelogram and opposite sides of a parallelogram are congruent.

19. Find point F on $\overline{AB}$ so that $AF = A'B'$ and point G on $\overline{AC}$ so $AC/AG = AB/AF$. Show that triangles AFG and $A'B'C'$ are congruent and then Eq. (2.14) follows.

21. Suppose $A'B' < AB$. Let D be on $\overline{AB}$ so that $AD = A'B'$ and E on $\overline{AC}$ so that $AE = A'C'$. Show that triangle ADE is similar to triangle ABC and congruent to triangle $A'B'C'$.

23. Triangles ABC and ACD are similar with correspondence $A \to A$, $B \to C$, $C \to D$.

25. Use Exercise 24.

Chapter 3

Chapter 3, Section 3.1

1. Take the right triangle we constructed with angle sum strictly less than 180° and make a second copy by flipping it over its hypotenuse. The two triangles together make a quadrilateral that has angle sum strictly less than 360°.

3. Let the larger triangle be ABC. Place the smaller triangle inside ABC with the vertex which corresponds to A in the similarity placed at A and with these angles coinciding. Thus, our large triangle is divided into a small triangle and a quadrilateral. By taking into account equal angles and supplements of angles we can see that the angle sum of the quadrilateral is 360°. But this contradicts Exercise 2.

5. Let the triangle be ABC with C the right angle. Extend the segment connecting A and C to D so that $AC = CD$. Triangles ABC and BCD are congruent by SAS. Therefore $\mathrm{d}(ABC) = \mathrm{d}(BCD)$. By Exercise 4, $\mathrm{d}(ABD) = \mathrm{d}(ABC) + \mathrm{d}(BCD) = 2\mathrm{d}(ABC)$.

7. Let $ABCD$ be a quadrilateral with diagonal AC. $\mathrm{d}(ABCD) = 360^\circ - (\mathrm{m}\angle A + \mathrm{m}\angle B + \mathrm{m}\angle C + \mathrm{m}\angle D) = 360^\circ - (\mathrm{m}\angle DAC + \mathrm{m}\angle CAB + \mathrm{m}\angle B + \mathrm{m}\angle BCA + \mathrm{m}\angle ACD + \mathrm{m}\angle D) = [180^\circ - (\mathrm{m}\angle DAC + \mathrm{m}\angle ACD + \mathrm{m}\angle D)] + [180^\circ - (\mathrm{m}\angle CAB + \mathrm{m}\angle B + \mathrm{m}\angle BCA)] = \mathrm{d}(ABC) + \mathrm{d}(ADC)$. If $ABCD$ is cut by a line that is not a diagonal, use the same style of proof to show that $\mathrm{d}(ABCD) =$ the sum of the defects of the pieces.

9. Triangles DAB and CBD are congruent by SAS since $DA = CB$, $\mathrm{m}\angle A = \mathrm{m}\angle B$, and $AB = AB$. $DB = CA$ by "corresponding parts."

11. $AB = A'B'$, $BC = B'C'$, $\mathrm{m}\angle B = \mathrm{m}\angle B'$ so triangles ABC and $A'B'C'$ are congruent by SAS. Thus $\alpha = \alpha'$ (corresponding parts). But $\beta = 90^\circ - \alpha = 90 - \alpha' = \beta'$.

13. By previous exercises, $\beta = \beta'$ and $AC = A'C'$. Then triangles ADC and $A'D'C'$ are congruent by SAS. $DC = D'C'$ by corresponding parts.

15. By Exercise 14, the summit cannot be congruent to the base. Suppose it is shorter, say, summit = base $- \varepsilon$ where $\varepsilon > 0$. Let the height be h. Now make a horizontal stack of Sachheri quadrilaterals. For example, in Figure 3.6 slide the second quadrilateral over until $A'D'$ meets BC and the two bases make one longer segment. Now add some more quadrilaterals, making n altogether where $n > 2h/\varepsilon$. Let the first quadrilateral be $A_1B_1C_1D_1$ and the last $A_nB_nC_nD_n$. $A_1B_n = n(\text{base})$ and we claim this is longer than the broken line path $A_1D_1C_1 \ldots C_nB_n$. The length of the broken line path is $2h + n(\text{base} - \varepsilon) < n(\text{base})$ since $\varepsilon n > 2h$.

Chapter 3, Section 3.2

1. Let C be the center, F the foot of the perpendicular from center to plane, and X any point on the intersection of the plane and sphere. Use the Pythagorean theorem on the right triangle CFX to show that FX has the same value regardless of which point in the intersection X is.

3. A cone whose point is at the origin.

5. A sphere centered at the origin with radius ρ.

7. (a) $R[\phi(1) - \phi(0)]$.

   (b) $m\angle OAA' = \phi$ and so $z = AA' = \rho \cos \phi$. Also $OA' = \rho \sin \phi$. Now drop perpendiculars from A' to x and y axes and reason about the right triangles to get the equations for x and y.

   (c) In Eqs. (3.4), replace x by $x(t)$, y by $y(t)$, z by $z(t)$, θ by $\theta(t)$, ϕ by $\phi(t)$ and ρ by R. Differentiate to get $x'(t)$ and $y'(t)$. Substitute into the length integral and simplify, making use of a trig identity.

   (d) Since our great circle has $\theta = \text{constant}$, $\theta'(t) = 0$, and we have length $(C) = R\int_0^1 \phi'(t)\,dt = R[\phi(1) - \phi(0)]$.

9. The angle at N must be $\leq 180^\circ$; otherwise, $\widehat{A_1A_2}$ won't be a shortest path. But even 180° is ruled out because then we would have A_1, A_2, and N all on one geodesic. Thus the angle sum is $< 180^\circ + 90^\circ + 90^\circ = 360^\circ$.

11. Let the triangle be ABC and let G be the geodesic containing $\widehat{AB}$. G divides the sphere into two hemispheres, one of which must contain the point C. The shortest paths from C to A and B must lie wholly in that hemisphere—going into the other hemisphere would be longer.

13. Divide the spherical quadrilateral $ABCD$ into two spherical triangles by a diagonal shortest path from A to C. The angle sum of the quadrilateral is the sum of the angle sums of the two spherical triangles. Likewise, the area of the quadrilateral is the sum of the areas of the two spherical triangles. Thus, adding two versions of the triangle formula,

$$\text{angle sum of quad} = 2\pi + \frac{\text{area of quad}}{R^2}.$$

15. The spherical triangle constructed in Example 3.2 has an exterior angle which is a right angle and a remote interior one which is also a right angle.

17. Since B lies on $\widehat{AC}$, length$(\widehat{AB})+$length$(\widehat{BC})=$length$(\widehat{AC})$. Since a perfect map would preserve lengths, we would have $A'B'+B'C'=A'C'$. But this implies that B' lies on the segment $A'C'$. (In the Euclidean plane, if three points are not collinear, their distances satisfy the triangle inequality: $AC < AB+BC$.)

Chapter 3, Section 3.3

1. Suppose we had the situation in Figure 3.17b. The distance from X to Y along the path on G which avoids A and B is (1/2)length(G) since X and Y are opposite. But since A and B are opposite, their distance along the part of G which includes X and Y is also (1/2)length(G). But it has to be longer than the path from X to Y since it contains that path. This contradiction proves that Figure 3.17b can't be right.

3. If $G_\perp$ contained A then A and M would be opposite by Theorem 3.17. But B is opposite A, and A can only have one opposite point.

5. In Euclidean geometry, a ray goes on indefinitely in one direction from a point. In spherical geometry, if we go on indefinitely in one direction, we create a geodesic. There seems to be nothing in spherical geometry that is "like" a Euclidean geometry ray, so there seems little point in trying to define a ray.

7. Suppose we have a bilateral whose sides are S_1 and S_2. Let G_1 and G_2 be the geodesics containing S_1 and S_2. By Pasch's axiom, G_1 has a side containing S_2. Call this side H_1. Let H_2 be the side of G_2 which contains S_1. The lune determined by S_1 and S_2 is defined to be $H_1 \cap H_2$.

9. (a) By ASA, the spherical triangle ABC is congruent to itself with this congruence: $A \to B, B \to A, C \to C$. Thus length$(\widehat{AC})=$length$(\widehat{BC})$ since these are corresponding parts.

   (b) This follows immediately from part (a).

Chapter 4

Chapter 4, Section 4.1

1. Just one. As described in Example 4.1, a translation is completely determined by the vector from any point to its image.

3. There are infinitely many possibilities. There are infinitely many circles that pass through the points. (Draw the segment from P to Q and then the perpendicular bisector of that segment. Any point on that perpendicular bisector is equidistant from both points, thereby serving as a center of a circle passing through both P and Q.) For any such circle, if you rotate around its center by the correct angle, you will rotate P to Q. (The rotation amounts will be different for different circles.)

5. Since the segments from P to P' and Q to Q' are equal and parallel, we can show that triangles $PP'Q$ and $Q'QP'$ are congruent. Thus $PQ = P'Q'$ since the segments are corresponding parts.

7. Connect P' to Q and find R, the intersection with L. Draw line $\overleftrightarrow{PR}$ and, along this line, measure from R away from P, a distance equal to RQ. The point you arrive at is Q'.

9. For a figure to illustrate this case, take Figure 4.5 but interchange the labels Q and Q'. Now in this figure, we can show that triangles $P'QF$ and $PQ'F$ are congruent by the same means as in Theorem 4.1. Thus, $P'Q = PQ'$ and $m\angle QP'P = m\angle Q'PP'$. Now we can use SAS to show that triangles $Q'P'P$ and QPP' are congruent. $P'Q' = PQ$ by corresponding parts.

11. (a) True.

    (b) False.

    (c) True. D is where the angle bisectors of triangle ABC meet. Because angles are invariant, the images of these angle bisectors are the angle bisectors of triangle $A'B'C'$. D' must be on each of these angle bisectors, so it is where they meet. But this is the center of the inscribed circle of triangle $A'B'C'$.

13. Reflection in any vertical or horizontal line. Any translation.

15. Let O be the origin, let P be another point on the y axis, and let Q be any point on the x axis other than O. Since angles are preserved by isometries, $m\angle Q'O'P' = m\angle QOP = 90°$. But since O' and P' are on the y axis, this means all points Q on the x axis map to points on a line perpendicular to the y axis (passing through O').

17. We give an indirect proof. If A', B', C' are collinear then one of the points is between the other two. With no loss of generality, we can assume B' is between A' and C'. Then $A'C' = A'B' + B'C'$. Since isometries preserve distance, this

equation implies $AC = AB + BC$. However, we know A, B, and C are not collinear so they obey the triangle inequality $AC < AB + BC$. This contradicts our previous equation.

19. (a) Let $S(O) = O$ and for each other point P let $S(P)$ be that point so that $OS(P) = 2OP$ and P is between O and $S(P)$. (See Section 7.1 of Chapter 7.)

    (b) The proof of Theorem 4.2 still works except from $AB + BC = AC$ we multiply by 2 to get $2AB + 2BC = 2AC$, which means $A'B' + B'C' = A'C'$.

    (c) If L and M are parallel lines, let T be any transversal. T makes alternate interior angles which are congruent. Since S preserves angles, L', M', and T' make a configuration where alternate interior angles are congruent and this means L' and M' are parallel.

21. No. The example asked for in Exercise 19 is a counterexample.

Chapter 4, Section 4.2

1. A rotation through 120° or through -120°.

3. $E'G' = EG$ since translation preserves distance and $EG = E^*G^*$ by congruence. These two equations imply $E'G' = E^*G^*$. Since $E' = E^*$, $E'G' = E'G^*$.

5. (a) False, it is either a rotation (if the lines cross) or a translation (if the lines are parallel).

   (b) False. If $S = R_O(180^\circ)$ and $S' = M_L$, then $S \circ S = S' \circ S'$.

   (c) False. We could have $S = R(180^\circ)$.

   (d) False if S is a reflection.

7. P' is under L and P'' is above L'. If the distance from P to L is x, then the distance from P to L' is $d - x$ and the distance from P' to L' is $d + x$. By the nature of reflection, the distance from P'' to L' is also $d + x$. The distance from P'' to P is $(d + x) + (d - x) = 2d$.

9. We replace L and L' by two more convenient lines. To get them, slide (translate) L and L' over until the image of L' is on top of L''. Let L''' be the image of L under this slide. By part (b) of Theorem 4.2, $M_{L'} \circ M_L = M_{L''} \circ M_{L'''}$. Thus, $M_{L''} \circ M_{L'} \circ M_L = M_{L''} \circ (M_{L''} \circ M_{L'''}) = (M_{L''} \circ M_{L''}) \circ M_{L'''} = M_{L'''}$, as we wished to prove.

11. (a) True. $ABCD$ is a parallelogram since it has a pair of sides which are congruent and parallel. Thus, $\overline{AB}$ and $\overline{DC}$ are congruent and parallel. Since $\overline{DE}$ is also congruent and parallel to $\overline{AB}$, we get $E = C$.

    (b) False.

    (c) True.

13. (a) No. Look at what happens to a point on the x axis.

    (b) No. $V \circ U = R_O(90^\circ)$ while $U \circ V = R_O(-90^\circ)$.

    (c) Yes, both give I.

    (d) No. Apply both to the origin. (Make a drawing.)

    (e) Yes.

15. The key is that if you reflect in the same line twice, you get I. $(M_L \circ M_{L'}) \circ (M_{L'} \circ M_L) = M_L \circ (M_{L'} \circ M_{L'}) \circ M_L = M_L \circ M_L = I$. In the same way, $(M_{L'} \circ M_L) \circ (M_L \circ M_{L'}) = I$.

17. Let L be $\overleftrightarrow{PQ}$, let L' be the line perpendicular to L through P and let L'' be the line perpendicular to L through Q. $R_Q(180^\circ) \circ R_P(180^\circ) = (M_{L''} \circ M_L) \circ (M_L \circ M_{L'}) = M_{L''} \circ (M_L \circ M_L) \circ M_{L'} = M_{L''} \circ M_{L'} =$ translation in the direction from L' to L'' (along L) by the amount $2PQ = PR$.

19. (a) L is the line obtained by rotating the x axis by θ around O.

    (b) Let L^* be the line obtained by rotating the x axis by $\theta/2$ around O. $R_O(\theta) = M_{L^*} \circ M_x$. $R_O(-\theta) = M_x \circ M_{L^*}$. Substituting these expressions and simplifying we get $R_O(\theta) \circ M_x \circ R_O(-\theta) = M_{L^*} \circ M_x \circ M_{L^*} = R_O(\theta) \circ M_{L^*} = M_L \circ M_{L^*} \circ M_{L^*} = M_L$.

21. Let L be the line you get by rotating the x axis by $\theta/2$ around O. Let T be any translation along the x axis and let H be the glide reflection $M_x \circ T$. $R_O(\theta) \circ H = (M_L \circ M_x) \circ (M_x \circ T) = M_L \circ T$, which is a glide reflection by Example 4.10.

23. (a) If S were not one-to-one, there would be two different points P and Q where $P' = Q'$. But in an isometry $P'Q' = PQ$, which is a contradiction since $P'Q' = 0$. Thus S is one-to-one.

    (b) Compare distances: $UV = S[S^{-1}(U)]S[S^{-1}(V)] = S^{-1}(U)S^{-1}(V)$, the second equality holding because S is an isometry.

Chapter 4, Section 4.3

1. Triangles F_1PF_2 and F_1QF_2 are congruent by SSS so $m\angle PF_1A = m\angle QF_1A$. Use this and SAS to conclude that triangles PF_1A and QF_1A are congruent. Then $AP = AQ$ and $m\angle F_1AP = m\angle F_1AQ$.

3. By Exercise 1, the perpendicular bisector of $\overline{PQ}$ passes through F_1, F_2, and F_3.

5. (a) False, it could be reflection.

   (b) True, unless it is rotation by some multiple of $360°$.

   (c) False. Suppose the reflections all use the same lines. The composition of all four reflections is the identity while the composition of any three is a single reflection in that line.

7. The identity, translation in a direction parallel to L, glide reflection in L, reflection in L.

9. If we let L^* be the line 1 unit from L'' in the opposite direction to that of the translation S then $S = M_{L''} \circ M_{L^*}$ so $M_{L''} \circ S = M_{L''} \circ (M_{L''} \circ M_{L^*}) = (M_{L''} \circ M_{L''}) \circ M_{L^*} = M_{L^*}$. All points on L^* are fixed.

11. (a) $[POP']$ is clockwise, $[OPP']$ is counterclockwise.

    (b) $[POP']$ is counterclockwise, $[OPP']$ is clockwise.

    (c) $[POP']$ is counterclockwise, $[OPP']$ is clockwise.

13. There is one possibility if S preserves sense and one if it reverses sense. Two altogether.

15. Let T be a translation that moves $S(A)$ back to A. Follow this by a rotation around A, say, $R_A(\theta)$ so the line $(T \circ S)(L)$ turns so as to coincide with L. Then $U = R_A(\theta) \circ T \circ S$ fixes A and the line L (although not necessarily every point on L). Now suppose U fixes some other point on L besides A. Then U is either I or a reflection in L. It can't be a reflection since U preserves sense. Thus $U = R_A(\theta) \circ T \circ S = I$ and $S = T^{-1}R_A(-\theta)$. This is one possibility for S. Now suppose U does not fix another point. Then if B is a point on L and we follow U by reflection in a line through A perpendicular to L, say, M, $M \circ U$ fixes A and B. Thus $M \circ U$ is either I or reflection M_L. It can't be I since it reverses sense. Thus $M \circ U = M_L$ and $M \circ R_A(\theta) \circ T \circ S = M_L$. This gives one more possibility for S.

17. Any reflection.

19. We know that a sense-preserving isometry is either the identity, a translation or a rotation. In case of the identity, use any line twice: $S = M_L \circ M_L$. For rotation, use Theorem 4.11 of Section 4.2. For translation, use Theorem 4.10 of Section 4.2.

21. (a) We can prove this by the same kind of argument as in Example 4.11.

    (b) Let C be the unknown center of the rotation. C is equidistant from A and A' and so it lies on the perpendicular bisector of $\overline{AA'}$. For similar reasons it lies on the perpendicular bisector of $\overline{BB'}$. Draw these perpendicular bisectors; C is where they cross.

23. (a) Two. Note that $m\angle S(A)S(B)S(C)$ is determined $(= m\angle ABC)$ and so there are only two rays through $S(B)$ on which $S(C)$ can lie.

    (b) One.

25. Examine each of the four types of isometry listed in Theorem 4.20. Reflection only preserves the directions of lines parallel or perpendicular to the reflection line, so reflection is not an example. We rule out glide reflection in the same way. That leaves rotation. The only rotation that works is a half-turn, $R_C(180°)$, about any center C.

27. S can be written $M_x \circ T$ where T is the translation by 2 units in the positive x direction. Then $(M_{L''} \circ M_x) \circ T = R \circ T$ where R is some rotation. But $R \circ T$ is also a rotation by Exercise 21.

Chapter 4, Section 4.4

1. Checklists for patterns in Figure 4.29.

	Translation	Reflections		Glide reflection	Half turn
		Vertical line	Horizontal line		
Pattern 1	yes	no	no	no	no
Pattern 2	yes	no	no	yes	no
Pattern 3	yes	yes	no	no	no
Pattern 4	yes	no	no	no	yes
Pattern 5	yes	no	yes	yes	no
Pattern 6	yes	yes	no	yes	yes
Pattern 7	yes	yes	yes	yes	yes

3. The shorter definition is not equivalent. Consider this pattern enclosed between border lines 1 inch apart. Place a ruler along the bottom horizontal border line and mark off the points at the inch marks. At each such point place a black vertical segment. The heights can be anything as long as they get bigger as you move to the right and smaller as you move to the left. (For example, at i inches, where $i = 0, 1, 2$, etc., have the height be $1 - [1/(i+2)]$. Now, as you move to the inch marks to the left of 0, cut the previous height in half at each next inch mark.) Translation by 1 inch to the right moves each black point onto a black point but does not move each white point onto a white point. This pattern is not a strip pattern. It has no symmetries.

5. Here is a way to construct a pattern that has a 60° rotational symmetry. Take a regular hexagon (all angles equal to 120°, all sides equal) and connect its center to each vertex, thereby dividing the hexagon into six equilateral triangles. Shade every other triangle. Now surround this hexagon with six identical hexagons, all shaded with the same pattern. Continue building on hexagons in all directions.

Chapter 4, Section 4.5

1. See Figure A.5.

3. See Figure A.6.

5. Example 4.14 is meant to show $T \circ M_V =$ some other vertical reflection. The original O could be moved to $T \circ M_V(\text{O})$ by translation T'. This does not prove that $T \circ M_V = T'$ because for this equation to be true it would have to hold for any motif: We would need $T \circ M_V$ (motif) $= T'$(motif) for any motif, not just the letter O. The problem arises because O has its own symmetry. Let $M_{V'}$ denote vertical reflection in the vertical line through the center of the original O. $\text{O} = M_{V'}(\text{O})$ so $T \circ M_V(\text{O}) = T \circ M_V[M_{V'}(\text{O})] = (T \circ T^*)(\text{O})$ for some horizontal translation T^*. But $T \circ T^*$ is itself a translation T', so $T \circ M_V(\text{O}) = T'(\text{O})$. But this only came out this way because $\text{O} = M_{V'}(\text{O})$.

7. $R = M_H \circ M_V$ so $R \circ M_V = (M_H \circ M_V) \circ M_V = M_H \circ (M_V \circ M_V) = M_H \circ I = M_H$. But M_H is not present according to the checklist. This is a contradiction, so this combination of symmetries cannot exist.

9. $T \circ M_H = G$, which is not present.

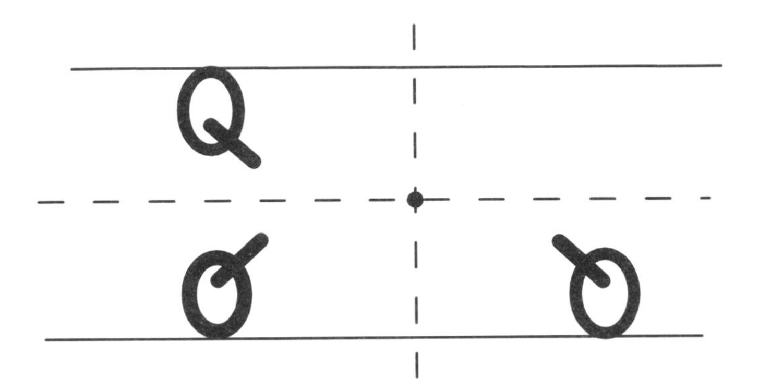

Figure A.5 $M_V \circ M_H = R$

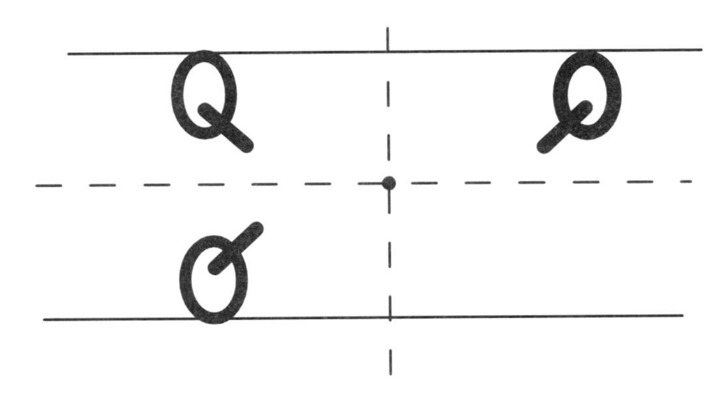

Figure A.6 $M_H \circ R = M_V$

Chapter 5

Chapter 5, Section 5.1

1. $\mathbf{b} - \mathbf{a} = \langle 1, -3, -2 \rangle$ and so $\mathbf{x} = \mathbf{a} + t(\mathbf{b} - \mathbf{a})$ means:

$$x = 2 + t,$$

$$y = -3 - 3t,$$

$$z = 4 - 2t.$$

3. If $(-3, 1)$ lies on the line, $\langle -3, 1\rangle = \mathbf{a} + t(\mathbf{b} - \mathbf{a}) = \langle 5, -2\rangle + t(\langle 1, 2\rangle - \langle 5, -2\rangle)$ for some t. This gives x and y component equations: $-3 = 5 - 4t$ and $1 = -2 + 4t$. There is no t that satisfies both equations so $(-3, 1)$ is not on the line.

5. $\langle -3, 5\rangle = \langle -5, 3\rangle + t\langle 6, 6\rangle$; $-3 = -5 + 6t$ implies $6t = 2$, which implies $t = 1/3$; $5 = 3 + 6t$ implies $6t = 2$, which implies $t = 1/3$. Since we get the same t values from each equation, $(-3, 5)$ lies on the line from S to G. Since $0 \leq 1/3 \leq 1$, it lies on the segment from S to G.

7. If $\langle x, y\rangle$ is the position vector of the midpoint we seek, $\langle 6, 2\rangle = \langle 0, 4\rangle + \frac{2}{3}(\langle x, y\rangle - \langle 0, 4\rangle)$. $x = 9$, $y = 1$.

9. $\mathbf{m}_1 = \frac{1}{2}\mathbf{a} + \frac{1}{2}\mathbf{b}$, $\mathbf{m}_2 = \frac{1}{2}\mathbf{b} + \frac{1}{2}\mathbf{c}$, $\mathbf{m}_3 = \frac{1}{2}\mathbf{c} + \frac{1}{2}\mathbf{d}$, $\mathbf{m}_4 = \frac{1}{2}\mathbf{d} + \frac{1}{2}\mathbf{a}$. From these, $\mathbf{m}_2 - \mathbf{m}_1 = \frac{1}{2}(\mathbf{b} + \mathbf{c} - \mathbf{a} - \mathbf{b}) = \frac{1}{2}(\mathbf{c} - \mathbf{a})$ and $\mathbf{m}_4 - \mathbf{m}_3 = \frac{1}{2}(\mathbf{d} + \mathbf{a} - \mathbf{c} - \mathbf{d}) = \frac{1}{2}(\mathbf{a} - \mathbf{c})$. We see that $\mathbf{m}_4 - \mathbf{m}_3 = -(\mathbf{m}_2 - \mathbf{m}_1)$, which means the vectors of the opposite sides are parallel and have the same length.

11. The ray from $(15, 3)$ pointing left is $\mathbf{x} = \langle 15, 3\rangle + t\langle -1, 0\rangle$, $t \geq 0$. Looking for an intersection of the ray with $\overline{AB}$: $x = 10 + 5s = 15 - t$, $y = 1 + 8s = 3$; this gives $s = 1/4$, $t = 15/4$, signifying a crossing of the ray and segment. Looking for an intersection of the ray with $\overline{BC}$: $x = 15 + 5s = 15 - t$, $y = 9 - 14s = 3$; this gives $s = 3/7$, $t = -15/7$. Since $t < 0$, there is no intersection with the segment. Looking for an intersection of the ray with $\overline{CA}$: $x = 10 + 10s = 15 - t$, $y = 1 - 6s = 3$; this gives $s = -(1/3)$. Since $s < 0$, there is no intersection with the segment.

13. The ray from $(14, 3)$ pointing up is $\mathbf{x} = \langle 14, 3\rangle + t\langle 0, 1\rangle$, $t \geq 0$. Looking for an intersection of the ray with $\overline{AB}$: $x = 10 + 5s = 14$, $y = 1 + 8s = 3 + t$; this gives $s = 4/5$, $t = 22/5$, signifying a crossing of the ray and segment. Looking for an intersection of the ray with BC: $x = 15 + 5s = 14$, $y = 9 - 14s = 3 + t$; this gives $s = -1/5$. Since $s < 0$ there is no crossing of ray and segment. Looking for an intersection of the ray with $\overline{CA}$: $x = 10 + 10s = 14$, $y = 1 - 6s = 3 + t$; this gives $s = 2/5$, $t = -22/5$. Since $t < 0$, there is no intersection of the ray with the segment. Since there is one crossing, and it is not at a vertex of the triangle, the point is inside the triangle.

15. For rays that don't cross the boundary of the polygon at any polygon vertices, the rule should be this: An even number of crossings means the point is outside; an odd number of crossings means the point is inside. But what if some vertices lie on the ray?

17. Equations for the two segments: $\mathbf{x} = \langle -6, 0\rangle + t\langle 12, 8\rangle$, $\mathbf{x} = \langle 2, 3\rangle + s\langle 2, 3\rangle$. Setting the two x components equal and the two y components equal gives $12t - 2s = 8$ and $8t - 3s = 3$. Solving simultaneously, $t = 9/10$ and $s = 14/10$. Since s is not in $[0, 1]$, the point of intersection does not lie in $\overline{AB}$.

19. Segment equations: $x = t$, $y = \sqrt{3}t$. To be on the circle, $(t - \sqrt{3})^2 + (\sqrt{3}t + 1)^2 = 2$, which yields $t^2 - 2\sqrt{3}t + 3 + 3t^2 + 2\sqrt{3}t + 1 = 2$, which gives $t^2 = -\frac{1}{2}$. This has no real solution for t.

21. If you use a small number of segments you will get the calculations done quickly but the approximation will be poor and your answer may come out wrong. There seems to be a trade-off between speed and accuracy.

23. When $t = \frac{2}{3}$, $x = 2 + 3(\frac{2}{3}) = 4$, $y = 4 - 6(\frac{2}{3}) = 0$, $z = 3 - 3(\frac{2}{3}) = 1$.

25. The danger zone consists of all points within 1 unit of any point on the circumference. Think about taking a point on the circumference and centering a unit circle there. Now slide it around the circumference. What you cover by the moving circle is the danger zone. This "grows" the obstacle by adding a ring of width 1 (the danger zone). Obstacle $\cup$ danger zone = a circle of radius 3 centered at (5, 5).

27. Equations for the lines are $\mathbf{x} = \langle 3, 0, 1\rangle + t_1\langle 2, 1, 1\rangle$, $\mathbf{x} = \langle 1, 9, 0\rangle + t_2\langle 2, -1, 1\rangle$. Setting components equal gives three equations: $3 + 2t_1 = 1 + 2t_2$, $t_1 = 9 - t_2$, $1 + t_1 = t_2$. These yield $t_1 = 4$, $t_2 = 5$. Substitute $t_1 = 4$ into the appropriate line equation: $\mathbf{x} = \langle 3, 0, 1\rangle + 4\langle 2, 1, 1\rangle = \langle 11, 4, 5\rangle$.

Chapter 5, Section 5.2

1. (a) $(1)(0) + (0)(1) + (0)(0) = 0$.

   (b) $(3)(1) + (-8)(-2) + (2)(3) = 25$.

   (c) $\langle 1, 2, -1\rangle \cdot \langle 1, 2, -1\rangle - t\langle 1, 2, -1\rangle \cdot \langle 0, 1, 2\rangle + t\langle 0, 1, 2\rangle \cdot \langle 1, 2, -1\rangle - t^2\langle 0, 1, 2\rangle \cdot \langle 0, 1, 2\rangle = 6 - 5t^2$.

3. (a) A and B lie on the plane: $\langle 1, 1, 2\rangle \cdot \langle 7, 3, -5\rangle = 7 + 3 - 10 = 0$, $\langle 2, -3, 1\rangle \cdot \langle 7, 3, -5\rangle = 14 - 9 - 5 = 0$. Point on the line lies in the plane: $[\langle 1, 1, 2\rangle + t(\langle 2, -3, 1\rangle - \langle 1, 1, 2\rangle)] \cdot \langle 9, 8, -23\rangle = \langle 1, 1, 2\rangle \cdot \langle 9, 8, -23\rangle + t(\langle 2, -3, 1\rangle \cdot \langle 9, 8, -23\rangle - \langle 1, 1, 2\rangle \cdot \langle 9, 8, -23\rangle) = 0 + t(0 - 0) = 0$.

   (b) $[\mathbf{a} + t(\mathbf{b} - \mathbf{a})] \cdot \mathbf{n} = \mathbf{a} \cdot \mathbf{n} + t(\mathbf{b} - \mathbf{a}) \cdot \mathbf{n} = \mathbf{a} \cdot \mathbf{n} + t(\mathbf{b} \cdot \mathbf{n} - \mathbf{a} \cdot \mathbf{n}) = k + t(k - k) = k$.

5. $(\mathbf{x}-\mathbf{y})\cdot(\mathbf{x}+\mathbf{y})=\mathbf{x}\cdot\mathbf{x}+\mathbf{x}\cdot\mathbf{y}-\mathbf{y}\cdot\mathbf{x}-\mathbf{y}\cdot\mathbf{y}=\mathbf{x}\cdot\mathbf{x}-\mathbf{y}\cdot\mathbf{y}=\|\mathbf{x}\|^2-\|\mathbf{y}\|^2$.

7. (a) $\mathbf{u}+\mathbf{v}=\mathbf{c}-\mathbf{a}$; $\mathbf{v}+(\mathbf{b}-\mathbf{d})=\mathbf{u}$ so $\mathbf{u}-\mathbf{v}=\mathbf{b}-\mathbf{d}$.

   (b) $AC^2=(\mathbf{c}-\mathbf{a})\cdot(\mathbf{c}-\mathbf{a})=(\mathbf{u}+\mathbf{v})\cdot(\mathbf{u}+\mathbf{v})=\|\mathbf{u}\|^2+2\mathbf{u}\cdot\mathbf{v}+\|\mathbf{v}\|^2$,
   $BD^2=(\mathbf{b}-\mathbf{d})\cdot(\mathbf{b}-\mathbf{d})=(\mathbf{u}-\mathbf{v})\cdot(\mathbf{u}-\mathbf{v})=\|\mathbf{u}\|^2-2\mathbf{u}\cdot\mathbf{v}+\|\mathbf{v}\|^2$.
   Therefore, $AC^2+BD^2=2\|\mathbf{u}\|^2+2\|\mathbf{v}\|^2$.

9. (a) $\mathbf{a}\cdot\mathbf{b}=\|\mathbf{a}\|\|\mathbf{b}\|\cos\theta$. $\langle 1,-3,5\rangle\cdot\langle 3,4,0\rangle=\|\langle 1,-3,5\rangle\|\|\langle 3,4,0\rangle\|\cos\theta$.
   $-9=5\sqrt{35}\cos\theta$. $\cos\theta=-0.3043$ so $90^\circ<\theta\le 180^\circ$.

   (b) $\mathbf{a}\cdot\mathbf{c}=\|\mathbf{a}\|\|\mathbf{c}\|\cos\theta$. $\langle 1,-3,5\rangle\cdot\langle 12,0,5\rangle=\|\langle 1,-3,5\rangle\|\|\langle 12,0,5\rangle\|\cos\theta$.
   $37=13\sqrt{35}\cos\theta$. $\cos\theta=0.4811$ so $0^\circ\le\theta<90^\circ$.

   (c) $\mathbf{b}\cdot\mathbf{c}=\|\mathbf{b}\|\|\mathbf{c}\|\cos\theta$. $\langle 3,4,0\rangle\cdot\langle 12,0,5\rangle=\|\langle 3,4,0\rangle\|\|\langle 12,0,5\rangle\|\cos\theta$.
   $36=65\cos\theta$. $\cos\theta=0.5538$ so $0^\circ\le\theta<90^\circ$.

11. Orthogonality at E: $(\mathbf{f}-\mathbf{e})\cdot(\mathbf{h}-\mathbf{e})=\langle 0,0,-1\rangle\cdot\langle 0.6,0.8,0\rangle=0$.
    Orthogonality at F: $(\mathbf{e}-\mathbf{f})\cdot(\mathbf{g}-\mathbf{f})=\langle 0,0,1\rangle\cdot\langle 0.6,0.8,0\rangle=0$.
    Orthogonality at G: $(\mathbf{h}-\mathbf{g})\cdot(\mathbf{f}-\mathbf{g})=\langle 0,0,1\rangle\cdot\langle -0.6,-0.8,0\rangle=0$.
    Orthogonality at H: $(\mathbf{e}-\mathbf{h})\cdot(\mathbf{g}-\mathbf{h})=\langle -0.6,-0.8,0\rangle\cdot\langle 0,0,-1\rangle=0$.

13. (a) $\mathbf{n}=(\mathbf{b}-\mathbf{a})\times(\mathbf{c}-\mathbf{a})=\begin{vmatrix}\mathbf{i}&\mathbf{j}&\mathbf{k}\\2&2&1\\7&1&-2\end{vmatrix}=\langle -5,11,-12\rangle$,

    $(\mathbf{x}-\langle 3,0,2\rangle)\cdot\langle -5,11,-12\rangle=0$.

    (b) $(\mathbf{x}-\langle 2,-4,5\rangle)\cdot\langle -3,-7,-4\rangle=0$.

    (c) $(\mathbf{x}-\langle 0,0,1\rangle)\cdot\langle -2,-6,-14\rangle=0$.

15. Face #1: outer normal is $\mathbf{n}=\langle -0.8,0.6,0\rangle$. $(\mathbf{s}-\mathbf{c})\cdot\mathbf{n}=\langle -1.8,2.6,10\rangle\cdot\langle -0.8,0.6,0\rangle>0$ so Face #1 is visible.
    Face #2: outer normal is $\mathbf{n}=\langle 0.8,-0.6,0\rangle$. $(\mathbf{s}-\mathbf{e})\cdot\mathbf{n}=\langle -2,4,9\rangle\cdot\langle 0.8,-0.6,0\rangle<0$ so Face #2 is hidden.
    Face #3: outer normal is $\mathbf{n}=\langle -0.6,-0.8,0\rangle$. $(\mathbf{s}-\mathbf{e})\cdot\mathbf{n}=\langle -2,4,9\rangle\cdot\langle -0.6,-0.8,0\rangle<0$ so Face #3 is hidden.
    Face #4: outer normal is $\mathbf{n}=\langle 0.6,0.8,0\rangle$. $(\mathbf{s}-\mathbf{c})\cdot\mathbf{n}=\langle -1.8,2.6,10\rangle\cdot\langle 0.6,0.8,0\rangle>0$ so Face #4 is visible.
    Face #5: outer normal is $\mathbf{n}=\langle 0,0,-1\rangle$. $(\mathbf{s}-\mathbf{c})\cdot\mathbf{n}=\langle -1.8,2.6,10\rangle\cdot\langle 0,0,-1\rangle<0$ so Face #5 is hidden.
    Face #6: outer normal is $\mathbf{n}=\langle 0,0,1\rangle$. $(\mathbf{s}-\mathbf{e})\cdot\mathbf{n}=\langle -2,4,9\rangle\cdot\langle 0,0,1\rangle>0$ so Face #6 is visible.

17. No faces would be calculated as visible. The algorithm seems inappropriate for an interior eyepoint.

19. Let the pyramid's base be $ABCD$ and let the fifth corner be E. Let F_1 be the base, let F_2 be face ABE, let F_3 be BCE, let F_4 be CDE and let F_5 be ADE. Let the slicing plane pass through A, E, and C, dividing P into two tetrahedra $P^* = ABCE$ and $P^{**} = ACDE$. After slicing, F_1 divides into two triangular pieces $F_1^* = ABC$ and $F_1^{**} = ACD$. Let F_6 be the face created by the slice where the two tetrahedra meet and let n_6^* be the unit outer normal of this face in P^* and let n_6^{**} be the unit outer normal in P^{**}. Note that $n_6^{**} = -n_6^*$. Then, applying Exercise 18 twice,

$$\text{Area}(F_1^*)\mathbf{n}_1 + \text{Area}(F_2)\mathbf{n}_2 + \text{Area}(F_3)\mathbf{n}_3 + \text{Area}(F_6)\mathbf{n}_6^* = 0,$$

$$\text{Area}(F_1^{**})\mathbf{n}_1 + \text{Area}(F_4)\mathbf{n}_4 + \text{Area}(F_5)\mathbf{n}_5 + \text{Area}(F_6)\mathbf{n}_6^{**} = 0.$$

Adding these, taking into account that $\text{Area}(F_1^*) + \text{Area}(F_1^{**}) = \text{Area}(F_1)$ and

$\mathbf{n}_6^* + \mathbf{n}_6^{**} = 0$ gives the result.

Chapter 5, Section 5.3

1. $\|\mathbf{x} - \langle 1, -2, 3\rangle\| = 2$; $(\mathbf{x} - \langle 1, -2, 3\rangle)\cdot(\mathbf{x} - \langle 1, -2, 3\rangle) = 4$; $\mathbf{x}\cdot\mathbf{x} - 2\mathbf{x}\cdot\langle 1, -2, 3\rangle + 14 = 4$.

3. $(\mathbf{x} - \mathbf{s}_i)\cdot(\mathbf{x} - \mathbf{s}_i) = r_i^2$, $\mathbf{x}\cdot\mathbf{x} - 2\mathbf{s}_i\cdot\mathbf{x} = r_i^2 - \mathbf{s}_i\cdot\mathbf{s}_i$.

   $i = 1$: $\mathbf{x}\cdot\mathbf{x} - 2\langle 1, 0, 0\rangle\cdot\mathbf{x} = 3 - 1 = 2$.
   $i = 2$: $\mathbf{x}\cdot\mathbf{x} - 2\langle 2, 0, 3\rangle\cdot\mathbf{x} = 5 - 13 = -8$.
   $i = 3$: $\mathbf{x}\cdot\mathbf{x} - 2\langle 3, 2, 3\rangle\cdot\mathbf{x} = 6 - 22 = -16$.
   $i = 4$: $\mathbf{x}\cdot\mathbf{x} - 2\langle 3, 0, 1\rangle\cdot\mathbf{x} = 2 - 10 = -8$.
   Radical plane of Sphere 1 and Sphere 4: $2\langle 2, 0, 1\rangle\cdot\mathbf{x} = 10$.
   Radical plane of Sphere 2 and Sphere 4: $2\langle 1, 0, -2\rangle\cdot\mathbf{x} = 0$.
   Radical plane of Sphere 3 and Sphere 4: $2\langle 0, -2, -2\rangle\cdot\mathbf{x} = -8$.
   Equations in x, y, z:

$$2x + z = 5$$
$$x - 2z = 0$$
$$-2y - 2z = 4.$$

   Solution: $x = 2$, $y = 1$, $z = 1$.

5. Using Eq. (5.17), $2\mathbf{x}\cdot(\langle 5, -6, 0\rangle - \langle 2, 3, -4\rangle) = 25 - 49 + 61 - 29 = 8$; $3x - 9y + 4z = 4$.

7. Using Eq. (5.17), $2\mathbf{x}\cdot\langle 3, 3, 3\rangle = 18$; $x+y+z=3$.

9. If x is any two-dimensional vector on the line, it must satisfy $(\mathbf{x}-\mathbf{p})\cdot\mathbf{n}=0$. This is the equation of a line in vector form in two dimensions.

11. $(\mathbf{x}-\langle 1, 2\rangle)\cdot(\mathbf{x}-\langle 1, 2\rangle)=4$; $(\mathbf{x}-\langle 4, 3\rangle)\cdot(\mathbf{x}-\langle 4, 3\rangle)=4$; $\mathbf{x}\cdot\mathbf{x}-2\mathbf{x}\cdot\langle 1, 2\rangle+\langle 1, 2\rangle\cdot\langle 1, 2\rangle=4$; $\mathbf{x}\cdot\mathbf{x}-2\mathbf{x}\cdot\langle 4, 3\rangle+\langle 4, 3\rangle\cdot\langle 4, 3\rangle=4$; $\mathbf{x}\cdot(\langle 4, 3\rangle-\langle 1, 2\rangle)-25+5=0$; $\mathbf{x}\cdot\langle 3, 1\rangle=20$; $3x+y=20$.

13. $\|\mathbf{x}-\mathbf{u}\|^2=(\mathbf{x}-\mathbf{u})\cdot(\mathbf{x}-\mathbf{u})=\mathbf{x}\cdot\mathbf{x}-2\mathbf{x}\cdot\mathbf{u}+\mathbf{u}\cdot\mathbf{u}=1-2+1=0$. Thus $\mathbf{x}-\mathbf{u}$ is the zero vector and $\mathbf{x}=\mathbf{u}$.

15. (a) $\mathbf{f}=\lambda\mathbf{n}$ for some scalar λ.

    (b) $\|\mathbf{x}-\mathbf{f}\|^2=(\mathbf{x}-\mathbf{f})\cdot(\mathbf{x}-\mathbf{f})=\mathbf{x}\cdot\mathbf{x}-2\mathbf{x}\cdot\mathbf{f}+\mathbf{f}\cdot\mathbf{f}=r^2-2\lambda\mathbf{n}\cdot\mathbf{x}+\mathbf{f}\cdot\mathbf{f}=r^2-2\lambda k+\mathbf{f}\cdot\mathbf{f}$. Note that λ depends only on $\mathbf{f}$ and $\mathbf{n}$.

17. If (a, b, c) lies on all four spheres, then applying Exercise 16 four times we see that $(a, b, -c)$ is also on all four spheres.

19. If P is on all the spheres, by part (a) of Exercise 18, the point P' whose position vector is $\mathbf{p}'=\mathbf{p}-2(\mathbf{p}\cdot\mathbf{u}-k)\mathbf{u}$ is also on all spheres. It is different from P as long as $\mathbf{p}\cdot\mathbf{u}-k\neq 0$, which will be true since P is not on the plane.

21. Let the equation of the line of centers be $\mathbf{x}=\mathbf{s}_1+t(\mathbf{s}_2-\mathbf{s}_1)$. Thus there is a constant t_3 so that $\mathbf{s}_3-\mathbf{s}_1=t_3(\mathbf{s}_2-\mathbf{s}_1)$. Now the radical plane of Sphere 1 and Sphere 3 has equation $(\mathbf{x}-\mathbf{p})\cdot(\mathbf{s}_3-\mathbf{s}_1)=0$ and the radical plane of Sphere 2 and Sphere 1 is $(\mathbf{x}-\mathbf{p})\cdot(\mathbf{s}_2-\mathbf{s}_1)$. Substituting from our previous equation into the radical plane equation of Sphere 1 and Sphere 3, $t_3(\mathbf{x}-\mathbf{p})\cdot(\mathbf{s}_2-\mathbf{s}_1)=0$, which gives $(\mathbf{x}-\mathbf{p})\cdot(\mathbf{s}_2-\mathbf{s}_1)=0$, the equation of the radical plane of Sphere 1 and Sphere 2. Thus the two radical planes are the same. In the same way we show that the radical planes of Sphere 2 and Sphere 3 are the same.

23. Since these points do not lie on a line, they determine one plane and this is the radical plane of the first two spheres (since P, Q, R are all in the intersection of those two spheres). But what we have said about the first two spheres can also be said about any other combination of two of the spheres.

Chapter 5, Section 5.4

1. (a) $\mathbf{p}(t)=\langle 2, -1\rangle t^3+\langle -3, 0\rangle t^2+\langle 0, 4\rangle t+\langle 1, 3\rangle$.

   (b) $\mathbf{p}'(t)=3\langle 2, -1\rangle t^2+2\langle -3, 0\rangle t+\langle 0, 4\rangle=\langle 6, -3\rangle t^2+\langle -6, 0\rangle t+\langle 0, 4\rangle$. $\mathbf{p}'(0)=\langle 0, 4\rangle$, $\mathbf{p}'(1)=\langle 0, 1\rangle$.

3. Following the outline of Example 5.22, $\mathbf{c}_2 = \langle 2, 4\rangle$, $\mathbf{c}_0 + \mathbf{c}_1 + \mathbf{c}_2 = \langle 3, 7\rangle$, $4\mathbf{c}_0 + 2\mathbf{c}_1 + \mathbf{c}_2 = \langle 4, 14\rangle$. Substituting for $\mathbf{c}_2$ in the other equations gives $\mathbf{c}_0 + \mathbf{c}_1 = \langle 1, 3\rangle$, $4\mathbf{c}_0 + 2\mathbf{c}_1 = \langle 2, 10\rangle$. Subtract twice the first from the second to get $2\mathbf{c}_0 = \langle 0, 4\rangle$, $\mathbf{c}_0 = \langle 0, 2\rangle$. Substituting this gives $\mathbf{c}_1 = \langle 1, 1\rangle$. $\mathbf{p}(t) = \langle 0, 2\rangle t^2 + \langle 1, 1\rangle t + \langle 2, 4\rangle$.

5. $\mathbf{c}_2 = \langle 2, 5\rangle$, $\mathbf{c}_0 + \mathbf{c}_1 + \mathbf{c}_2 = \langle 3, 9\rangle$, $4\mathbf{c}_0 + 2\mathbf{c}_1 + \mathbf{c}_2 = \langle 4, 13\rangle$. Substituting for $\mathbf{c}_2$ in the other equations gives: $\mathbf{c}_0 + \mathbf{c}_1 = \langle 1, 4\rangle$, $4\mathbf{c}_0 + 2\mathbf{c}_1 = \langle 2, 8\rangle$. Subtract twice the first from the second to get $2\mathbf{c}_0 = \langle 0, 0\rangle$, $\mathbf{c}_0 = \langle 0, 0\rangle$. Substituting this gives $\mathbf{c}_1 = \langle 1, 4\rangle$. $\mathbf{p}(t) = \langle 0, 0\rangle t^2 + \langle 1, 4\rangle t + \langle 2, 5\rangle$.

7. Yes, but the t^3 coefficient is $\langle 0, 0\rangle$ so our cubic is actually a quadratic. Examples like this lead mathematicians to regard a quadratic as a "degenerate" kind of cubic.

9. $\mathbf{c}_0 = 2\langle 1, 1\rangle - 2\langle 2, 3\rangle + \langle 0, 1\rangle + \langle 1, 0\rangle = \langle -1, -3\rangle$.
   $\mathbf{c}_1 = -3\langle 1, 1\rangle + 3\langle 2, 3\rangle - 2\langle 0, 1\rangle - \langle 1, 0\rangle = \langle 2, 4\rangle$.
   $\mathbf{c}_2 = \langle 0, 1\rangle$.
   $\mathbf{c}_3 = \langle 1, 1\rangle$.
   Thus $\mathbf{p}(t) = \langle -1, -3\rangle t^3 + \langle 2, 4\rangle t^2 + \langle 0, 1\rangle t + \langle 1, 1\rangle$.

11. First spline:
    $\mathbf{c}_0 = 2\langle 2, 5\rangle - 2\langle 3, 7\rangle + \langle 1, 1\rangle + \langle 1, 4\rangle = \langle 0, 1\rangle$.
    $\mathbf{c}_1 = -3\langle 2, 5\rangle + 3\langle 3, 7\rangle - 2\langle 1, 1\rangle - \langle 1, 4\rangle = \langle 0, 0\rangle$.
    $\mathbf{c}_2 = \langle 1, 1\rangle$.
    $\mathbf{c}_3 = \langle 2, 5\rangle$.
    Thus $\mathbf{p}_1(t) = \langle 0, 1\rangle t^3 + \langle 1, 1\rangle t + \langle 2, 5\rangle$.
    Second spline:
    $\mathbf{c}_0 = 2\langle 3, 7\rangle - 2\langle 4, 13\rangle + \langle 1, 4\rangle + \langle 1, 6\rangle = \langle 0, -2\rangle$.
    $\mathbf{c}_1 = -3\langle 3, 7\rangle + 3\langle 4, 13\rangle - 2\langle 1, 4\rangle - \langle 1, 6\rangle = \langle 0, 4\rangle$.
    $\mathbf{c}_2 = \langle 1, 4\rangle$.
    $\mathbf{c}_3 = \langle 3, 7\rangle$.
    Thus $\mathbf{p}_2(t) = \langle 0, -2\rangle t^3 + \langle 0, 4\rangle t^2 + \langle 1, 4\rangle t + \langle 3, 7\rangle$.

13. (a) First spline: $\mathbf{p}_1''(t) = 6\langle -1, -4\rangle t + 2\langle 3, 6\rangle$. Thus $\mathbf{p}_1''(1) = \langle 0, -12\rangle$.

    (b) Second spline: $\mathbf{p}_2''(t) = 6\langle 5, -9\rangle t + 2\langle -10, 13\rangle$. Thus $\mathbf{p}_2''(0) = \langle -20, 26\rangle$.

    (b) Since $\mathbf{p}(t) = \mathbf{c}_0 t^3 + \mathbf{c}_1 t^2 + \mathbf{c}_2 t + \mathbf{c}_3$, we have $\mathbf{p}''(t) = 6\mathbf{c}_0 t + 2\mathbf{c}_1$. Using Eq. (5.34):

$$\mathbf{p}_1''(t) = 6(2\mathbf{p}_0 - 2\mathbf{p}_1 + \mathbf{v}_0 + \mathbf{v}_1)t + 2(-3\mathbf{p}_0 + 3\mathbf{p}_1 - 2\mathbf{v}_0 - \mathbf{v}_1),$$

$$\mathbf{p}_2''(t) = 6(2\mathbf{p}_1 - 2\mathbf{p}_2 + \mathbf{v}_1 + \mathbf{v}_2)t + 2(-3\mathbf{p}_1 + 3\mathbf{p}_2 - 2\mathbf{v}_1 - \mathbf{v}_2).$$

(c) In order to have $\mathbf{p}_1''(1) = \mathbf{p}_2''(0)$ we must have (by substitution into previous equations):

$$6(2\mathbf{p}_0 - 2\mathbf{p}_1 + \mathbf{v}_0 + \mathbf{v}_1) + 2(-3\mathbf{p}_0 + 3\mathbf{p}_1 - 2\mathbf{v}_0 - \mathbf{v}_1) \\ = 2(-3\mathbf{p}_1 + 3\mathbf{p}_2 - 2\mathbf{v}_1 - \mathbf{v}_2).$$

This simplifies to $3(\mathbf{p}_0 - \mathbf{p}_2) + \mathbf{v}_0 + 4\mathbf{v}_1 + \mathbf{v}_2 = 0$.

15. There are four requirements on the first polynomial, so a polynomial with four coefficients that can be freely assigned should do. Thus let $\mathbf{p}_{02}(t) = \mathbf{c}_0 t^3 + \mathbf{c}_1 t^2 + \mathbf{c}_2 t + \mathbf{c}_3$. $\mathbf{p}_{02}(0) = \mathbf{p}_0$ so $\mathbf{c}_3 = \mathbf{p}_0$, $\mathbf{p}_{02}(1) = \mathbf{p}_1$ so $\mathbf{c}_0 + \mathbf{c}_1 + \mathbf{c}_2 + \mathbf{c}_3 = \mathbf{p}_1$, $\mathbf{p}_{02}(2) = \mathbf{p}_2$ so $8\mathbf{c}_0 + 4\mathbf{c}_1 + 2\mathbf{c}_2 + \mathbf{c}_3 = \mathbf{p}_2$. $\mathbf{p}_{02}'(t) = 3\mathbf{c}_0 t^2 + 2\mathbf{c}_1 t + \mathbf{c}_2$. Since $\mathbf{p}_{02}'(2) = \mathbf{v}_2$ we have $12\mathbf{c}_0 + 4\mathbf{c}_1 + \mathbf{c}_2 = \mathbf{v}_2$. Substituting the value for $\mathbf{c}_3$ into the equations involving the other $\mathbf{c}_i$ we get three equations in three unknowns whose solution is $\mathbf{c}_0 = \frac{1}{4}[2\mathbf{v}_2 + 3\mathbf{p}_1 - 3\mathbf{p}_2 - \mathbf{p}_0]$,

$$\begin{aligned} \mathbf{c}_1 &= \tfrac{1}{2}[\mathbf{p}_2 - 2\mathbf{p}_1 + \mathbf{p}_0 - \tfrac{3}{2}(2\mathbf{v}_2 + 3\mathbf{p}_1 - 3\mathbf{p}_2 - \mathbf{p}_0)] \\ &= \tfrac{1}{4}[-7\mathbf{p}_2 - 13\mathbf{p}_1 + 5\mathbf{p}_0 - 6\mathbf{v}_2], \\ \mathbf{c}_2 &= \tfrac{1}{4}[10\mathbf{p}_2 + 10\mathbf{p}_1 + -8\mathbf{p}_0 + \mathbf{v}_2]. \end{aligned}$$

Chapter 6

Chapter 6, Section 6.1

1. (a) For P': $x' = \cos 90° + 2 \sin 90° = 2$, $y' = \sin 90° - 2 \cos 90° = 1$. $P' = (2, 1)$.

   For Q': $x' = \cos 90° - \sin 90° = -1$, $y' = \sin 90° + \cos 90° = 1$. $Q' = (-1, 1)$.

   (b) For P': $x' = \cos 30° + 2 \sin 30° = 1 + \frac{\sqrt{3}}{2}$, $y' = \sin 30° - 2 \cos 30° = \frac{1}{2} - \sqrt{3}$.

   For Q': $x' = \cos 30° - \sin 30° = \frac{\sqrt{3}-1}{2}$, $y' = \sin 30° + \cos 30° = \frac{\sqrt{3}+1}{2}$.

   (c) $P' = (1, 2)$, $Q' = (1, -1)$.

   (d) $P' = (-1, -2)$, $Q' = (-1, 1)$.

3. Take the origin of the new frame to be at the point $(2, 0)$ of $\mathcal{W}$. Have the new x axis point up (in the direction of $\mathcal{W}$'s y-vector) and the new x axis point to the left (opposite to the direction of $\mathcal{W}$'s x-vector).

5. Let $P(x, y)$ be a point that is reflected to $P'(x', y')$. For simplicity let's consider just the case where P is in the first quadrant beneath the reflection line. Let $\overline{PP'}$ meet the reflection line at Q. Drop a perpendicular from P to the x axis, meeting it at F. Drop a perpendicular from P' to the y axis, meeting it at

G. We need to show $GP' = PF$ and $OG = OF$. To do this, we need triangles OGP' and OFP congruent. As a stepping stone toward that goal, show triangles $OP'Q$ and OPQ congruent (by SAS). This will give $OP' = OP$ and $m\angle P'OQ = m\angle POQ$. Subtracting these congruent angles from the congruent angles $\angle GOQ$ and $\angle FOQ$ we get $m\angle GOP' = m\angle FOP$.

7. (a) $OQ/OP = OQ/OP' = \cos\alpha$.

   (b) $\mathbf{q} = (\cos\alpha)\mathbf{p}$.

   (c) $\mathbf{p}' - \mathbf{q}$ is a $90°$ rotation of vector $\langle x, y\rangle$ so, by Exercise 6, $\mathbf{p}' - \mathbf{q}$ has the same direction as $\langle -y, x\rangle$. But if a vector has the same direction as a second, it is a positive scalar multiple of the second. Thus $\mathbf{p}' - \mathbf{q} = k\langle -y, x\rangle$ for some k. To find k: $k = \|\mathbf{p}' - \mathbf{q}\|/\|\langle -y, x\rangle\| = \|\mathbf{p}' - \mathbf{q}\|/r = \sin\alpha$.

   (d) $\mathbf{p}' = \mathbf{q} + \mathbf{p}' - \mathbf{q} = \cos\alpha\ \mathbf{p} + \sin\alpha\langle -y, x\rangle = \cos\alpha\langle x, y\rangle + \sin\alpha\langle -y, x\rangle = \langle x\cos\alpha - y\sin\alpha, x\sin\alpha + y\cos\alpha\rangle$. In other words, $x' = x\cos\alpha - y\sin\alpha$; $y' = x\sin\alpha + y\cos\alpha$.

9. $m\angle QPQ' = m\angle P'Q'P$ since these are alternate interior angles of parallel lines. Thus triangles QPQ' and $P'Q'P$ are congruent (SAS). $\angle QQ'P$ and $\angle P'QQ'$ are corresponding parts and thus congruent. But these are also alternate interior angles for the lines $\overleftrightarrow{QQ'}$ and $\overleftrightarrow{PP'}$ cut by the transversal $\overleftrightarrow{PQ'}$.

11. $\begin{pmatrix} 3 & 7 \\ 4 & 8 \end{pmatrix}$.

13. $\begin{pmatrix} \sqrt{2}/2 & -\sqrt{2}/2 \\ \sqrt{2}/2 & -\sqrt{2}/2 \end{pmatrix}$, $\begin{pmatrix} \sqrt{2}/2 & -\sqrt{2}/2 & 0 \\ \sqrt{2}/2 & \sqrt{2}/2 & 0 \\ 0 & 0 & 1 \end{pmatrix}$, $\begin{pmatrix} \sqrt{2}/2 & -\sqrt{2}/2 \\ \sqrt{2}/2 & \sqrt{2}/2 \end{pmatrix}\begin{pmatrix} 4 \\ -3 \end{pmatrix}$

    $= \begin{pmatrix} 7\sqrt{2}/2 \\ \sqrt{2}/2 \end{pmatrix}$.

15. If (x, y) is to the right of $x = a$, then the distance to the line is $(x - a)$. The image point has distance $(a - x')$ to the line. These must be equal so $x' = 2a - x$. In addition $y' = y$. The matrix version of these equations is

$$\begin{pmatrix} x' \\ y' \\ 1 \end{pmatrix} = \begin{pmatrix} -1 & 0 & 2a \\ 0 & 1 & 0 \\ 0 & 0 & 1 \end{pmatrix}\begin{pmatrix} x \\ y \\ 1 \end{pmatrix}.$$

If the point is to the left of the line a similar argument gives the same result.

17. The standard frame has its x- and y-vectors meeting at (0, 0, 1), the tip of the x vector at (1, 0, 1) and the tip of the y-vector at (0, 1, 1). This is a frame with counterclockwise sense. Apply the matrix to these points and plot the resulting frame and check its sense.

Chapter 6, Section 6.2

1. Multiply the matrices for the two reflections and compare the product to $\mathbf{R}_O(180^\circ)$. An alternative approach, using substitution, amounts to the same thing:

$$\mathbf{p}' = \begin{pmatrix} -1 & 0 & 0 \\ 0 & 1 & 0 \\ 0 & 0 & 1 \end{pmatrix} \mathbf{p} \quad \text{and} \quad \mathbf{p}'' = \begin{pmatrix} 1 & 0 & 0 \\ 0 & -1 & 0 \\ 0 & 0 & 1 \end{pmatrix} \mathbf{p}'.$$

By substitution,

$$\mathbf{p}'' = \begin{pmatrix} -1 & 0 & 0 \\ 0 & 1 & 0 \\ 0 & 0 & 1 \end{pmatrix} \begin{pmatrix} 1 & 0 & 0 \\ 0 & -1 & 0 \\ 0 & 0 & 1 \end{pmatrix} \mathbf{p} = \begin{pmatrix} -1 & 0 & 0 \\ 0 & -1 & 0 \\ 0 & 0 & 1 \end{pmatrix} \mathbf{p}.$$

But this last matrix is $\mathbf{R}_O(180^\circ)$.

3. $$\mathbf{T}_1 = \begin{pmatrix} 1 & 0 & a \\ 0 & 1 & 0 \\ 0 & 0 & 1 \end{pmatrix}, \mathbf{T}_2 = \begin{pmatrix} 1 & 0 & 0 \\ 0 & 1 & b \\ 0 & 0 & 1 \end{pmatrix} \text{ so } \mathbf{T}_2\mathbf{T}_1 = \begin{pmatrix} 1 & 0 & 0 \\ 0 & 1 & b \\ 0 & 0 & 1 \end{pmatrix} \begin{pmatrix} 1 & 0 & a \\ 0 & 1 & 0 \\ 0 & 0 & 1 \end{pmatrix}$$

$$= \begin{pmatrix} 1 & 0 & a \\ 0 & 1 & b \\ 0 & 0 & 1 \end{pmatrix}.$$

$$(T_2 \circ T_1)(1, 0) = \mathbf{T}_2\mathbf{T}_1 \begin{pmatrix} 1 \\ 0 \\ 1 \end{pmatrix} = \begin{pmatrix} 1 + a \\ b \\ 1 \end{pmatrix} = \langle 1 + a, b, 1 \rangle.$$

Likewise we compute $(T_2 \circ T_1)(0, 1) = \langle a, 1 + b, 1 \rangle$ and $(T_2 \circ T_1)(2, 3) = \langle 2 + a, 3 + b, 1 \rangle$. T_1 and T_2 do commute. If you mutiply their matrices in either order, you get the same matrix.

5. Let α and $\mathbf{g} = \langle x_g, y_g \rangle$ be given and let $C = (x_C, y_C)$ be the unknown center. By Theorem 6.1, $T_g \circ R_O(\alpha)$ has matrix

$$\begin{pmatrix} \cos\alpha & -\sin\alpha & x_g \\ \sin\alpha & \cos\alpha & y_g \\ 0 & 0 & 1 \end{pmatrix}.$$

By Theorem 6.2 and Eq. (6.9), this matrix must equal

$$\begin{pmatrix} \cos\alpha & -\sin\alpha & -x_C\cos\alpha + y_C\sin\alpha + x_C \\ \sin\alpha & \cos\alpha & -x_C\sin\alpha + y_C\cos\alpha + y_C \\ 0 & 0 & 1 \end{pmatrix}.$$

Setting the corresponding entries in the last columns of these matrices equal gives:

$$x_C(1-\cos\alpha) + y_C\sin\alpha = x_g,$$

$$-x_C\sin\alpha + y_C(1+\cos\alpha) = y_g.$$

Now solve these equations simultaneously to find x_C and y_C in terms of α and the components of **g**.

7. For 100 points: Equation (6.7) needs 1800 multiplications. Equation (6.8) needs 927. Computing **TR** is about 3% of the work ($27/927 = 0.029$). For 1000 points: Equation (6.7) needs 18,000 multiplications. Equation (6.8) needs 9027. Computing **TR** is about 0.3% of the work. For 10,000 points: Equation (6.7) needs 180,000 multiplications. Equation (6.8) needs 90,027. Computing **TR** is about 0.03% of the work.

9. For 100 points: Equation (6.7) needs 3000 operations. Equation (6.8) needs 1545. Computing **TR** is about 3% of the work ($45/1545 = 0.029$). For 1000 points: Equation (6.7) needs 30,000 multiplications. Equation (6.8) needs 15,045. Computing **TR** is about 0.3% of the work. For 10,000 points: Equation (6.7) needs 300,000 multiplications. Equation (6.8) needs 150,045. Computing **TR** is about 0.03% of the work.

11. Apply Theorem 6.1 to each composition, obtaining the following matrix in both cases:

$$\begin{pmatrix} 1 & 0 & a+c \\ 0 & 1 & b+d \\ 0 & 0 & 1 \end{pmatrix}.$$

13. Let S have matrix

$$\mathbf{S} = \begin{pmatrix} a & b & c \\ d & e & f \\ 0 & 0 & 1 \end{pmatrix}.$$

Then $S(\mathbf{g})$ has components

$$\begin{pmatrix} a & b & c \\ d & e & f \\ 0 & 0 & 1 \end{pmatrix}\begin{pmatrix} u \\ v \\ 1 \end{pmatrix} = \begin{pmatrix} au+bv+c \\ du+ev+f \\ 1 \end{pmatrix}.$$

By Theorem 6.1, $T_{S(\mathbf{g})} \circ S$ has matrix

$$\begin{pmatrix} a & b & au+bv+2c \\ d & e & du+ev+2f \\ 0 & 0 & 1 \end{pmatrix}.$$

$S \circ T_{\mathbf{g}}$ has matrix

$$\begin{pmatrix} a & b & c \\ d & e & f \\ 0 & 0 & 1 \end{pmatrix}\begin{pmatrix} 1 & 0 & u \\ 0 & 1 & v \\ 0 & 0 & 1 \end{pmatrix} = \begin{pmatrix} a & b & au+bv+c \\ d & e & du+ev+f \\ 0 & 0 & 1 \end{pmatrix}.$$

The matrices for $S \circ T_{\mathbf{g}}$ and $T_{S(\mathbf{g})} \circ S$ differ in their last columns, so the equality does not hold unless $c = f = 0$.

15. $$M_x + M_y = \begin{pmatrix} 1 & 0 & 0 \\ 0 & -1 & 0 \\ 0 & 0 & 1 \end{pmatrix} + \begin{pmatrix} -1 & 0 & 0 \\ 0 & 1 & 0 \\ 0 & 0 & 1 \end{pmatrix} = \begin{pmatrix} 0 & 0 & 0 \\ 0 & 0 & 0 \\ 0 & 0 & 2 \end{pmatrix}.$$

This is not the matrix of any isometry. If you multiply it by any homogeneous representation of a point we get

$$\begin{pmatrix} 0 \\ 0 \\ 2 \end{pmatrix},$$

which does not represent a point.

Chapter 6, Section 6.3

1. $$\mathbf{T}_{\langle 2,1\rangle}\mathbf{R}_O(30^\circ)\mathbf{M}_x = \mathbf{T}_{\langle 2,1\rangle}\begin{pmatrix} \sqrt{3}/2 & -1/2 & 0 \\ 1/2 & \sqrt{3}/2 & 0 \\ 0 & 0 & 1 \end{pmatrix}\begin{pmatrix} 1 & 0 & 0 \\ 0 & -1 & 0 \\ 0 & 0 & 1 \end{pmatrix}$$

$$= \mathbf{T}_{\langle 2,1\rangle}\begin{pmatrix} \sqrt{3}/2 & 1/2 & 0 \\ 1/2 & -\sqrt{3}/2 & 0 \\ 0 & 0 & 1 \end{pmatrix} = \begin{pmatrix} \sqrt{3}/2 & 1/2 & 2 \\ 1/2 & -\sqrt{3}/2 & 1 \\ 0 & 0 & 1 \end{pmatrix}.$$

3. (a) Rotation by 90° around the origin of $\mathcal{W}$ followed by translation 15 units to the right and 6 units up.

   (b) Case (b).

   (c) $\mathbf{T} = \begin{pmatrix} 1 & 0 & 15 \\ 0 & 1 & 6 \\ 0 & 0 & 1 \end{pmatrix}, \mathbf{R} = \begin{pmatrix} 0 & -1 & 0 \\ 1 & 0 & 0 \\ 0 & 0 & 1 \end{pmatrix}.$

5. $\mathbf{T}_{(5,4)}\mathbf{R}_O(45°) = \begin{pmatrix} \sqrt{2}/2 & -\sqrt{2}/2 & 5 \\ \sqrt{2}/2 & \sqrt{2}/2 & 4 \\ 0 & 0 & 1 \end{pmatrix}.$

7. Let A be the point at the tip of the x-vector of a frame and let B be the tip of the y-vector. The frame has counterclockwise sense if and only if the sense of $[OAB]$ is counterclockwise.

9. Every isometry is either translation, rotation, reflection, or glide reflection and each of these has a matrix.

11. It is easy to see from Figure A.7 how to get the $\mathcal{W}$ coordinates for the corner — just interchange the coordinates obtained for P on the screen (in the screen frame $\mathcal{S}$).

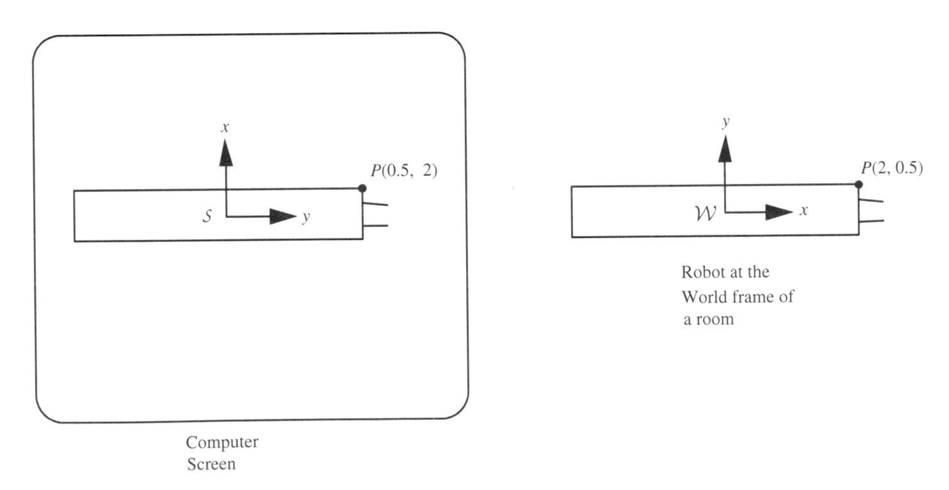

Figure A.7

13. Multiply the matrix of Exercise 12 by $\begin{pmatrix} -2 \\ 7 \\ 1 \end{pmatrix}$ to obtain $\begin{pmatrix} 0.11 \\ 5.44 \\ 1 \end{pmatrix}$.

15. Find the inverse of $\mathbf{F}$ and compute $\mathbf{F}^{-1}\begin{pmatrix} a \\ b \\ 1 \end{pmatrix}$.

17. Think of the plane embedded in three-dimensional space and let $\mathbf{u} = \langle a, b, 0\rangle$ and $\mathbf{v} = \langle c, d, 0\rangle$. If $\mathbf{u} \times \mathbf{v}$ has a positive z component we have a frame of counter-clockwise sense.

Chapter 6, Section 6.4

1. $\begin{pmatrix} 12/13 & -5/13 & 4 \\ -5/13 & -12/13 & -1 \\ 0 & 0 & 1 \end{pmatrix}$.

3. See Figure A.8.

5. (a) Last column is $\begin{pmatrix} 0 \\ 0 \\ 1 \end{pmatrix}$.

   (b) First two entries of first column make a vector $\mathbf{u}$ with $\|\mathbf{u}\| = 1$.

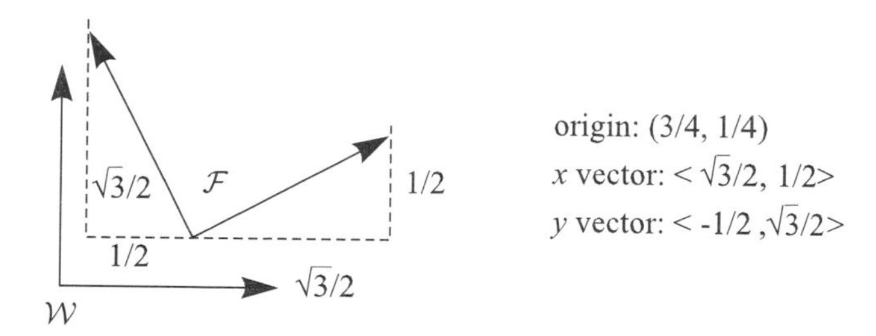

Figure A.8

(c) First two entries of second column make a vector $\mathbf{v}$ with $\|\mathbf{v}\| = 1$.

(d) $\mathbf{u} \cdot \mathbf{v} = 1$.

7. The x-vector $\langle a, b\rangle$ is a unit vector so $\|\langle a, b\rangle\| = \sqrt{a^2 + b^2} = 1$.

9. $$\mathbf{F} = \mathbf{T}_{\langle 5,\ 10\sqrt{3}/2\rangle}\mathbf{R}_O(60^\circ) = \begin{pmatrix} 1/2 & -\sqrt{3}/2 & 5 \\ \sqrt{3}/2 & 1/2 & 15\sqrt{3} \\ 0 & 0 & 1 \end{pmatrix}.$$

11. $$\begin{pmatrix} 5/13 & 12/13 & 3 \\ 12/13 & -5/13 & 2 \\ 0 & 0 & 1 \end{pmatrix}\begin{pmatrix} \sqrt{2}/2 & -\sqrt{2}/2 & \sqrt{3}/2 \\ \sqrt{2}/2 & \sqrt{2}/2 & -1/2 \\ 0 & 0 & 1 \end{pmatrix}$$

$$= \begin{pmatrix} 17\sqrt{2}/26 & 7\sqrt{2}/26 & 5\sqrt{3}/26 + 33/13 \\ 7\sqrt{2}/26 & -17\sqrt{2}/26 & 12\sqrt{3}/26 + 57/26 \\ 0 & 0 & 1 \end{pmatrix}.$$

13. $$\begin{pmatrix} 0 & -1 & 15 \\ 1 & 0 & 6 \\ 0 & 0 & 1 \end{pmatrix}\begin{pmatrix} 4/5 & -3/5 & -2 \\ 3/5 & 4/5 & 1 \\ 0 & 0 & 1 \end{pmatrix} = \begin{pmatrix} -3/5 & -4/5 & 14 \\ 4/5 & -3/5 & 4 \\ 0 & 0 & 1 \end{pmatrix}.$$

15. New origin is

$$\begin{pmatrix} \sqrt{3}/2 & -1/2 & 3/4 \\ 1/2 & \sqrt{3}/2 & 1/4 \\ 0 & 0 & 1 \end{pmatrix}\begin{pmatrix} 15 \\ 16 \\ 1 \end{pmatrix} = \begin{pmatrix} 10.74 \\ 12.95 \\ 1 \end{pmatrix}.$$

This point is within the left–right limits of the room, but we can't tell about whether it is beyond the top border. The labeling on the diagram does not reveal where that border is.

17. (a) The last column of $\mathbf{F}$ is the last column of $\mathbf{T}_{\langle a,\ b\rangle}\mathbf{R}_O(\theta)$, but by Theorem 6.1, this is the same as the last column of $\mathbf{T}_{\langle a,\ b\rangle}$, whose first two entries are a and b. Thus a and b can be read from $\mathbf{F}$ directly. Now we "solve for" $\mathbf{R}_O(\theta)$: $\mathbf{T}_{\langle -a,\ -b\rangle}\mathbf{F} = \mathbf{T}_{\langle -a,\ -b\rangle}\mathbf{T}_{\langle a,\ b\rangle}\mathbf{R}_O(\theta) = \mathbf{R}_O(\theta)$. The entries of $\mathbf{R}_O(\theta)$ reveal the sine and cosine of θ and from these we can solve for θ.

(b) According to Theorem 6.8, we are looking for $\mathbf{G}$ where $\mathbf{GF} = \mathbf{I}$, that is, $\mathbf{GT}_{\langle a,\ b\rangle}\mathbf{R}_O(\theta) = \mathbf{I}$. Take $\mathbf{G} = \mathbf{R}_O(-\theta)\mathbf{T}_{\langle -a,\ -b\rangle}$.

Chapter 6, Section 6.5

1. See Figure A.9.

3. $x_g = 20 \cos 90^\circ + 10 \cos(90^\circ - 45^\circ) = 5\sqrt{2}$.
   $y_g = 20 \sin 90^\circ + 10 \sin(90^\circ - 45^\circ) = 20 + 5\sqrt{2}$.

5. It is the ring between circles of radius 10 and 30 centered at the base point of the robot.

7. It is a circle of radius $2l_1$ centered at the base point of the robot.

9. Using the notation of Figure 6.19, let α be the angle between $\overline{OG}$ and the positive x axis, and let β be m$\angle BOG$ and let $\gamma = $ m$\angle OBG$. From (x_G, y_G) we can get α (for example, $\alpha = \tan^{-1}(y_G/x_G)$ if G is in the first quadrant). Use the law of cosines (twice) to find β and γ then find the two possibilities for θ_1 (if E is under $\overline{OG}$, $\theta_1 = \alpha - \beta$; otherwise $\theta_1 = \alpha + \beta$). θ_2 is either $180^\circ - \gamma$ or $\gamma - 180^\circ$.

11. The configurations shown in the solution for Exercise 9 are ones where there is no other pair of joint angles that get the end effector point to the same place. (If there were a different configuraton, with the links forming a true triangle, the triangle inequality would give a contradiction.)

13. $$\begin{pmatrix} \cos 90^\circ & -\sin 90^\circ & x_g \\ \sin 90^\circ & \cos 90^\circ & y_g \\ 0 & 0 & 1 \end{pmatrix} = \begin{pmatrix} 0 & -1 & x_g \\ 1 & 0 & y_g \\ 0 & 0 & 1 \end{pmatrix}$$

    where $x_G = 30 \cos 45^\circ + 10 \cos 90^\circ = 15\sqrt{2}$, $y_G = 30 \sin 45^\circ + 10 \sin 90^\circ = 15\sqrt{2} + 10$.

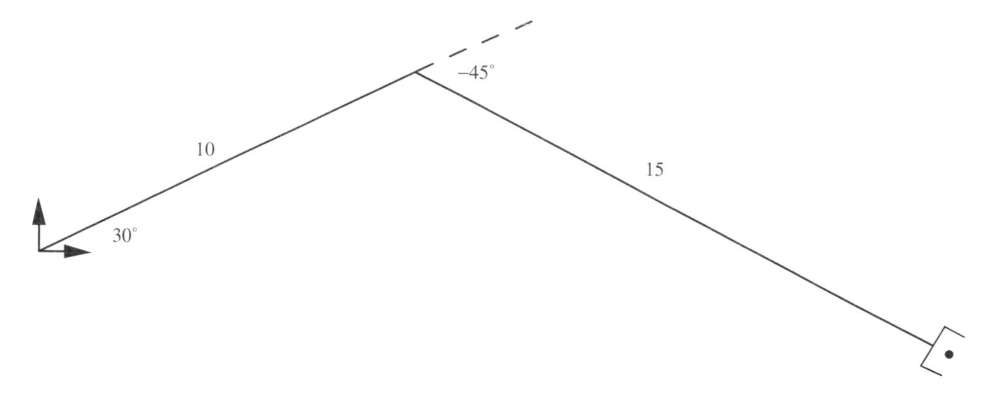

Figure A.9

15. $\mathbf{M}_1 = \mathbf{R}_O(45^\circ) = \begin{pmatrix} \sqrt{2}/2 & -\sqrt{2}/2 & 0 \\ \sqrt{2}/2 & \sqrt{2}/2 & 0 \\ 0 & 0 & 1 \end{pmatrix}.$

17. $\mathbf{M}_2 = \mathbf{T}_{\langle 6\cos 45^\circ,\ 6\sin 45^\circ \rangle}\mathbf{M}_1$. Use $\mathbf{M}_1$ from Exercise 15 to get

$$\mathbf{M}_2 = \begin{pmatrix} \sqrt{2}/2 & -\sqrt{2}/2 & 3\sqrt{2} \\ \sqrt{2}/2 & \sqrt{2}/2 & 3\sqrt{2} \\ 0 & 0 & 1 \end{pmatrix}.$$

19. $\mathbf{M}_3 = \mathbf{R}_b(45^\circ)\mathbf{T}_b\mathbf{R}_O(45^\circ) = \mathbf{T}_b\mathbf{R}_O(45^\circ)\mathbf{R}_O(45^\circ) = \mathbf{T}_b\mathbf{R}_O(90^\circ)$

$$= \begin{pmatrix} 0 & -1 & 3\sqrt{2} \\ 1 & 0 & 3\sqrt{2} \\ 0 & 0 & 1 \end{pmatrix}.$$

21. The isometry matrix **C** for the camera frame is obtained just as we obtained **E** except that the last translation is by an amount $l_2/4$ instead of l_2. Just replace l_2 by $l_2/4$ in the matrix of Equation (6.17).

23. Precede the isometries used to derive Theorem 6.10 by reflection in the x axis of $\mathcal{W}$. We get

$$\begin{pmatrix} \cos(\theta_1+\theta_2) & \sin(\theta_1+\theta_2) & x_g \\ \sin(\theta_1+\theta_2) & -\cos(\theta_1+\theta_2) & y_g \\ 0 & 0 & 1 \end{pmatrix}.$$

25. If the last column of **E** represents a point **G** which is so far from the origin the robot can't reach it, there is no solution. Even for a point the robot can reach, there are at most two sets of angles which will achieve that point, and for each of these angle sets the end effector frame is determined. So if you pick the first two columns so that they don't specify either of those frame orientations, there will be no solution.

Chapter 7

Chapter 7, Section 7.1

1. $P' = (2, -6)$, $Q' = (6, -8)$, $R' = (0, 4)$.

3. In Figure A.10, let P' be the foot of the perpendicular from P to the horizontal line through C. Let Q' be the image of Q. $CQ'/CQ = CP'/CP = r$. Thus, $(x^* - x_C)/(x - x_C) = r$ and $x^* = r(x - x_C) + x_C$. By part (b) of Theorem 7.1, $P'Q'$ is parallel to PQ and so perpendicular to the x axis and so $x^* = x'$. Substitute into the previous equation.

5. (a) $P' = (1.5, 1)$, $Q' = (2.5, -4)$, $R' = (2.5, -1)$. Each image is one-half unit to the left of the original point. This transformation is translation by the vector $\langle -0.5, 0 \rangle$.

   (b) (x, y) transforms to $(x - \frac{1}{2}, y)$.

7. The scaling plus scan conversion method is more accurate.

9. The center C is the intersection of lines $\overleftrightarrow{AA'}$ and $\overleftrightarrow{BB'}$. $CA'/CA = r$.

11. The given curve is a circle with radius 4, centered at the origin, so the image is a circle of radius 2, centered at the origin.

13. For $S_{O,r}$ we have $x' = rx, y' = ry$; for $R_O(\theta)$ we have $x'' = x' \cos\theta - y' \sin\theta$, $y'' = x' \sin\theta + y' \cos\theta$. Making substitutions from the first set of equations into the second gives $x'' = r(x \cos\theta - y \sin\theta)$, $y'' = r(x \sin\theta + y \sin\theta)$.

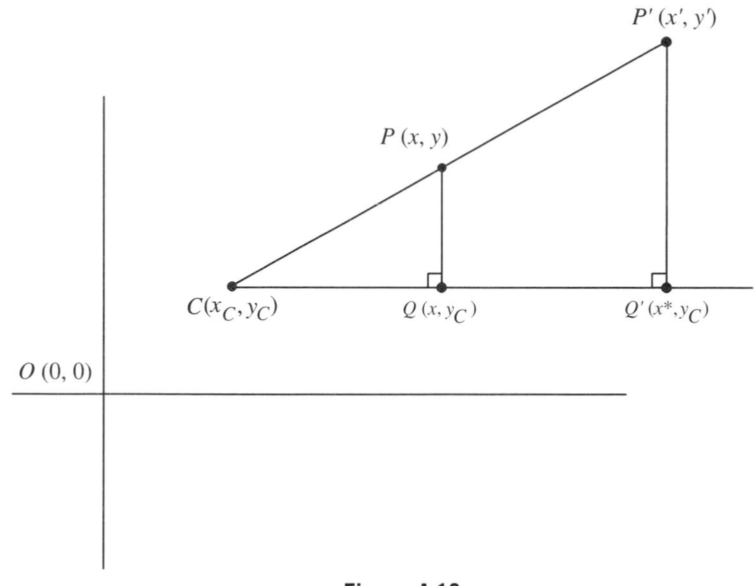

Figure A.10

15. Triangle $P'Q'R'$ is similar to PQR and therefore each of its angles is equal to the corresponding angle in triangle PQR.

17. $R_O(\theta) \circ S_{O,r}$ has O as a fixed point but unless θ is a multiple of 180° it does not preserve direction. Thus, by Theorem 7.1, it cannot be a central similarity.

19. (a) The composite transformation preserves direction and scales each segment by the same factor, pq. By the Exercise 18, it is a central similarity if $pq \neq 1$. Since it keeps C fixed and a central similarity with $pq \neq 1$ has a single fixed point, C is its center.

    (b) Let A be some point and A' its image. $S_{c,q} \circ S_{c,p}$ is the translation that moves A to A'. To see this, let B be any other point. $\overline{A'B'}$ has the same length and direction as $\overline{AB}$ so $AA'B'B$ is a parallelogram. Thus $\overline{BB'}$ has the same lenth and direction as $\overline{AA'}$.

21. Use Exercise 18. Its ratio is pq.

23. Let $U = S_{C,p} \circ E$ where E is an isometry. The transformation $E^{-1} \circ U$ will preserve direction (whatever E does to change a direction is left unaffected by $S_{C,p}$ but then reversed by E^{-1}) and scales segment lengths by p. By Exercise 18, $E^{-1} \circ U = S_{D,q}$ for some D and q. Thus $U = E \circ S_{D,q}$ from which we deduce $U^{-1} = S_{D,1/q} \circ E^{-1}$, which is a similarity transformation.

25. Let PQR be any triangle and $P'Q'R'$ its image under U. Since PQR and $P'Q'R'$ are similar, by Theorem 7.5 there is a similarity S that maps $P'Q'R'$ to PQR. $S \circ U$ fixes P, Q, and R and maps every triangle to one similar to it. By Exercise 24, $S \circ U = I$ and then $U = S^{-1}$. But by Exercise 23, S^{-1} is a similarity.

Chapter 7, Section 7.2

1. (a) 1/5, (b) $\sqrt{2}/2$, (c) 1.

3. Let M be the infinite line obtained by extending the top of the square and let A be the point where it meets the circle of inversion. M' is a circle internally tangent to the circle of inversion at A, and passing through the center of inversion. Each of the other boundary lines transforms into a similar circle. But we are not inverting entire infinite lines, only the sides of the square. Thus we take an arc on each circle, extending from its intersection with one neighboring circle, through the point of tangency, to its intersection with the other neighboring circle. This gives a cross-shaped figure with circular boundaries.

5. Take the circle of inversion to have center at $O(0, 0)$ and $r=1$. Then let $P=(-1, 0)$, $Q=(0, \frac{1}{2})$, $R=(1, 0)$. P and R are fixed but $R'=(0, 2)$ is outside the circle of inversion. Thus it is visually evident that $m\angle PQR \neq m\angle P'Q'R'$.

7. Inside the circle of inversion, close to the center of inversion (the closer you go, the farther the inverses move to infinity).

9. Say P and Q are on the x axis with $P=(a, 0)$ and $Q=(b, 0)$. Suppose we imagine such a circle of inversion centered at $(x, 0)$ and having radius of inversion r. Then we must have $(a - x)(b - x)=r^2$. We can pick many combinations of x and r so as to make this work. For example, pick any x so that $x < a$ and $x < b$. Then the left side is >0 and it will be possible to find r as its square root.

11. First observe that $(\gamma')'=\gamma$. Since P lies on γ, P' lies on γ'. Since P lies on γ', P' lies on $(\gamma')'=\gamma$. Thus P' is on both γ and γ'.

13. (a) $\rho' = r^2/\rho$, $\theta'=\theta$. (b) $\rho' = r^2/\rho$, $\theta'=\theta$, $\phi'=\phi$.

15. By similar triangles $CP'/CT = CT/CP$.

17. Triangles CPT and CTP' are similar (since $\overline{CT}$ is perpendicular to $\overline{TP'}$ and $\angle C$ is in both triangles). Thus $CP'/CT = CT/CP$.

19. Let Q be any point on the circle other than P and let the other intersection point of $\overleftrightarrow{CT}$ and γ be denoted Q^*. Draw a tangent to γ meeting it at T. By the secant–tangent theorem, $(CQ)(CQ^*)=CT^2=(CP)(CP')=r^2$. Thus Q^* is the inverse of Q.

21. Base angles of an isosceles triangle are congruent so $m\angle PST=m\angle PRT$ and $m\angle QST=m\angle QRT$. Adding these equations gives $m\angle PSQ=m\angle PRQ$. Then triangles PSQ and PRQ are congruent by SAS. Using the corresponding parts principle and since base angles of isosceles triangles are congruent, the following four angles are all congruent: $\angle SPT$, $\angle RPT$, $\angle SQT$, $\angle RQT$. Now it is possible to show that all four triangles which touch at T are congruent. Thus the angles at T are all equal and all are right angles.

23. As S moves around the circle, the distance SC gets smaller. The closest S can be to C is when $m\angle PCQ=0$ and S lies on $\overline{CP}$. So find $CP - PS$ and find that point U on the circle where $CU=CP - PS$. U is as far as S can go around the circle.

Chapter 7, Section 7.3

1. Take the four boundary segments of the rectangle that is the image surface and project these boundary segments through C. This creates a sort of infinite pyramid extending out in front of the camera. The edges of this solid are rays starting at C. (Each ray, if extended backward into the camera, would pass through a corner of the image surface.) Every point inside or on the boundary of this unbounded pyramid is visible.

3. (a) The image enlarges as the line segment gets closer to C.

   (b) There is no limit.

   (c) If the line segment is parallel to the image plane, there will be no image. Otherwise, the image will be a single point.

5. The image will be larger after the stretch.

7. Your right hand.

9. If P, Q, and R determine a plane that is parallel to the projection direction, all three projector lines through P, Q, and R lie in the plane. Thus the images P, Q, and R will lie in the intersection of this plane with the image plane. This intersection is a line.

11. If the polygon lies in a plane containing the center of projection.

13. Let P, Q, R determine a plane containing C, the center of projection, and let S be outside this plane. The quadrilateral $PQRS$ projects into a triangle because P', Q', and R' are collinear.

15. If C lies on the line determined by that segment.

17. If we are looking at a scene which has a variety of solid objects, in any direction we look, our eye can see only the nearest object. Points on other objects are hidden even though they are in our line of sight.

19. $x' = \dfrac{4(2)}{4-1} = \dfrac{8}{3},\ y' = \dfrac{4(-3)}{(4-1)} = -4,\ z' = 0.$

21. First we get two points on the line. For $t = 0$, get $\mathbf{p}_0 = \langle 1, 2, -5\rangle$; for $t = 1$, get $\mathbf{p}_1 = \langle 1, 5, -12\rangle$. The line determined by $\mathbf{p}_0$ and $\mathbf{c}$ is $\mathbf{x} = \langle 1, 2, -5\rangle + t(\langle -2, 6, 3\rangle - \langle 1, 2, -5\rangle) = \langle 1, 2, -5\rangle + t\langle -3, 4, 8\rangle$. Intersection of this line with the image plane: $(\langle 1, 2, -5\rangle + t\langle -3, 4, 8\rangle) \cdot \langle -1, 3, 0\rangle = 4$; $\langle -1+6\rangle + t\langle 3+12\rangle = 4$; $t = -1/15$. Thus $\mathbf{p}_{0'} = \langle 1, 2, -5\rangle + (-1/15)\langle -3, 4, 8\rangle$

$= \langle 18/15,\ 26/15,\ -83/15\rangle$. The line determined by $\mathbf{p}_1$ and $\mathbf{c}$ is $x = \langle 1, 5, -12\rangle + t(\langle -2, 6, 3\rangle - \langle 1, 5, -12\rangle) = \langle 1, 5, -12\rangle + t\langle -3, 1, 15\rangle$. Intersection of this line with the image plane: $(\langle 1, 5, -12\rangle + t\langle -3, 1, 15\rangle)\cdot\langle -1, 3, 0\rangle = 4$; $14 + 6t = 4$; $t = -5/3$. Thus $\mathbf{p}_{1'} = \langle 1, 5, -12\rangle + (-5/3)\langle -3, 1, 15\rangle = \langle 6, 10/3, -37\rangle$. The line connecting the two image points: $\mathbf{x} = \mathbf{p}_{0'} + t(\mathbf{p}_{1'} - \mathbf{p}_{0'}) = \langle 18/15, 26/15, -83/15\rangle + t\langle 72/15,\ 24/15,\ -472/15\rangle$.

23. Take noncollinear points A, B, and C in the x-z plane. $m\angle ABC \neq 0$ but $m\angle A'B'C' = 0$.

25. The projection equations are $x' = \frac{-1}{2}x$, $y' = \frac{-1}{2}y$ so $x = -2x'$ and $y = -2y'$. Substitute these into the equation of the circle: $(-2x' - 6)^2 + (-2y' - 2)^2 = 25$. Equivalently, $4[(x' + 3)^2 + (y' + 1)^2] = 25$ or $(x' + 3)^2 + (y' + 1)^2 = (\frac{5}{2})^2$. This, together with $z' = 0$, tells us we have a circle in the plane $z = 0$ with center $(-3, -1)$ and radius $\frac{5}{2}$.

27. $\mathbf{p}' = \left\langle \dfrac{cx_P}{c-k}, \dfrac{cy_P}{c-k}, 0\right\rangle = \dfrac{c}{c-k}\langle x_P, y_P, 0\rangle$. $\mathbf{q}' = \dfrac{c}{c-k}\langle x_Q, y_Q, 0\rangle$. Thus

$$\mathbf{p}' - \mathbf{q}' = \frac{c}{c-k}\langle x_P - x_Q, y_P - y_Q, 0\rangle = \frac{c}{c-k}(\langle x_P, y_P, k\rangle - \langle x_Q, y_Q, k\rangle) = \frac{c}{c-k}(\mathbf{p} - \mathbf{q}).$$

Thus $\|\mathbf{p}' - \mathbf{q}'\| = \left\|\dfrac{c}{c-k}(\mathbf{p} - \mathbf{q})\right\| = \dfrac{c}{c-k}\|(\mathbf{p} - \mathbf{q})\|$. $c/(c-k)$ is the r asked for in the exercise.

29. Let P be any world point and P' its image. The projector line has equation $\mathbf{x} = \mathbf{p} + t(\mathbf{c} - \mathbf{p})$. Substitute this expression for $\mathbf{x}$ in the plane equation and solve for t to obtain $t = \dfrac{k - \mathbf{p}\cdot\mathbf{n}}{(\mathbf{c} - \mathbf{p})\cdot\mathbf{n}}$. Thus $\mathbf{p}' = \mathbf{p} + \dfrac{k - \mathbf{p}\cdot\mathbf{n}}{(\mathbf{c} - \mathbf{p})\cdot\mathbf{n}}(\mathbf{c} - \mathbf{p})$.

31. No, it does not lie in the plane $z = 6$.

33. We need to check if $(\mathbf{p} - \mathbf{c})\cdot\mathbf{n} = 0$. $(\langle -1, -1, 4\rangle - \langle 5, 1, 2\rangle)\cdot\langle -1, 3, 0\rangle = \langle -6, -2, 2\rangle\cdot\langle -1, 3, 0\rangle = (-6)(-1) + (-2)(3) + (2)(0) = 0$. Yes, it is exceptional.

Chapter 7, Section 7.4

1. (a) This could be. The edges of the box which are parallel are depicted as converging to a vanishing point.

(b) No. Parallel edges of the box which project to parallel segments in the image must be parallel to the image plane. But we can't have adjacent faces both parallel to the image plane.

(c) No. For this to be the image of a box, some pairs of box edges would have to project to segments on the same line in the image. This would mean the plane determined by those box edges contains C. But then there are four different face planes of the box which would have to contain C, an impossibility.

3. (a) Identical.

   (b) Parallel.

   (c) Different and not parallel.

5. (a) $\mathbf{x} = \langle 1, 0, 0 \rangle + t\langle 7, 3, -8 \rangle$.

   (b) $\mathbf{x} = \langle 2, -3, 4 \rangle + t\langle 1, 1, -1 \rangle$.

   (c) $\mathbf{x} = \langle 3, 6, -3 \rangle + t\langle 3, 4, 5 \rangle$.

7. Let E_{13} be the plane determined by lines L_1 and L_3, and let E_{23} be the plane determined by lines L_2 and L_3. Let P be a point on L_1. Let E be the plane determined by P and L_2. We'd like to prove L_1 is contained in E. If not, E and E_{13} intersect in a line L different from L_1 but passing through P. Since L_2 does not meet E_{13} (if it did, it would have to do so on L_3, which is impossible), and since L is contained in E_{13}, it follows that L_2 does not meet L. This, together with the fact that L and L_2 are in E, means L and L_2 are parallel. L meets L_3 since L_1 is parallel to L_3 and we can't have two parallels to L_3 in plane E_{13}, both passing through P. Let Q be the intersection of L and L_3. L and L_3 are two lines through Q parallel to L_2, a contradiction. This shows L_1 is contained in E. If L_1 and L_2 were to cross, it would have to be at a point in E_{13} (since L_1 is in E_{13}) and it would have to be in E_{23} (since L_2 is in E_{23}). The only points in both planes are on L_3. Thus L_1 and L_3 cross, contrary to the hypothesis. Thus L_1 and L_2 don't cross.

9. Let A, B and P, Q be four points on line L. We need to show direction vectors $\mathbf{b} - \mathbf{a}$ and $\mathbf{q} - \mathbf{p}$ are proportional. Since P and Q are both on L, their position vectors must satisfy the parametric equations for some t's:

$$\mathbf{q} = \mathbf{a} + t_Q(\mathbf{b} - \mathbf{a}) \qquad \text{and} \qquad \mathbf{p} = \mathbf{a} + t_P(\mathbf{b} - \mathbf{a}).$$

Subtracting gives $\mathbf{q} - \mathbf{p} = (t_Q - t_P)(\mathbf{b} - \mathbf{a})$. Thus the direction vectors are proportional with proportionality factor $(t_Q - t_P)$.

11. $\mathbf{a}_1 + t_1\mathbf{d} = \mathbf{a}_2 + t_2\mathbf{d}$ so $\mathbf{a}_2 = \mathbf{a}_1 + (t_1 - t_2)\mathbf{d}$ and this means A_2 is on L_2. Now suppose P is any other point on L_2. There is a t_P so that $\mathbf{p} = \mathbf{a}_2 + t_P\mathbf{d} = \mathbf{a}_1 + (t_1 - t_2)\mathbf{d} + t_P\mathbf{d} = \mathbf{a}_1 + (t_1 - t_2 + t_P)\mathbf{d}$. Thus P is on L_1.

13. (a) $\mathbf{d} \cdot \mathbf{n} = \langle 2, -1, 0\rangle \cdot \langle 1, 2, -4\rangle = 0$ so the line is either parallel to or in the plane.

    (b) $\mathbf{d} \cdot \mathbf{n} = \langle -4, 2, 3\rangle \cdot \langle 1, 2, -4\rangle = -12 \neq 0$ so there is one intersection point. To get V, intersect $\mathbf{x} = \langle 1, -2, 1\rangle + t\langle -4, 2, 3\rangle$ with the plane: $(\langle 1, -2, 1\rangle + t\langle -4, 2, 3\rangle) \cdot \langle 1, 2, -4\rangle = 3$. $-7 + t(1 - 12) = 3$. $t = -5/6$. Substitute into line equation: $\mathbf{v} = \langle 4, 3, -2\rangle - (5/6)\langle -4, 2, 3\rangle = \langle 22/3, 4/3, -9/2\rangle$.

    (c) $\mathbf{d} \cdot \mathbf{n} = \langle 5, -2, 3\rangle \cdot \langle 1, 2, -4\rangle = -11 \neq 0$ so there is one intersection point.

15. We proceed indirectly. If the images lie in a line, then they lie in the plane determined by that line and C. Since the points are not collinear, there is just one plane containing all three, so there can't be a plane containing them which does not contain C. This is a contradiction to the hypothesis.

17. Any direction orthogonal to $\langle -2, 4, 3\rangle$, that is, **d**'s components should satisfy $-2x_d + 4y_d + 3z_d = 0$. Pick $x_d = 0$, $y_d = 1$, and solve for $z_d = -4/3$.

19. Since L and L_C are parallel, there is a plane M containing both and it intersects plane I in a line L'. Now L' can't be L since L does not lie in I. Thus L_C intersects L' because otherwise there would be two parallels in M to L_C which both contain P. L_C does not lie within I since C does not.

21. For Q to project to U, we would need to know $Q \neq C$ (because C has no image). Since Q is on L, if C is not on L then $Q \neq C$.

23. M contains L, which intersects the image plane.

Chapter 7, Section 7.5

1. (a) No. The parallel edges of the box are not parallel in the projection shown.

   (b) Yes. If the projection direction is parallel to the top and bottom faces of the box and if the front and back vertical edges "line up" along the projection direction.

   (c) Yes, if the projection direction is the same as one of the edges of the box.

   (d) No. The parallel edges of the box are not parallel in the projection shown.

3. $x' = 4 + y - 2z = 4 + 1 - 2(-3) = 11$, $y' = y = 1$, $z' = z = -3$.

5. In place of $\mathbf{d} = \langle x_d, y_d, z_d\rangle$ we would have $\langle kx_d, ky_d, kz_d\rangle$. Following the same steps, the equations would be $x' = x - (kx_d/kz_d)\,z$ and $y' = y - (ky_d/kz_d)\,z$. Cancel the k's.

7. I is $\mathbf{x}\cdot\langle 1, 2, 1\rangle = k$. Projector line $\mathbf{p}' = \mathbf{p} + t\mathbf{d}$. Intersection with I: $[\mathbf{p} + t\mathbf{d}]\cdot\langle 1, 2, 1\rangle = k$. This yields $\mathbf{p}\cdot\langle 1, 2, 1\rangle + t\mathbf{d}\cdot\langle 1, 2, 1\rangle = k$, which yields

$$t = \frac{k - x - 2y - z}{\mathbf{d}\cdot\langle 1, 2, 1\rangle}.$$

Substitute into the equation for the projector line.

9. Suppose L' were the image of L_1 and L_2. The points mapping into L' are on the projector lines through L' which make up a plane by Corollary 7.24. L_1 and L_2 would have to be in this plane. But there is no plane containing L_1 and L_2 since they are skew.

11. All the projector lines coincide with the given line.

13. (a) $$\mathbf{M} = \mathbf{d}^t\mathbf{n} = \begin{pmatrix} x_d \\ y_d \\ z_d \end{pmatrix} \langle x_n, y_n, z_n\rangle = \begin{pmatrix} x_d x_n & x_d y_n & x_d z_n \\ y_d x_n & y_d y_n & y_d z_n \\ z_d x_n & z_d y_n & z_d z_n \end{pmatrix}$$

(b) $\langle x, y, z\rangle\cdot\mathbf{n} = xx_n + yy_n + zz_n$ so the left side is the column vector

$$\begin{pmatrix} (xx_n + yy_n + zz_n)x_d \\ (xx_n + yy_n + zz_n)y_d \\ (xx_n + yy_n + zz_n)z_d \end{pmatrix}.$$

To calculate the right side, insert parentheses as follows: $\mathbf{d}^t(\mathbf{np})$, where $\mathbf{np}$ is clearly $xx_n + yy_n + zz_n$ and so $\mathbf{d}^t(\mathbf{np})$ comes out to the column vector above.

(c) Setting $k = 0$ in Eq. (7.15) and using part (b) we get

$$\mathbf{p}' = \mathbf{p} - \frac{\mathbf{p}\cdot\mathbf{n}}{\mathbf{d}\cdot\mathbf{n}}\mathbf{d}^t = \left(\mathbf{I} - \frac{1}{\mathbf{d}\cdot\mathbf{n}}\mathbf{M}\right)\mathbf{p}.$$

15. Theorem 7.28 tells us that the direction of projection has no effect on the size or shape of the image of the face in question.

17. The line equation, broken into components: $x = 1 + 2t$; $y = 3 - t$; $z = -4$. Substituting these into the projection equations: $x' = 4 + y - 2z = 4 + (3 - t) - 2(-4) = 15 - t$, $y' = 3 - t$, $z' = 4$.

19. Redo of 17: $\mathbf{p}_0 = \langle 1, 3, -4\rangle$, $\mathbf{p}_1 = \langle 1, 3, -4\rangle + 1\langle 2, -1, 0\rangle = \langle 3, 2, -4\rangle$. Then $\mathbf{p}_0' = \langle 15, 3, -4\rangle$, $\mathbf{p}_1' = \langle 14, 2, -4\rangle$. The line connecting them is $x = \langle 15, 3, -4\rangle + t(\langle 14, 2, -4\rangle - \langle 15, 3, -4\rangle) = \langle 15, 3, -4\rangle + t\langle -1, -1, 0\rangle$.

Redo of 18: Take the same points $\mathbf{p}_0$ and $\mathbf{p}_1$. Then $\mathbf{p}'_0 = \langle 1, 3, 0\rangle$, $\mathbf{p}'_1 = \langle 3, 2, 0\rangle$. The line connecting them is $\mathbf{x} = \langle 1, 3, 0\rangle + t(\langle 3, 2, 0\rangle - \langle 1, 3, 0\rangle) = \langle 1, 3, 0\rangle + t\langle 2, -1, 0\rangle$.

21. $\mathbf{n} = \langle 0, 0, 1\rangle$, $\mathbf{d} = \langle 0, 0, 1\rangle$, $\mathbf{e} = \langle 2, 3, 4\rangle$. Direction vector of the image is

$$\left(\mathbf{e} - \frac{\mathbf{e}\cdot\mathbf{n}}{\mathbf{d}\cdot\mathbf{n}}\mathbf{d}\right) = \langle -2, 3, 4\rangle - 4\langle 0, 0, 1\rangle = \langle -2, 3, 0\rangle.$$

23. No, the images converge to the vanishing point.

25. Suppose the points are A, B, C and they project onto line L' in the image plane. The projector plane of L' (see Corollary 7.24) must contain A', B', C'. But the projector plane, by its very nature, contains the direction of projection.

27. (a) $x' = \dfrac{cx}{c-z} = \dfrac{x}{1-z}$. $x'' = -\dfrac{x}{10}$.

(b) $\dfrac{(x'-x'')}{x'} = \dfrac{x/(1-z) + x/10}{x/(1-z)} = \dfrac{10x + x(1-z)}{10x} = \dfrac{11-z}{10}$.

To show $-1/10 < (11-z)/10$ is equivalent to showing $-1 < 11 - z$, which follows from $z < 12$. To show $(11-z)/10 < 1/10$ is equivalent to showing $11 - z < 1$, which follows from $10 < z$.

Chapter 8

Chapter 8, Section 8.1

1. (a) See Figure A.11.

   (b) See Figure A.12.

   (c) The graph in part (a).

3. See Figure A.13.

5. See Figure A.14.

7. First flip e and g and the edges touching them inside $bcdf$. Then flip edge ab inside the three-sided region fbg.

9. See the tetrahedron graph in Figure 8.23.

11. (a) and (b) See Figure A.15.

13. It suffices to remove one edge (see Figure A.16).

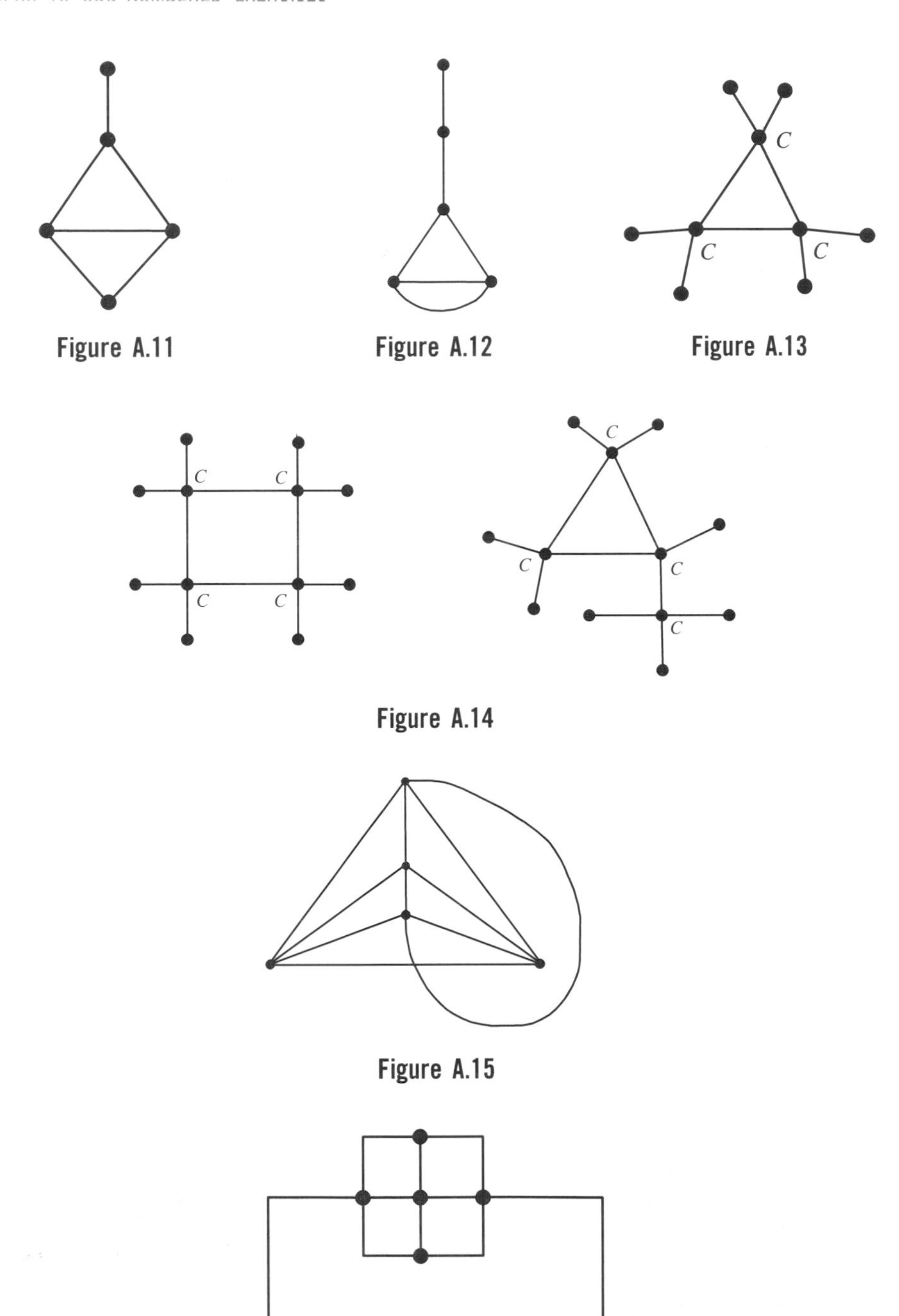

Figure A.11

Figure A.12

Figure A.13

Figure A.14

Figure A.15

Figure A.16

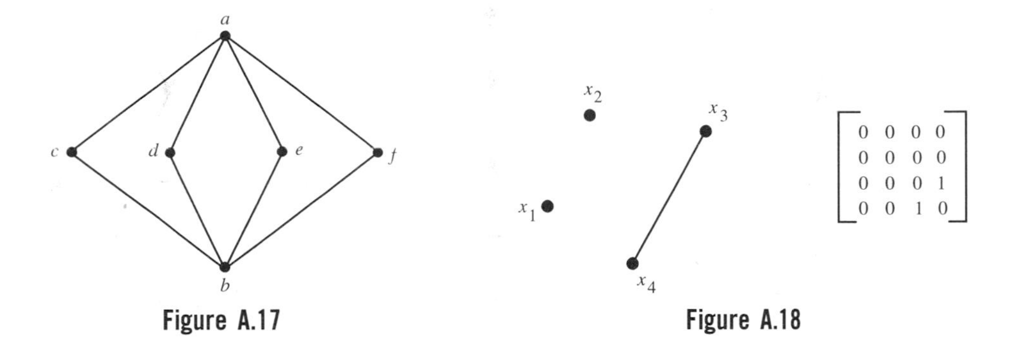

Figure A.17

$$\begin{bmatrix} 0 & 0 & 0 & 0 \\ 0 & 0 & 0 & 0 \\ 0 & 0 & 0 & 1 \\ 0 & 0 & 1 & 0 \end{bmatrix}$$

Figure A.18

15. In Figure A.17, a and b are the "top" vertices.

17. See Figure A.18.

19. Let x_1, x_2, x_3 (labeling first three rows and columns) be the top vertices and x_4, x_5, x_6 be the bottom vertices:

$$\begin{pmatrix} 0 & 0 & 0 & 1 & 1 & 1 \\ 0 & 0 & 0 & 1 & 1 & 1 \\ 0 & 0 & 0 & 1 & 1 & 1 \\ 1 & 1 & 1 & 0 & 0 & 0 \\ 1 & 1 & 1 & 0 & 0 & 0 \\ 1 & 1 & 1 & 0 & 0 & 0 \end{pmatrix}.$$

21. See Figure A.19.

23. See Figure A.20.

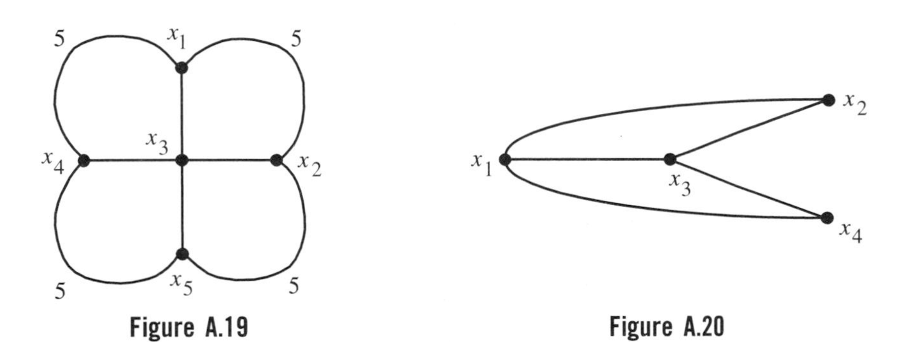

Figure A.19

Figure A.20

25. Add the 1's in a column to get the valence for the corresponding vertex. Therefore, adding all the 1's in the whole matrix gives the sum of the valences. Divide by 2 to get the number of edges.

27. (a) *a*-Water-*b*-Gas-*c*-Electricity-*a*.

    (b) Since Figure 8.6 is contained in $K_{3,3}$, use the circuit of part (a).

    (c) Start at any vertex. Now visit vertices at random, making sure you do not revisit one you have been to until you have covered them all, at which point you go back to the starting one. Since all pairs of vertices are joined, no matter which vertex you are at, you can get to any of the vertices you have not been to before. Furthermore, when you have gotten to the fifth vertex, there will be an edge to get you back to the starting vertex.

    (d) Start at the top and go down and to the left. Then across to the ones colored 3 and 4. Now down to the bottom vertex and then back up to the rightmost vertex and finally back up to the top.

29. (a) See the answer to Exercise 27(c).

    (b) If there were a Hamilton circuit, it would alternate between top (T) and bottom (B) vertices: for example *T-B-T-B-T-B* ... *T-B-T*. Except for the last, all of these vertices are different. Then there must be as many T vertices as B vertices. Thus $m = n$. Conversely, if $m = n$, there is a Hamilton circuit and you can get it by the "random" method used in the solution to Exercise 27(c).

31. (a) If there is an edge *ab*, then *aba* is a path that comes back to its starting vertex, a circuit. (But it is not a simple circuit.)

    (b) In all of your examples, you should notice $v_1 = 2 + v_3 + 2v_4 + 3v_5 + \ldots + (m-2)v_m$. This could be proved by induction.

33. Color the center face color #1. Alternate #2 and #3 around the center face. Use #1 again on the infinite face.

35. (a) See Figure A.21.

    (b) See Figure A.22.

37. Three colors will do.

39. mn.

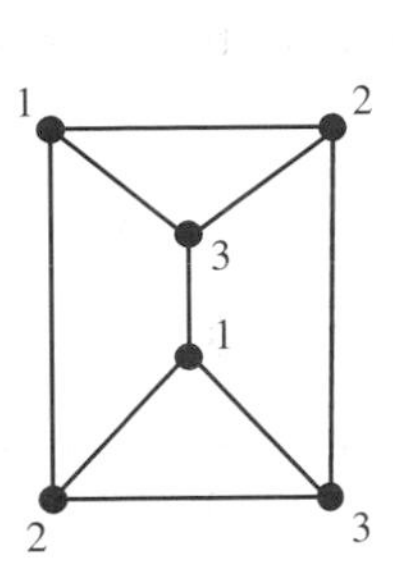

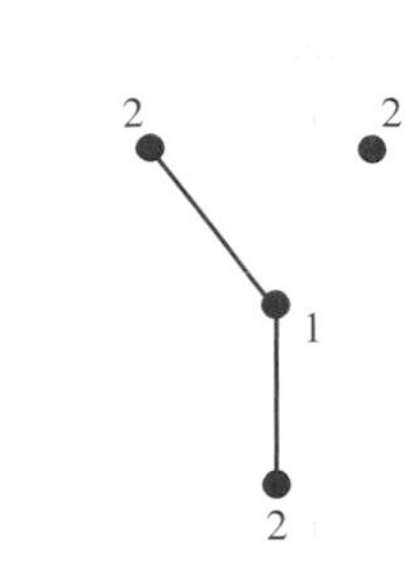

Figure A.21

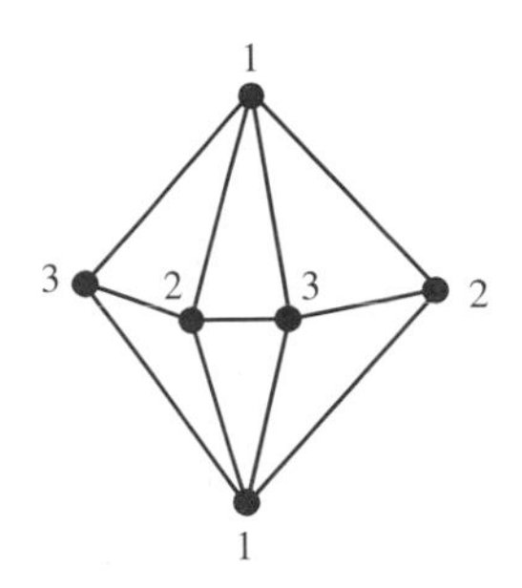

Figure A.22

41. $\dfrac{n(n-1)}{2}$.

Chapter 8, Section 8.2

1. Theorem 8.7 applies. Since each face has three edges, all $f_i = 0$ except for f_3. Thus $3f_3 = 2e$. Since $f_3 = f$, the result follows.

3. (a) A graph consisting of a single edge. This is the only example.

   (b) A graph consisting of two edges that are either not connected to one another or have one vertex in common. These are the only examples.

5. $v_1 + 2v_2 + \ldots + mv_m = 2e$. But all $v_i = 0$ except for $i = 3$. $v = v_3$ and $3v = 3v_3 = 2e$.

7. From Exercise 6, $3f \geq 6 + e$. Multiply by 2: $6f \geq 12 + 2e$. By Exercise 2, $2e \geq 3f$ so $12 + 2e \geq 12 + 3f$. Thus $6f \geq 12 + 3f$, which gives $3f \geq 12$, $f \geq 4$.

9. (a) $v + f = e + 2$. Multiply by 3 and use Exercise 3 to get $2e + 3f = 3e + 6$. This gives $3f - e = 6$.

   (b) From Theorem 8.7.

   (c) Replacing f by $f_4 + f_6$ in part (a): $3f_4 + 3f_6 = e + 6$. Double this and subtract the equation in part (b) to deduce $f_4 = 6$.

   (d) The author knows two examples. To get the smaller of the two, draw a hexagon inside another hexagon and then connect each vertex of the outer hexagon to a vertex of the inner hexagon, making a ring of four-sided faces between the hexagons. For the second example, draw a four-sided face and surround it with four 6-sided faces (keeping all valences 3). Now surround that ring of six-sided faces by four-sided faces alternating with six-sided faces in such a way that each four-sided face is joined by an edge to the four-sided face you started with. If you stop now you should have an infinite face which is the sixth four-sided face.

11. Let *ab* be any edge of such a map. There is a simple circuit that includes *ab*. Thus there are two ways to get from *a* to *b*: directly using edge *ab*, and the "other way around" on the circuit. Thus a path connecting two vertices *u* and *v* which used *ab* could still connect *u* and *v* if *ab* is removed by using the "other way around." This means the graph is still connected after *ab* is removed. Since *ab* was any edge, no edge disconnects and Eq. (8.7) holds.

13. We know K_5 has $v=5$, $e=10$ so if it were planar then $f=1+e+c-v=1+10+1-5=7$. K_5 has no loops or multiple edge sets and no edge disconnects it. Thus the inequality of Exercise 2 applies and we have $3f \leq 2e$. This imples $3(7) \leq 2(10)$, which is a contradiction, so K_5 is not planar.

15. $K_{3,3}$ and K_5 can both be drawn on the torus with no crossings. To see this in the case of $K_{3,3}$, draw as much of the graph as you can on a long skinny piece of paper. This should be all but one edge of $K_{3,3}$. Now make a torus in two steps: Tape together the bottom and top edges, making a cylindrical tube; bend the tube so that the two circular ends come together and tape them together. In this way you have created a torus and you have also provided routes for the missing edge. Try the same with K_5.

17. (a) See Figure A.23.

    (b) One piece, one hole [the (2,2) pixel], Euler number = 0.

19.
```
e := 0;                                        /* comment: start tally */
for k going from 1 to m * n repeat the following steps
        { r := k mod m;                        /* compute row index */

          c := (k-r)/m + 1;                    /* compute column index */

          /* next step is edge update, exactly as in text */

          if pixel (r, c) is black then
              { e := e+4;
                  if pixel (r-1, c) is black then e := e-1;
                  if pixel (r, c-1) is black then e := e-1 }
              }
```

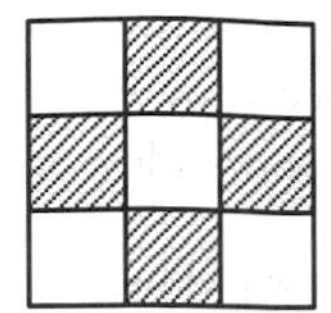

Figure A.23

21. (a) Since the pieces add up and the holes add up, $E(I_1 \cup I_2) = E(I_1) + E(I_2)$. If they have just a point in common, the pieces of the union are one less than the sum of the pieces, but the holes add up, so $E(I_1 \cup I_2) = E(I_1) + E(I_2) - 1$. In general, we have $E(I_1 \cup I_2) = E(I_1) + E(I_2) - E(I_1 \cap I_2)$. This is not obvious. If you want to prove it you need first to see whether Theorem 8.10 holds for intersections of digital images (which might contain isolated pixel edges and isolated pixel corners). Then apply the set theory theorem cardinality$(A \cup B)$ = cardinality(A) + cardinality(B) − cardinality$(A \cap B)$ to vertices, edges, and black pixels separately.

Chapter 8, Section 8.3

1. It is not a solid.

3. It is a polyhedron.

5. Suppose A and B are in the circle of radius r, whose center is C, and D is in $\overline{AB}$ but not in the circle. Then $CD >$ both AC and BC. By Exercise 12 of Section 2.1 in Chapter 2, $m\angle CAD > m\angle CDA$ and $m\angle CBD > m\angle CDB$. Adding these inequalities, $m\angle CAD + m\angle CBD > m\angle CDA + m\angle CDB = 180°$. This is a contradiction.

7. A segment is convex. The union of two of them making an "*L*" is not convex.

9. (b) The point closest to you and the rightmost point.

   (c) Take any points in the two solid parts except for the points where the connecting segment touches a solid part.

   (d) Any points on the two separate parallelograms as long as they are not on the segment where the parallelograms meet.

   (e) A and Z for example.

11. Let S be any plane containing U and V. It intersects the polyhedron in a convex set (Exercise 6). This set has straight boundaries because the intersection of S with the faces of the polyhedron must be line segments (planes intersect in a line). Thus it is a convex polygon in the plane S. U and V divide the polygon into two arcs, either of which will do as the sequence of line segments we are looking for.

13. See Figure A.24.

    The vertex correspondence is $A \to A'$, $B \to B'$, etc. (each vertex to its primed version.)

    The edge correspondence is $AB \to A'B'$, $BC \to B'C'$, etc.

    The face correspondence is $ABC \to A'B'C'$, $ABED \to A'B'E'D'$.

15. See Figure A.25.

17. When S lies on the plane determined by face $ABCD$.

19. Let v_o, e_o, f_o be the vertices, edges, and faces of the original polyhedron (before drilling) and let v_h, e_h, f_h be the vertices, edges, and faces of the polyhedron which is the hole, and let v, e, f be the vertices, edges, and faces of the toroidal polyhedron obtained by drilling. $v = v_o + v_h$, $e = e_o + e_h$, $f = f_o + f_h - 2$. Euler's formula applies to the original and to the hole: $v_o - e_o + f_o = 2$, $v_h - e_h + f_h = 2$. Adding these gives $v - e + f + 2 = 4$ and so $v - e + f = 2$.

21. There must be at least one face and it must have at least three edges. Each of these edges touches another face—and these are all different. Thus there must be at least four.

23. See Figure A.26.

25. All planar maps with those statistics are combinatorially equivalent. An easy way to draw one is to draw a square and a point inside it. Join the point to each vertex of the square. (This is the Schlegel diagram of a pyramid.)

Chapter 8, Section 8.4

1. Figure 8.23 shows one example. All other examples are combinatorially equivalent because when you try to draw an example, the next thing to add on at a given part of your drawing is always determined.

3. Same answer as Exercise 1.

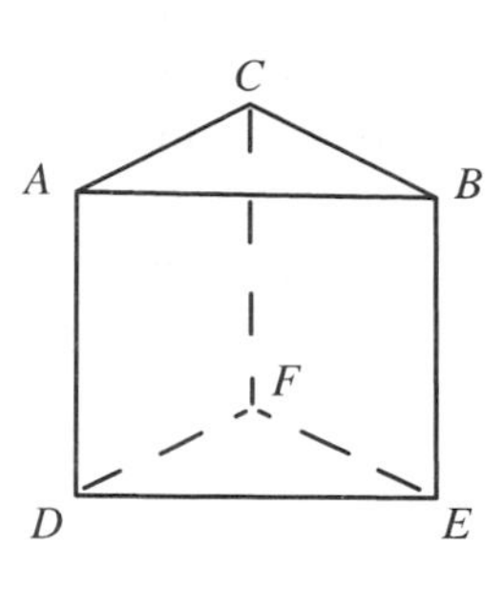

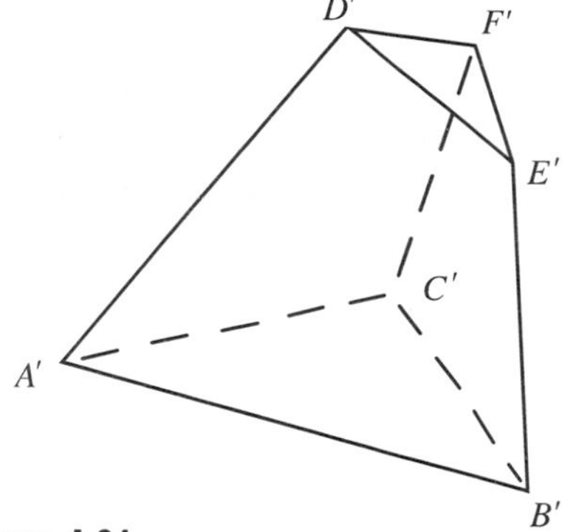

Figure A.24

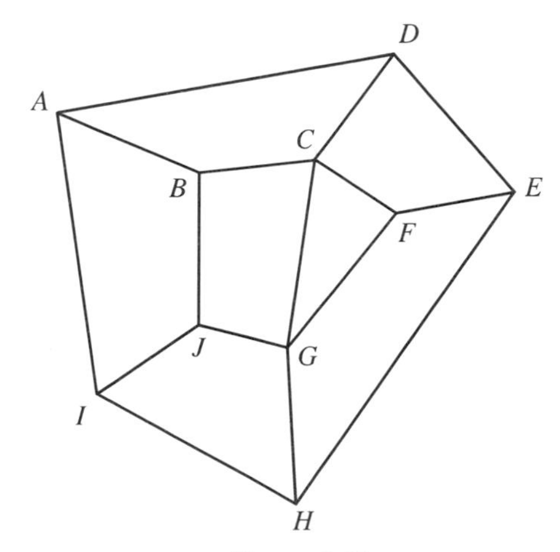

Figure A.25

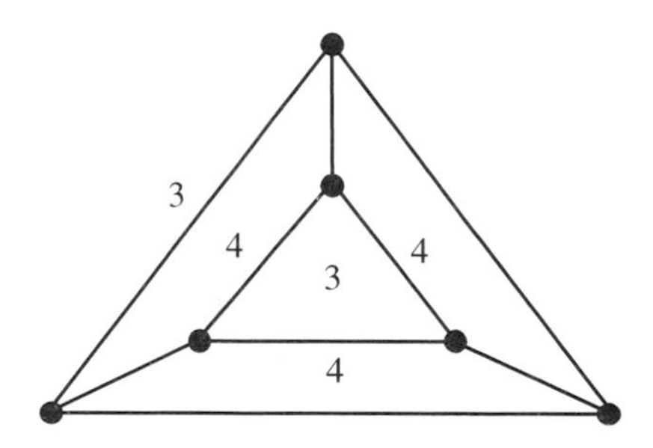

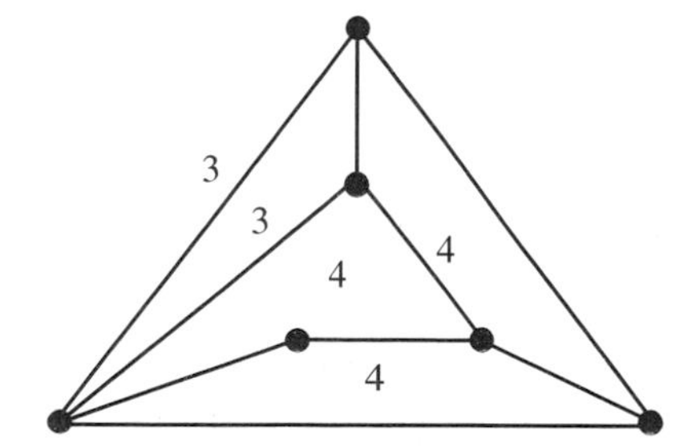

Figure A.26

5. $pf = qv = 2e$ so $f = 2e/p$ and $v = 2e/q$. We need to express e in terms of p and q in the last two equations. To do this, solve Eq. (8.12) for e:

$$\begin{aligned} e &= \frac{1}{1/p + 1/q - 1/2} \\ &= \frac{2pq}{2q + 2p - pq}. \end{aligned}$$

Thus

$$f = \frac{4q}{2q + 2p - pq} \quad \text{and} \quad v = \frac{4p}{2q + 2p - pq}.$$

7. Take two polygons that touch along an edge and therefore have the same side length. By part (b) of Exercise 6, they are congruent. Now add a third polygon that meets one of the first two on an edge. It will also be congruent to them. We can continue like this until we have connected all of the polygons.

9. For $p = 3$, $q = 6$, extend the pattern of Figure A.27 indefinitely. For $p = 6$, $q = 3$, extend the pattern of Figure A.28 indefinitely.

11. Let A and B be points on opposite faces where the sphere is tangent to those faces. If we could show that segment $\overline{AB}$ is perpendicular to the two faces then $AB = s$. (Here we are relying on the fact that parallel planes are everywhere equidistant.) If we could also show that $\overline{AB}$ passes through the center of the

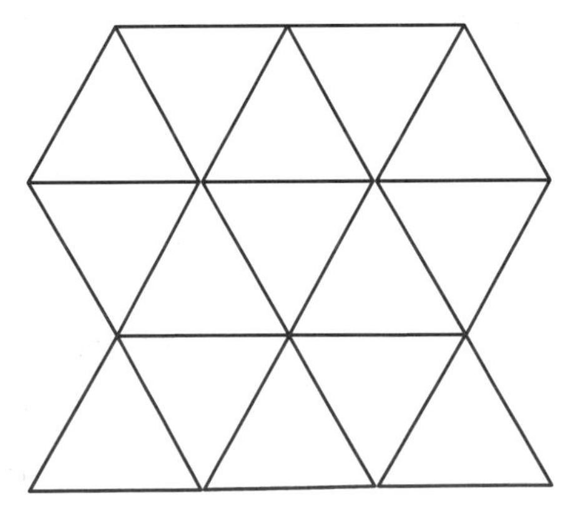

Figure A.27

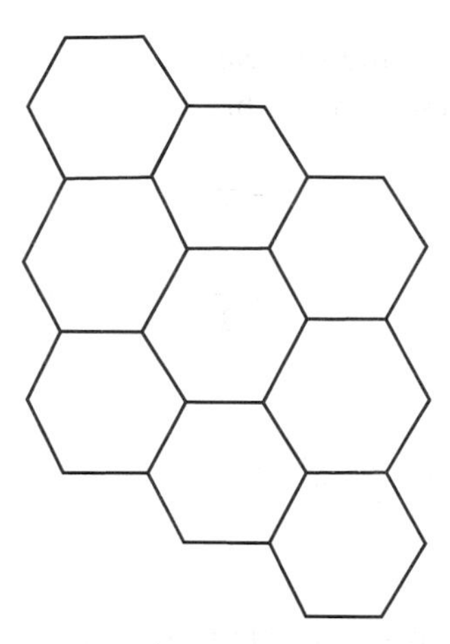

Figure A.28

circle, then diameter $= AB$, and we have diameter $= s$. The key fact is this: The segment from the center of a sphere to the point of tangency with a plane is perpendicular to the plane at the point of tangency.

13. Start by drawing two hexagons as required (Figure A.29a). Now add edges (Figure A.29b) so as to make all valences 3. Some of these added edges have to meet since the only remaining faces are pentagons. The point is that at every stage of building, you have no real choice about what to do. (You might draw the edges longer or shorter or angle them differently, but the combinatorial type of what you build is forced on you.) Eventually you will find you cannot avoid another hexagon.

15. Since $f = f_5 + f_6$, $f = 12 + f_6$. In the proof of Theorem 8.14 we found $2e = 5f_5 + 6f_6$, so $2e = 60 + 6f_6$. We also have $3v = 2e$ so $v = (1/3)(60 + 6f_6) = 20 + 2f_6$.

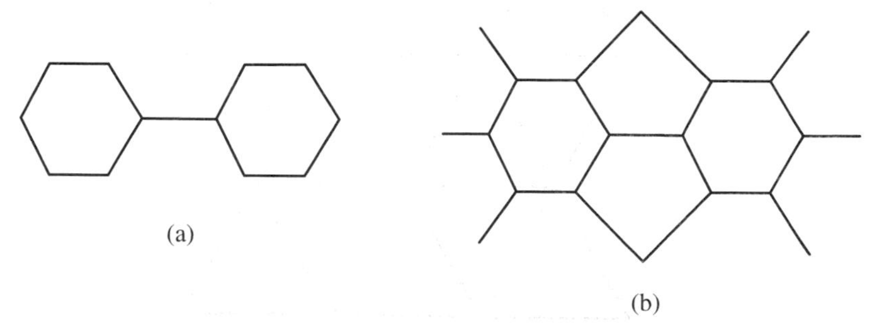

Figure A.29

Chapter 8, Section 8.5

1. See Figure A.30.

3. One of the colors available for x_2 will be different from the one chosen for x_1, so use it. Every time you come to a new vertex, there will be a color in its list that is different from the one just used, so you can color all the vertices.

5. Alternate colors as you go around the circuit.

7. Alternate colors as you go around the circuit, just as in Exercise 5.

9. A simple circuit of three edges is the simplest example.

11. See Figure A.31.

13. This is wrong. If you carry out the alteration in Figure A.32, you will not be able to list color it.

15. See Figure A.33.

17. The map of the octahedron (see Figure 8.23 of Section 8.3 in Chapter 8).

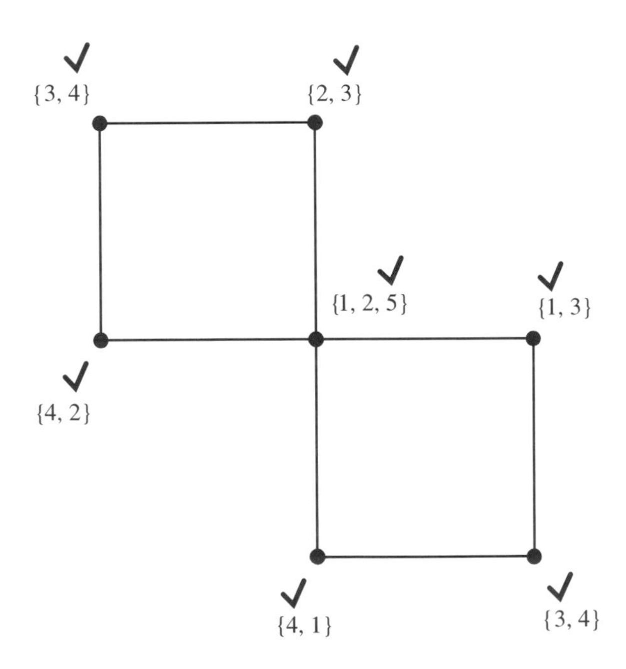

Figure A.30

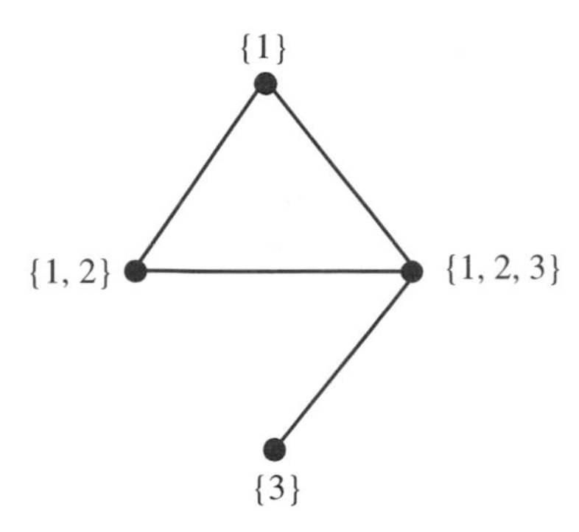

Figure A.31

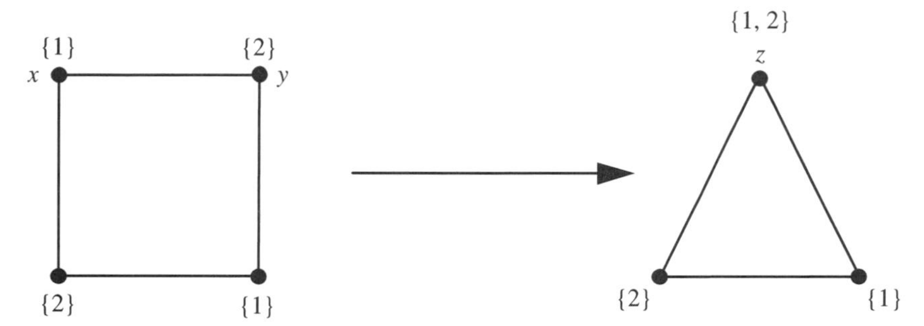

Figure A.32

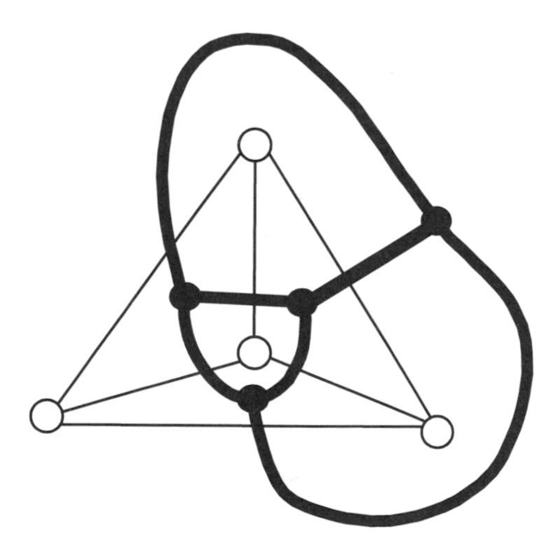

Figure A.33

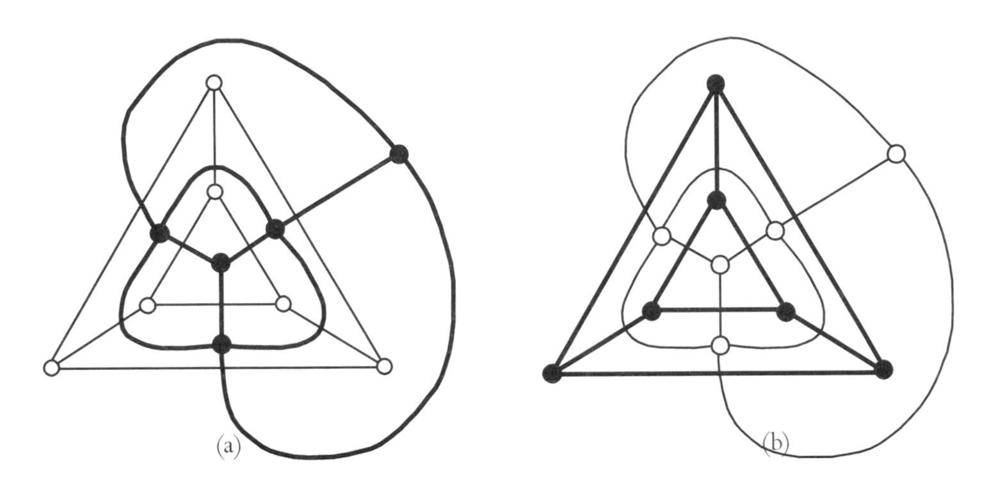

Figure A.34

19. The dual of the dual (bold map in Figure A.34b) is isomorphic to the original map.

21. Take a triangle ($d = 2$) where each vertex has the color list $\{1, 2\}$. This can't be list colored. This is a counterexample.

23. Let M be a planar map that is not necessarily 3-valent. Truncate every vertex as in Figure 8.47. (Reread the proof of Theorem 8.19.) By hypothesis we can face color the truncated map with four colors. Now shrink the new faces so they disappear, bringing back the original map. Keep the colors on the faces which do not disappear and we have a face coloring of the original map using just four colors.

Index